THE MATHEMATICAL THEORY OF NON-UNIFORM GASES

AN ACCOUNT OF THE KINETIC THEORY OF VISCOSITY, THERMAL CONDUCTION AND DIFFUSION IN GASES

SYDNEY CHAPMAN, F.R.S.

Geophysical Institute, College, Alaska
National Center for Atmospheric Research, Boulder, Colorado

AND

T. G. COWLING, F.R.S.

Professor of Applied Mathematics
Leeds University

THIRD EDITION
PREPARED IN CO-OPERATION WITH
D. BURNETT

CAMBRIDGE UNIVERSITY PRESS
Cambridge, New York, Melbourne, Madrid, Cape Town,
Singapore, São Paulo, Delhi, Tokyo, Mexico City

Cambridge University Press
The Edinburgh Building, Cambridge CB2 8RU, UK

Published in the United States of America by Cambridge University Press, New York

www.cambridge.org
Information on this title: www.cambridge.org/9780521408448

First published 1939
Second edition 1952
Third edition 1970
Reissued as a paperback with a Foreword by Carlo Cercignani
in the Cambridge Mathematical Library Series 1990
Reprinted 1995

A catalogue record for this publication is available from the British Library

ISBN 978-0-521-07577-0 Hardback
ISBN 978-0-521-40844-8 Paperback

CONTENTS

FOREWORD

The atomic theory of matter asserts that material bodies are made up of small particles. This theory was founded in ancient times by Democritus and expressed in poetic form by Lucretius. This view was challenged by the opposite theory, according to which matter is a continuous expanse. As quantitative science developed, the study of nature brought to light many properties of bodies which appear to depend on the magnitude and motions of their ultimate constituents, and the question of the existence of these tiny, invisible, and immutable particles became conspicuous among scientific enquiries.

As early as 1738 Daniel Bernoulli advanced the idea that gases are formed of elastic molecules rushing hither and thither at large speeds, colliding and rebounding according to the laws of elementary mechanics. The new idea, with respect to the Greek philosophers, was that the mechanical effect of the impact of these moving molecules, when they strike against a solid, is what is commonly called the pressure of the gas. In fact, if we were guided solely by the atomic hypothesis, we might suppose that pressure would be produced by the repulsions of the molecules. Although Bernoulli's scheme was able to account for the elementary properties of gases (compressibility, tendency to expand, rise of temperature in a compression and fall in an expansion, trend toward uniformity), no definite opinion could be formed until it was investigated quantitatively. The actual development of the kinetic theory of gases was, accordingly, accomplished much later, in the nineteenth century.

Although the rules generating the dynamics of systems made up of molecules are easy to describe, the phenomena associated with this dynamics are not so simple, especially because of the large number of particles: there are about $2 \times 7 \times 10^{19}$ molecules in a cubic centimeter of a gas at atmospheric pressure and a temperature of 0 °C.

Taking into account the enormous number of particles to be considered, it would of course be a perfectly hopeless task to attempt to describe the state of the gas by specifying the so-called microscopic state, i.e. the position and velocity of every individual particle, and we must have recourse to statistics. This is possible because in practice all that our observation can detect is changes in the macroscopic state of the gas, described by quantities such as density, velocity, temperature, stresses, heat flow, which are related to the suitable averages of quantities depending on the microscopic state.

J. P. Joule appears to have been the first to estimate the average velocity of a molecule of hydrogen. Only with R. Clausius, however, the kinetic theory of gases entered a mature stage, with the introduction of the concept of mean free-path (1858). In the same year, on the basis of this concept, J. C. Maxwell developed a preliminary theory of transport processes and gave an heuristic derivation of the velocity distribution function that bears his name. However, he almost immediately realized that the mean free-path method was inadequate as a foundation for kinetic theory and in 1866 developed a much more accurate method, based on the transfer equations, and discovered the particularly simple properties of a model, according to which the molecules interact at distance with a force inversely proportional to the fifth power of the distance (nowadays these are commonly called Maxwellian molecules). In the same paper he gave a better justification of his formula for the velocity distribution function for a gas in equilibrium.

With his transfer equations, Maxwell had come very close to an evolution equation for the distribution, but this step must be credited to L. Boltzmann. The equation under consideration is usually called the Boltzmann equation and sometimes the Maxwell–Boltzmann equation (to acknowledge the important role played by Maxwell in its discovery).

In the same paper, where he gives an heuristic derivation of his equation, Boltzmann deduced an important consequence from it, which later came to be known as the H-theorem. This theorem attempts to explain the irreversibility of natural processes in a gas, by showing how molecular collisions tend to increase entropy. The theory was attacked by several physicists and mathematicians in the 1890s, because it appeared to produce paradoxical results. However, a few years after Boltzmann's suicide in 1906, the existence of atoms was definitely established by experiments such as those on Brownian motion and the Boltzmann equation became a practical tool for investigating the properties of dilute gases.

In 1912 the great mathematician David Hilbert indicated how to obtain approximate solutions of the Boltzmann equation by a series expansion in a parameter, inversely proportional to the gas density. The paper is also reproduced as Chapter XXII of his treatise entitled *Gründzüge einer allgemeinen Theorie der linearen Integralgleichungen*. The reasons for this are clearly stated in the preface of the book ('Neu hinzugefügt habe ich zum Schluss ein Kapitel über kinetische Gastheorie. [. . .] erblicke ich in der Gastheorie die gläzendste Anwendung der die Auflösung der Integralgleichungen betreffenden Theoreme').

In 1917, S. Chapman and D. Enskog simultaneously and independently obtained approximate solutions of the Boltzmann equation, valid for a sufficiently dense gas. The results were identical as far as practical applications were concerned, but the methods differed widely in spirit and detail. Enskog presented a systematic technique generalizing Hilbert's idea, while Chapman simply extended a method previously indicated by Maxwell to obtain transport coefficients. Enskog's method was adopted by S. Chapman and T. G. Cowling when writing *The Mathematical Theory of Non-uniform Gases* and thus came to be known as the Chapman–Enskog method.

This is a reissue of the third edition of that book, which was the standard reference on kinetic theory for many years. In fact after the work of Chapman and Enskog, and their natural developments described in this book, no essential progress in solving the Boltzmann equation came for many years. Rather the ideas of kinetic theory found their way into other fields, such as radiative transfer, the theory of ionized gases, the theory of neutron transport and the study of quantum effects in gases. Some of these developments can be found in Chapters 17 and 18.

In order to appreciate the opportunity afforded by this reissue, we must enter into a detailed description of what was the kinetic theory of gases at the time of the first edition and how it has developed. In this way, it will be clear that the subsequent developments have not diminished the importance of the present treatise.

The fundamental task of statistical mechanics is to deduce the macroscopic observable properties of a substance from a knowledge of the forces of interaction and the internal structure of its molecules. For the equilibrium states this problem can be considered to have been solved in principle; in fact the method of Gibbs ensembles provides a starting point for both qualitative understanding and quantitative approximations to equilibrium behaviour. The study of nonequilibrium states is, of course, much more difficult; here the simultaneous consideration of matter in all its phases – gas, liquid and solid – cannot yet be attempted and we have to use different kinetic theories, some more reliable than others, to deal with the great variety of nonequilibrium phenomena occurring in different systems.

A notable exception is provided by the case of gases, particularly monatomic gases, for which Boltzmann's equation holds. For gases, in fact, it is possible to obtain results that are still not available for general systems, i.e. the description of the thermomechanical properties of gases in the pressure and temperature ranges for which the description suggested by continuum mechanics also holds. This is the object of the approximations associated with the names Maxwell, Hilbert, Chapman, Enskog and Burnett, as well as of the systematic treatment presented in this volume. In these approaches, out of all the distribution functions f which could be assigned to given values of the velocity, density and temperature, a single one is chosen. The precise method by which this is done is rather subtle and is described in Chapters 7 and 8. There exists, of course, an exact set of equations which the basic continuum variables, i.e. density, bulk velocity (as opposed to molecular velocity) and temperature, satisfy, i.e., the full conservation equations. They are a consequence of the Boltzmann equation but do not form a closed system, because of the appearance of additional variables, i.e. stresses and heat flow. The same situation occurs, of course, in ordinary continuum mechanics, where the system is closed by adding further relations known as 'constitutive equations'. In the method described in this book, one starts by assuming a special form for f depending only on the basic variables (and their gradients); then the explicit form of f is determined and, as a consequence, the stresses and heat flow are evaluated in terms of the basic variables, thereby closing the system of conservation equations. There are various degrees of approximation possible within this scheme, yielding the Euler equations, the Navier–Stokes equations, the Burnett equations, etc. Of course, to any degree of approximation, these solutions approximate to only one part of the manifold of solutions of the Boltzmann equation; but this part turns out to be the one needed to describe the behaviour of the gas at ordinary temperatures and pressures. A byproduct of the calculations is the possibility of evaluating the transport coefficients (viscosity, heat conductivity, diffusivity, . . .) in terms of the molecular parameters. The calculations are by no means simple and are presented in detail in Chapters 9 and 10. These results are also compared with experiment (Chapters 12, 13 and 14).

In 1949, H. Grad wrote a paper which became widely known because it contained a systematic method of solving the Boltzmann equation by expanding the solution into a series of orthogonal polynomials. Although the solutions which could be obtained by means of Grad's 13-moment equations (see section 15.6) were more general than the 'normal solutions' which could be obtained by the Chapman–Enskog method, they failed to be sufficiently general to cover the new applications of the Boltzmann equation to the study of upper atmosphere flight. In the late 1950s and in the 1960s, under the impact of the problems related to space research, the main interest was in the direction of finding approximate solutions of the Boltzmann equation in regions having a thickness of the order of a mean free-path. These new solutions were, of course, beyond the reach of the methods described in this book. In fact, at the time when the book was written, the next step was to go beyond the Navier–Stokes level in the Chapman–Enskog expansion. This leads to the so-called Burnett equations briefly described in Chapter 15 of this book. These equations, generally speaking, are not so good in describing departures from the Navier–Stokes model, because their corrections are usually of the same order of magnitude as the difference between the normal solutions and the solutions of interest in practical problems. However, as pointed out by several Russian authors in the early 1970s, there are certain flows, driven by temperature gradients, where the Burnett terms are of importance. For this reason as well for his historical interest, the chapter on the Burnett equations still retains some importance.

Let us now briefly comment on the chapters of the book, which have not been mentioned so far in this foreword. Chapters 1–6 are of an introductory nature; they describe the heavy apparatus that anybody dealing with the kinetic theory of gases must know, as well as the results which can be obtained by simpler, but less accurate tools. Chapter 11 describes a classical model for polyatomic gases, the rough sphere molecule; this model, although not so accurate when compared with experiments, retains an important role from a conceptual point of view, because it offers a simple example of what one should expect from a model describing a polyatomic molecule. Chapter 16 describes the kinetic theory of dense gases; although much has been done in this field, the discussion by Chapman and Cowling is still useful today.

Where is kinetic theory going today? The main recent developments are in the direction of developing a rigorous mathematical theory: existence and uniqueness of the solutions to initial and boundary value problems and their asymptotic trends, but also rigorous justification of the approximate methods of solution. Among these is the method described in this book. It is unfair, however, to criticise, in the light of the standards and achievement of today, the approach described in this book, as is sometimes done. In addition to still being a good description of an important part of the kinetic theory of gases, this book has played the important role of transmitting the solved and unsolved problems of kinetic theory to generations of students and scholars. Thus it is not only useful, but also historically important.

Carlo Cercignani

Milano

EXTRACT FROM PREFACE TO FIRST EDITION

In this book an account is given of the mathematical theory of gaseous viscosity, thermal conduction, and diffusion. This subject is complete in itself, and possesses its own technique; hence no apology is needed for separating it from related subjects such as statistical mechanics.

The accurate theory originated with Maxwell and Boltzmann, who established the fundamental equations of the subject. The general solution of these equations was first given more than forty years later, when within about a year (1916–1917) Chapman and Enskog independently obtained solutions by methods differing widely in spirit and detail, but giving identical results. Although Chapman's treatment of the general theory was fully effective, its development was intuitive rather than systematic and deductive; the work of Enskog showed more regard for mathematical form and elegance. His treatment is the one chosen for presentation here, but with some differences, including the relatively minor one of vector and tensor notation.* A more important change is the use of expansions of Sonine polynomials, following Burnett (1935). We have also attempted to expound the theory more simply than is done in Enskog's dissertation, where the argument is sometimes difficult to follow.

The later chapters describe more recent work, on dense gases, on the quantum theory of collisions (so far as it affects the theory of the transport phenomena in gases), and on the theory of conduction and diffusion in ionized gases, in the presence of electric and magnetic fields.

Although most of the book is addressed to the mathematician and theoretical physicist, an effort has been made to serve the needs of laboratory workers in chemistry and physics by collecting and stating, as clearly as possible, the chief formulae derived from the theory, and discussing them in relation to the best available data.

S. C.

1939 T. G. C.

* The notation used in this book for three-dimensional Cartesian tensors was devised jointly by E. A. Milne and S. Chapman in 1926, and has since been used by them in many branches of applied mathematics.

PREFACE TO THIRD EDITION

Until now, this book has appeared in substantially its 1939 form, apart from certain corrections and the addition, in 1952, of a series of notes indicating advances made in the intervening years. A more radical revision has been made in the present edition.

Chapter 11 has been wholly rewritten, and discusses general molecular models with internal energy. The discussion is primarily classical, but in a form readily adaptable to a quantum generalization. This generalization is made in Chapter 17, which also discusses (in rather more detail than before) quantum effects on the transport properties of hydrogen and helium at low temperatures. The theory is applied to additional molecular models in Chapter 10, and these are compared with experiment in Chapters 12–14; the discussion in these chapters aims for the maximum simplicity consistent with reasonable accuracy. Chapter 16 now includes a short summary of the BBGKY theory of a dense gas, with comments on its difficulties. A new Chapter 18 discusses mixtures of several gases. Chapter 19 (the old Chapter 18) discusses phenomena in ionized gases, on which an enormous amount of work has been done in recent years. This chapter has been much extended, even though attention is confined to aspects related to the transport phenomena. Finally, in Chapter 6 and elsewhere, a more detailed account is given of approximate theories, especially those that illuminate some feature of the general theory.

To accommodate the new material, some cuts have been necessary. These include the earlier approximate discussion of the electron-gas in a metal, and the Appendices A and B. The Historical Summary, and the discussion of the Lorentz approximation have been curtailed. The discussion of certain other topics has been modified, especially in the light of the work of Kihara, of Waldmann, of Grad and of Hirschfelder, Curtiss and Bird. A few minor changes of notation have been made; these are set out at the end of the list of symbols on pp. xxv and xxvi.

The third edition has been prepared throughout with the co-operation of Professor D. Burnett. We are deeply indebted to him for numerous valuable improvements, and for his continuous attention to details that might otherwise have been overlooked. He has given unstinted assistance over a long period.

Our thanks are due to many others for their interest and encouragement. Special mention should be made of Professors Waldmann and Mason for their helpful interest. Our thanks are also due, as earlier, to the officials of the Cambridge University Press for their willing and expert help both before and during the printing of this edition.

S. C.
T. G. C.

1969

NOTE REGARDING REFERENCES

The chapter-sections are numbered decimally.

The equations in each section are numbered consecutively and are preceded by the section number, (3.41, 1), (3.41, 2), ...

References to equations are also preceded by the section number and where a series of numbers occur they are elided (3.41, 2, 3, ...) or (3.41, 1–16).

References to periodicals give first (in italic type) the name of the periodical, next (in Clarendon type) the volume-number, then the number of the page or pages referred to, and finally the date in parenthesis.

CHAPTER AND SECTION TITLES

Chapter 4. Boltzmann's *H*-theorem and the Maxwellian velocity-distribution

Chapter 5. The free path, the collision-frequency and persistence of velocities

Chapter 6. Elementary theories of the transport phenomena

Chapter 7. The non-uniform state for a simple gas

Chapter 8. The non-uniform state for a binary gas-mixture

Chapter 9. Viscosity, thermal conduction, and diffusion: general expressions

Chapter 10. Viscosity, thermal conduction, and diffusion: theoretical formulae for special molecular models

Chapter 11. Molecules with internal energy

Chapter 12. Viscosity: comparison of theory with experiment

Chapter 17. Quantum theory and the transport phenomena

Chapter 18. Multiple gas mixtures

Chapter 19. Electromagnetic phenomena in ionized gases

DIAGRAMS

LIST OF SYMBOLS

Clarendon type is used for vectors, **roman clarendon type** for unit vectors, and sans serif type for tensors.

The bracket symbols [,] and { , } are defined on page 83.

In general, symbols which occur only in a few consecutive pages are not included in this list. Greek symbols are placed at the end of the list.

The *italic figures* indicate the pages on which the symbols are introduced.

a_p, *128, 145* a_{pq}, *128, 146* a'_{pq}, a''_{pq}, *162* $A_l(\nu)$, *171* $A(\mathcal{C})$, *123*
 $\mathcal{A}^{(m)}$, $\mathcal{A}^{(m)}_{pq}$, *129, 146* $\mathcal{A}'^{(m)}$, $\mathcal{A}'^{(m)}_{pq}$, *146* $\boldsymbol{a}^{(p)}$, *128* $\boldsymbol{a}_1^{(p)}$, $\boldsymbol{a}_2^{(p)}$, *145*
 $\boldsymbol{A}$, *123* $\boldsymbol{A}_1$, $\boldsymbol{A}_2$, *139, 211* $\bar{\boldsymbol{A}}_1$, $\bar{\boldsymbol{A}}_2$, *143* $\boldsymbol{A}_s$, *344* $\bar{\boldsymbol{A}}_s$, *347*
 A, *163* A_{st}, *350*

b, *55* b_p, *130, 147* b_{pq}, *130, 148* b'_{pq}, b''_{pq}, *162* $B(\mathcal{C})$, *124* B_1, B_2, *211*
 $\mathcal{B}^{(m)}$, $\mathcal{B}^{(m)}_{pq}$, *131, 148* $\mathsf{b}^{(p)}$, *130* $\mathsf{b}_1^{(p)}$, $\mathsf{b}_2^{(p)}$, *147* B_1, B_2, *139, 211*
 B_s, *344* B, *163* B_{st}, *350*

$\mathbf{c}$, $\boldsymbol{c}$, *25* $\mathbf{c}_0$, $\boldsymbol{c}_0$, *27, 44* c_v, *39* c_p, *40* $\boldsymbol{c}_s$, *44* $\mathbf{c}_1$, $\mathbf{c}_2$, $\mathbf{c}'_1$, $\mathbf{c}'_2$, *53*
 c_1, c_2, c'_1, c'_2, *54* c'_v, c''_v, *222* $(c'_v)_s$, $(c''_v)_s$, *207* C_v, C_p, *40* $\mathbf{C}$, $\mathbf{C}'$, C, *27*
 C', *28* $\boldsymbol{C}_s$, *44* C_s, *45* $\mathcal{C}$, $\mathcal{C}$, *122* $\mathcal{C}_1$, $\mathcal{C}_2$, $\mathcal{C}_1$, $\mathcal{C}_2$, *138* c, *163* c_{st}, *351*

d_p, *145* D_{12}, *102* D_{11}, *103* D_T, *141* D_{int}, *224* D_{Ts}, *345*
 D_{st}, *109, 346* $\mathcal{D}f$, $\mathcal{D}_s f_s$, *47* $\mathcal{D}_1 f_1$, $\mathcal{D}_2 f_2$, *134* $\mathcal{D}_1^{(r)}$, $\mathcal{D}_2^{(r)}$, *134*
 $\dfrac{\partial_e f}{\partial t}$, *46* $\dfrac{\partial_e f_s}{\partial t}$, *47* $\left(\dfrac{\partial_e f_1}{\partial t}\right)_1$, $\left(\dfrac{\partial_e f_1}{\partial t}\right)_2$, *61* $\dfrac{\partial_r}{\partial t}$, *115* $\dfrac{D}{Dt}$, *48*
 $\dfrac{D_0}{Dt}$, *116* $d\boldsymbol{r}$, *13* $d\boldsymbol{c}$, *25* $d\mathbf{k}$, *14* $d\mathbf{e}$, *62* $d\mathbf{e}'$, *59*
 $d\Omega_s$, *199* $\dfrac{\partial}{\partial \boldsymbol{r}}$, *12* $\dfrac{\partial}{\partial \boldsymbol{C}}$, *13* $\dfrac{\partial}{\partial \boldsymbol{c}}$, *25* $\dfrac{\partial}{\partial \boldsymbol{P}_s}$, $\dfrac{\partial}{\partial \boldsymbol{Q}_s}$, *199*
 $\dfrac{\partial(\boldsymbol{u})}{\partial(\boldsymbol{v})}$, *20* $\boldsymbol{d}_{12}$, $\boldsymbol{d}_{21}$, *138* $\boldsymbol{d}_r$, *210, 343* $\boldsymbol{D}_1$, $\boldsymbol{D}_2$, *139, 211* $\boldsymbol{D}_s^{(t)}$, *344*

e_1, e_2, *176* e_s, *358* $\mathbf{e}$, $\mathbf{e}'$, *59* E, $\bar{E}$, *37* $\boldsymbol{E}$, *359* e, $\mathcal{E}$, *19*
 $\mathcal{E}_s$, $\mathcal{E}_s^{(i)}$, *207* E, *163*

$f(\boldsymbol{c}, \boldsymbol{r}, t)$, $f(\boldsymbol{C}, \boldsymbol{r}, t)$, *28* $f_s(\boldsymbol{c}_s, \boldsymbol{r}, t)$, *44* $f^{(0)}, f^{(1)}, f^{(2)}, \ldots$, *110* $f_1^{(r)}, f_2^{(r)}$, *134*
 $f^{(s)}(\boldsymbol{c}_1, \ldots \boldsymbol{c}_s, \boldsymbol{r}_1, \ldots \boldsymbol{r}_s, t)$, *312* f, *101* $\boldsymbol{F}$, *46* $\boldsymbol{F}_s$, *47*

$\boldsymbol{g}_{12}$, $\boldsymbol{g}_{21}$, $\boldsymbol{g}'_{12}$, $\boldsymbol{g}'_{21}$, $\boldsymbol{G}$, *53* g_{12}, g_{21}, g'_{12}, g'_{21}, g, g', G, *54* $\mathcal{g}$, $\mathcal{g}'$, $\mathcal{g}$, $\mathcal{g}'$, *150*
 $\boldsymbol{G}_0$, *149* G_0, *150* $\mathcal{G}_0$, $\mathcal{G}_0$, *150* g (gravity), *227*

GREEK SYMBOLS

The notation r_q is used to denote the product $r(r-1)\ldots(r-q+1)$, e.g. on pages 127 and 173.

Differences in notation from earlier editions are noted in the following list, which gives the new equivalents of earlier combinations.

Old	$\mathrm{C}-1$	$k_{12}d\mathbf{k}, k_1 d\mathbf{k}$	$n_1 n_2\{F, G\}$	n_{10}, n_{20}
New	C	$g\alpha_{12}d\mathbf{e}', g\alpha_1 d\mathbf{e}'$	$n^2\{F, G\}$	$\mathbf{x}_1, \mathbf{x}_2$

Old	B, B_1, B_2	$n\boldsymbol{D}_1$, $n\boldsymbol{D}_2$	$\boldsymbol{a}_1^{(0)}$, $\boldsymbol{a}_2^{(0)}$	$2\pi\phi_{12}^{(l)}$
New	2B, 2B_1, 2B_2	$\boldsymbol{D}_1$, $\boldsymbol{D}_2$	$m_0^{\frac{1}{2}}\boldsymbol{a}_1^{(0)}$, $m_0^{\frac{1}{2}}\boldsymbol{a}_2^{(0)}$	$g\phi_{12}^{(l)}$

Old	P_1, P_2
New	$M_1^2 P_1$, $M_2^2 P_2$

The symbols R_1, R_2, R_{12}, R'_{12} used here are unrelated to those of earlier editions.

INTRODUCTION

1. The molecular hypothesis

The purpose of this book is to elucidate some of the observed properties of the natural objects called gases. The method used is a mathematical one.

The foundation on which our work is based is the molecular hypothesis of matter. This postulates that matter is not continuous and indefinitely divisible, but is composed of a finite number of small bodies called molecules. These in any particular case may be all of one kind, or of several kinds: the number of kinds is usually far less than the number of molecules. Free atoms, ions and electrons are considered merely as special types of molecule. The individual molecules are too small to be seen individually even with the most powerful ultra-microscope.

The joint labours of experimental and theoretical physicists have suggested certain hypotheses regarding the structure and interaction of molecules: the precise details, however, are known for only a few kinds of molecule. The mathematician has therefore to consider ideal systems, chosen as illustrating the particular features of actual gas-molecules that are to be studied, and to work out their properties as accurately as possible. The difficulty of this undertaking imposes limitations on the models that can be used. For example, if the systems are not spherically symmetrical, the investigation of their interactions includes the solution of some difficult dynamical problems: the mass-distribution and field of force of a molecule are therefore usually taken to be spherically symmetrical. As this book shows, the investigations even then are very complicated; the complexity is enormously enhanced when the condition of spherical symmetry is relaxed in the least degree. The special models of molecules that are considered in this book are described in 3.3 and in Chapter 11.

2. The kinetic theory of heat

The molecular hypothesis is of great importance in chemistry as well as in physics. For some purposes, particularly in chemistry and crystallography, the molecules can be considered statically; but usually it is essential to take account of the molecular motions. These are not individually visible, but there is evidence that they are extremely rapid. An important extension of the molecular hypothesis is the theory (called the kinetic theory of heat) that the molecules move more or less rapidly, the hotter or colder the body of which they form part; and that the heat energy of the body is in reality mechanical energy, kinetic and potential, of the unseen molecular motions, relative to the body as a whole. The heat energy is thus taken to include the translatory kinetic energy of the molecules, relative to axes moving with the element of the body of which at the time these molecules form part; it includes also kinetic energy of rotation, and kinetic and potential energy of vibration, if the molecular constitution permits of these motions.

Since heat energy is regarded as hidden mechanical energy, it must be expressible in terms of mechanical units. Joule, in fact, showed that the ordinary measure of a given amount of heat energy is proportional to the amount of mechanical energy that can be converted, for example by friction, into the given quantity of heat. The ratio

$$\frac{\text{measure of heat energy in heat units}}{\text{measure of the same energy in mechanical units}}$$

is therefore called Joule's 'Mechanical equivalent of heat'—usually denoted by J.

3. The three states of matter

The molecular hypothesis and the kinetic theory of heat are applicable to matter in general. The three states of matter—solid, liquid, and gaseous—are distinguished merely by the degree of proximity and the intensity of the motions of the molecules. In a solid the molecules are supposed to be packed closely, each hemmed in by its neighbours so that only by a rare chance can it slip between them and get into a new set. If the solid is heated, the motions of the molecules become more violent, and their impacts in general produce a slight thermal expansion of the body. At a certain point, depending on the pressure to which the body is subjected, the motions are sufficiently intense for the molecules, though still close-packed, to be able to pass from one set of neighbours to another set: the liquid state has then been attained. Further application of heat will ultimately lead to a state in which the molecules break the bonds of their mutual attractions, so that they expand to fill any volume available to them; the matter has then attained the gaseous state. At certain pressures and temperatures two states of matter (liquid and gas, solid and liquid, or solid and gas) can coexist in equilibrium; all three states can coexist at a particular pressure and temperature.

4. The theory of gases

In a solid or liquid the mutual forces between pairs of neighbouring molecules are considerable, strong enough, in fact, to hold the mass of molecules together, at least for a time, even if all external pressure is removed. A static picture of a solid is obtained if the molecules are imagined to be rigid bodies in contact: a molecule can be supposed to possess a size, equal to the size of such a rigid body.

The density of a gas is ordinarily low compared with that of the same substance in the liquid or solid form. The molecules in a gas are therefore separated by distances large compared with their sizes, and they move hither and thither, influencing each other only slightly except when two or more happen to approach closely, when they will sensibly deflect each other's paths. In this case the molecules are said to have *encountered* one another; expressed otherwise, an *encounter* has occurred. Obviously an encounter is a less definite event than a collision between two rigid bodies; definiteness can be imparted to the conception of an encounter only by specifying a minimum deflection which must result from the approach of two molecules, if the event is to qualify for the name encounter.

When the molecules are regarded as rigid bodies not surrounded by fields of force, their motion between successive impacts is quite free from any mutual influences: each molecule is said to traverse a *free path* between its successive collisions. The average or *mean* free path will be greater or less, the rarer or denser the gas.

The conception of the free path loses some if its definiteness when the molecules, though still rigid, are surrounded by fields of force. The loss of definiteness is greater still if the molecules are non-rigid. The conception can, however, be applied to gases composed of such molecules, by giving to encounters, in the manner described above, the definiteness that attaches to collisions.

Collisions or encounters in a gas of low density are mainly between pairs of molecules, whereas in a solid or liquid each molecule is usually near or in contact with several neighbours. The legitimate neglect of all but binary encounters in a gas is one of the important simplifications that have enabled the theory of gases to attain its present high development.

5. Statistical mechanics

In ordinary mechanics our aim is usually to determine the events that follow from prescribed initial conditions. Our approach to the theory of a gas must be different from this, for two reasons. Firstly, we never know the detailed initial conditions, that is, the situation and state of motion of every molecule at a prescribed initial instant; secondly, even if we did, our powers are quite unequal to the task of following the subsequent motions of all the many molecules that compose the gas. Hence we do not even attempt to consider the fate of the individual molecules, but interest ourselves only in statistical properties—such as the mean number, momentum or energy of the molecules within an element of volume, averaged over a short time interval, or the average distribution of linear velocities or other motions among these molecules.

It is not only *necessary*, for mathematical reasons, to restrict our aims in this way: it is also physically *adequate*, because experiments on a mass of gas measure only such 'averaged' properties of the gas. Thus our aim is to find out how, for example, the distribution of the 'averaged' or 'mass' motion of a gas, supposed known at one instant, will vary with the time; or again, how a non-uniform mixture of two sets of molecules of different kinds will vary, by the process known as diffusion.

In such attempts, we consider not only the dynamics of the molecular encounters, but also the statistics of the encounters. In this we must use probability assumptions, such, for example, as that the molecules are in general distributed 'at random', or evenly, throughout a small volume, and moreover, that this is true also for the molecules having velocities in a certain range.

The pioneers in the development of the kinetic theory of gases employed such probability considerations intuitively. Their work laid the foundations of a now very extensive branch of theoretical physics, known as statistical mechanics, which deals with systems much more general than gases. This applies probability methods to mechanical problems, and as regards its underlying principles it shares some of the obscurities that attach to the theory of probability itself. These philosophical difficulties were glimpsed already by the founders of the subject, and have been partly though not completely clarified by subsequent discussion.

In one aspect, the theory of probability is merely a definite mathematical theory of arrangements. The simplest problem in that subject is to find in how many different ways m different objects can be set out in n rows ($m > n$), account being taken of the order of the objects in the rows. A great variety of problems of this and more complicated types can be solved, in a completely definite way.

One such problem throws some light on the uniformity of density in a gas. Consider all possible arrangements of m molecules in a certain volume, supposed divided into n cells of equal extent, m being very large compared with n. The number of different arrangements, if regard is paid only to the presence, and not to the order or disposition, of individual molecules in each cell, is n^m. Among these arrangements there will be many in which the total numbers of molecules in the respective cells 1 to n have the same particular set of values $a_1, a_2, \ldots, a_n$, where of course

$$a_1+a_2+\ldots+a_n = m.$$

It is not difficult to show that, when m/n is large, the great majority of the n^m arrangements correspond to distributions for which every number a_1 to a_n differs by a very small fraction from the average number m/n per cell. Hence, if we regard the original n^m arrangements as all equally probable (on the ground, for example, that all the cells are equal in volume, and that there is no reason why any particular molecule should be placed in one cell rather than in another),* we are led to conclude that in any arbitrarily chosen mass of gas the density of the molecules will almost certainly be very nearly uniform throughout the volume.

It needs little consideration to recognize that this somewhat vague statement is very different from the original results about the arrangements

* This, of course, implies that the volume of the molecules is negligible: if the volume of one cell is already largely occupied by molecules, another molecule may be supposed less likely to find a place in this cell than in a relatively empty one.

of the molecules; it is less precise (though it can be expressed in the form of inequalities with narrow limits): moreover, it depends on an assumption as to *a priori* probability. Every statement about probability depends, in a similar way, on some assumption as to *a priori* probability, and is less definite than the results of the arrangement theory.

Similar considerations as to arrangements can be applied to the distribution of a given total amount of translatory kinetic energy between the molecules of a gas when the mass-centre of the whole set is at rest. Here it is assumed that all velocities of a given molecule are *a priori* equally probable. The result obtained is that the velocities of the molecules are almost certainly distributed in a manner agreeing very nearly with a formula first inferred (from intuitive and unjustifiable probability considerations) by Maxwell. The *a priori* assumption cannot be verified: but it can be shown, using a purely dynamical theorem due to Liouville, that as the state of the gas varies with the passage of time, the 'arrangements' which are found initially to be most abundant, as regards both space and velocity-distribution, will always remain most abundant. Hence it is concluded that the uniform density and the Maxwellian velocity-distribution will always be the most probable, though a particular mass of gas may, very rarely (with a degree of improbability that can be estimated), pass through a state which departs to some extent from these usual or *normal* conditions.

These results of statistical mechanics, and others of a like kind, illustrate the use made of probability in the kinetic theory of gases. The results obtained in this theory are usually stated in a quite definite form, but the validity of the conclusions cannot be rated higher than that of the arguments leading thereto. Since in these arguments we appeal to probability, the results of the kinetic theory remain only probable. But the study of statistical mechanics suggests that statements of probability about systems containing a very large number of independent units, such as molecules, usually have a degree of probability so high as to be equivalent, for all practical purposes, to certainty: results which statistical mechanics asserts to be extremely probable are usually taken as rigorously true in experimental work and in thermodynamic theory. Hence though in theory we cannot exclude the rare possibility of a fleeting departure from the most probable states, in practice there need be no question whether the results of kinetic theory will agree with those of experiment.

By statistical mechanics we are led to certain conclusions about the equilibrium states of systems, independent of the mode whereby these equilibrium states are attained; but statistical mechanics does not show how, or at what rate, a system will attain an equilibrium state. This can be determined only if we know certain details about the molecules or other units composing the system, details which, for the purposes of statistical mechanics, can be ignored.

It is the province of a detailed kinetic theory to study the problems of non-equilibrium states, and such investigations occupy the greater part of this book. The probability methods of the kinetic theory are also, however, applied in the earlier chapters (3 and 4) to determine the equilibrium states; the results thus obtained are merely special cases of much more general results of statistical mechanics.

6. The interpretation of kinetic-theory results

The methods of the kinetic theory are successful in giving results of practical interest, although the molecular models chosen are not believed to correspond at all closely with actual molecules. By comparing results obtained for different models, we are able to gain some idea as to how far any particular kind of result depends on this or that feature of the molecular model. It appears that the assumption that the centres of molecules approach each other more closely, the greater their speed of mutual approach, leads to quantitative results for various properties of gases more in accordance with those actually observed than the assumption that the molecules are rigid. Thus a molecule surrounded by a field of force is a better model for quantitative treatment, if not for simple illustrative discussions, than a rigid molecule.

In actual gases, at moderate temperatures, in all but a very small fraction of the molecular encounters the least distance between the centres of the molecules is still distinctly greater than would correspond to an overlapping of the normal detailed structures of the molecules. These structures are therefore not of immediate concern in the kinetic theory of gases; they determine the exterior fields of force, which form the outworks of the molecule, and it is only the nature of the outworks that is here important. It can be adequately specified, for our purpose, by a formula expressing the approximate rate of variation of the force-intensity with distance from the centre of the molecule, over the range of distance outwards from that corresponding to close encounters. At smaller distances the field might have any value without affecting the kinetic-theory calculations; the actual structure of the molecule within this minimum distance can be ignored, and the molecule may without detriment be regarded as a point-centre of force.

The restriction of kinetic-theory calculations to molecules that are spherically symmetrical is also not of such importance as might appear from the practical certainty that many actual molecules are not at all spherically symmetrical. This is because the molecules in general rotate, and at encounters they may be oriented relative to each other in any manner: the detailed consequences of a particular encounter depend on the orientations, but such consequences averaged over a large number of encounters are probably not very different from the corresponding averaged results of the encounters of a set of spherically symmetrical molecules, the force between pairs of which, at any distance, is equal to the average, over all orientations, of the force between pairs of the actual molecules whose centres are at that distance apart. Such averaging of the consequences of encounters is of the essence of kinetic-theory calculations, so that many of the results obtained in this book should be correct qualitatively, and not far from correct quantitatively, for gases whose molecules are non-spherical. The chief exceptions are in problems involving the total heat energy, since the actual molecules may possess an average amount of internal energy different from that of the spherically symmetric models.

7. The interpretation of some macroscopic concepts

Our aim is to explain things that are seen and directly measurable by means of imagined things that are not seen and not directly measurable. The general lines along which we are to proceed have already been indicated, in describing the molecular hypothesis and the kinetic theory of heat. There remain, however, further points on which there is room for freedom of interpretation. The criterion by which our choice is to be judged is whether the relations found between the quantities we identify with measurable macroscopic quantities do or do not approximate to the observed relations between those macroscopic quantities. Success in this test affords ground for a reasonable expectation that any hitherto unknown macroscopic relation suggested by the kinetic theory on its own basis of interpretation will be confirmed on experimental trial.

It should be emphasized that only approximate agreement is to be expected between observed macroscopic relations and those inferred from kinetic theory, because some divergence between the two sets of relations may reasonably be attributed to the imperfect representation of the actual molecules by the 'model' molecules with which the mathematician works.

The kinetic-theory interpretations of some typical macroscopic properties are briefly summarized here.

The combined masses of the molecules of a set are taken as giving the macroscopically observed mass of the set.

The heat energy of a small portion of matter is identified with the translatory kinetic energy of the molecular motions relative to the element as a whole, together with the total of such other forms of molecular energy as are interchangeable with translatory kinetic energy at encounters. Thus in diatomic and polyatomic molecules the relative motion of the atomic nuclei may contribute kinetic and potential energy to the heat energy; but energy like the kinetic and potential energy of electrons in an inert gas-molecule, which is normally unaffected by encounters, is neglected. Correspondingly, certain of our molecular models, such as the *smooth* rigid spherical model, make no provision for a possible interchange of translatory and rotatory kinetic energy: we can ignore the energy of rotation in discussing these.

The pressure of a gas on a bounding surface is identified with the mean time-rate of communication of momentum to the surface, per unit area, by molecular impacts; the momentum is imparted in a more or less discontinuous manner, but the individual impulses are so small, frequent, and numerous as to simulate a continuous pressure. In addition to this momentum-pressure there is a much smaller stress due to the action at a distance between the molecules of the gas and the wall; this is normally not taken into account in the present book.

The mean translatory kinetic energy per molecule, relative to the general motion of the gas, is taken to be proportional to the thermodynamic temperature, the constant of proportionality depending on the units of energy and temperature, but not on the gas. Such an identification is permissible (on the understanding that it is to be justified by its results), so long as we are concerned with the phenomena of a single portion of gas; but the question arises whether this identification will be valid for different portions of gas, composed of molecules of different kinds. So long as our discussion is confined purely to the phenomena of gases, the question seems to depend on whether, when we mix two different gases, to which, according to this definition, we ascribe equal temperatures, the same temperature will characterize the mixture, as is observed to be the case when we mix actual rare gases whose *thermodynamic* temperatures are equal. The kinetic theory is able to give a fairly satisfactory affirmative answer to this question (4.3), to this extent justifying its procedure as regards temperature definition. But questions as to the thermal equilibrium of two different gases with another body (say a diathermanous wall between the compartments of a vessel containing the two gases) are outside the scope of the kinetic theory of gases: they lie within the domain of statistical mechanics, which considers assemblies much more general than gases.

8. Quantum theory

A wise conservatism, rather than reasons valid *a priori*, prompted the pioneers of the kinetic theory to attribute to their imagined molecules the same rules of behaviour—or, in technical language, the same mechanical laws of motion—as those that characterize the objects of our ordinary experience. Their rigid spherical molecules were idealizations of ordinary billiard balls, while their point-centres of force were suggested by planets viewed, from the large astronomical standpoint, as point-centres of gravitational attraction.

The consequences of this assumed behaviour of molecules correspond closely in general to the observed behaviour of gases; this supports the view that molecules do behave in the supposed way. It is not a matter for surprise, however, that the kinetic-theory consequences of the assumption do not fit the whole range of observed facts. The discrepancies are of a nature to be explained by attributing to the molecules rules or laws of behaviour (including statistical laws) that deviate from the classical laws in a way suggested by the study of many other phenomena of matter, particularly spectroscopic phenomena. The new laws and the body of science dealing with them, known as the quantum laws and quantum theory, are only briefly touched on in this book (Chapter 17), whose main aim is to record the basic development of the kinetic theory of gases, founded on classical mechanics.

1

VECTORS AND TENSORS

1.1. Vectors

The notation and calculus of vectors, and also of three-dimensional Cartesian tensors, are largely used in this book. In this chapter we summarize the vector and tensor notation and calculus that we adopt.

Any physical quantity possessing both magnitude and direction is called a vector quantity, or, briefly, a vector. Such quantities will be denoted by symbols in heavy (Clarendon) type, in various founts, as, for example,*

$$\boldsymbol{a}, \boldsymbol{A}, \boldsymbol{\mathscr{C}}, \boldsymbol{\omega}, \mathbf{n}.$$

The (positive) magnitude of a vector denoted by a Clarendon symbol will usually be denoted by the same symbol in the corresponding ordinary type, e.g. for $\boldsymbol{a}$, $\boldsymbol{A}$, $\boldsymbol{\mathscr{C}}$, $\boldsymbol{\omega}$ by a, A, $\mathscr{C}$, ω (of course in the case of a unit vector no such magnitude symbol is needed).

The component of a vector $\boldsymbol{A}$ along a direction inclined to $\boldsymbol{A}$ at an angle θ $(0 \leqslant \theta \leqslant \pi)$ is defined to be $A \cos \theta$; this may be positive or negative. Any vector is completely specified when its components in three mutually perpendicular directions are given. When these directions are those of the axes Ox, Oy, Oz of a Cartesian system,† the components are called the rectangular Cartesian components relative to these axes. They may be denoted by adding the suffices x, y, z to the symbol denoting the magnitude of the vector (e.g. a_x, a_y, a_z denote the x, y, z components of $\boldsymbol{a}$), or by special symbols (as in 1.2 for $\boldsymbol{r}$ and $\boldsymbol{C}$, and as in 1.33 for $\boldsymbol{c}$). The magnitude of a vector is given in terms of its rectangular components by an equation of the form

$$a^2 = a_x^2 + a_y^2 + a_z^2. \qquad (1.1, 1)$$

Let $\boldsymbol{a}$ be a vector whose components relative to the axes Ox, Oy, Oz are

* For the convenience of the reader certain special conventions regarding such types will be made in this book, as follows:

(i) vectors whose magnitude is unity (or, briefly, *unit vectors*), will be denoted by ordinary small upright letters in Clarendon type, namely

$$\mathbf{e}, \mathbf{h}, \mathbf{i}, \mathbf{j}, \mathbf{k}, \mathbf{l}, \mathbf{n};$$

(ii) script Clarendon capitals, such as

$$\boldsymbol{\mathscr{C}}, \boldsymbol{\mathscr{G}}_0,$$

will denote certain non-dimensional vectors associated with vectors represented by the corresponding Clarendon italic capitals, namely

$$\boldsymbol{C}, \boldsymbol{G}_0.$$

† Throughout this book all Cartesian axes of reference are understood to be mutually perpendicular (or *orthogonal*) and right-handed.

a_x, a_y, a_z, and let Ox', Oy', Oz' be a second set of orthogonal axes whose direction cosines relative to the first set are (l_1, m_1, n_1), (l_2, m_2, n_2), (l_3, m_3, n_3). Then the components $a_{x'}, a_{y'}, a_{z'}$ of $\boldsymbol{a}$ relative to the second set of axes are given by

$$a_{x'} = l_1 a_x + m_1 a_y + n_1 a_z \tag{1.1, 2}$$

and two similar equations. Similarly

$$a_x = l_1 a_{x'} + l_2 a_{y'} + l_3 a_{z'}, \tag{1.1, 3}$$

and so on. These equations take a simpler form if in place of $l_1, l_2, l_3, m_1, m_2, \ldots$, we write $t_{xx'}, t_{xy'}, t_{xz'}, t_{yx'}, t_{yy'}, \ldots$. The nine symbols $t_{\alpha\beta'}$, where α and β may stand for x or y or z, define a matrix which we call the transformation matrix; the typical element $t_{\alpha\beta'}$ of this matrix is the cosine of the angle between the axes $O\alpha$, $O\beta'$. In this notation, the equations of transformation may be written

$$a_{\beta'} = \sum_{\alpha} a_\alpha t_{\alpha\beta'}, \tag{1.1, 4}$$

$$a_\alpha = \sum_{\beta'} t_{\alpha\beta'} a_{\beta'}. \tag{1.1, 5}$$

1.11. Sums and products of vectors

The sum of two vectors is defined as the vector whose components are the sums of the corresponding components of the vectors. Thus the rule for the addition of vectors is the same as the parallelogram law for the composition of forces or velocities.

Let $\boldsymbol{a}$, $\boldsymbol{b}$ be two vectors inclined at an angle θ ($\leqslant \pi$). Then $ab\cos\theta$ is a scalar quantity (i.e. a quantity possessing magnitude but not direction). It is called the *scalar product* of $\boldsymbol{a}$ and $\boldsymbol{b}$, and is denoted by $\boldsymbol{a}.\boldsymbol{b}$. In terms of the components of $\boldsymbol{a}$ and $\boldsymbol{b}$,

$$\boldsymbol{a}.\boldsymbol{b} = a_x b_x + a_y b_y + a_z b_z = \boldsymbol{b}.\boldsymbol{a}. \tag{1.11, 1}$$

From this it follows that

$$(\boldsymbol{a}+\boldsymbol{b}).(\boldsymbol{c}+\boldsymbol{d}) = \boldsymbol{a}.\boldsymbol{c}+\boldsymbol{b}.\boldsymbol{c}+\boldsymbol{a}.\boldsymbol{d}+\boldsymbol{b}.\boldsymbol{d},$$

of which the following are important special cases

$$(\boldsymbol{a}+\boldsymbol{b}).(\boldsymbol{a}+\boldsymbol{b}) = a^2+2\boldsymbol{a}.\boldsymbol{b}+b^2,$$

$$(\boldsymbol{a}-\boldsymbol{b}).(\boldsymbol{a}-\boldsymbol{b}) = a^2-2\boldsymbol{a}.\boldsymbol{b}+b^2,$$

$$(\boldsymbol{a}+\boldsymbol{b}).(\boldsymbol{a}-\boldsymbol{b}) = a^2-b^2.$$

The *vector product* of the vectors $\boldsymbol{a}$, $\boldsymbol{b}$ is defined to be the vector of magnitude $ab\sin\theta$, perpendicular to both $\boldsymbol{a}$ and $\boldsymbol{b}$, and in the direction of translation of a right-handed screw, rotated in the sense *from* $\boldsymbol{a}$ *to* $\boldsymbol{b}$, through the angle θ ($\leqslant \pi$) between $\boldsymbol{a}$ and $\boldsymbol{b}$. It is here denoted by $\boldsymbol{a}\wedge\boldsymbol{b}$. Its Cartesian components are

$$a_y b_z - a_z b_y, \quad a_z b_x - a_x b_z, \quad a_x b_y - a_y b_x. \tag{1.11, 2}$$

Using these expressions, it may readily be proved that

$$\boldsymbol{a}\wedge(\boldsymbol{b}\wedge\boldsymbol{c}) = (\boldsymbol{a}.\boldsymbol{c})\boldsymbol{b}-(\boldsymbol{a}.\boldsymbol{b})\boldsymbol{c}. \qquad (1.11, 3)$$

In connection with vector products it is of interest to distinguish a special class of vectors, associated with rotation about an axis: typical vectors of this class are the angular velocity of a body, and the moment of a force. The direction of such a 'rotation-vector' is taken to be along the axis, in the direction of translation of a right-handed screw rotated in the sense of the quantity considered. Thus the sign of a rotation-vector depends on a convention as to the relation between the positive directions of translation along, and rotation about, a given axis, and would be reversed if this convention were altered. Since the same convention is used in the definition of a vector product, the vector product of two ordinary vectors is an example of a rotation-vector: the vector product of an ordinary vector and a rotation-vector, in whose definition the convention is used twice, does not have its sign altered if the convention is changed, and so is an ordinary vector.

In mechanical equations rotation-vectors can be equated only to other rotation-vectors, and not to vectors of other types.

1.2. Functions of position

Any point in space may be specified either by the '*position-vector*' $\boldsymbol{r}$ giving its displacement from some origin O, or by its Cartesian coordinates x, y, z (the components of $\boldsymbol{r}$), referred to a set of rectangular axes with O as origin. For brevity, the phrase 'at the point $\boldsymbol{r}$ at time t' will usually be contracted to 'at $\boldsymbol{r}, t$'.

A function ϕ of position may be denoted by $\phi(\boldsymbol{r})$ or $\phi(x,y,z)$, if scalar; if it is a vector function, the functional symbol will be printed in heavy type, as $\boldsymbol{\phi}(\boldsymbol{r})$, and its Cartesian components will be denoted by $\phi_x(\boldsymbol{r})$, $\phi_y(\boldsymbol{r})$, $\phi_z(\boldsymbol{r})$, or, more briefly, by ϕ_x, ϕ_y, ϕ_z.

The equations of transformation of the operator whose components are $\partial/\partial x$, $\partial/\partial y$, $\partial/\partial z$, from one set of axes Ox, Oy, Oz to another set Ox', Oy', Oz', are the same as for a vector: for, in the notation of 1.1,

$$\frac{\partial}{\partial x'} = \frac{\partial x}{\partial x'}\frac{\partial}{\partial x}+\frac{\partial y}{\partial x'}\frac{\partial}{\partial y}+\frac{\partial z}{\partial x'}\frac{\partial}{\partial z}$$

$$= l_1\frac{\partial}{\partial x}+m_1\frac{\partial}{\partial y}+n_1\frac{\partial}{\partial z}.$$

Thus the operator in question may be treated as a vector; it will be denoted by $\partial/\partial\boldsymbol{r}$ or ∇.

The result of the operation of $\partial/\partial\boldsymbol{r}$ on a scalar function $\phi(\boldsymbol{r})$ is called the gradient of the function; it is a vector with components $\partial\phi/\partial x$, $\partial\phi/\partial y$, $\partial\phi/\partial z$.

When $\phi(r)$ is a function of the magnitude r alone, it is readily seen that

$$\frac{\partial\phi}{\partial \boldsymbol{r}} = \frac{\boldsymbol{r}}{r}\frac{\partial\phi}{\partial r}; \tag{1.2, 1}$$

in particular,

$$\frac{\partial r^2}{\partial \boldsymbol{r}} = 2\boldsymbol{r}. \tag{1.2, 2}$$

The scalar product of $\partial/\partial\boldsymbol{r}$ and a vector function $\boldsymbol{\phi}(\boldsymbol{r})$, i.e. $\partial/\partial\boldsymbol{r}.\boldsymbol{\phi}$ or $\nabla.\boldsymbol{\phi}$, is called the divergence of the vector (sometimes written as $\operatorname{div}\boldsymbol{\phi}$); it is, of course, invariant for a change of axes. Clearly

$$\frac{\partial}{\partial \boldsymbol{r}}.\boldsymbol{\phi} = \frac{\partial\phi_x}{\partial x} + \frac{\partial\phi_y}{\partial y} + \frac{\partial\phi_z}{\partial z}. \tag{1.2, 3}$$

Similarly, if $\boldsymbol{C}$ is a vector whose x, y, z components are U, V, W, and $\boldsymbol{\phi}(\boldsymbol{C})$ is any vector function of $\boldsymbol{C}$,

$$\frac{\partial}{\partial \boldsymbol{C}}.\boldsymbol{\phi}(\boldsymbol{C}) = \frac{\partial\phi_x}{\partial U} + \frac{\partial\phi_y}{\partial V} + \frac{\partial\phi_z}{\partial W}, \tag{1.2, 4}$$

where ϕ_x, ϕ_y, ϕ_z are the x, y, z components of $\boldsymbol{\phi}(\boldsymbol{C})$. Likewise if $\phi(\boldsymbol{C})$ is any scalar function of $\boldsymbol{C}$, an associated vector is

$$\frac{\partial\phi}{\partial \boldsymbol{C}}, \tag{1.2, 5}$$

with components $\frac{\partial\phi}{\partial U}, \frac{\partial\phi}{\partial V}, \frac{\partial\phi}{\partial W}$. In particular, if $\phi(\boldsymbol{C}) = C^2 = U^2 + V^2 + W^2$, it is readily seen that

$$\frac{\partial C^2}{\partial \boldsymbol{C}} = 2\boldsymbol{C}; \tag{1.2, 6}$$

more generally, if $\phi(\boldsymbol{C}) = F(C^2)$, where F is any function, it is easy to verify that

$$\frac{\partial\phi}{\partial \boldsymbol{C}} = \frac{\partial F(C^2)}{\partial \boldsymbol{C}} = 2\boldsymbol{C}\frac{\partial F}{\partial C^2}. \tag{1.2, 7}$$

Again, if $\boldsymbol{A}$ is any vector independent of $\boldsymbol{C}$, it is easy to verify that

$$\frac{\partial}{\partial \boldsymbol{C}}(\boldsymbol{C}.\boldsymbol{A}) = \boldsymbol{A}. \tag{1.2, 8}$$

1.21. Volume elements and spherical surface elements

An element of volume enclosing the point $\boldsymbol{r}$ or (x, y, z) will be denoted by the symbol $d\boldsymbol{r}$. This must be distinguished from $\boldsymbol{dr}$, which denotes the small vector joining $\boldsymbol{r}$ to an adjacent point, and from dr, which denotes a small increment in the length r. If Cartesian coordinates are employed, it is convenient to take $d\boldsymbol{r}$ as the parallelepiped $dx\,dy\,dz$; using polar coordinates r, θ, φ, we take $d\boldsymbol{r} = r^2 \sin\theta\, dr\, d\theta\, d\varphi$, and so on.

If $\mathbf{k}$ denotes a unit vector, then the point whose position vector, relative to an origin O, is $\mathbf{k}$, lies on a sphere of unit radius (or 'unit sphere') with centre O. Thus $d\mathbf{k}$ must be interpreted not as an element of volume, but as an element of the surface of the unit sphere, or, what is equivalent, as the element of solid angle subtended by this element of surface at O; the element $d\mathbf{k}$ will be supposed to include the point $\mathbf{k}$. The element may be of any form; if $\mathbf{k}$ is specified by its polar angles θ, φ, it is appropriate to take $d\mathbf{k} = \sin\theta\, d\theta\, d\varphi$.

1.3. Dyadics and tensors

Any two vectors $\boldsymbol{a}$, $\boldsymbol{b}$ determine, relative to the set of axes chosen, a matrix, each of whose components is the product of one component of $\boldsymbol{a}$ with one of $\boldsymbol{b}$, namely

$$\left.\begin{matrix} a_x b_x, & a_x b_y, & a_x b_z, \\ a_y b_x, & a_y b_y, & a_y b_z, \\ a_z b_x, & a_z b_y, & a_z b_z. \end{matrix}\right\} \tag{1.3, 1}$$

Such a matrix gives the ordered components, relative to the given axes, of an entity called a *dyadic*, which will be denoted by $\boldsymbol{ab}$.* It is to be noted that the dyadic $\boldsymbol{ba}$ differs from $\boldsymbol{ab}$ unless the vectors $\boldsymbol{a}$, $\boldsymbol{b}$ are parallel. The order of the suffixes in the matrix may be remembered by aid of the symbol $a_{\underset{\downarrow}{x}} b_{x\rightarrow}$, indicating how the suffices succeeding x are disposed in (1.3, 1).

The components of the dyadic $\boldsymbol{ab}$ relative to a second set of axes Ox', Oy', Oz' are given, in the notation of 1.1, by

$$\begin{aligned} a_{\alpha'} b_{\beta'} &= (\sum_\gamma a_\gamma t_{\gamma\alpha'})(\sum_\delta b_\delta t_{\delta\beta'}) \\ &= \sum_\gamma \sum_\delta a_\gamma b_\delta t_{\gamma\alpha'} t_{\delta\beta'}. \end{aligned} \tag{1.3, 2}$$

Any 3×3 matrix (related to a set of axes Ox, Oy, Oz) of the type

$$\left.\begin{matrix} w_{xx}, & w_{xy}, & w_{xz}, \\ w_{yx}, & w_{yy}, & w_{yz}, \\ w_{zx}, & w_{zy}, & w_{zz}, \end{matrix}\right\} \tag{1.3, 3}$$

of which the general term may be denoted by $w_{\alpha\beta}$, is said to constitute the array of components (relative to those axes) of a second-order *tensor* (which will be denoted by the symbol w), provided that the components, $w_{\alpha'\beta'}$ say, relative to any other set of axes Ox', Oy', Oz', are such that

$$w_{\alpha'\beta'} = \sum_\gamma \sum_\delta w_{\gamma\delta} t_{\gamma\alpha'} t_{\delta\beta'}. \tag{1.3, 4}$$

This set of equations of transformation is the same as the set (1.3, 2) for the components of a dyadic, so that every dyadic is a tensor.

* This symbol must be carefully distinguished from $\boldsymbol{a}.\boldsymbol{b}$. The insertion of the dot changes the symbol for the dyadic to that for the scalar product of two vectors.

The matrix of components of a dyadic or tensor must be carefully distinguished from the *determinant* which might be formed from the matrix; the matrix is an *ordered set* of numbers, and the determinant is a certain sum of *products* of these numbers.

The sum of two tensors is defined as the tensor whose components are equal to the sums of the corresponding components of the two tensors.

The product of a tensor and a scalar magnitude k is defined as the tensor whose components are each k times the corresponding components of the original tensor.

If the rows and columns of the matrix (1.3, 3) are interchanged, a new tensor is derived, which is known as the tensor conjugate to w, and denoted by $\overline{\mathsf{w}}$. When this is identical with w, w is said to be *symmetrical*. If w is not symmetrical, a symmetrical tensor denoted by $\overline{\overline{\mathsf{w}}}$ can be derived from it, whose components are the means of the corresponding components of w and $\overline{\mathsf{w}}$, so that

$$\overline{\overline{\mathsf{w}}} = \tfrac{1}{2}(\overline{\mathsf{w}}+\mathsf{w}). \tag{1.3, 5}$$

The components of $\overline{\overline{\mathsf{w}}}$ are

$$\begin{matrix} w_{xx}, & \tfrac{1}{2}(w_{xy}+w_{yx}), & \tfrac{1}{2}(w_{xz}+w_{zx}), \\ \tfrac{1}{2}(w_{yx}+w_{xy}), & w_{yy}, & \tfrac{1}{2}(w_{yz}+w_{zy}), \\ \tfrac{1}{2}(w_{zx}+w_{xz}), & \tfrac{1}{2}(w_{zy}+w_{yz}), & w_{zz}. \end{matrix}$$

The simplest symmetrical tensor is the unit tensor U, whose components relative to any set of orthogonal axes are given by

$$U_{xx} = U_{yy} = U_{zz} = 1, \quad U_{xy} = U_{yx} = \text{etc.} = 0; \tag{1.3, 6}$$

it is easy to show that they are unaltered by transformation of orthogonal axes.

The sum of the diagonal terms of the dyadic $\boldsymbol{ab}$ is $a_x b_x + a_y b_y + a_z b_z$ or $\boldsymbol{a}.\boldsymbol{b}$, which is invariant for change of axes. Thus the sum $w_{xx}+w_{yy}+w_{zz}$ of the diagonal terms of any tensor w will also be an invariant; it is known as the *divergence* of the tensor. If the divergence of a tensor vanishes, it is said to be *non-divergent*.

From any tensor w a non-divergent tensor, denoted by $\mathring{\mathsf{w}}$, can be derived, by subtraction of one-third of the divergence from each of the diagonal terms: thus

$$\mathring{\mathsf{w}} = \mathsf{w} - \tfrac{1}{3}(w_{xx}+w_{yy}+w_{zz})\,\mathsf{U}. \tag{1.3, 7}$$

The components of $\mathring{\mathsf{w}}$ are

$$\begin{matrix} \tfrac{1}{3}(2w_{xx}-w_{yy}-w_{zz}), & w_{xy}, & w_{xz}, \\ w_{yx}, & \tfrac{1}{3}(2w_{yy}-w_{xx}-w_{zz}), & w_{yz}, \\ w_{zx}, & w_{zy}, & \tfrac{1}{3}(2w_{zz}-w_{xx}-w_{yy}). \end{matrix}$$

The symbols $\circ$ and $=$ may both be placed above a tensor symbol, as in $\mathring{\overline{\overline{\mathsf{w}}}}$, which in accordance with (1.3, 7) signifies

$$\overline{\overline{\mathsf{w}}} - \tfrac{1}{3}(w_{xx}+w_{yy}+w_{zz})\,\mathsf{U}. \tag{1.3, 8}$$

Clearly the components of $\overset{\circ}{\overline{\overline{\mathsf{w}}}}$ are

$$\begin{array}{lll} \tfrac{1}{3}(2w_{xx}-w_{yy}-w_{zz}), & \tfrac{1}{2}(w_{xy}+w_{yx}), & \tfrac{1}{2}(w_{xz}+w_{zx}), \\ \tfrac{1}{2}(w_{yx}+w_{xy}), & \tfrac{1}{3}(2w_{yy}-w_{xx}-w_{zz}), & \tfrac{1}{2}(w_{yz}+w_{zy}), \\ \tfrac{1}{2}(w_{zx}+w_{xz}), & \tfrac{1}{2}(w_{zy}+w_{yz}), & \tfrac{1}{3}(2w_{zz}-w_{xx}-w_{yy}). \end{array}$$

If $\mathbf{h}, \mathbf{i}, \mathbf{j}$ are three mutually perpendicular unit vectors (1.1, footnote),

$$\mathbf{hh}+\mathbf{ii}+\mathbf{jj} = \mathsf{U}, \tag{1.3, 9}$$

as is evident if the elements of the tensors are written out in full. Hence also

$$\overset{\circ}{\mathbf{hh}}+\overset{\circ}{\mathbf{ii}}+\overset{\circ}{\mathbf{jj}} = \overset{\circ}{\mathsf{U}} = 0. \tag{1.3, 10}$$

1.31. Products of vectors or tensors with tensors

The product $\mathsf{w}.\boldsymbol{a}$ of the tensor w and a vector $\boldsymbol{a}$ is defined as the vector whose components are given by

$$(\mathsf{w}.\boldsymbol{a})_\alpha = \sum_\beta w_{\alpha\beta}a_\beta. \tag{1.31, 1}$$

The product $\boldsymbol{a}.\mathsf{w}$ (which is in general not equal to $\mathsf{w}.\boldsymbol{a}$) is similarly defined by the relation

$$(\boldsymbol{a}.\mathsf{w})_\alpha = \sum_\beta a_\beta w_{\beta\alpha}.$$

Clearly

$$\mathsf{w}.\boldsymbol{a} = \boldsymbol{a}.\overline{\mathsf{w}}, \quad \mathsf{U}.\boldsymbol{a} = \boldsymbol{a}.\mathsf{U} = \boldsymbol{a}, \tag{1.31, 2}$$

and if p is any symmetrical tensor, $\mathsf{p}.\boldsymbol{a} = \boldsymbol{a}.\mathsf{p}$.

The *simple product* $\mathsf{w}.\mathsf{w}'$ of two tensors w, w' is defined as the tensor with components

$$(\mathsf{w}.\mathsf{w}')_{\alpha\beta} = \sum_\gamma w_{\alpha\gamma}w'_{\gamma\beta}. \tag{1.31, 3}$$

The *double*, or *scalar product* $\mathsf{w}:\mathsf{w}'$ is defined as the scalar equal to the divergence of $\mathsf{w}.\mathsf{w}'$; thus

$$\mathsf{w}:\mathsf{w}' = \sum_\alpha\sum_\beta w_{\alpha\beta}w'_{\beta\alpha} = \mathsf{w}':\mathsf{w}, \tag{1.31, 4}$$

that is, it is equal to the sum of the products of corresponding components of w and $\overline{\mathsf{w}}'$. In particular, $\mathsf{w}:\overline{\mathsf{w}}$ is the sum of the squares of the components of w, and $\mathsf{U}:\mathsf{U} = 3$.

From these definitions it follows that each of the above products satisfies the distributive law of ordinary algebra; but the commutative law is not in general satisfied, since, except in the case of the double product of two tensors, the terms of the product cannot be interchanged without altering the value of the expression.

An important particular case of (1.31, 4) is

$$\mathsf{U}:\mathsf{w} = w_{xx}+w_{yy}+w_{zz}, \tag{1.31, 5}$$

which gives the divergence of w. Thus $\mathring{\mathsf{w}}:\mathsf{U}$ or $\mathsf{U}:\mathring{\mathsf{w}}$ is zero (by the definition of $\mathring{\mathsf{w}}$). Also (1.3, 7) may be written

$$\mathring{\mathsf{w}} = \mathsf{w} - \tfrac{1}{3}\mathsf{U}(\mathsf{U}:\mathsf{w}), \tag{1.31, 6}$$

and so
$$\begin{aligned}\mathring{\mathsf{w}}:\mathring{\mathsf{w}}' &= \mathring{\mathsf{w}}:\{\mathsf{w}' - \tfrac{1}{3}\mathsf{U}(\mathsf{U}:\mathsf{w}')\}\\ &= \mathring{\mathsf{w}}:\mathsf{w}' - \tfrac{1}{3}(\mathring{\mathsf{w}}:\mathsf{U})(\mathsf{U}:\mathsf{w}')\\ &= \mathring{\mathsf{w}}:\mathsf{w}',\end{aligned}$$

whence, by symmetry, $\quad \mathring{\mathsf{w}}:\mathring{\mathsf{w}}' = \mathring{\mathsf{w}}:\mathsf{w}' = \mathsf{w}:\mathring{\mathsf{w}}'.$ (1.31, 7)

Again, it follows from (1.31, 4) that

$$\overline{\mathsf{w}}:\overline{\mathsf{w}}' = \sum_{\alpha}\sum_{\beta} w_{\beta\alpha} w'_{\alpha\beta} = \mathsf{w}:\mathsf{w}', \tag{1.31, 8}$$

and so
$$\begin{aligned}\overline{\overline{\mathsf{w}}}:\overline{\overline{\mathsf{w}}}' &= \tfrac{1}{2}\mathsf{w}:\overline{\overline{\mathsf{w}}}' + \tfrac{1}{2}\overline{\mathsf{w}}:\overline{\overline{\mathsf{w}}}'\\ &= \tfrac{1}{2}\mathsf{w}:\overline{\overline{\mathsf{w}}}' + \tfrac{1}{2}\mathsf{w}:\overline{\overline{\mathsf{w}}}'\\ &= \mathsf{w}:\overline{\overline{\mathsf{w}}}',\end{aligned}$$

whence, by symmetry,
$$\overline{\overline{\mathsf{w}}}:\overline{\overline{\mathsf{w}}}' = \mathsf{w}:\overline{\overline{\mathsf{w}}}' = \overline{\overline{\mathsf{w}}}:\mathsf{w}'. \tag{1.31, 9}$$

1.32. Theorems on dyadics

Since dyadics form a special class of tensors, the above notations and results for tensors also apply to dyadics.

If $\boldsymbol{ab}$ is symmetrical, $\boldsymbol{ab} = \boldsymbol{ba}$, and $\boldsymbol{a}$ must be a scalar multiple of $\boldsymbol{b}$; further, whatever $\boldsymbol{a}$ and $\boldsymbol{b}$,

$$\overline{\boldsymbol{ab}} = \boldsymbol{ba}, \quad \overline{\overline{\boldsymbol{ab}}} = \tfrac{1}{2}(\boldsymbol{ab}+\boldsymbol{ba}) = \overline{\overline{\boldsymbol{ba}}}. \tag{1.32, 1}$$

The notation $\mathring{\boldsymbol{ab}}$ may be illustrated by the special case $\mathring{\boldsymbol{CC}}$, where $\boldsymbol{C}$ is a vector with amplitude C and components (U, V, W); the components of this tensor are

$$\left.\begin{matrix} U^2 - \tfrac{1}{3}C^2, & UV, & UW,\\ VU, & V^2-\tfrac{1}{3}C^2, & VW,\\ WU, & WV, & W^2-\tfrac{1}{3}C^2.\end{matrix}\right\} \tag{1.32, 2}$$

The product of a dyadic $\boldsymbol{ab}$ by a vector $\boldsymbol{d}$ has a specially simple form; for

$$\{(\boldsymbol{ab}).\boldsymbol{d}\}_\alpha = \sum_\beta (\boldsymbol{ab})_{\alpha\beta} d_\beta = \sum_\beta a_\alpha b_\beta d_\beta = a_\alpha(\boldsymbol{b}.\boldsymbol{d}),$$

and so
$$(\boldsymbol{ab}).\boldsymbol{d} = \boldsymbol{a}(\boldsymbol{b}.\boldsymbol{d}). \tag{1.32, 3}$$

Similarly
$$\boldsymbol{d}.(\boldsymbol{ab}) = (\boldsymbol{d}.\boldsymbol{a})\boldsymbol{b}. \tag{1.32, 4}$$

The scalar product of the vector $\mathsf{w}.\boldsymbol{a}$ by a vector $\boldsymbol{b}$ is equal to the double product of the tensors w, $\boldsymbol{ab}$; for

$$(\mathsf{w}.\boldsymbol{a}).\boldsymbol{b} = \sum_\alpha (\mathsf{w}.\boldsymbol{a})_\alpha b_\alpha = \sum_\alpha\sum_\beta w_{\alpha\beta} a_\beta b_\alpha = \sum_\alpha\sum_\beta w_{\alpha\beta}(\boldsymbol{ab})_{\beta\alpha} = \mathsf{w}:\boldsymbol{ab}. \tag{1.32, 5}$$

Similarly
$$\boldsymbol{b}.(\boldsymbol{a}.\mathsf{w}) = \boldsymbol{ba}:\mathsf{w}. \tag{1.32, 6}$$

When w is itself of the form $\boldsymbol{cd}$, it follows that

$$\boldsymbol{ab}:\boldsymbol{cd} = \boldsymbol{a}.(\boldsymbol{b}.\boldsymbol{cd}) = \boldsymbol{a}.\{(\boldsymbol{b}.\boldsymbol{c})\boldsymbol{d}\} = (\boldsymbol{a}.\boldsymbol{d})(\boldsymbol{b}.\boldsymbol{c}), \tag{1.32, 7}$$

whence also

$$\boldsymbol{ab}:\boldsymbol{cd} = \boldsymbol{ac}:\boldsymbol{bd}. \tag{1.32, 8}$$

From these results and (1.31, 6, 7) it follows that

$$\begin{aligned}\overset{\circ}{\boldsymbol{C}_1\boldsymbol{C}_1}:\overset{\circ}{\boldsymbol{C}_2\boldsymbol{C}_2} &= \overset{\circ}{\boldsymbol{C}_1\boldsymbol{C}_1}:\boldsymbol{C}_2\boldsymbol{C}_2\\ &= \boldsymbol{C}_1\boldsymbol{C}_1:\boldsymbol{C}_2\boldsymbol{C}_2 - \tfrac{1}{3}C_1^2(\mathsf{U}:\boldsymbol{C}_2\boldsymbol{C}_2)\\ &= (\boldsymbol{C}_1.\boldsymbol{C}_2)^2 - \tfrac{1}{3}C_1^2C_2^2.\end{aligned} \tag{1.32, 9}$$

Again, if w is independent of $\boldsymbol{C}$,

$$\frac{\partial}{\partial \boldsymbol{C}}(\mathsf{w}:\overset{\circ}{\boldsymbol{CC}}) = \frac{\partial}{\partial \boldsymbol{C}}(\mathring{\overline{\overline{\mathsf{w}}}}:\boldsymbol{CC}) = 2\mathring{\overline{\overline{\mathsf{w}}}}.\boldsymbol{C}. \tag{1.32, 10}$$

1.33. Dyadics involving differential operators

One of the vectors in a dyadic may be a vector differential operator such as $\partial/\partial\boldsymbol{r}$ ($\equiv\nabla$). If, for example, $\boldsymbol{c}$ is a vector with components u, v, w, the components of $\frac{\partial}{\partial\boldsymbol{r}}\boldsymbol{c}$ ($\equiv\nabla\boldsymbol{c}$) are

$$\left.\begin{matrix}\dfrac{\partial u}{\partial x}, & \dfrac{\partial v}{\partial x}, & \dfrac{\partial w}{\partial x},\\[2ex] \dfrac{\partial u}{\partial y}, & \dfrac{\partial v}{\partial y}, & \dfrac{\partial w}{\partial y},\\[2ex] \dfrac{\partial u}{\partial z}, & \dfrac{\partial v}{\partial z}, & \dfrac{\partial w}{\partial z},\end{matrix}\right\} \tag{1.33, 1}$$

and the components of $\mathring{\overline{\overline{\frac{\partial}{\partial\boldsymbol{r}}}}}\boldsymbol{c}$ ($\equiv\mathring{\overline{\overline{\nabla\boldsymbol{c}}}}$) are

$$\left.\begin{matrix}\dfrac{1}{3}\left(2\dfrac{\partial u}{\partial x}-\dfrac{\partial v}{\partial y}-\dfrac{\partial w}{\partial z}\right), & \dfrac{1}{2}\left(\dfrac{\partial v}{\partial x}+\dfrac{\partial u}{\partial y}\right), & \dfrac{1}{2}\left(\dfrac{\partial w}{\partial x}+\dfrac{\partial u}{\partial z}\right),\\[2ex] \dfrac{1}{2}\left(\dfrac{\partial u}{\partial y}+\dfrac{\partial v}{\partial x}\right), & \dfrac{1}{3}\left(2\dfrac{\partial v}{\partial y}-\dfrac{\partial u}{\partial x}-\dfrac{\partial w}{\partial z}\right), & \dfrac{1}{2}\left(\dfrac{\partial w}{\partial y}+\dfrac{\partial v}{\partial z}\right),\\[2ex] \dfrac{1}{2}\left(\dfrac{\partial u}{\partial z}+\dfrac{\partial w}{\partial x}\right), & \dfrac{1}{2}\left(\dfrac{\partial v}{\partial z}+\dfrac{\partial w}{\partial y}\right), & \dfrac{1}{3}\left(2\dfrac{\partial w}{\partial z}-\dfrac{\partial u}{\partial x}-\dfrac{\partial v}{\partial y}\right).\end{matrix}\right\} \tag{1.33, 2}$$

If (as in 2.2) $\boldsymbol{c}_0$ denotes the velocity of a medium, the tensor $\frac{\partial}{\partial\boldsymbol{r}}\boldsymbol{c}_0$ ($\equiv\nabla\boldsymbol{c}_0$) will be called the *velocity-gradient* tensor. Its symmmetrical part $\overline{\overline{\nabla\boldsymbol{c}_0}}$ and its non-divergent symmetrical part $\mathring{\overline{\overline{\nabla\boldsymbol{c}_0}}}$ will be called respectively the *rate-of-*

strain and the *rate-of-shear* tensors; for these we use the notation

$$\mathsf{e} \equiv \overline{\overline{\nabla \boldsymbol{c}_0}}, \quad \overset{\circ}{\mathsf{e}} \equiv \overset{\circ}{\overline{\overline{\nabla \boldsymbol{c}_0}}}. \tag{1.33, 3}$$

When the operator $\partial/\partial\boldsymbol{r}$ appears in the product of two tensors, or of a vector and a tensor, attention must be paid to the order in which the terms occur, so that in each case the terms on which the operator acts may be made clear. For example, when a dyadic $\boldsymbol{ab}$ is multiplied by $\partial/\partial\boldsymbol{r}$, both $\boldsymbol{a}$ and $\boldsymbol{b}$ being functions of $\boldsymbol{r}$, the operator, being supposed to act on the components of the tensor, should be written before it; thus

$$\left(\frac{\partial}{\partial \boldsymbol{r}}.\boldsymbol{ab}\right)_\alpha = \sum_\beta \left(\frac{\partial}{\partial \boldsymbol{r}}\right)_\beta a_\beta b_\alpha = \sum_\beta a_\beta \left(\frac{\partial}{\partial \boldsymbol{r}}\right)_\beta b_\alpha + \sum_\beta b_\alpha \left(\frac{\partial}{\partial \boldsymbol{r}}\right)_\beta a_\beta$$

or

$$\nabla.\boldsymbol{ab} = (\boldsymbol{a}.\nabla)\,\boldsymbol{b} + \boldsymbol{b}(\nabla.\boldsymbol{a}). \tag{1.33, 4}$$

If, on the other hand, in the product $\mathsf{w}.\boldsymbol{a}$ or $\boldsymbol{a}.\mathsf{w}$, the tensor w is of the form $\dfrac{\partial}{\partial \boldsymbol{r}}\boldsymbol{b}$, these products should be written as follows

$$\boldsymbol{a}.\mathsf{w} = \boldsymbol{a}.\frac{\partial}{\partial \boldsymbol{r}}\boldsymbol{b} = (\boldsymbol{a}.\nabla)\,\boldsymbol{b}, \quad \mathsf{w}.\boldsymbol{a} = \boldsymbol{a}.\overline{\mathsf{w}} = \boldsymbol{a}.(\overline{\nabla \boldsymbol{b}}). \tag{1.33, 5}$$

Similarly in obtaining the double product of $\boldsymbol{ab}$ and $\nabla\boldsymbol{c}$, (1.32, 7) should be put in the form

$$\boldsymbol{ab}:\nabla\boldsymbol{c} = \boldsymbol{a}.(\boldsymbol{b}.\nabla)\,\boldsymbol{c}. \tag{1.33, 6}$$

As further examples of the notation,

$$\nabla\{\boldsymbol{a}.(\nabla\boldsymbol{b})\} = (\nabla\boldsymbol{a}).(\nabla\boldsymbol{b}) + (\boldsymbol{a}.\nabla)(\nabla\boldsymbol{b}), \tag{1.33, 7}$$

and, if T is a scalar function of $\boldsymbol{r}$,

$$\nabla(\boldsymbol{a}.\nabla T) = (\nabla\boldsymbol{a}).\nabla T + (\boldsymbol{a}.\nabla)\nabla T. \tag{1.33, 8}$$

We may also note here the form of the components of the product $\nabla.\mathsf{p}$, where p is a tensor function of position; the x-component is given by

$$(\nabla.\mathsf{p})_x \equiv \left(\frac{\partial}{\partial \boldsymbol{r}}.\mathsf{p}\right)_x = \frac{\partial p_{xx}}{\partial x} + \frac{\partial p_{yx}}{\partial y} + \frac{\partial p_{zx}}{\partial z}. \tag{1.33, 9}$$

SOME RESULTS ON INTEGRATION

1.4. Integrals involving exponentials

Consider the integral

$$\int_0^\infty e^{-\alpha C^2} C^r dC.$$

In this write $s = \alpha C^2$; then it becomes equal to

$$\tfrac{1}{2}\alpha^{-\frac{1}{2}(r+1)} \int_0^\infty e^{-s} s^{\frac{1}{2}(r-1)} ds = \tfrac{1}{2}\alpha^{-\frac{1}{2}(r+1)}\,\Gamma\left(\frac{r+1}{2}\right) \tag{1.4, 1}$$

if $r > -1$. In particular, if r is an integer,

$$\int_0^\infty e^{-\alpha C^2} C^r dC = \frac{\sqrt{\pi}}{2} \cdot \frac{1}{2} \cdot \frac{3}{2} \cdot \frac{5}{2} \cdots \frac{r-1}{2} \alpha^{-\frac{1}{2}(r+1)}, \tag{1.4, 2}$$

or

$$\int_0^\infty e^{-\alpha C^2} C^r dC = \tfrac{1}{2}\alpha^{-\frac{1}{2}(r+1)} \left(\frac{r-1}{2}\right)!, \tag{1.4, 3}$$

according as r is even or odd.

1.41. Transformation of multiple integrals

Consider the multiple integral

$$\iiint \ldots F(u_1, u_2, \ldots, u_n)\, du_1\, du_2 \ldots du_n$$

extended over any range of values of the variables u. As is well known from integration theory, if the variables of integration are changed to a set $v_1, v_2, \ldots, v_n$, then the integral is transformed to

$$\iiint \ldots \mathscr{F}(v_1, v_2, \ldots, v_n)\, |J|\, dv_1\, dv_2 \ldots dv_n,$$

where $\mathscr{F}(v_1, v_2, \ldots, v_n) \equiv F(u_1, u_2, \ldots, u_n)$, and J denotes the Jacobian determinant

$$\frac{\partial(u_1, u_2, \ldots, u_n)}{\partial(v_1, v_2, \ldots, v_n)} = \begin{vmatrix} \dfrac{\partial u_1}{\partial v_1}, & \dfrac{\partial u_2}{\partial v_1}, & \ldots, & \dfrac{\partial u_n}{\partial v_1} \\ \dfrac{\partial u_1}{\partial v_2}, & \dfrac{\partial u_2}{\partial v_2}, & \ldots, & \dfrac{\partial u_n}{\partial v_2} \\ \ldots & \ldots & \ldots & \ldots \\ \dfrac{\partial u_1}{\partial v_n}, & \dfrac{\partial u_2}{\partial v_n}, & \ldots, & \dfrac{\partial u_n}{\partial v_n} \end{vmatrix}.$$

The new integral extends over the range of values of the variables v that corresponds to the range of values of the original variables u.

A special case of this result was used in 1.21, when it was pointed out that the volume $d\mathbf{r}$, whose expression in Cartesian coordinates is $dx\,dy\,dz$, is taken to be equal to $r^2 \sin\theta\, dr\, d\theta\, d\varphi$ in terms of polar coordinates r, θ, φ, where $x = r\cos\varphi\sin\theta$, $y = r\sin\varphi\sin\theta$, $z = r\cos\theta$. It may easily be verified that in this case

$$J \equiv \frac{\partial(x, y, z)}{\partial(r, \theta, \varphi)} = -r^2 \sin\theta,$$

and so $|J|\, dr\, d\theta\, d\varphi = r^2 \sin\theta\, dr\, d\theta\, d\varphi$.

1.411. Jacobians

The general Jacobian of 1.41 may conveniently be denoted by $\partial(\mathbf{u})/\partial(\mathbf{v})$, regarding $u_1, u_2, \ldots, u_n$ and $v_1, v_2, \ldots, v_n$ as components of vectors $\mathbf{u}$ and $\mathbf{v}$ in n-dimensional spaces. Alternatively, if each set is divided into two groups,

$(u_1, u_2, \ldots, u_m)$, $(u_{m+1}, u_{m+2}, \ldots, u_n)$ and $(v_1, \ldots, v_r)$, $(v_{r+1}, \ldots, v_n)$, regarded as components of vectors of dimensions m and $n-m$ or r and $n-r$, namely $\boldsymbol{u}'$, $\boldsymbol{u}''$ and $\boldsymbol{v}'$, $\boldsymbol{v}''$, the Jacobian may be denoted by $\partial(\boldsymbol{u}', \boldsymbol{u}'')/\partial(\boldsymbol{v}', \boldsymbol{v}'')$. This notation can obviously be extended to the case of division of the n components into more than two groups. For example, consider the case when $n = 6$, $m = 3$, $r = 3$, so that $\boldsymbol{u}'$, $\boldsymbol{u}''$, $\boldsymbol{v}'$, $\boldsymbol{v}''$ are all three-dimensional vectors. If we write out in full the determinants in the equations

$$\frac{\partial(\boldsymbol{u}' + \mathrm{k}\boldsymbol{u}'', \boldsymbol{u}'')}{\partial(\boldsymbol{v}', \boldsymbol{v}'')} = \frac{\partial(\boldsymbol{u}', \boldsymbol{u}'')}{\partial(\boldsymbol{v}', \boldsymbol{v}'')}, \tag{1.411, 1}$$

$$\frac{\partial(\boldsymbol{u}', \boldsymbol{u}'' + \mathrm{k}'\boldsymbol{u}')}{\partial(\boldsymbol{v}', \boldsymbol{v}'')} = \frac{\partial(\boldsymbol{u}', \boldsymbol{u}'')}{\partial(\boldsymbol{v}', \boldsymbol{v}'')}, \tag{1.411, 2}$$

where k, k′ are any constants, the truth of these equations is readily seen. The notation adopted here is convenient and suggestive.

1.42. Integrals involving vectors or tensors

Let $\boldsymbol{C}$ be a vector with components U, V, W, and let $d\boldsymbol{C}$ denote an element of volume in a space in which $\boldsymbol{C}$ denotes the vector of displacement from an origin. Consider integrals (supposed convergent) of the type

$$\int \phi(\boldsymbol{C})\, d\boldsymbol{C},$$

taken over the whole of the $\boldsymbol{C}$-space.

If ϕ is a function of odd degree in U or V or W, the part of the integral for which ϕ is positive cancels the part for which ϕ is negative, and the integral vanishes.

If $\phi(\boldsymbol{C}) = U^2 F(C)$, then by symmetry

$$\int U^2 F(C)\, d\boldsymbol{C} = \int V^2 F(C)\, d\boldsymbol{C} = \int W^2 F(C)\, d\boldsymbol{C}$$
$$= \tfrac{1}{3}\int (U^2 + V^2 + W^2) F(C)\, d\boldsymbol{C} = \tfrac{1}{3}\int C^2 F(C)\, d\boldsymbol{C}. \tag{1.42, 1}$$

Thus
$$\int F(C)\, \boldsymbol{C}\boldsymbol{C}\, d\boldsymbol{C} = \tfrac{1}{3}\mathsf{U}\int F(C)\, C^2\, d\boldsymbol{C} \tag{1.42, 2}$$

(the integrals involving the non-diagonal terms of $\boldsymbol{C}\boldsymbol{C}$ vanish, since these terms are odd functions of U, V, or W). Hence also

$$\int F(C)\, \overset{\circ}{\boldsymbol{C}\boldsymbol{C}}\, d\boldsymbol{C} = 0, \tag{1.42, 3}$$

and if $\boldsymbol{A}$ is any constant vector

$$\int F(C)(\boldsymbol{A}.\boldsymbol{C})\boldsymbol{C}\, d\boldsymbol{C} = \boldsymbol{A}.\int F(C)\, \boldsymbol{C}\boldsymbol{C}\, d\boldsymbol{C}$$
$$= \tfrac{1}{3}\boldsymbol{A}\cdot\mathsf{U}\int F(C)\, C^2\, d\boldsymbol{C}$$
$$= \tfrac{1}{3}\boldsymbol{A}\int F(C)\, C^2\, d\boldsymbol{C}. \tag{1.42, 4}$$

Again, let $\phi(\boldsymbol{C}) = U^4 F(C)$. Using polar coordinates C, θ, φ such that

$U = C\cos\theta$, $V = C\sin\theta\cos\varphi$, $W = C\sin\theta\sin\varphi$, it is found that

$$\begin{aligned}\int F(C)\,U^4 d\mathbf{C} &= \iiint F(C)\,C^4\cos^4\theta . C^2\sin\theta\, dC\, d\theta\, d\varphi \\ &= \tfrac{1}{5}\iiint F(C)\,C^4 . C^2 \sin\theta\, dC\, d\theta\, d\varphi \\ &= \tfrac{1}{5}\int F(C)\,C^4 d\mathbf{C}, \end{aligned} \qquad (1.42, 5)$$

since
$$\int_0^\pi \cos^4\theta\sin\theta\, d\theta = \frac{1}{5}\int_0^\pi \sin\theta\, d\theta.$$

Similarly it may be proved that

$$\int F(C)\,U^2V^2 d\mathbf{C} = \tfrac{1}{15}\int F(C)\,C^4 d\mathbf{C}. \qquad (1.42, 6)$$

1.421. An integral theorem

Let w be any tensor independent of $\mathbf{C}$. Then the five integrals

(i) $\int F(C)\,\mathbf{CC}(\overset{\circ}{\mathbf{CC}}:\mathsf{w})\,d\mathbf{C}$, (ii) $\int F(C)\,\overset{\circ}{\mathbf{CC}}(\overset{\circ}{\mathbf{CC}}:\mathsf{w})\,d\mathbf{C}$,

(iii) $\int F(C)\,\overset{\circ}{\mathbf{CC}}(\mathbf{CC}:\mathsf{w})\,d\mathbf{C}$, (iv) $\tfrac{1}{5}\overset{\circ}{\overline{\overline{\mathsf{w}}}}\int F(C)\,(\overset{\circ}{\mathbf{CC}}:\overset{\circ}{\mathbf{CC}})\,d\mathbf{C}$,

(v) $\tfrac{2}{15}\overset{\circ}{\overline{\overline{\mathsf{w}}}}\int F(C)\,C^4 d\mathbf{C}$,

represent identical tensors, if $F(C)$ is any function of C such that the integrals converge.

For, if (ii) is subtracted from (i), the result is

$$\tfrac{1}{3}\int F(C)\,\mathsf{U}C^2(\overset{\circ}{\mathbf{CC}}:\mathsf{w})\,d\mathbf{C}$$

or
$$\tfrac{1}{3}\mathsf{U}(\mathsf{w}:\int F(C)\,C^2\overset{\circ}{\mathbf{CC}}\,d\mathbf{C}),$$

which vanishes, by (1.42, 3). Similarly the result of subtracting (ii) from (iii) is

$$\tfrac{1}{3}\int F(C)\,\overset{\circ}{\mathbf{CC}}\,C^2(\mathsf{U}:\mathsf{w})\,d\mathbf{C},$$

which likewise vanishes. Thus the equality of (i), (ii) and (iii) is established. Again, by (1.31, 7, 9) the integral (i) is equal to

$$\int F(C)\,\mathbf{CC}(\mathbf{CC}:\overset{\circ}{\overline{\overline{\mathsf{w}}}})\,d\mathbf{C}.$$

A typical diagonal element of this tensor is

$$\int F(C)\,U^2(\mathbf{CC}:\overset{\circ}{\overline{\overline{\mathsf{w}}}})\,d\mathbf{C}.$$

Neglecting terms in the integrand which involve functions odd in U, V or W, this may be written

$$\int F(C)\,U^2(U^2\overset{\circ}{\overline{\overline{w}}}_{xx} + V^2\overset{\circ}{\overline{\overline{w}}}_{yy} + W^2\overset{\circ}{\overline{\overline{w}}}_{zz})\,d\mathbf{C},$$

or, using (1.42, 5, 6),

$$(\tfrac{1}{5}\overset{\circ}{\overline{\overline{w}}}_{xx}+\tfrac{1}{15}\overset{\circ}{\overline{\overline{w}}}_{yy}+\tfrac{1}{15}\overset{\circ}{\overline{\overline{w}}}_{zz})\int F(C)\,C^4 d\boldsymbol{C}$$

$$=(\tfrac{2}{15}\overset{\circ}{\overline{\overline{w}}}_{xx}+\tfrac{1}{15}\mathsf{U}:\overset{\circ}{\overline{\overline{\mathsf{w}}}})\int F(C)\,C^4 d\boldsymbol{C}$$

$$=\tfrac{2}{15}\overset{\circ}{\overline{\overline{w}}}_{xx}\int F(C)\,C^4 d\boldsymbol{C}.$$

Similarly the typical non-diagonal element

$$\int F(C)\,UV(\boldsymbol{CC}:\overset{\circ}{\overline{\overline{\mathsf{w}}}})\,d\boldsymbol{C}$$

reduces to the form $$2\int F(C)\,U^2V^2\overset{\circ}{\overline{\overline{w}}}_{xy}\,d\boldsymbol{C},$$

which by (1.42, 6) is equal to

$$\tfrac{2}{15}\overset{\circ}{\overline{\overline{w}}}_{xy}\int F(C)\,C^4 d\boldsymbol{C}.$$

Thus the integral (i) is equal to

$$\tfrac{2}{15}\overset{\circ}{\overline{\overline{\mathsf{w}}}}\int F(C)\,C^4 d\boldsymbol{C},$$

i.e. to (v). Finally, the equality of the integrals (iv) and (v) follows from (1.32, 9). Thus the theorem is proved.

By a similar argument, if W is a symmetric non-divergent tensor,

$$\int F(C)\,\overset{\circ}{\boldsymbol{C}\boldsymbol{C}}(\boldsymbol{CC}:\mathsf{W})^2 d\boldsymbol{C}=\frac{8}{105}\overset{\circ}{\overline{\mathsf{W}.\mathsf{W}}}\int F(C)\,C^6 d\boldsymbol{C}. \quad (1.421, 1)$$

1.5. Skew tensors

Let w be a tensor with conjugate $\overline{\mathsf{w}}$. If $\mathsf{w}=-\overline{\mathsf{w}}$, the tensor w is said to be antisymmetric, or skew. In this case

$$w_{\alpha\beta}=-w_{\beta\alpha},\quad w_{\alpha\alpha}=0, \quad (1.5, 1)$$

so that w has only three independent components.

If $\mathsf{w}\neq-\overline{\mathsf{w}}$, the skew tensor $\overset{\times}{\mathsf{w}}$ is defined by

$$\overset{\times}{\mathsf{w}}\equiv\tfrac{1}{2}(\mathsf{w}-\overline{\mathsf{w}}), \quad (1.5, 2)$$

so that $$\overset{\times}{w}_{\alpha\beta}=\tfrac{1}{2}(w_{\alpha\beta}-w_{\beta\alpha})=-\overset{\times}{w}_{\beta\alpha}.$$

Since $\overline{\overline{\mathsf{w}}}=\tfrac{1}{2}(\mathsf{w}+\overline{\mathsf{w}})$, clearly $$\mathsf{w}=\overline{\overline{\mathsf{w}}}+\overset{\times}{\mathsf{w}}. \quad (1.5, 3)$$

The two tensors on the right of (1.5, 3) are the symmetric and skew parts of w.

The tensor $\overset{\times}{\mathsf{w}}$ has components

$$\left.\begin{matrix} 0, & \tfrac{1}{2}(w_{xy}-w_{yx}), & \tfrac{1}{2}(w_{xz}-w_{zx}),\\ \tfrac{1}{2}(w_{yx}-w_{xy}), & 0, & \tfrac{1}{2}(w_{yz}-w_{zy}),\\ \tfrac{1}{2}(w_{zx}-w_{xz}), & \tfrac{1}{2}(w_{zy}-w_{yz}), & 0. \end{matrix}\right\} \quad (1.5, 4)$$

Thus it can be specified by the three components $\breve{w}_{yz}, \breve{w}_{zx}, \breve{w}_{xy}$, which are

$$\tfrac{1}{2}(w_{yz}-w_{zy}),\quad \tfrac{1}{2}(w_{zx}-w_{xz}),\quad \tfrac{1}{2}(w_{xy}-w_{yx}).$$

These can be shown to be the components of a vector $\breve{\boldsymbol{w}}$ which is of the rotational type (1.1). This is called *the vector* of the tensor w or $\breve{\mathsf{w}}$.

In the special case when w is a dyadic $\boldsymbol{AB}$, the vector is clearly $\tfrac{1}{2}\boldsymbol{A}\wedge\boldsymbol{B}$; in the case of a differential dyadic, such as $\nabla\boldsymbol{c}$, the vector is $\tfrac{1}{2}\nabla\wedge\boldsymbol{c}$, or $\tfrac{1}{2}\operatorname{curl}\boldsymbol{c}$.

It is easy to show that $\breve{\mathsf{w}}.\boldsymbol{a} = \boldsymbol{a}\wedge\breve{\boldsymbol{w}}$, so that

$$\mathsf{w}.\boldsymbol{a} = \overline{\overline{\mathsf{w}}}.\boldsymbol{a}+\boldsymbol{a}\wedge\breve{\boldsymbol{w}}. \tag{1.5, 5}$$

An example occurring later (cf. 15.3) is

$$(\nabla\boldsymbol{c}_0).\nabla T = \overline{\overline{\nabla\boldsymbol{c}_0}}.\nabla T+\tfrac{1}{2}\nabla T\wedge\operatorname{curl}\boldsymbol{c}_0$$
$$= \mathsf{e}.\nabla T+\tfrac{1}{2}\nabla T\wedge\operatorname{curl}\boldsymbol{c}_0.$$

2

PROPERTIES OF A GAS: DEFINITIONS AND THEOREMS

2.1. Velocities, and functions of velocity

The linear velocity of the mass-centre of a molecule will be denoted vectorially by the letter $\mathbf{c}$, and its components relative to Cartesian coordinates by (u, v, w); its magnitude c will be called the molecular *speed*. The velocity-vector $\mathbf{c}$ may be regarded as the position-vector, or vector of displacement from an origin, of a point in a *velocity-space* or *velocity-domain*: this point is called the *velocity-point* of the molecule.

This representation of a velocity by a point in an auxiliary velocity-space suggests the analogy used in 1.2 between scalar or vector functions of $\mathbf{r}$ and similar functions of $\mathbf{c}$. Thus a scalar function of velocity, $\phi(\mathbf{c})$ or $\phi(u, v, w)$, has an associated gradient function $\partial\phi/\partial\mathbf{c}$ or $\partial\phi/\partial u$, $\partial\phi/\partial v$, $\partial\phi/\partial w$; similarly with a vector function of velocity, $\boldsymbol{\phi}(\mathbf{c})$, there is associated a scalar function

$$\frac{\partial}{\partial \mathbf{c}}\cdot\boldsymbol{\phi} = \frac{\partial\phi_x}{\partial u}+\frac{\partial\phi_y}{\partial v}+\frac{\partial\phi_z}{\partial w}$$

corresponding to the divergence. Again, a triple integration with respect to u, v, w corresponds to a volume integration in the velocity-space; this will be denoted by

$$\int \ldots d\mathbf{c},$$

the symbol $d\mathbf{c}$ denoting an element of volume (of any shape) in the velocity-space, surrounding the point $\mathbf{c}$. Unless the contrary is expressly stated, such an integration is supposed to extend over the whole velocity-space.

The phrase 'velocities in a range $d\mathbf{c}$ about the value $\mathbf{c}$' will be contracted to 'velocities in the range $\mathbf{c}$, $d\mathbf{c}$', or, more briefly, to 'velocities in the range $d\mathbf{c}$'. Similarly 'during a time-interval dt including the instant t' will be contracted to 'during a time t, dt' or 'during a time dt'. Likewise 'the volume-element $d\mathbf{r}$ containing the point $\mathbf{r}$' will be contracted to 'the volume-element $\mathbf{r}$, $d\mathbf{r}$' or, more briefly, to 'the volume-element $d\mathbf{r}$'.

The position and velocity of a molecule can both together be represented by a point in a space of six dimensions, whose coordinates in that space are the three components of $\mathbf{r}$ and the three components of $\mathbf{c}$.

If the molecule is not merely a mass-point, but is of finite size, it will in general possess rotatory motion; and if it is not rigid, it may also possess vibratory or other internal motion. The condition of such a molecule may be represented by a point in a space of n dimensions, where n is the number of independent positional and velocity (or momentum) variables needed to specify the configuration and motion of the molecule.

For example, if the molecule is rigid, it has six positional variables (three of location and three of orientation), and six velocity variables (three translatory and three rotatory); in this case $n = 12$. Some of these may be unimportant in particular cases: thus if the molecule be spherically symmetrical, its three variables of orientation will have no dynamical interest. If it is also smooth, its angular velocity will be unalterable by collisions, and its three angular-velocity variables are also without further interest; in this case $n = 6$ as for a point-molecule. But if the molecule, though spherical, be rough, its three angular-velocity variables are important, affecting and being affected by collisions; in this case $n = 9$.

If a gas is composed of molecules of more than one kind, we may associate a separate velocity-domain with each kind; each domain will have the appropriate number of dimensions for that kind of molecule, corresponding to the number of independent variables of position and velocity for such molecules.

2.2. Density and mean motion

In a non-uniform, variable, *continuous* medium, the density at the point $\boldsymbol{r}$ at time t is defined as the limit of the mean density (mass/volume) in a small volume $d\boldsymbol{r}$ surrounding the point $\boldsymbol{r}$, as the dimensions of $d\boldsymbol{r}$ diminish indefinitely, the limit being supposed independent of the shape of $d\boldsymbol{r}$. This definition cannot usefully be applied to a medium like a gas, composed of discrete molecules, especially when these are separated by distances large compared with molecular dimensions; it leads to a value of the density which varies rapidly from point to point, and with passage of time, and does not correspond to any ordinary measurable quantity. Hence some other definition is necessary.

Consider in the first instance a 'simple' gas, that is, a gas composed of molecules all of which are alike. Let the mass of any molecule be m. Let $d\boldsymbol{r}$ denote a small volume surrounding the point $\boldsymbol{r}$, which is large enough to contain a great number of molecules, while still possessing dimensions small compared with the scale of variation of such macroscopic quantities as the pressure, temperature, or mass-velocity of the gas. (For example, a cube of edge one-hundredth of a millimetre contains about $2{\cdot}687 \times 10^{10}$ molecules in a gas 'at standard temperature and pressure'.*) Let the mass contained in $d\boldsymbol{r}$ be averaged over a time t, dt which is long compared with the average time that would be taken by a molecule to cross $d\boldsymbol{r}$ if undeflected, yet short compared with the scale of time-variation of the macroscopic properties of the gas. (For example, since the mean speed of molecules in a gas at S.T.P. is

* This phrase is commonly abbreviated to 'at S.T.P.'; it signifies at a temperature of 0 °C., and a pressure of 760 mm. of mercury ($1{\cdot}013 \times 10^{6}$ dynes/cm.^{2}).

several hundred metres per second, a molecule would move one-hundredth of a millimetre in less than 10^{-7} of a second.) Then the averaged value of the mass contained by $d\mathbf{r}$ will be proportional only to its volume, and will not depend on its shape. It will be denoted by $\rho\, d\mathbf{r}$; ρ is termed the *mass-density* or *the density* of the gas at $\mathbf{r}$, t.

Similarly, the *number* of molecules in $d\mathbf{r}$ averaged over dt is proportional to $d\mathbf{r}$. It will be denoted by $n\,d\mathbf{r}$; n is called the *number-density* of the molecules. The quantities ρ and n are connected by the relation

$$\rho = nm.$$

Both ρ and n are functions of position and time; when it is desired to indicate this, they may be denoted by $\rho(\mathbf{r}, t)$, $n(\mathbf{r}, t)$.

The *mean molecular velocity* at $\mathbf{r}$, t in a simple gas, denoted by $\mathbf{c}_0$, is defined by the vector equation

$$(n\,d\mathbf{r})\,\mathbf{c}_0 = \Sigma\mathbf{c},$$

where the summation on the right is extended over the $n\,d\mathbf{r}$ molecules in the small volume $\mathbf{r}$, $d\mathbf{r}$, both $n\,d\mathbf{r}$ and $\Sigma\mathbf{c}$ being averaged over a small time-element t, dt. Similarly we obtain any other mean value; for example, the mean speed is $(\Sigma c)/(n\,d\mathbf{r})$. In particular, the mean momentum of a molecule at $\mathbf{r}$, t is equal to $m\mathbf{c}_0$; like $\mathbf{c}_0$ itself, it is, in general, a function of $\mathbf{r}$ and t.

The translational motions of the individual molecules in $d\mathbf{r}$ may be specified either by their 'actual' velocities $\mathbf{c}$ (i.e. their velocities relative to some standard frame of reference) or by their velocities $\mathbf{C}'$ relative to axes moving with some velocity $\mathbf{c}'$, so that $\mathbf{C}' = \mathbf{c} - \mathbf{c}'$. Generally $\mathbf{c}_0$, the mean velocity of the gas at the point, will be adopted as the velocity $\mathbf{c}'$; $\mathbf{C}'$ is then written $\mathbf{C}$, and is termed the *peculiar velocity* of the molecule; also C is the *peculiar speed*. The mean peculiar velocity of molecules at $\mathbf{r}$, t is $\mathbf{c}_0 - \mathbf{c}_0$, that is, zero. The components of $\mathbf{C}'$, $\mathbf{c}'$ and $\mathbf{C}$, $\mathbf{c}_0$ are denoted respectively by (U', V', W'), (u', v', w') and (U, V, W), (u_0, v_0, w_0).

2.21. The distribution of molecular velocities

The distribution of velocities among the large number $n\,d\mathbf{r}$ of molecules in $d\mathbf{r}$ can be represented by the distribution of their velocity-points $\mathbf{c}$ in the velocity-space. Owing to the continual changes of velocity by molecular encounters, and to the appearance and disappearance of velocity-points, as molecules pass into or out of $d\mathbf{r}$, the distribution of velocity-points will vary with time. As, however, the number $n\,d\mathbf{r}$ of velocity-points is very large, it will be assumed that, just as there is a number-density of molecules in the actual space occupied by the gas, so there is a statistically definite number-density of the $n\,d\mathbf{r}$ velocity-points in the velocity-space. This number-density is supposed to be proportional to the volume of the element $d\mathbf{r}$, but not to depend on its shape. It will in general be a function of $\mathbf{r}$ and of t as well as of position in the velocity-space; it will therefore be denoted by $f(\mathbf{c}, \mathbf{r}, t)\,d\mathbf{r}$. The definition implies that the probable number of molecules which, at the time t, are situated in the volume element $\mathbf{r}$, $d\mathbf{r}$, and have velocities lying in the range $\mathbf{c}$, $d\mathbf{c}$, is equal to

$$f(\mathbf{c}, \mathbf{r}, t)\, d\mathbf{c}\, d\mathbf{r}.$$

This does not mean that the given element $d\mathbf{r}$ actually contains this number of molecules having velocities in the range $\mathbf{c}$, $d\mathbf{c}$ at the time t; this is the average number of such molecules when the fluctuations which occur in a short time dt are, as it were, smoothed out. The function $f(\mathbf{c}, \mathbf{r}, t)$ or, briefly, f is termed the *velocity-distribution function*. Its definition involves probability concepts; any result in which it appears will be a result as to the probable, or average, behaviour of the gas.

The distribution of velocity-points is clearly unaffected if the origin in the velocity-space is changed to the point $\mathbf{c}'$. In this case the position-vector of a velocity-point is changed to $\mathbf{c}-\mathbf{c}'$ or $\mathbf{C}'$; the volume-element $d\mathbf{c}$ is now denoted by $d\mathbf{C}'$, and contains the same number of molecules, i.e.

$$f(\mathbf{C}' + \mathbf{c}', \mathbf{r}, t)\, d\mathbf{C}'\, d\mathbf{r}.$$

This will often be written as $f(\mathbf{C}', \mathbf{r}, t)\, d\mathbf{C}'\, d\mathbf{r}$, with an appropriate change in the nature of the function f. The change of variable from $\mathbf{c}$ to $\mathbf{C}'$ can be made in any integral without altering its value. Usually $\mathbf{c}'$ is taken to be $\mathbf{c}_0$, so that $\mathbf{C}'$ becomes $\mathbf{C}$, the peculiar velocity, and f becomes $f(\mathbf{C}, \mathbf{r}, t)$.

The whole number of molecules in the element $d\mathbf{r}$ is obtained by integrating $f\,d\mathbf{c}\,d\mathbf{r}$ or $f\,d\mathbf{C}'\,d\mathbf{r}$ throughout the whole velocity-space; this number is, by hypothesis, $n\,d\mathbf{r}$. Hence

$$n = \int f(\mathbf{c}, \mathbf{r}, t)\, d\mathbf{c} = \int f(\mathbf{C}' + \mathbf{c}', \mathbf{r}, t)\, d\mathbf{C}'.$$

Clearly the function f is never negative, and it must tend to zero as c or C' becomes infinite. It is assumed to be finite and continuous for all values of t.

The function f gives the number-density of points in the six-dimensional space of 2.1, in which the coordinates of a point are the components of $\mathbf{c}$ and $\mathbf{r}$ for a molecule. Similarly, if the molecule also possesses rotational or vibratory motion, account can be taken of this by employing a space of more dimensions, as in 2.1; the number-density f of representative points in this space is the generalized velocity-distribution function. The function f then represents the distribution of density and of translatory, rotatory and internal motions.

2.22. Mean values of functions of the molecular velocities

Let $\phi(\mathbf{c})$ be any function of the molecular velocity $\mathbf{c}$. The function ϕ may be a scalar, vector or tensor; for example, it may be $\mathbf{c}$ itself, or $\mathbf{C}'$, or a combination of components such as uv^2 or uW', or again a function of the speed c or C'. It may also be a function of position and time, and will therefore be denoted by $\phi(\mathbf{c}, \mathbf{r}, t)$ or, in terms of $\mathbf{C}'$, by $\phi(\mathbf{C}' + \mathbf{c}', \mathbf{r}, t)$ or, with an appropriate change in the function ϕ, by $\phi(\boldsymbol{C}', \boldsymbol{r}, t)$. Any such function ϕ we call a molecular property.

Let $\Sigma\phi$ denote the time-average during t, dt of the sum of the values of ϕ for the $n\,d\boldsymbol{r}$ molecules in $\boldsymbol{r}$, $d\boldsymbol{r}$, and write

$$\Sigma\phi = n\bar{\phi}\,d\boldsymbol{r}. \tag{2.22, 1}$$

Then $\bar{\phi}$ is the mean value* of ϕ for the molecules at (or near) the point $\boldsymbol{r}$. It is a function of $\boldsymbol{r}$, t, even if ϕ itself does not explicitly involve position and time. It can be expressed in terms of the velocity-distribution function f; for each of the $f(\boldsymbol{c}, \boldsymbol{r}, t)\,d\boldsymbol{c}\,d\boldsymbol{r}$ molecules in $d\boldsymbol{r}$, whose velocities are in the range $\boldsymbol{c}$, $d\boldsymbol{c}$, contributes $\phi(\boldsymbol{c}, \boldsymbol{r}, t)$ to $\Sigma\phi$, and their aggregate contribution is $\phi f\,d\boldsymbol{c}\,d\boldsymbol{r}$. By integration over the whole velocity-space, we obtain

$$\Sigma\phi = d\boldsymbol{r}\int \phi f\,d\boldsymbol{c},$$

whence

$$n\bar{\phi} = \int \phi f\,d\boldsymbol{c} = \int \phi f\,d\boldsymbol{C}'.$$

In particular, since $\boldsymbol{c}_0$, by definition, is the mean molecular velocity at $\boldsymbol{r}$, t,

$$n\boldsymbol{c}_0 = \int \boldsymbol{c} f\,d\boldsymbol{c}; \tag{2.22, 2}$$

clearly

$$\bar{\boldsymbol{C}} = 0, \quad \bar{U} = \bar{V} = \bar{W} = 0. \tag{2.22, 3}$$

2.3. Flow of molecular properties

Consider the passage of molecules across a small element of surface dS, moving in the gas with any velocity $\boldsymbol{c}'$. The surface element is supposed to have a positive and a negative side. Let $\mathbf{n}$ be a unit vector drawn normal to the element in the direction from the negative to the positive side. The passage of a molecule across dS is regarded as positive or negative according as the molecule crosses to the positive or negative side of dS. The velocity $\boldsymbol{C}'$ of a molecule relative to dS is equal to $\boldsymbol{c} - \boldsymbol{c}'$, or, if $\boldsymbol{C}$ is the peculiar velocity of the molecule, to $\boldsymbol{c}_0 + \boldsymbol{C} - \boldsymbol{c}'$.

Consider the molecules whose peculiar velocities lie in the range $\boldsymbol{C}$, $d\boldsymbol{C}$.† If one such molecule crosses the element dS in a time dt so short that we may ignore the possibility of the molecule encountering another during dt, then at the beginning of dt the molecule must lie somewhere inside the cylinder on dS as base,‡ with generators specified in length and direction by

* The significance of the bar placed over ϕ is here totally different from that of the bar placed over the symbol for a tensor, in 1.3. When, as may happen, ϕ denotes a tensor, or when a single bar occurs over a dyadic (and therefore tensorial) expression (as in (2.31, 3) for example), it is necessary to know in which of the two possible senses it is used. In the remainder of this book a single bar placed over the symbol for a tensor will indicate that the *mean value* is to be taken. The double bar, on the other hand, always has the same meaning as in 1.3.

† For brevity, the phrase 'molecules whose peculiar velocities lie in the range $\boldsymbol{C}$, $d\boldsymbol{C}$' is contracted to 'molecules $\boldsymbol{C}$, $d\boldsymbol{C}$'.

‡ Here, as frequently, it is convenient to regard molecules as mass-points. Thus the exact position of a molecule can be specified, and also the exact time at which it crosses the element dS.

$-\mathbf{C}'\,dt$ (see Fig. 1). Thus, if $d\mathbf{r}$ denotes the volume of this cylinder, the number of molecules $\mathbf{C}$, $d\mathbf{C}$ crossing dS during dt is $f d\mathbf{C} d\mathbf{r}$.

Now $d\mathbf{r} = \pm C' \cos\theta\, dt\, dS$, where θ is the angle between $\mathbf{C}'$ and $\mathbf{n}$, the sign + or − being chosen so as to make the expression for $d\mathbf{r}$ positive. But θ is acute or obtuse (and so $\cos\theta$ is positive or negative), according as the flow is positive or negative. Thus the flow is expressed, both in magnitude and sign, by

$$f(\mathbf{C})\,d\mathbf{C}.C'\cos\theta\,dt\,dS.$$

But $C'\cos\theta$ is the component of $\mathbf{C}'$ normal to dS, which is equal to $\mathbf{C}'.\mathbf{n}$; we denote it by C'_n.* Then the flow of molecules $\mathbf{C}$, $d\mathbf{C}$ across dS in time dt is

$$C'_n f(\mathbf{C})\,d\mathbf{C}\,dt\,dS. \qquad (2.3, 1)$$

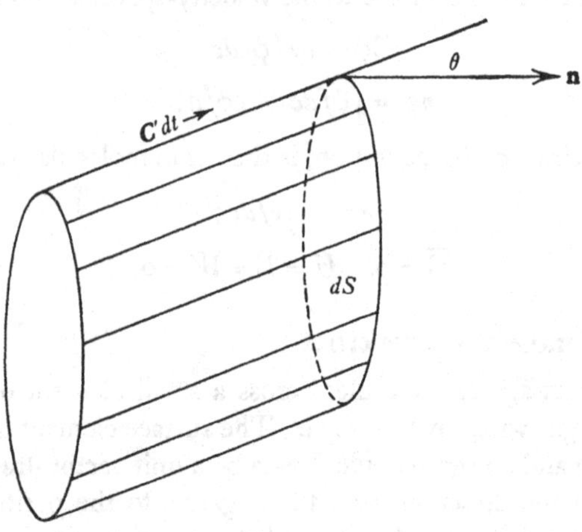

Fig. 1

The net flow of molecules across dS during time dt is found by summing over all velocity groups, i.e. by integrating over the whole range of $\mathbf{C}$, which gives

$$dS\,dt\int C'_n f(\mathbf{C})\,d\mathbf{C} = dS\,dt\,n\overline{C'_n}. \qquad (2.3, 2)$$

The number of molecules crossing from the negative to the positive side is similarly

$$dS\,dt\int_{C'_n>0} C'_n f(\mathbf{C})\,d\mathbf{C}.$$

The molecules that cross the element carry with them their energy, momentum, and so on. The net rate of transport of such quantities across dS can be found by methods similar to those just used. Thus let $\phi(\mathbf{C})$ denote any scalar molecular property; each of the molecules $\mathbf{C}$, $d\mathbf{C}$ that cross dS carries an amount $\phi(\mathbf{C})$ of ϕ with it. Hence the contribution of the group to the flow of ϕ across dS during dt is, by (2.3, 1),

$$\phi(\mathbf{C}).C'_n f(\mathbf{C})\,d\mathbf{C}\,dt\,dS,$$

* The suffix n in this symbol, and in p_n and C_n (2.31), has reference to $\mathbf{n}$, the normal to dS; it has no relation to n the number-density.

and the total net flow of ϕ across dS during dt is

$$dS\,dt\int C'_n\phi(\boldsymbol{C})f(\boldsymbol{C})\,d\boldsymbol{C} = dS\,dt\,n\overline{C'_n\phi(\boldsymbol{C})}. \tag{2.3, 3}$$

The expression (2.3, 2) for the flow of molecules across dS is a special case of this result, corresponding to $\phi(\boldsymbol{C}) = 1$.

The *rate* of flow of ϕ across dS per unit area is obtained by dividing (2.3, 3) by $dS\,dt$, giving

$$n\overline{C'_n\phi(\boldsymbol{C})}. \tag{2.3, 3'}$$

Since $C'_n = \boldsymbol{C}'.\mathbf{n}$, this is the component along $\mathbf{n}$ of the vector

$$n\overline{\boldsymbol{C}'\phi(\boldsymbol{C})}.$$

Now $\boldsymbol{C}' = \boldsymbol{C}+\boldsymbol{c}_0-\boldsymbol{c}'$, so that

$$n\overline{\boldsymbol{C}'\phi(\boldsymbol{C})} = n\overline{\boldsymbol{C}\phi(\boldsymbol{C})}+n(\boldsymbol{c}_0-\boldsymbol{c}')\overline{\phi(\boldsymbol{C})}. \tag{2.3, 4}$$

Hence the component of this vector in any direction $\mathbf{n}$ represents the rate of flow of the property $\phi(\boldsymbol{C})$ per unit area across a surface normal to this direction, and moving with the velocity $\boldsymbol{c}'$.

The *number-flow* is given by the vector $n(\boldsymbol{c}_0-\boldsymbol{c}')$, since in this special case $\phi(\boldsymbol{C}) = 1$, $\overline{\boldsymbol{C}\phi(\boldsymbol{C})} = \overline{\boldsymbol{C}} = 0$. This vector is the product of the number-density and the mean velocity of the molecules relative to the surface element. If $\boldsymbol{c}' = \boldsymbol{c}_0$ the number-flow is zero, whatever the orientation of the surface.

This enables us to interpret the second term on the right-hand side of (2.3, 4). Its component normal to dS represents the contribution to the rate of flow of $\phi(\boldsymbol{C})$ due to the net number-flow of the molecules across dS, each carrying, on the average, the quantity $\overline{\phi(\boldsymbol{C})}$ of ϕ. The first term, on the other hand, is independent of the number-flow, and its component normal to dS represents the rate of flow of ϕ when the number-flow vanishes, i.e. when the element shares the mean motion of the gas at the point.

The vector $n\overline{\boldsymbol{C}\phi(\boldsymbol{C})}$ may conveniently be termed the '*flux-vector*' for the property ϕ. The rate of flow of ϕ across unit area of a surface which moves with the gas is the component of the flux-vector normal to the surface. If, however, the surface is in motion relative to the gas, the rate of flow is increased by the normal component of the relative velocity, multiplied by $n\overline{\phi(\boldsymbol{C})}$. In general, when the flow of some molecular property across a surface is considered, it will be assumed that the surface is moving with the gas.

In the case of a *vector* property $\boldsymbol{\phi}(\boldsymbol{C})$ of the molecular velocities it is convenient to consider the flux-vector of each component of $\boldsymbol{\phi}$, which will be a scalar quantity, as in the preceding discussion. Thus the flux-vector of the component ϕ_α, where α stands for any one of x, y, and z, is $n\overline{\boldsymbol{C}\phi_\alpha(\boldsymbol{C})}$.

These results may be generalized so as to apply to molecules that are free to rotate, or that possess other internal degrees of freedom; ϕ may then depend on the variables specifying the orientation, angular velocity, and internal state, as well as on the translational velocity. The flow will again be represented by an expression of the form (2.3, 4), where averages are now to be taken over all values of the velocities and also of the other variables specifying the motion.

2.31. Pressure and the pressure tensor

The case in which $\phi(\mathbf{C})$ is equal to some component of the molecular momentum $m\mathbf{c}$ is of great importance, because of its connection with the pressure distribution.

At the boundary of the containing vessel, every molecule that impinges and rebounds exerts an impulse equal to the difference between its momenta before and after impact. When such impacts are sufficiently numerous and sufficiently uniform in distribution, they simulate a continuous force on the boundary, equal in magnitude and direction to the rate at which momentum is being communicated to the surface by impacts. The force per unit area of the surface is called the pressure (or 'boundary pressure') on the surface at the point. The surface clearly exerts an equal and opposite pressure on the gas. The pressure is a vector, whose direction is not necessarily normal to the surface at the point considered.

Suppose that dS is an element of the surface of the containing vessel, moving with the velocity $\mathbf{c}'$, and let the internal face be taken as the negative face. Let the direction of the outward normal to dS be that of the unit vector $\mathbf{n}$, and let $\mathbf{p}_n$ denote the pressure on the wall at this point. Then by the definition of $\mathbf{p}_n$, the momentum communicated to dS in the time dt is $\mathbf{p}_n dS\,dt$.

As in 2.3, we may show that the total momentum of the molecules impinging on the element dS in time dt is equal, before impact, to

$$dS\,dt\int_{+} C'_n m\mathbf{c}f(\mathbf{C})\,d\mathbf{C},$$

where the suffix (+) signifies that the integration is extended only over that part of the velocity-range for which C'_n, the $\mathbf{n}$-component of the velocity of a molecule relative to dS, is positive (since only molecules for which C'_n is positive can impinge on the surface). Similarly the total momentum of the molecules rebounding from dS during dt is

$$dS\,dt\int_{-} (-C'_n)\,m\mathbf{c}f(\mathbf{C})\,d\mathbf{C},$$

the suffix (−) signifying that the range of integration is over all values of $\mathbf{C}$ for which C'_n is negative: the minus sign before C'_n is introduced because C'_n enters into the integrand through the expression for the number of molecules $\mathbf{C}, d\mathbf{C}$ leaving dS during dt, and this number is essentially positive. Thus the total momentum communicated to dS during dt, which is the difference between the momentum of the impinging molecules and that of those rebounding from the surface, is equal to

$$dS\,dt\left[\int_+ C'_n mc f(\mathbf{C})\,d\mathbf{C} - \int_- (-C'_n) mc f(\mathbf{C})\,d\mathbf{C}\right]$$

$$= dS\,dt \int C'_n mc f(\mathbf{C})\,d\mathbf{C}$$

$$= dS\,dt\,.\,nm\overline{C'_n \mathbf{c}}.$$

Hence $$p_n = nm\overline{C'_n \mathbf{c}} = \rho\overline{C'_n \mathbf{c}}. \qquad (2.31, 1)$$

The velocity $\mathbf{c}'$ of the wall will not in general equal the mean velocity $\mathbf{c}_0$ of the neighbouring gas. Experiment shows that the behaviour of a gas in the neighbourhood of a wall may be rather complicated; some molecules do not immediately rebound off the wall, but enter it or adhere to it for a time before they return to the gas. If the gas is neither condensing upon nor evaporating from the surface, the total number of impinging molecules, namely,

$$dS\,dt \int_+ C'_n f(\mathbf{C})\,d\mathbf{C},$$

must equal the number rebounding, which is

$$dS\,dt \int_- (-C'_n) f(\mathbf{C})\,d\mathbf{C};$$

hence $$dS\,dt \int C'_n f(\mathbf{C})\,d\mathbf{C} = 0$$

or $$\overline{C'_n} = 0.$$

That is, the mean velocity relative to the wall, for the molecules in its neighbourhood, has no component normal to the wall; the gas may, however, have a mean motion relative to the wall, in a direction parallel to the surface.

Using this result and (2.22, 3), we have

$$\overline{C'_n \mathbf{c}} = \overline{C'_n(\mathbf{c}_0 + \mathbf{C})} = \overline{C'_n}\mathbf{c}_0 + \overline{C'_n \mathbf{C}} = \overline{C'_n \mathbf{C}}$$

$$= \overline{(\mathbf{n}\,.\,\mathbf{C}')\mathbf{C}} = \overline{\{\mathbf{n}\,.\,(\mathbf{c} - \mathbf{c}')\}\mathbf{C}}$$

$$= \overline{\{\mathbf{n}\,.\,(\mathbf{C} + \mathbf{c}_0 - \mathbf{c}')\}\mathbf{C}}$$

$$= \overline{(\mathbf{n}\,.\,\mathbf{C})\mathbf{C}} + \{\mathbf{n}\,.\,(\mathbf{c}_0 - \mathbf{c}')\}\overline{\mathbf{C}}$$

$$= \overline{(\mathbf{n}\,.\,\mathbf{C})\mathbf{C}} = \overline{C_n \mathbf{C}}.$$

Hence from (2.31, 1), using (1.32, 3), we obtain the following alternative forms for $\mathbf{p}_n$:

$$\mathbf{p}_n = \rho\overline{C_n \mathbf{C}} = \rho\overline{(\mathbf{n}\,.\,\mathbf{C})\mathbf{C}} = \mathbf{n}\,.\,\rho\overline{\mathbf{C}\mathbf{C}}$$

$$= \mathbf{n}\,.\,\mathsf{p} = \mathsf{p}\,.\,\mathbf{n}, \qquad (2.31, 2)$$

where p is the symmetrical tensor defined by the equation

$$\mathsf{p} = \rho\overline{\mathbf{CC}} = \begin{Bmatrix} \rho\overline{U^2}, & \rho\overline{UV}, & \rho\overline{UW}, \\ \rho\overline{VU}, & \rho\overline{V^2}, & \rho\overline{VW}, \\ \rho\overline{WU}, & \rho\overline{WV}, & \rho\overline{W^2}. \end{Bmatrix} \tag{2.31, 3}$$

This tensor depends only on the distribution of the peculiar velocities; its components are given by

$$p_{xx} = \rho\overline{U^2}, \quad p_{yy} = \rho\overline{V^2}, \quad p_{zz} = \rho\overline{W^2}, \tag{2.31, 4}$$

$$p_{yz} = p_{zy} = \rho\overline{VW}, \quad p_{zx} = p_{xz} = \rho\overline{WU}, \quad p_{xy} = p_{yx} = \rho\overline{UV}. \tag{2.31, 5}$$

The pressure distribution at any point P *within* the gas is defined as follows. As in 2.3, let dS be any surface element containing P, and let $\mathbf{n}$ be its unit positive normal vector. Let dS share the mean motion of the gas at P, so that $\mathbf{c}' = \mathbf{c}_0$, $\overline{\mathbf{C}}' = \overline{\mathbf{C}} = 0$, and therefore $\overline{C}'_n = 0$. Then $\boldsymbol{p}_n$, the *pressure across* dS, *towards its positive side*, is defined as the rate of flow of molecular momentum $m\mathbf{c}$ across dS, per unit area, in the positive direction. This is given by (2.3, 3'), if ϕ is taken to be the vector function $m\mathbf{c}$. Consequently $\boldsymbol{p}_n = n\overline{C'_n m\mathbf{c}} = \rho\overline{C'_n \mathbf{c}}$, as in (2.31, 1); since $\overline{C}'_n = 0$, this is equivalent to (2.31, 2), which therefore holds in the interior as well as at the boundary of the gas. In the interior, however, $\mathbf{n}$ may have any direction.

The distribution of pressure across planes in all directions through P is therefore determined by the *pressure tensor* p.

When $\mathbf{n}$ is $\mathbf{x}$, the unit vector in the direction of Ox, $\boldsymbol{p}_n = \boldsymbol{p}_x = \mathsf{p}\cdot\mathbf{x}$, of which the components are p_{xx}, p_{xy}, p_{xz}; the other components of p are components of the similarly defined vectors $\boldsymbol{p}_y, \boldsymbol{p}_z$. Thus the components of p are the components of the pressures across surfaces parallel to the three coordinate planes.

The above results are valid whether the molecules possess only energy of translation, or have internal energy of rotation, vibration, or any other form.

2.32. The hydrostatic pressure

The normal component of the pressure on a surface normal to the unit vector $\mathbf{n}$ is

$$\mathbf{n}\cdot\boldsymbol{p}_n = \mathbf{n}\cdot\rho\overline{C_n\mathbf{C}} = \rho\overline{C_n^2}. \tag{2.32, 1}$$

Thus the normal component of the pressure on any surface is essentially positive; that is, the normal force exerted on any surface by the gas is always a pressure, and never a traction.

The sum of the normal pressures across three planes through any point P, parallel to the coordinate planes, is

$$\begin{aligned} p_{xx}+p_{yy}+p_{zz} &= \rho(\overline{U^2+V^2+W^2}) \\ &= \rho\overline{C^2}. \end{aligned} \tag{2.32, 2}$$

Thus the *mean* of the normal pressures across any three orthogonal planes is $\frac{1}{3}\rho\overline{C^2}$. This is called the *mean hydrostatic pressure*, or *the pressure* at P; we denote it by p. By (1.31, 5)

$$p = \tfrac{1}{3}\mathsf{p}:\mathsf{U}. \qquad (2.32, 3)$$

If the non-diagonal elements of the tensor p vanish, and the diagonal elements are equal, then

$$p = p_{xx} = p_{yy} = p_{zz},$$

and

$$\mathsf{p} = p\mathsf{U}.$$

In this case (cf. (1.31, 2))

$$\boldsymbol{p}_n = \mathsf{p}.\mathbf{n} = p\mathsf{U}.\mathbf{n} = p\mathbf{n},$$

so that the pressure on any surface element through the point $\boldsymbol{r}$ is normal to the surface: its magnitude is independent of the orientation of the surface, and equal to the hydrostatic pressure. These are the conditions satisfied by the pressure in hydrostatic problems; hence such a pressure system, in which p is a scalar multiple of U, is said to be hydrostatic.

2.33. Intermolecular forces and the pressure

In the above discussion the whole of the pressure of the gas on the walls of the containing vessel, or across a hypothetical internal surface, was tacitly assumed to be due to the transfer of momentum. In actual gases, intermolecular forces also contribute to these pressures. Since at distances large compared with the molecular diameters these forces are usually attractive, they add an attractive component to the total pressure. This component is relatively small for gases at ordinary temperatures: but in liquids or solids the importance of intermolecular forces may equal or exceed that of the momentum transfer.

Attractions between the gas molecules operate to reduce the pressure on the walls in the following way. The average resultant force exerted on a molecule in the interior of the gas by the other molecules of the gas is in general zero, because only the adjacent molecules exert any appreciable attraction, and so, unless there is a very steep density gradient at the point, the attractions are approximately equal in all directions; but at the walls the gas lies on one side only, so that there is a resultant attraction inwards, which is roughly proportional to the number of attracting molecules in the neighbourhood, i.e. to the density of the gas. Consequently the momentum imparted to the wall by each molecule impinging thereon will be smaller than if there were no attractive force, by an amount proportional to the density ρ. The rate at which molecules strike the wall is also proportional to ρ; hence the correction to the momentum pressure is proportional to ρ^2, whereas the momentum pressure itself varies as ρ. Thus the correction becomes of greater importance as the density increases; it is small for gases at ordinary pressures, and at temperatures well above their critical temperatures.*

* For a detailed discussion of the effects of these factors on the equation of state of a gas, see R. H. Fowler's *Statistical Mechanics*, chapters 8 and 9 (1928, 1936).

In this book attention will be directed more particularly to the deviation of the actual pressure system from the hydrostatic system $p\mathsf{U}$. Intermolecular forces at distances large compared with the molecular dimensions have little effect on the pressure deviations, except for very dense gases; accordingly we ignore them. The finite size of molecules, which results in a reduction of the effective volume of the vessel containing the gas is also of importance mainly for dense gases; this is taken into account in Chapter 16.

2.34. Molecular velocities: numerical values

The quantities p and ρ appearing in the equation

$$p = \tfrac{1}{3}\rho\overline{C^2} \qquad (2.34, 1)$$

are directly measurable. From their experimental values the corresponding values of $\overline{C^2}$ can be found. For example, at standard temperature and pressure ($1\cdot013 \times 10^6$ dynes/cm.^{2}) the densities of hydrogen and nitrogen are respectively $8\cdot99 \times 10^{-5}$ g./cm.^{3} and $1\cdot25 \times 10^{-3}$ g./cm.^{3}. The corresponding values for $\sqrt{(\overline{C^2})}$ are 1839 and 493 m./sec. For a gas in a uniform steady state,

$$\bar{C} = \sqrt{(\overline{C^2})}/1\cdot086$$

(cf. (4.11, 4)); thus the corresponding values of $\bar{C}$ are 1694 and 454 m./sec.

These mean speeds are very large, and appear at first sight startling. Evidence confirming their order of magnitude is, however, supplied by other phenomena. If the basic assumptions of the kinetic theory of gases are valid, sound must be transmitted by the motions of individual molecules, and the velocities of sound in these gases are known to be similar in order of magnitude to the above molecular mean speeds. Again, the speed of effusion of a gas from a vessel into a vacuum through a small aperture should be of the same order of magnitude as the mean molecular speed; experiment shows that the speeds of effusion for different gases are, in fact, of this order.

2.4. Heat

The amount of translatory kinetic energy possessed by the molecules in the element $\mathbf{r}$, $d\mathbf{r}$ at time t is $n\,d\mathbf{r}\,\tfrac{1}{2}m\overline{c^2}$. Writing $\mathbf{c} = \mathbf{c}_0 + \mathbf{C}$, we may express the energy in the form

$$n\,d\mathbf{r}\,.\,\tfrac{1}{2}m(c_0^2 + 2\mathbf{c}_0\,.\,\overline{\mathbf{C}} + \overline{C^2})$$

or

$$\tfrac{1}{2}\rho\,d\mathbf{r}\,.\,c_0^2 + n\,d\mathbf{r}\,.\,\tfrac{1}{2}m\overline{C^2}.$$

Since $\rho\,d\mathbf{r}$ is the mass of the gas contained in $d\mathbf{r}$, the first term in the last expression represents the kinetic energy of the visible or mass motion of the gas. The second term is the kinetic energy of the invisible peculiar motion: its ratio to the first is $\overline{C^2}/c_0^2$. In 2.34 it was shown that $\overline{C^2}$ is very large for ordinary gases at S.T.P., the value of $\sqrt{(\overline{C^2})}$ being several hundred metres per second. Hence unless c_0 is much greater than is usual, $\overline{C^2}/c_0^2$ is very large, and there is much more hidden energy of peculiar motion than visible kinetic energy; for example, in the case of hydrogen at S.T.P., if $c_0 = 10$ cm./sec., the ratio is $3{\cdot}4 \times 10^8$. In addition, there may be further hidden molecular energy, kinetic and perhaps also potential, corresponding to molecular rotations, vibrations, and so on.

In the kinetic theory this hidden molecular energy, or rather that part which is communicable between molecules at encounters, is identified with the heat energy of the gas. Thus the heat energy per unit volume, or the *heat-density*, in a gas whose molecules are point-centres of force and therefore possess only translatory kinetic energy, is $\frac{1}{2}nm\overline{C^2}$ or $\frac{1}{2}\rho\overline{C^2}$. This is true also for a gas whose molecules are smooth rigid elastic spheres, for though these may also possess rotatory energy, this is not communicable between molecules at collision. In general, however, the molecules will possess other kinds of communicable energy, whose amounts vary from one molecule to another; the total heat energy E of a molecule is the sum of this communicable energy and the peculiar kinetic energy $\frac{1}{2}mC^2$, and the heat-density is $n\bar{E}$.

Here $\frac{1}{2}mC^2$ and E are supposed expressed in mechanical units, and the heat-density $n\bar{E}$ will be in the same units. If expressed in thermal units, the heat-density is $n\bar{E}/J$, where J is Joule's mechanical equivalent of heat ($4{\cdot}185 \times 10^7$ ergs/cal., or $4{\cdot}185$ joules/cal.)

2.41. Temperature

Two systems of temperature-reckoning are in common use among physicists. In experimental work they generally use the empirical temperature measured by expansion (mercury or gas) thermometers: in theoretical work they use the absolute temperature of thermodynamics. In the kinetic theory, on the other hand, the temperature T of a gas in a uniform steady state at rest or in uniform translation is defined directly in terms of the peculiar speeds of the molecules, by the relation

$$\tfrac{1}{2}m\overline{C^2} = \tfrac{3}{2}kT, \qquad (2.41, 1)$$

where k is a constant, the same for all gases, whose value will be assigned later (2.431); it is called the Boltzmann constant.

At a given density and temperature, the interchange of energy between the translatory and internal motions at molecular encounters establishes a balance between the mean translatory and internal energies. Thus, for a gas in a uniform steady state, the total mean thermal energy $\bar{E}$ of a molecule is a function of the T defined by (2.41, 1). For a gas not in a uniform steady state, the temperature T at any point is defined as that for which the same gas, when in a uniform steady state at the same density, would have the same mean thermal energy $\bar{E}$ per molecule at that point.

The kinetic-theory definition of temperature, being applicable whether or not the gas is in a uniform or steady state, is more general than that of thermodynamics and statistical mechanics, where only equilibrium states are considered. It is of importance to examine, however, whether the kinetic-theory definition is in agreement with that of thermodynamics if the gas is in equilibrium. Before so doing we proceed to deduce certain relations from the definition (2.41, 1).

2.42. The equation of state

An immediate consequence of the definition of temperature is that the hydrostatic pressure p of a gas in equilibrium is given by

$$p = \tfrac{1}{3}nm\overline{C^2} = knT. \tag{2.42, 1}$$

This formula applies also to a gas not in equilibrium if the molecules possess only translatory energy; for a gas with internal energy it need not be exact, though still correct to a close approximation. The formula is in agreement with the well-known hypothesis of Avogadro, according to which equal volumes of different gases, at the same pressure and temperature, contain equal numbers of molecules.

Consider now a mass M of gas contained in a volume V. The number of molecules in the mass M is M/m; the number-density n is therefore $M/m\text{V}$. On substituting in (2.42, 1), this takes the form

$$p\text{V} = k(M/m)\,T. \tag{2.42, 2}$$

This important relation embodies the well-known experimental laws of Boyle and Charles, which are closely followed by many gases, at moderate or low densities, and at temperatures well above their critical temperatures. These laws may be stated as follows:

Boyle's Law: For a given mass of gas at constant temperature the product of the pressure and the volume is constant.

Charles's Law: For a given mass of gas at constant pressure the volume varies directly as the *absolute* temperature.

The deduction of these laws from kinetic-theory principles and definitions affords some measure of justification for the latter. By itself, however, it does not suffice to show that the kinetic-theory definition of T is in accord with the thermodynamic definition for equilibrium states.

The relation between p, V and T for a given mass M of gas in equilibrium is called the equation of state of the gas. The above simple form of this equation is only an approximation to the equation of state as found for actual gases; the error of the simple formula $p\text{V} \propto T$ becomes considerable at high pressures and low temperatures. This is to be ascribed, not to any fault in the above kinetic-theory definition of temperature, but to the neglect, in deriving the expression $\tfrac{1}{3}\rho\overline{C^2}$ for p, of such factors as the finite size of molecules, their fields of force at large distances, and, when the gas is near the point of liquefaction, their tendency to aggregate into clusters. The effect of these neglected factors becomes considerable in precisely those conditions of high pressure and low temperature under which large deviations from (2.42, 2) are observed.

From (2.42, 2) other important relations can be obtained. The (chemical) molecular weight W of a gas is defined as $12m/m_c$, where m is the mass of a molecule of the gas, and m_c that of an atom of carbon, $= 1{\cdot}993 \times 10^{-23}$ g. A mass W grams of the gas is called a gram-molecule or mole of the gas; naturally it is different for different gases. It is a convenient mass to consider, because for all gases it contains the same number of molecules, W/m or $12/m_c$. This number is Loschmidt's number ($6{\cdot}022 \times 10^{23}$). Some writers call it Avogadro's number; but this term should strictly be applied only to the number of molecules in one c.c. of gas at S.T.P., $2{\cdot}687 \times 10^{19}$.

Suppose that the mass M used in (2.41, 2) is a gram-molecule W. Then

$$p\mathrm{V} = RT, \tag{2.42, 3}$$

where

$$R = k\mathrm{W}/m = 12k/m_c. \tag{2.42, 4}$$

Clearly R has the same value for all gases. It is called the gas-constant per mole.

Again, from (2.42, 1, 4)

$$p = \frac{R}{\mathrm{W}}\rho T. \tag{2.42, 5}$$

2.43. Specific heats

Let unit mass of a gas in equilibrium be enclosed in a constant volume. To increase its temperature from T to $T+\delta T$, a certain amount of heat must be added, which, if δT is small, will be proportional to δT; we write it as

$$c_v \delta T.$$

The coefficient c_v is called the specific heat of the gas at constant volume. Since the gas does no mechanical work against external pressure during the process, the added heat $c_v\delta T$ must go entirely to increase the heat energy of the gas. The number of molecules in unit mass is $(1/m)$, and so the initial heat energy is $\bar{E}/m$; this is increased by $c_v\delta T$ when T is increased by δT; hence

$$\delta\bar{E}/m = c_v\delta T,$$

or, proceeding to the limit,

$$c_v = \frac{1}{m}\left(\frac{d\bar{E}}{dT}\right)_{\mathrm{V}}, \tag{2.43, 1}$$

where $(d\bar{E}/dT)_{\mathrm{V}}$ denotes the rate of increase of $\bar{E}$ with respect to T, when V is kept constant.

If the thermal energy consists only of energy of translation of the molecules, $E = \frac{1}{2}mC^2$, and $\bar{E} = \frac{3}{2}kT$, by (2.41, 1); hence

$$c_v = \frac{3k}{2m} \tag{2.43, 2}$$

in mechanical units, or
$$c_v = \frac{3k}{2Jm} \tag{2.43, 2'}$$
in thermal units.

If, instead of the volume, it is the pressure of the gas which is kept constant while T is increased by δT, the volume V will increase by δV, and on putting $M = 1$ in (2.42, 2), we find
$$p\delta V = (k/m)\delta T.$$

In the expansion mechanical work of amount $p\delta$V will be done, and the heat supplied to raise T must provide this energy $p\delta$V as well as the increase $\delta\bar{E}/m$ in the heat energy of the gas. Writing $c_p\delta T$ for the required amount of heat, we have
$$c_p\delta T = p\delta V + \delta\bar{E}/m$$
$$= (k\delta T + \delta\bar{E})/m.$$

Hence
$$c_p = \frac{k}{m} + \frac{1}{m}\left(\frac{d\bar{E}}{dT}\right)_p \tag{2.43, 3}$$
in mechanical units, the suffix p denoting that the pressure is kept constant.

An increase in pressure, or density, of a gas of assigned temperature can affect $\bar{E}$ only by increasing the number of pairs of molecules whose fields of force overlap, and which in consequence possess mutual potential energy. However, in the relatively rare gases for which Boyle's and Charles's laws are valid, at any given moment all save a negligible fraction of the molecules are independent systems, and $\bar{E}$ may be taken as depending only on T, and not on p and V. Thus for such a gas
$$c_p = \frac{k}{m} + c_v \tag{2.43, 4}$$
in mechanical units, or
$$c_p = \frac{k}{Jm} + c_v \tag{2.43, 4'}$$
in thermal units.

If the specific heats c_v and c_p are multiplied by the molecular weight W, we obtain the specific heats C_v and C_p per mole of the gas. Thus in mechanical units
$$C_v = \frac{W}{m}\frac{d\bar{E}}{dT}, \tag{2.43, 5}$$
$$C_p = \frac{W}{m}\left(k + \frac{d\bar{E}}{dT}\right)$$
$$= R + C_v, \tag{2.43, 6}$$
by (2.42, 4). It is an experimental fact that $C_p - C_v$ has nearly the same value for all actual gases under moderate conditions of pressure and temperature, a fact which further supports the principles and interpretations here used.

The ratio c_p/c_v or C_p/C_v of the specific heats is denoted by γ. Thus

$$\gamma = \frac{c_p}{c_v} = \frac{C_p}{C_v},$$

and (2.43, 4, 6) can be written in the forms

$$c_v(\gamma - 1) = k/m, \qquad (2.43, 7)$$

$$C_v(\gamma - 1) = R. \qquad (2.43, 8)$$

For a gas possessing no communicable internal energy, it is clear from (2.43, 2, 7) that $\gamma = \frac{5}{3} = 1{\cdot}6\dot{6}$.

2.431. The kinetic-theory temperature and thermodynamic temperature

It is now possible to establish the consistency of the kinetic-theory definition of temperature and the thermodynamic definition. Let a mass M of a gas undergo a small change of state, such that the temperature increases by δT and the volume by $\delta \mathrm{V}$. Then the energy it receives is

$$Mc_v\delta T + p\delta \mathrm{V} = T\left\{\frac{Mc_v}{T}\delta T + \frac{Mk}{m}\frac{\delta \mathrm{V}}{\mathrm{V}}\right\},$$

by (2.42, 2). Since c_v, like $\bar{E}$, is a function of T alone, the expression in the bracket denotes the increment of a function S of T and V. If the gas undergoes an adiabatic change, i.e. a change in which no energy is supplied, the change in S must be zero, that is, S is constant; in an isothermal change, at temperature T, the energy received is T multiplied by the change in S.

Suppose now that the gas is taken round a Carnot cycle working between lower and upper temperatures T_1 and T_2; then S returns to its original value when the cycle is completed. Since it is unaltered during the adiabatic processes, its increase ΔS at temperature T_1 must be equal and opposite to its decrease at temperature T_2, and the heats gained and lost at temperatures T_1 and T_2 respectively are $T_1\Delta$S, $T_2\Delta$S. Thus the efficiency of the cycle is

$$\frac{T_2\Delta \mathrm{S} - T_1\Delta \mathrm{S}}{T_2\Delta \mathrm{S}} = \left(1 - \frac{T_1}{T_2}\right).$$

This proves that our T is proportional to the temperature on the thermodynamic scale: the function S is the *entropy*.

It remains to consider the constant k introduced in (2.41, 1). This was taken to be the same for all gases: this implies the assumption that the mean peculiar kinetic energy of translation is the same for molecules of different gases at the same temperature. We shall show in 4.3 that in a gas-mixture in equilibrium the mean peculiar kinetic energies of molecules of the different constituent gases are the same, which is a similar result: but to establish the actual result it is necessary to consider the equilibrium of two gases separated by a diathermanous wall. This problem lies in the domain of statistical mechanics rather than of kinetic theory, and the reader is referred to books on that subject for a proof of the result in question. Alternatively we may regard the result as established by experiment, since Avogadro's hypothesis and the law $C_p - C_v = \text{const.}$ both follow directly from it.

In practice, k is chosen such as to make the kinetic-theory temperature coincide with the thermodynamic temperature measured in degrees K. That is, k is such that the difference between the temperatures of melting ice and boiling water at standard atmospheric pressure is $100°$; the zero of temperature is found to be approximately equal to $-273{\cdot}15°$ C. The value of k can be determined if we know either the number-density of molecules in a gas at given p and T, or (cf. (2.42, 4)) the mass of a molecule. The determination of either of these is a matter of some difficulty, but the values of each have been found for several gases. The different determinations agree in giving*

$$k = 1{\cdot}3806 \times 10^{-16} \text{ ergs/degree.}$$

The determination of the gas-constant R is much easier. It is found that

$$R = 8{\cdot}314 \times 10^7 \text{ ergs/degree,}$$

or, in thermal units,

$$\frac{R}{J} = 1{\cdot}9865 \text{ cal./degree.}$$

2.44. Specific heats: numerical values

The values of γ, C_p, C_v, c_v, $C_p - C_v$ are given for several gases in Table 1. The values refer to a temperature of $15°$ C. and a pressure of 1 atmosphere; departures from the Boyle–Charles law affect the values appreciably, even at this pressure. The units of C_p and C_v are calories (at $15°$ C.) per degree per mole; for c_v the units are calories per degree per gram.†

As Table 1 shows, γ has very nearly the value $5/3$ for the monatomic gases helium, neon and argon, which are on many grounds believed to possess no internal energy communicable at ordinary encounters.‡ This is a further confirmation of the kinetic-theory interpretations.

For other gases, let us write

$$\frac{d\bar{E}}{dT} = \tfrac{1}{2}Nk, \tag{2.44, 1}$$

* B. N. Taylor, W. H. Parker and D. N. Langenberg, *Rev. Mod. Phys.* **41**, 375 (1969).

† The values given in this table are based on the following sources: J. R. Partington and W. G. Shilling, *The Specific Heats of Gases*, p. 201 (Benn, 1924); J. Hilsenrath *et al.*, U.S. National Bureau of Standards, Circular 564 (1955); F. Din, *Thermodynamic Functions of Gases* (Butterworth, 1956). The values given for D_2 and Xe are values *calculated* by a method that gives good agreement with experiment for other gases (A. Michels, W. de Graaff and G. J. Wolkers, *Physica*, **25**, 1097 (1959), for D_2; A. Michels, T. Wassenaar, G. J. Wolkers and J. Dawson, *Physica*, **22**, 17 (1956), for Xe). The value of γ for Ne was found by W. H. Keesom and J. A. van Lammeren, *Physica*, **1**, 1161 (1934).

‡ The deviation from $\gamma = 5/3$ for xenon is due to departures from the perfect-gas laws.

Table 1. *Specific heats*

Gas	γ	C_p	C_v	c_v	$C_p - C_v$
Air	1·402	6·960	4·965	0·172	1·995
He	1·668	4·97	2·98	0·745	1·99
A	1·669	4·975	2·98	0·0747	1·995
Ne	1·668	·	·	·	·
Xe	[1·679]	[5·032]	[2·996]	[0·0228]	[2·036]
H_2	1·408	6·87	4·88	2·42	1·99
D_2	[1·399]	[6·98]	[4·99]	[1·238]	[1·99]
N_2	1·403	6·96	4·96	0·177	2·00
CO	1·402	6·97	4·97	0·177	2·00
NO	1·400	7·00	5·00	0·167	2·00
O_2	1·398	7·03	5·03	0·157	2·00
Cl_2	1·356	8·04	5·93	0·084	2·11
H_2S	1·340	8·15	6·08	0·178	2·07
CO_2	1·303	8·85	6·79	0·154	2·06
N_2O	1·300	8·85	6·81	0·155	2·04
SO_2	1·284	9·62	7·49	0·117	2·13
NH_3	1·318	8·75	6·64	0·390	2·11
CH_4	1·310	8·49	6·48	0·404	2·01
C_2H_4	1·250	10·25	8·20	0·293	2·05
C_2H_6	1·20	12·42	10·36	0·345	2·06

so that for monatomic gases $N = 3$, while for other gases we expect that $N > 3$. Then

$$c_v = \frac{kN}{2m}, \quad c_p = \frac{k}{m}\left(1 + \frac{N}{2}\right), \tag{2.44, 2}$$

$$\gamma = 1 + \frac{2}{N}. \tag{2.44, 3}$$

If c_v is independent of T, as is found to be approximately the case for many gases (monatomic and otherwise) over a considerable range of temperature, then N is independent of T. Consequently, apart from a possible additive constant,

$$\bar{E} = \tfrac{1}{2}NkT.$$

For many diatomic gases, as the above table shows, $\gamma = 1{\cdot}4$ very approximately; this corresponds to $N = 5$, indicating that the communicable internal energy is two-thirds the peculiar kinetic energy of translation. For polyatomic molecules the values of γ are less than $1{\cdot}4$, and the corresponding values of N are greater than 5; this implies a still larger proportion of non-translatory energy.

2.45. Conduction of heat

An important flux-vector (cf. 2.3, p. 31) is that giving the rate of flow of heat energy, corresponding to $\phi(\boldsymbol{C}) = E$. We denote this flux-vector by $\boldsymbol{q}$, so that

$$\boldsymbol{q} = n\overline{E\boldsymbol{C}}. \tag{2.45, 1}$$

Thus the rate of flow of heat across a surface through the point $\boldsymbol{r}$, normal to the unit vector $\mathbf{n}$, is equal to

$$\boldsymbol{q}.\mathbf{n} = n\overline{EC_n}$$

per unit area. The vector $\boldsymbol{q}$ is termed the thermal flux-vector.

2.5. Gas-mixtures

All the above definitions and results may be generalized to apply to a mixture of gases. The definitions of the number-density and velocity-distribution function of each of the constituent gases are analogous to those employed for a simple gas in 2.2 and 2.21. The velocity-distribution function $f_s(\boldsymbol{c}_s, \boldsymbol{r}, t)$ and the number-density n_s of the sth constituent are connected by the relation

$$n_s = \int f_s(\boldsymbol{c}_s, \boldsymbol{r}, t)\, d\boldsymbol{c}_s. \tag{2.5, 1}$$

The number-density n of the whole gas is given by the relation

$$n = \sum_s n_s. \tag{2.5, 2}$$

The different masses and mean molecular speeds of the different constituents render it pointless to consider the velocity-distribution function of the whole gas.

If the mass of a molecule of the sth constituent is m_s, the partial density of this constituent is ρ_s, where

$$\rho_s = n_s m_s, \tag{2.5, 3}$$

and the density ρ of the whole gas is given by

$$\rho = \sum_s \rho_s = \sum_s n_s m_s. \tag{2.5, 4}$$

We shall frequently refer to molecules of the sth constituent as molecules m_s.

If ϕ is any function of the velocities of the molecules, its mean value $\overline{\phi}_s$ at any point, for molecules m_s, is given by

$$n_s\overline{\phi}_s = \int f_s \phi_s d\boldsymbol{c}_s, \tag{2.5, 5}$$

and the mean value $\overline{\phi}$ for all molecules of the mixture is given by

$$n\overline{\phi} = \sum_s n_s\overline{\phi}_s = \sum_s \int f_s \phi_s d\boldsymbol{c}_s. \tag{2.5, 6}$$

The *mass-velocity* $\boldsymbol{c}_0$ of the gas at any point is defined by the equation

$$\rho\boldsymbol{c}_0 = \sum_s \int f_s m_s \boldsymbol{c}_s d\boldsymbol{c}_s = \sum_s \rho_s \overline{\boldsymbol{c}}_s; \tag{2.5, 7}$$

it is not the mean velocity $\overline{\boldsymbol{c}}$ of the molecules, but a weighted mean, giving to each molecule a weight proportional to its mass. The momentum of the gas per unit volume is the same as if every molecule moved with the mass-velocity $\boldsymbol{c}_0$.

The peculiar velocity $\boldsymbol{C}_s$ of a molecule m_s in a gas-mixture is defined by the equation

$$\boldsymbol{C}_s = \boldsymbol{c}_s - \boldsymbol{c}_0. \tag{2.5, 8}$$

Clearly

$$\sum_s \rho_s \overline{\boldsymbol{C}}_s = 0. \tag{2.5, 9}$$

The temperature T of the gas at any point is defined, as for a simple gas, by the equation

$$\tfrac{1}{2}\overline{mC^2} \equiv \frac{1}{n}\sum_s \int f_s \tfrac{1}{2}m_s C_s^2 \, d\boldsymbol{c}_s = \tfrac{3}{2}kT, \qquad (2.5, 10)$$

for a gas in a uniform steady state. For a gas in a more general state it is defined by $\bar{E} = \bar{E}(T)$, where $\bar{E}(T)$ is the value of $\bar{E}$ for a uniform gas in a steady state, with the density and composition of the actual gas at the point considered, at the temperature T.

The partial pressure of any constituent, on the surface of a containing vessel or on an internal surface moving with the mass-velocity of the gas, is defined as the mean rate at which momentum of that constituent is communicated to, or transferred across, unit area of the surface; the total pressure of the gas on the surface is the sum of the partial pressures of the constituents. It follows as in 2.31 that the pressure-tensor p is defined by the equation

$$\mathsf{p} \equiv \sum_s \mathsf{p}_s \equiv \sum_s n_s m_s \overline{\boldsymbol{C}_s \boldsymbol{C}_s} \equiv n\overline{m\boldsymbol{C}\boldsymbol{C}}. \qquad (2.5, 11)$$

As before, the pressure on a surface normal to the unit vector $\mathbf{n}$ is $\mathsf{p}.\mathbf{n}$.

The mean hydrostatic pressure p of the gas at any point is $\tfrac{1}{3}\mathsf{p}:\mathsf{U}$, as in (2.32, 3); hence for a gas in equilibrium, by (2.5, 10, 11),

$$p = \tfrac{1}{3}n\overline{mC^2} = knT, \qquad (2.5, 12)$$

which is equivalent to Boyle's and Charles's laws. By (2.5, 2), this implies that

$$p = \sum_s (kn_s T).$$

Thus the hydrostatic pressure of the mixture at a given temperature is equal to the sum of the hydrostatic pressures p_s which would be exerted by the constituents if each separately occupied the same volume and were at the same temperature. This is Dalton's law.

Finally, the vector $\boldsymbol{q}$ of thermal flow is connected with the molecular energy E by the equation

$$\boldsymbol{q} = n\overline{E\boldsymbol{C}}, \qquad (2.5, 13)$$

as for a simple gas.

3

THE EQUATIONS OF BOLTZMANN AND MAXWELL

3.1. Boltzmann's equation derived

In Chapter 2 it was shown that the macroscopic properties of a gas can be calculated from the velocity-distribution function f. This function can be determined from a certain integral equation first given by Boltzmann. In order to indicate Maxwell's association with the ideas thus formulated by Boltzmann, Hilbert* termed it the Maxwell–Boltzmann equation.

In deriving the equation it is assumed that encounters with other molecules occupy a very small part of the lifetime of a molecule. This implies that only binary encounters are important.

Consider a gas in which each molecule is subject to an external force $m\boldsymbol{F}$, which may be a function of $\boldsymbol{r}$ and t but not† of $\boldsymbol{c}$. Between the times t and $t+dt$ the velocity $\boldsymbol{c}$ of any molecule that does not collide with another will change to $\boldsymbol{c}+\boldsymbol{F}dt$, and its position-vector $\boldsymbol{r}$ will change to $\boldsymbol{r}+\boldsymbol{c}dt$. There are $f(\boldsymbol{c}, \boldsymbol{r}, t)\, d\boldsymbol{c}\, d\boldsymbol{r}$ molecules which at time t lie in the volume-element $\boldsymbol{r}$, $d\boldsymbol{r}$, and have velocities in the range $\boldsymbol{c}$, $d\boldsymbol{c}$. After the interval dt, if the effect of encounters could be neglected, the same molecules, and no others, would compose the set that occupy the volume $\boldsymbol{r}+\boldsymbol{c}dt$, $d\boldsymbol{r}$, and have velocities in the range $\boldsymbol{c}+\boldsymbol{F}dt$, $d\boldsymbol{c}$: the number in this set is

$$f(\boldsymbol{c}+\boldsymbol{F}dt, \boldsymbol{r}+\boldsymbol{c}dt, t+dt)\, d\boldsymbol{c}\, d\boldsymbol{r}.$$

The number of molecules in the second set will, however, in general differ from that in the first, since molecular encounters will have deflected some molecules of the initial set from their course, and will have deflected other molecules so that they become members of the final set. The net gain of molecules to the second set must be proportional to $d\boldsymbol{c}\, d\boldsymbol{r}\, dt$, and will be denoted by $(\partial_e f/\partial t)\, d\boldsymbol{c}\, d\boldsymbol{r}\, dt$. Consequently

$$\{f(\boldsymbol{c}+\boldsymbol{F}dt, \boldsymbol{r}+\boldsymbol{c}dt, t+dt) - f(\boldsymbol{c}, \boldsymbol{r}, t)\}\, d\boldsymbol{c}\, d\boldsymbol{r} = \frac{\partial_e f}{\partial t}\, d\boldsymbol{c}\, d\boldsymbol{r}\, dt.$$

On dividing by $d\boldsymbol{c}\, d\boldsymbol{r}\, dt$, and making dt tend to zero, Boltzmann's equation for f is obtained, namely

$$\frac{\partial f}{\partial t}+u\frac{\partial f}{\partial x}+v\frac{\partial f}{\partial y}+w\frac{\partial f}{\partial z}+F_x\frac{\partial f}{\partial u}+F_y\frac{\partial f}{\partial v}+F_z\frac{\partial f}{\partial w} = \frac{\partial_e f}{\partial t}, \qquad (3.1, 1)$$

* D. Hilbert, *Grundzüge einer allgemeinen Theorie der linearen Integralgleichungen*, p. 269 (Teubner, 1912).

† A special case in which the force on a molecule depends on its velocity is considered in Chapter 19.

or

$$\mathscr{D}f = \frac{\partial_e f}{\partial t}, \tag{3.1, 2}$$

where $\mathscr{D}f$ denotes the left-hand side of (3.1, 1); in vector notation

$$\mathscr{D}f \equiv \frac{\partial f}{\partial t} + \boldsymbol{c}.\frac{\partial f}{\partial \boldsymbol{r}} + \boldsymbol{F}.\frac{\partial f}{\partial \boldsymbol{c}}. \tag{3.1, 3}$$

The quantity $\partial_e f/\partial t$ defined above is equal to the rate of change, owing to encounters, in the velocity-distribution function f at a *fixed* point. It will appear later that $\partial_e f/\partial t$ is expressible as an integral involving the unknown function f. Thus Boltzmann's equation is an integral (or integro-differential) equation.

The generalization for a mixture of gases is

$$\mathscr{D}_s f_s \equiv \frac{\partial f_s}{\partial t} + \boldsymbol{c}_s.\frac{\partial f_s}{\partial \boldsymbol{r}} + \boldsymbol{F}_s.\frac{\partial f_s}{\partial \boldsymbol{c}_s} = \frac{\partial_e f_s}{\partial t}, \tag{3.1, 4}$$

where $m_s \boldsymbol{F}_s$ denotes the force on a molecule m_s at $\boldsymbol{r}$, t, and $\partial_e f_s/\partial t$ denotes the rate at which the velocity-distribution function f_s is being altered by encounters. The equation can also be modified to apply to more general molecular models (2.1, 2.21); in the case of rotating molecules possessing spherical symmetry, f depends only on $\boldsymbol{c}$, $\boldsymbol{r}$, t and the angular velocity $\boldsymbol{\omega}$, and the equation for f has the same form as (3.1, 1). For more general models f will involve further variables, specifying the orientation and other properties of a molecule; terms corresponding to these variables must in general appear in Boltzmann's equation.

3.11. The equation of change of molecular properties

Another important equation may be derived from Boltzmann's equation as follows. Consider first a simple gas. Let ϕ be any molecular property as defined in 2.22. Multiply Boltzmann's equation by $\phi d\boldsymbol{c}$ and integrate throughout the velocity-space; it is supposed that all the integrals obtained are convergent, and that products such as ϕf tend to zero as $\boldsymbol{c}$ tends to infinity in any direction. The result may be written as

$$\int \phi \mathscr{D}f d\boldsymbol{c} = n\Delta\bar{\phi}, \tag{3.11, 1}$$

where

$$n\Delta\bar{\phi} \equiv \int \phi \frac{\partial_e f}{\partial t} d\boldsymbol{c}. \tag{3.11, 2}$$

The significance of $\Delta\bar{\phi}$ is readily seen; $(\partial_e f/\partial t) d\boldsymbol{c}$ measures the rate of change in the number of molecules with velocities in the range $\boldsymbol{c}$, $d\boldsymbol{c}$, per unit volume at $\boldsymbol{r}$, t, owing to encounters. Consequently $\phi(\partial_e f/\partial t) d\boldsymbol{c}$ represents the rate of change in the sum $\Sigma\phi$ extended over all the molecules of this set, due to the same cause. Similarly $\int \phi(\partial_e f/\partial t) d\boldsymbol{c}$ is the rate of change by encounters in $\Sigma\phi$, summed over all the molecules in unit volume. But $\Sigma\phi = n\bar{\phi}$, and, since encounters do not in themselves modify the number-density n, the rate of change of $\overline{\phi}$ by encounters is equal to

$$\frac{1}{n}\int \phi \frac{\partial_e f}{\partial t} d\mathbf{c} = \Delta\overline{\phi}.$$

Equation (3.11, 1) can be generalized to apply to a mixture of gases; for the sth constituent

$$\int \phi_s \mathscr{D}_s f_s d\mathbf{c}_s = n_s \Delta\overline{\phi}_s, \tag{3.11, 3}$$

where $\Delta\overline{\phi}_s$ is equal to the rate of change of $\overline{\phi}_s$ by molecular encounters. A modified form of (3.11, 1) is, moreover, satisfied by the generalized velocity-distribution function of 2.21, ϕ being then a function of further variables besides $\mathbf{c}$.

3.12. $\mathscr{D}f$ expressed in terms of the peculiar velocity

If the peculiar velocity $\mathbf{C}(\equiv \mathbf{c}-\mathbf{c}_0)$ is used as an independent variable instead of $\mathbf{c}$, the meanings of $\partial/\partial t$ and $\partial/\partial\mathbf{r}$ are changed, since now $\mathbf{C}$, not $\mathbf{c}$, is to be kept constant while performing the differentiation. Hence in (3.1, 3) $\partial f/\partial t$ and $\partial f/\partial\mathbf{r}$ have to be replaced respectively by

$$\frac{\partial f}{\partial t}-\frac{\partial \mathbf{c}_0}{\partial t}\cdot\frac{\partial f}{\partial \mathbf{C}} \quad\text{and}\quad \frac{\partial f}{\partial \mathbf{r}}-\left(\frac{\partial}{\partial \mathbf{r}}\mathbf{c}_0\right)\cdot\frac{\partial f}{\partial \mathbf{C}}$$

in order to take account of the implicit dependence of f on t and $\mathbf{r}$ through the dependence of $\mathbf{C}$ on $\mathbf{c}_0$. Also $\partial f/\partial\mathbf{c}$ becomes $\partial f/\partial\mathbf{C}$; hence the expression for $\mathscr{D}f$ becomes

$$\frac{\partial f}{\partial t}-\frac{\partial \mathbf{c}_0}{\partial t}\cdot\frac{\partial f}{\partial \mathbf{C}}+(\mathbf{c}_0+\mathbf{C})\cdot\left\{\frac{\partial f}{\partial \mathbf{r}}-\left(\frac{\partial}{\partial \mathbf{r}}\mathbf{c}_0\right)\cdot\frac{\partial f}{\partial \mathbf{C}}\right\}+\mathbf{F}\cdot\frac{\partial f}{\partial \mathbf{C}}.$$

Let

$$\frac{D}{Dt}=\frac{\partial}{\partial t}+\mathbf{c}_0\cdot\frac{\partial}{\partial \mathbf{r}}, \tag{3.12, 1}$$

so that D/Dt is the 'mobile operator', or time-derivative following the motion, as in hydrodynamics. Then

$$\mathscr{D}f=\frac{Df}{Dt}+\mathbf{C}\cdot\frac{\partial f}{\partial \mathbf{r}}+\left(\mathbf{F}-\frac{D\mathbf{c}_0}{Dt}\right)\cdot\frac{\partial f}{\partial \mathbf{C}}-\frac{\partial f}{\partial \mathbf{C}}\mathbf{C}:\frac{\partial}{\partial \mathbf{r}}\mathbf{c}_0. \tag{3.12, 2}$$

3.13. Transformation of $\int \phi \mathscr{D} f d\mathbf{c}$

Suppose that ϕ, like f, is expressed as a function of $\mathbf{C}$, $\mathbf{r}$, t. The various terms obtained on substituting for $\mathscr{D}f$ from (3.12, 2) into $\int\phi\mathscr{D}f d\mathbf{c}$ (or $\int\phi\mathscr{D}f d\mathbf{C}$) can be transformed by means of relations such as

$$\int \phi\frac{Df}{Dt}d\mathbf{C}=\frac{D}{Dt}\int\phi f d\mathbf{C}-\int\frac{D\phi}{Dt}f d\mathbf{C}=\frac{D(n\overline{\phi})}{Dt}-n\overline{\frac{D\phi}{Dt}}, \tag{3.13, 1}$$

$$\int \phi\mathbf{C}\cdot\frac{\partial f}{\partial \mathbf{r}}d\mathbf{C}=\frac{\partial}{\partial \mathbf{r}}\cdot\int\phi\mathbf{C}f d\mathbf{C}-\int\frac{\partial\phi}{\partial \mathbf{r}}\cdot\mathbf{C}f d\mathbf{C}=\frac{\partial}{\partial \mathbf{r}}\cdot n\overline{\phi\mathbf{C}}-n\overline{\mathbf{C}\cdot\frac{\partial\phi}{\partial \mathbf{r}}}, \tag{3.13, 2}$$

$$\int\phi\frac{\partial f}{\partial U}d\mathbf{C}=\iint\Big[\phi f\Big]_{U=-\infty}^{U=\infty}dV\,dW-\int\frac{\partial\phi}{\partial U}f d\mathbf{C}=-n\overline{\frac{\partial\phi}{\partial U}}. \tag{3.13, 3}$$

In (3.13, 1, 2) the variable $\boldsymbol{C}$ is not included in the differentiations D/Dt, $\partial/\partial\boldsymbol{r}$, since $\boldsymbol{C}$ is now regarded as an independent variable. In (3.13, 3) the term involving ϕf, obtained by integrating by parts, vanishes because, by hypothesis, $\phi f \to 0$ as $U \to \pm\infty$. Clearly (3.13, 3) is one component of the equation

$$\int \phi \frac{\partial f}{\partial \boldsymbol{C}} d\boldsymbol{c} = -n \overline{\frac{\partial \phi}{\partial \boldsymbol{C}}}. \qquad (3.13, 4)$$

By a precisely similar argument

$$\int \phi \frac{\partial f}{\partial \boldsymbol{C}} \boldsymbol{C} d\boldsymbol{C} = -n \overline{\frac{\partial}{\partial \boldsymbol{C}} \phi \boldsymbol{C}} = -n\overline{\phi}\mathsf{U} - n \overline{\frac{\partial \phi}{\partial \boldsymbol{C}} \boldsymbol{C}}, \qquad (3.13, 5)$$

where (cf. 1.3) U denotes the unit tensor.

Using these results in (3.11, 1), we obtain the equation

$$n\Delta\overline{\phi} = \int \phi \mathscr{D} f d\boldsymbol{C} = \frac{Dn\overline{\phi}}{Dt} + n\overline{\phi}\frac{\partial}{\partial \boldsymbol{r}}.\boldsymbol{c}_0 + \frac{\partial}{\partial \boldsymbol{r}}.n\overline{\phi \boldsymbol{C}}$$
$$-n\left\{\overline{\frac{D\phi}{Dt}} + \overline{\boldsymbol{C}.\frac{\partial \phi}{\partial \boldsymbol{r}}} + \left(\boldsymbol{F} - \frac{D\boldsymbol{c}_0}{Dt}\right).\overline{\frac{\partial \phi}{\partial \boldsymbol{C}}} - \overline{\frac{\partial \phi}{\partial \boldsymbol{C}}\boldsymbol{C}} : \frac{\partial}{\partial \boldsymbol{r}}\boldsymbol{c}_0\right\}. \qquad (3.13, 6)$$

This is called the equation of change of $\overline{\phi}$; it is a generalization by Enskog of an equation of transfer due to Maxwell. Maxwell's equation refers to a function $\phi(\boldsymbol{c})$ of $\boldsymbol{c}$ alone, and does not introduce the peculiar velocity.

3.2. Molecular properties conserved after encounter; summational invariants

Some important results can be deduced from the equation of change without actually evaluating $\Delta\overline{\phi}$, because certain functions of the velocities of the molecules are conserved during encounters; that is, their sum for the molecules participating in an encounter is unaltered by the encounter, so that $\Delta\overline{\phi} = 0$. Such functions, which may be termed *summational invariants* for encounters, are of fundamental importance in the theory of gases, both because they can be measured throughout the fluctuations in the condition of a gas, and also because, as will appear later, their mean values, if completely specified at any instant as functions of position, in general determine the whole condition and future course of the gas.

For every gas, of whatever kind, three such summational invariants are

$$\psi^{(1)} = 1, \quad \boldsymbol{\psi}^{(2)} = m\boldsymbol{C}, \quad \psi^{(3)} = E, \qquad (3.2, 1)$$

where, as in 2.4, E denotes the total thermal energy of a molecule. For special types of gas there may be one or more additional summational invariants (cf. 11.24).

The statement $\Delta\overline{\psi^{(1)}} = 0$ merely implies that the number-density of molecules is unaltered by encounters. Similarly $\Delta\overline{\boldsymbol{\psi}^{(2)}} = 0$ expresses the principle of conservation of momentum (relative to axes moving with velocity $\mathbf{c}_0$), while $\Delta\overline{\psi^{(3)}} = 0$ expresses the principle of conservation of energy. Although $\boldsymbol{\psi}^{(2)}$ is a vector it is convenient to refer to the three conserved functions by the single symbol $\psi^{(i)}$.

It is sometimes more convenient to refer the momentum $\boldsymbol{\psi}^{(2)}$ and the energy $\psi^{(3)}$ to axes fixed in space, instead of to axes moving with the gas. Thus, for example, if the molecules possess only energy of translation the summational invariants can be taken as

$$\psi^{(1)} = 1, \quad \boldsymbol{\psi}^{(2)} = m\mathbf{c}, \quad \psi^{(3)} = \tfrac{1}{2}mc^2. \tag{3.2, 2}$$

In the case of binary encounters, the conservation of $\psi^{(i)}$ is expressible in the form

$$\psi^{(i)\prime} + \psi_1^{(i)\prime} - \psi^{(i)} - \psi_1^{(i)} = 0, \tag{3.2, 3}$$

where $\psi^{(i)}$, $\psi_1^{(i)}$ refer to the two molecules before encounter, and $\psi^{(i)\prime}$, $\psi_1^{(i)\prime}$ to these molecules after encounter.

Any linear combination of the three conserved functions $\psi^{(i)}$ is also a summational invariant, but no further summational invariant, linearly independent of $\psi^{(1)}$, $\boldsymbol{\psi}^{(2)}$ and $\psi^{(3)}$ as given by (3.2, 2), can exist for molecules whose energy is purely translatory. This is because an encounter between two such molecules involves two disposable geometrical variables (for example, if the molecules are rigid elastic spheres, these two variables may be the polar angles θ, φ of the line of centres at collision; cf. 3.43); when these two variables are eliminated from the six scalar relations which express the six components of $\mathbf{c}'$, $\mathbf{c}_1'$, the velocities after encounter, in terms of the six components of the initial velocities $\mathbf{c}$, $\mathbf{c}_1$, only four general scalar relations between the two sets of six components are obtainable. But we already have four such relations, expressing the conservation of energy and of the three components of momentum; hence no additional independent relation, valid for all encounters, is possible.

3.21. Special forms of the equation of change of molecular properties

By substituting each of the functions $\psi^{(i)}$ for ϕ in (3.13, 6), important special forms of the equation of change are obtained.

Case I. Let $\phi = \psi^{(1)} = 1$; then $\bar{\phi} = 1$, $\overline{\phi\mathbf{C}} = 0$, $\partial\phi/\partial\mathbf{C} = 0$, $D\phi/Dt = 0$, $\partial\phi/\partial\mathbf{r} = 0$, $\Delta\bar{\phi} = 0$. Thus (3.13, 6) becomes

$$\int \psi^{(1)}\mathscr{D}f\,d\mathbf{c} \equiv \frac{Dn}{Dt} + n\frac{\partial}{\partial\mathbf{r}}\cdot\mathbf{c}_0 = 0; \tag{3.21, 1}$$

this is the equation of *continuity*, expressing the conservation of number of molecules (or mass) in the gas. It has the alternative forms

$$\left.\begin{aligned} \frac{D\ln n}{Dt} + \frac{\partial}{\partial\mathbf{r}}\cdot\mathbf{c}_0 = 0, \\ \frac{D\rho}{Dt} + \rho\frac{\partial}{\partial\mathbf{r}}\cdot\mathbf{c}_0 = 0, \quad \frac{D\ln\rho}{Dt} + \frac{\partial}{\partial\mathbf{r}}\cdot\mathbf{c}_0 = 0. \end{aligned}\right\} \tag{3.21, 2}$$

Case II. Let $\phi = \psi_x^{(2)} = mU$; then $\overline{\phi} = 0$, $n\overline{\phi \boldsymbol{C}} = \rho\overline{U\boldsymbol{C}} = \boldsymbol{p}_x$ (cf. 2.31), $\partial\phi/\partial\boldsymbol{r} = 0$, $D\phi/Dt = 0$, $\partial\phi/\partial\boldsymbol{C} = (m, 0, 0)$, $\overline{(\partial\phi/\partial\boldsymbol{C})\boldsymbol{C}} = 0$, $\Delta\overline{\phi} = 0$. Hence the equation of change becomes

$$\int \psi_x^{(2)} \mathscr{D} f\, d\boldsymbol{c} \equiv \frac{\partial}{\partial \boldsymbol{r}} \cdot \boldsymbol{p}_x - \rho\left(F_x - \frac{Du_0}{Dt}\right) = 0;$$

this (cf. (1.33, 9)) is one component of the equation of *momentum* for the gas, namely

$$\int \boldsymbol{\psi}^{(2)} \mathscr{D} f\, d\boldsymbol{c} \equiv \frac{\partial}{\partial \boldsymbol{r}} \cdot \mathsf{p} - \rho\left(\boldsymbol{F} - \frac{D\boldsymbol{c}_0}{Dt}\right) = 0. \qquad (3.21, 3)$$

Equations (3.21, 2, 3) are identical with the equations of continuity and momentum derived for a continuous fluid in hydrodynamics; they provide a justification for the hydrodynamical treatment of a gas.

Case III. Let $\phi = \psi^{(3)} = E$; then $n\overline{\phi\boldsymbol{C}} = \boldsymbol{q}$ (cf. (2.45, 1)), $\partial\phi/\partial\boldsymbol{r} = 0$, $D\phi/Dt = 0$ and, since E depends on $\boldsymbol{C}$ only through the contribution of the kinetic energy of translation, $\partial\phi/\partial\boldsymbol{C} = \frac{1}{2}m\,\partial C^2/\partial\boldsymbol{C} = m\boldsymbol{C}$; thus $\overline{\partial\phi/\partial\boldsymbol{C}} = 0$, and $n\overline{(\partial\phi/\partial\boldsymbol{C})\boldsymbol{C}} = \rho\overline{\boldsymbol{C}\boldsymbol{C}} = \mathsf{p}$. Since $\Delta\overline{\phi} = 0$, the equation of change becomes

$$\int \psi^{(3)} \mathscr{D} f\, d\boldsymbol{c} \equiv \frac{D(n\overline{E})}{Dt} + n\overline{E}\frac{\partial}{\partial \boldsymbol{r}} \cdot \boldsymbol{c}_0 + \frac{\partial}{\partial \boldsymbol{r}} \cdot \boldsymbol{q} + \mathsf{p} : \frac{\partial}{\partial \boldsymbol{r}} \boldsymbol{c}_0 = 0. \qquad (3.21, 4)$$

Since $d\overline{E}/dT = \frac{1}{2}Nk$ (cf. (2.44, 1)) equation (3.21, 4) may be transformed to

$$\frac{DT}{Dt} = -\frac{2}{Nkn}\left\{\mathsf{p} : \frac{\partial}{\partial \boldsymbol{r}} \boldsymbol{c}_0 + \frac{\partial}{\partial \boldsymbol{r}} \cdot \boldsymbol{q}\right\}, \qquad (3.21, 5)$$

using (3.21, 1). This, or more properly (3.21, 4), is the equation of *thermal energy* for the gas. Let equation (3.21, 4) be multiplied by $d\boldsymbol{r}\,dt$: then $D(n\overline{E})/Dt \,.\, d\boldsymbol{r}\,dt$ represents the increase of thermal energy during the time dt in a volume $d\boldsymbol{r}$ moving with the gas. This may be interpreted from a macroscopic standpoint as the sum of (i) the energy brought into $d\boldsymbol{r}$ by the net inflow of molecules into the element, (ii) the gain due to the greater energy, as distinct from the greater number, of the inflowing as compared with the outflowing molecules, and (iii) the work done on the element by the pressures on its surface as it varies in shape and volume during the time dt; these three quantities are represented by the last three terms on the left of the equation, with their signs reversed.

Equations (3.21, 2, 3, 5) represent the maximum information which can be derived from the equation of change without determining the form of the velocity-distribution function. To determine it we must first find explicit forms for the expressions $\partial_e f/\partial t$ and $\Delta\overline{\phi}$. This involves an investigation of the statistical effect of encounters.

3.3. Molecular encounters

Exact expressions for $\partial_e f/\partial t$ and $\Delta\overline{\phi}$ can be given only when the nature of the interaction between molecules at encounter is known. Only in relatively few cases can atomic theory describe the process of encounter exactly. It is therefore necessary to assume some law of interaction: the appropriateness of the assumed law can be tested by comparing the results deduced from it with experimental results.

Physical data for gases, relating to the deviations of the equation of state from Boyle's law, show that, at distances large compared with molecular dimensions, molecules may exert a weak attractive force on each other, whereas at distances of the order of molecular dimensions they repel each other strongly. Moreover, at an encounter between complex molecules possessing internal energy some interchange of this energy with energy of translation may occur. Assumptions as to the nature of the forces between molecules at encounter must take these facts into account.

Various special models, chosen for their physical simplicity, or for the mathematical simplicity of their laws of interaction, have been studied. One of the earliest and simplest molecular models is a rigid, smooth, and perfectly elastic sphere. The impulse between two spheres at collision here represents the repulsive force between molecules at a close encounter. The representation can, however, only be approximate, since molecules, being complicated electronic structures, cannot closely resemble rigid spheres; the interaction between them varies continuously as they approach one another. This fact is better represented by treating a molecule as a point-centre of force, the force depending on the nature of the interacting molecules and on their distance apart. One simple assumption is that the force is always repulsive and varies inversely as some power of the distance. A better representation of the facts is afforded, however, if the force is supposed to change sign at a certain distance, beyond which it is an attraction. The elastic sphere model may also be improved by supposing the spheres to attract one another weakly, with a force depending on the distance.

If the molecule is represented either as a smooth sphere or as a point-centre of force, no provision is made for a possible interchange between internal energy and energy of translation. Such molecules may be termed *smooth*: their internal energy can be neglected, as it does not vary with the temperature. In most of this book, only smooth molecules are considered: but models permitting interchange between internal and translational energy are considered in Chapter 11.

All the smooth models we consider possess the property of spherical symmetry. The molecules of a monatomic gas closely approximate to such symmetry, but diatomic and polyatomic molecules diverge widely from it, by reason of the concentration of mass in the atomic nuclei. Thus our theoretical results do not apply strictly to diatomic and polyatomic gases;

since, however, in our calculations we average over all possible orientations of pairs of molecules at encounter, many of our results may be expected to apply approximately to such gases, if we ascribe to our spherically symmetrical molecules a field which is the average of the true field over all possible orientations of the molecules.

3.4. The dynamics of a binary encounter

Consider the encounter of two molecules of masses m_1, m_2. Since only smooth and spherically symmetrical molecules are considered, the force which either exerts on the other is directed along the line joining their centres, A, B; it may arise only at contact, or may act when the molecules are at any distance from each other, and be equal to some function of the distance AB. It is supposed that any external forces (gravitational, electric,...) which act on the molecules are so small compared with those brought into play during the encounter that their effect can be neglected in a consideration of the dynamical effect of an encounter.

The phrase 'before the encounter' will refer to the time before the molecules have begun to influence one another appreciably, so that each is moving in a straight line, or (more accurately) close to the asymptote of the orbit which it describes under the influence of the other; the phrase 'after the encounter' is to be interpreted in a similar way. With these conventions the velocities before and after the encounter have definite values, which will be denoted by $\boldsymbol{c}_1$, $\boldsymbol{c}_2$ (before) and $\boldsymbol{c}_1'$, $\boldsymbol{c}_2'$ (after).* It is desired to express either pair of velocities in terms of the other pair, and of any geometrical variables required to complete the specification of the encounter. Since the motion is reversible, the relation between the two pairs of velocities must be reciprocal. The details of the encounter are of no importance for our purpose; we wish only to know the relation between the initial and final velocities.

3.41. Equations of momentum and of energy for an encounter

Let $$m_0 \equiv m_1 + m_2, \quad M_1 \equiv m_1/m_0, \quad M_2 \equiv m_2/m_0, \qquad (3.41, 1)$$

so that $$M_1 + M_2 = 1. \qquad (3.41, 2)$$

The mass-centre of the two molecules will move uniformly throughout the encounter; its constant velocity $\boldsymbol{G}$ is given by

$$m_0 \boldsymbol{G} = m_1 \boldsymbol{c}_1 + m_2 \boldsymbol{c}_2 = m_1 \boldsymbol{c}_1' + m_2 \boldsymbol{c}_2'. \qquad (3.41, 3)$$

Let $\boldsymbol{g}_{21}$, $\boldsymbol{g}_{21}'$ and $\boldsymbol{g}_{12}$, $\boldsymbol{g}_{12}'$ denote respectively the initial and final velocities of the second molecule relative to the first, and of the first relative to the second, so that

$$\boldsymbol{g}_{21} = \boldsymbol{c}_2 - \boldsymbol{c}_1 = -\boldsymbol{g}_{12}, \quad \boldsymbol{g}_{21}' = \boldsymbol{c}_2' - \boldsymbol{c}_1' = -\boldsymbol{g}_{12}'. \qquad (3.41, 4)$$

* In 3.52 $\boldsymbol{c}_1'$, $\boldsymbol{c}_2'$ are used to denote the *initial* velocities, and $\boldsymbol{c}_1$, $\boldsymbol{c}_2$ the final velocities, in an *inverse* encounter. These uses of the symbol $\boldsymbol{c}'$ must be distinguished from that in 2.2 (p. 27), where $\boldsymbol{c}'$ denotes the velocity of moving axes of reference.

The magnitudes of $\boldsymbol{g}_{21}$ and $\boldsymbol{g}_{12}$ are equal, and can both be denoted by g; likewise for the final relative velocities; thus

$$g_{21} = g_{12} = g, \quad g'_{21} = g'_{12} = g'. \tag{3.41, 5}$$

By means of (3.41, 3, 4) we can express $\boldsymbol{c}_1$, $\boldsymbol{c}_2$, $\boldsymbol{c}'_1$, $\boldsymbol{c}'_2$ in terms of $\boldsymbol{G}$, $\boldsymbol{g}_{21}$, and $\boldsymbol{g}'_{21}$; thus

$$\boldsymbol{c}_1 = \boldsymbol{G} - M_2\boldsymbol{g}_{21}, \quad \boldsymbol{c}_2 = \boldsymbol{G} + M_1\boldsymbol{g}_{21}, \tag{3.41, 6}$$

$$\boldsymbol{c}'_1 = \boldsymbol{G} - M_2\boldsymbol{g}'_{21}, \quad \boldsymbol{c}'_2 = \boldsymbol{G} + M_1\boldsymbol{g}'_{21}. \tag{3.41, 7}$$

Hence a knowledge of $\boldsymbol{G}$ and $\boldsymbol{g}_{21}$ or of $\boldsymbol{G}$ and $\boldsymbol{g}'_{21}$ is equivalent to a knowledge of $\boldsymbol{c}_1$ and $\boldsymbol{c}_2$ or of $\boldsymbol{c}'_1$ and $\boldsymbol{c}'_2$, that is, of the initial or final state of motion.

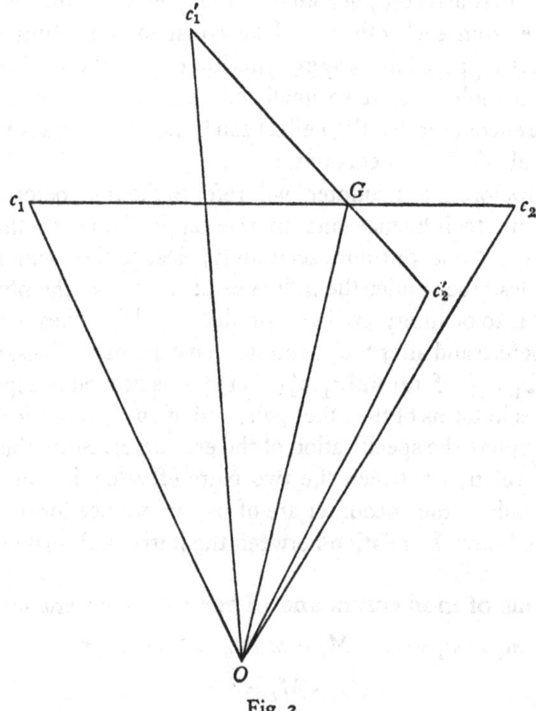

Fig. 2

The mutual potential energy of the two molecules is zero both before and after the encounter; thus the equation of energy gives

$$\tfrac{1}{2}(m_1 c_1^2 + m_2 c_2^2) = \tfrac{1}{2}(m_1 c_1'^2 + m_2 c_2'^2).$$

Using (3.41, 6, 7), it is readily shown that

$$\left.\begin{aligned} \tfrac{1}{2}(m_1 c_1^2 + m_2 c_2^2) &= \tfrac{1}{2}m_0(G^2 + M_1 M_2 g^2), \\ \tfrac{1}{2}(m_1 c_1'^2 + m_2 c_2'^2) &= \tfrac{1}{2}m_0(G^2 + M_1 M_2 g'^2). \end{aligned}\right\} \tag{3.41, 8}$$

Hence $$g = g',$$

so that the relative velocity is changed only in *direction*, and not in *magnitude*, by the encounter. The dynamical effect of the encounter is therefore known when the change in *direction* of $\mathbf{g}_{21}$ is determined.

These facts are illustrated in Fig. 2. The initial velocities $\mathbf{c}_1$, $\mathbf{c}_2$ are represented by Oc_1, Oc_2, and $\mathbf{G}$ by OG, where G divides c_1c_2 in the ratio $m_2:m_1$. The ends of the lines Oc'_1, Oc'_2 representing $\mathbf{c}'_1$, $\mathbf{c}'_2$ are likewise collinear with G, which divides $c'_1c'_2$ in the same ratio $m_2:m_1$. The lines c_1c_2 and $c'_1c'_2$ represent $\mathbf{g}_{21}$ and $\mathbf{g}'_{21}$. Thus $c_1c_2 = c'_1c'_2$, $c_1G = c'_1G$, $c_2G = c'_2G$.

3.42. The geometry of an encounter

Considerations of momentum and energy alone do not suffice to determine the direction of $\mathbf{g}'_{21}$. As will now appear, this direction depends not only on the initial velocities $\mathbf{c}_1, \mathbf{c}_2$ (or on $\mathbf{G}, \mathbf{g}_{21}$) but also on two geometric variables which complete the specification of the encounter.

Consider the motion of the centre B of the second molecule relative to the centre A of the first (or to axes moving with A). Since the force between the molecules is directed along AB, this motion will be confined to a plane through A; let the curve described by B be LMN (Fig. 3 *a*). The asymptotes PO, OQ of this curve are in the directions of the initial and final relative velocities, $\mathbf{g}_{21}$ and $\mathbf{g}'_{21}$, and so the plane of LMN is parallel to the plane $c_1Gc'_1$ of Fig. 2. Let $P'A$ be a line parallel to PO, so that it is in the direction of $\mathbf{g}_{21}$. The direction of AP' is then fixed by the initial velocities $\mathbf{c}_1$, $\mathbf{c}_2$; the orientation of the plane LMN about AP' is, however, independent of these velocities, and is thus *one* of the additional variables of the encounter. We specify it by the angle ϵ between the plane LMN and a plane containing AP' and a direction fixed in space, such as that of Oz.

The angle χ through which $\mathbf{g}_{21}$ is deflected depends, in general, on the magnitude g of the initial relative velocity, and on the distance b of A from either of the asymptotes. This distance b is the *second* of the additional geometric variables of the encounter.

The functional relation between χ, b and g depends on the law of interaction between the molecules. This law is involved in the following discussion solely through the dependence of χ on b and g. Hence, both for generality and brevity, χ will be retained as an unspecified function of b and g as long as possible.

3.43. The apse-line and the change of relative velocity

The orbit LMN of the second molecule relative to the first is symmetrical about the *apse-line*, or line joining the two molecules when at the points of closest approach. This apse-line passes through O, the intersection of the two asymptotes, and bisects the angle between them. In Fig. 3 *a* the direction of the apse-line is represented by OAK, K being the point in which OA produced cuts the unit sphere of centre A. The unit vector AK is denoted

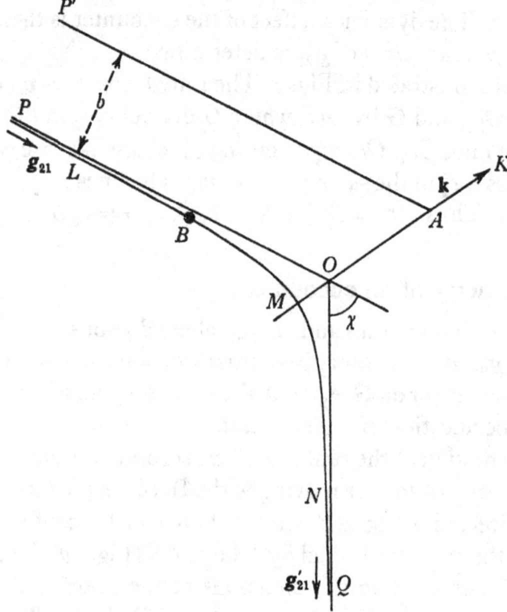

Fig. 3(*a*). Direct encounter.

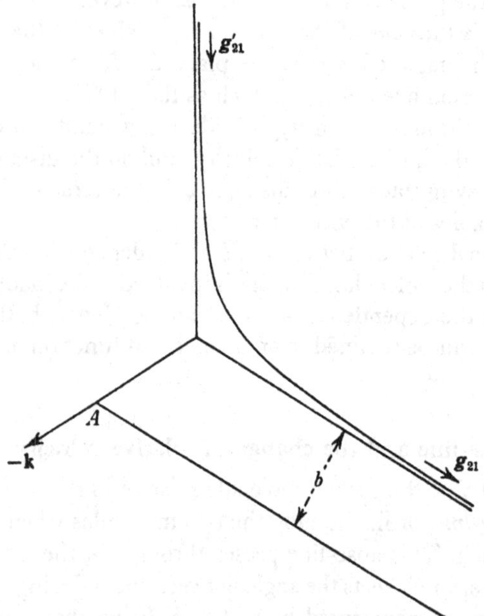

Fig. 3(*b*). Inverse encounter (see p. 62).

by $\mathbf{k}$. The components of $\boldsymbol{g}_{21}$ and $\boldsymbol{g}'_{21}$ in the direction of $\mathbf{k}$ are equal in magnitude, but opposite in sign, so that $\boldsymbol{g}'_{21}.\mathbf{k} = -\boldsymbol{g}_{21}.\mathbf{k}$; the components perpendicular to $\mathbf{k}$ are equal. Hence $\boldsymbol{g}_{21}$ and $\boldsymbol{g}'_{21}$ differ by twice the component of $\boldsymbol{g}_{21}$ in the direction of $\mathbf{k}$, so that

$$\boldsymbol{g}_{21}-\boldsymbol{g}'_{21} = 2(\boldsymbol{g}_{21}.\mathbf{k})\mathbf{k} = -2(\boldsymbol{g}'_{21}.\mathbf{k})\mathbf{k}. \tag{3.43, 1}$$

Combining this with (3.41, 6, 7) it follows that

$$\left.\begin{aligned} \boldsymbol{c}'_1-\boldsymbol{c}_1 &= 2M_2(\boldsymbol{g}_{21}.\mathbf{k})\mathbf{k} = -2M_2(\boldsymbol{g}'_{21}.\mathbf{k})\mathbf{k},\\ \boldsymbol{c}'_2-\boldsymbol{c}_2 &= -2M_1(\boldsymbol{g}_{21}.\mathbf{k})\mathbf{k} = 2M_1(\boldsymbol{g}'_{21}.\mathbf{k})\mathbf{k}. \end{aligned}\right\} \tag{3.43, 2}$$

Thus when $\mathbf{k}$, $\boldsymbol{c}_1$, $\boldsymbol{c}_2$ are given, the velocities after encounter are determinate, and knowledge of $\mathbf{k}$ is equivalent to knowledge of the geometrical variables b and ϵ.

If the force between molecules is always repulsive or always attractive then for a given direction of $\boldsymbol{g}_{21}$ the point K can range over one or other of the unit hemispheres having AP' as axis; if the molecules repel one another, the pole of this hemisphere is in the same direction as $\boldsymbol{g}_{21}$, while if they attract it is in the opposite direction. The possible positions of K are similarly related to $-\boldsymbol{g}'_{21}$: for a repulsive force $\boldsymbol{g}_{21}.\mathbf{k} > 0$, and $\boldsymbol{g}'_{21}.\mathbf{k} < 0$; for an attractive force these inequalities are reversed.

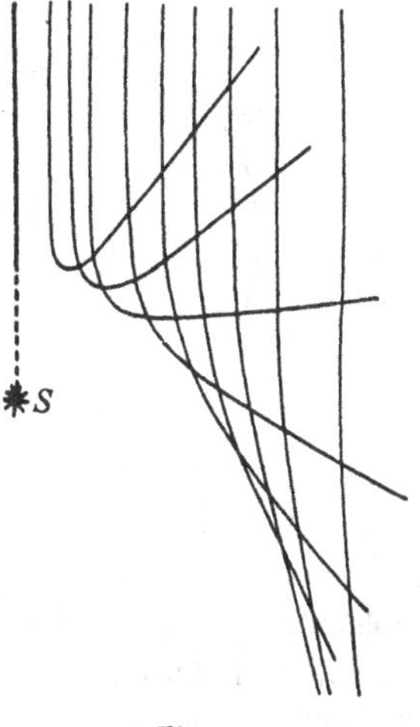

Fig. 4

3.44. Special types of interaction

Figure 4, drawn by Maxwell,* shows a number of the orbits described by one molecule, relative to another molecule represented by S, when they exert a mutual repulsive force varying as the inverse fifth power of their distance apart; they correspond to equal values of g, but different values of b. With this law of force, or, more generally, when the force varies as any inverse power of the distance, the families of paths for different values of g differ only in scale; with more general laws of force this is not true.

When the molecules are rigid elastic spheres the apse-line becomes identical with the line of centres at collision. In this case the distance σ_{12} between the centres of the spheres at collision is connected with their diameters σ_1, σ_2 by the relation

$$\sigma_{12} = \tfrac{1}{2}(\sigma_1+\sigma_2), \tag{3.44, 1}$$

and (cf. Fig. 5)

$$b = \sigma_{12}\sin\psi = \sigma_{12}\cos\tfrac{1}{2}\chi, \tag{3.44, 2}$$

* J. C. Maxwell, *Collected Papers*, vol. 2, 42; *Phil. Trans. R. Soc.* **157**, 49 (1867).

where ψ is the angle between $\boldsymbol{g}_{21}$ and $\mathbf{k}$; clearly $\psi = \frac{1}{2}(\pi - \chi)$. This model is unique in that χ depends only on b, and not on g.

3.5. The statistics of molecular encounters

In evaluating $\partial_e f/\partial t$ and $\Delta\bar{\phi}$, we suppose that encounters in which more than two molecules take part are negligible in number and effect, compared with binary encounters. This implies that the gas is of low density, so that encounters occupy only a small fraction of the life of a molecule.

The probability is zero that at a given instant, in a finite volume of gas, $d\boldsymbol{r}$, there shall be any molecule whose velocity is exactly equal to any

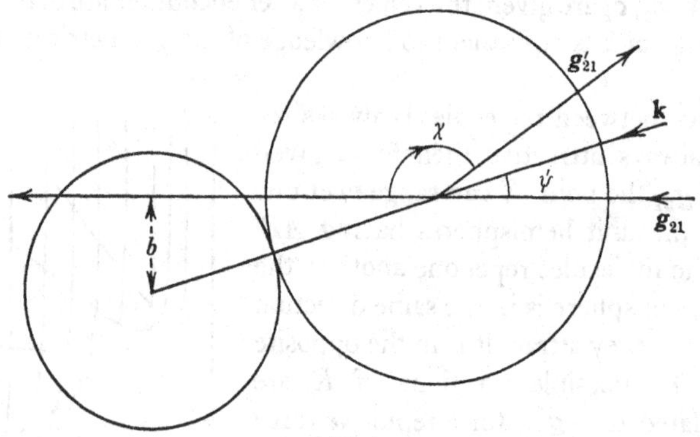

Fig. 5

specified value $\boldsymbol{c}$ out of the whole continuous range; it is necessary to consider a small but finite range of velocity, $d\boldsymbol{c}$. Thus the probable number of molecules of the first kind, in $d\boldsymbol{r}$, having velocities within the small range $\boldsymbol{c}_1$, $d\boldsymbol{c}_1$, is $f_1 d\boldsymbol{c}_1 d\boldsymbol{r}$, where f_1 stands for $f_1(\boldsymbol{c}_1, \boldsymbol{r}, t)$; likewise the probable number of molecules of the second kind, in $d\boldsymbol{r}$, having velocities within the range $\boldsymbol{c}_2$, $d\boldsymbol{c}_2$, is $f_2 d\boldsymbol{c}_2 d\boldsymbol{r}$, where f_2 stands for $f_2(\boldsymbol{c}_2, \boldsymbol{r}, t)$.

The probable number of *encounters* in $d\boldsymbol{r}$, during a small interval dt, between molecules in the velocity-ranges $d\boldsymbol{c}_1$, $d\boldsymbol{c}_2$, will in the same way be zero if the geometric encounter-variables b, ϵ are exactly assigned; it is necessary to suppose that b, ϵ also lie in small finite ranges db, $d\epsilon$. The ranges $d\boldsymbol{c}_1$, $d\boldsymbol{c}_2$, db, $d\epsilon$ are regarded as positive quantities; as they are small, the average number of encounters of the type considered is proportional to the product $d\boldsymbol{c}_1 d\boldsymbol{c}_2 db\, d\epsilon\, d\boldsymbol{r}\, dt$.

In considering such encounters between molecules having velocities within assigned ranges, it is assumed that both sets of molecules are distributed at random, and without any correlation between velocity and position, in the neighbourhood of the point $\boldsymbol{r}$.* Also the interval dt which

* This 'assumption of molecular chaos' was considered by J. H. Jeans, *Dynamical Theory of Gases* (4th ed.), chapter 4 (1925). See also H. Grad, *Handbuch der Physik*, vol. 12, 205–94 (1958).

we consider is supposed short compared with the scale of time-variation of macroscopic properties, but large compared with the duration of an encounter.

In an encounter between two molecules as specified, the velocity of the second relative to the first, before encounter, is $\boldsymbol{c}_2 - \boldsymbol{c}_1$, or $\boldsymbol{g}_{21}$.

Consider the motion of the centre B of the second molecule relative to the centre A of the first, or relative to axes moving with A. For such an encounter to occur, the line PO of Fig. 3 must cut a plane through A, perpendicular to AP', within an area, of magnitude $b\,db\,d\epsilon$, bounded by circles of radii b, $b+db$ and centre A, and by radii from A including an angle $d\epsilon$. Also, since the relative velocity is $\boldsymbol{g}_{21}$, and dt is large compared with the duration of an

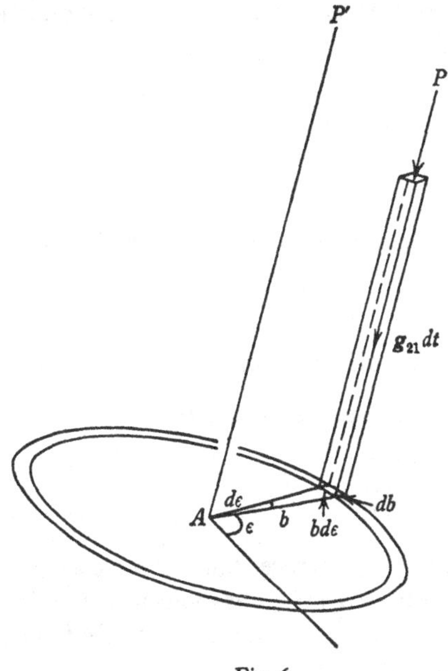

Fig. 6

encounter, it follows by an argument similar to the one used in 2.3 that at the beginning of dt the point B must lie within the cylinder indicated in Fig. 6, having the area $b\,db\,d\epsilon$ as base, and generators equal to $-\boldsymbol{g}_{21}\,dt$; that is, it must lie in a volume $(g\,dt)(b\,db\,d\epsilon)$ or $gb\,db\,d\epsilon\,dt$.

We can imagine such a cylinder to be associated with each of the $f_1\,d\boldsymbol{c}_1\,d\boldsymbol{r}$ molecules of the first kind, within the specified velocity range, in $d\boldsymbol{r}$. If db and $d\epsilon$ are small, it can safely be assumed that the cylinders do not overlap to any significant extent, so that the total volume dv of all the cylinders is given by

$$dv = f_1 gb\,db\,d\epsilon\,d\boldsymbol{c}_1\,d\boldsymbol{r}\,dt.$$

In many of these tiny cylinders there will be no molecule of the second kind, having a velocity within the range $\boldsymbol{c}_2$, $d\boldsymbol{c}_2$; and if db, $d\epsilon$ and $d\boldsymbol{c}_2$ are sufficiently small, we can ignore the possibility that in any one cylinder there are two such molecules. The total number of such molecules in the whole combined volume dv is $f_2 d\boldsymbol{c}_2 dv$, which is therefore the number of 'occupied' cylinders in which such a molecule occurs. Each occupied cylinder corresponds to an encounter of the specified type, occurring within $d\boldsymbol{r}$ during the time dt. Inserting the above expression for dv in $f_2 d\boldsymbol{c}_2 dv$, the number of encounters is found to be

$$f_1 f_2 g b \, db \, d\epsilon \, d\boldsymbol{c}_1 d\boldsymbol{c}_2 d\boldsymbol{r} \, dt. \tag{3.5, 1}$$

Let $\mathbf{e}$, $\mathbf{e}'$ be unit vectors in the directions of $\boldsymbol{g}_{21}$, $\boldsymbol{g}'_{21}$, so that $\boldsymbol{g}_{21} = g\mathbf{e}$, $\boldsymbol{g}'_{21} = g\mathbf{e}'$. Then (following Waldmann*) (3.5, 1) may be expressed in terms of the element $d\mathbf{e}'$, which (since $\mathbf{e}'$ is a unit vector) represents an element of solid angle (cf. 1.21). The angles χ, ϵ may be regarded as polar angles specifying the orientation of $\mathbf{e}'$ relative to an axis in the direction of $\mathbf{e}$; thus

$$\begin{aligned} d\mathbf{e}' &= \sin\chi \, d\chi \, d\epsilon \\ &= \left(\sin\chi \Big/ \left|\frac{\partial b}{\partial \chi}\right|\right) db \, d\epsilon. \end{aligned} \tag{3.5, 2}$$

Hence we may write†

$$b \, db \, d\epsilon = \alpha_{12} d\mathbf{e}' \tag{3.5, 3}$$

where the positive scalar α_{12} is a function of g and b, or of g and χ, given by

$$\alpha_{12} = b \left|\frac{\partial b}{\partial \chi}\right| \Big/ \sin\chi. \tag{3.5, 4}$$

On substituting $\alpha_{12} d\mathbf{e}'$ for $b \, db \, d\epsilon$ in (3.5, 1), we obtain the alternative expression

$$f_1 f_2 g \alpha_{12} d\mathbf{e}' d\boldsymbol{c}_1 d\boldsymbol{c}_2 d\boldsymbol{r} \, dt \tag{3.5, 5}$$

for this number of encounters.

3.51. An expression for $\Delta\bar{\phi}$

If there are several gases in a mixture, the rate of change, by molecular encounters, of the mean value ($\bar{\phi}_1$) of ϕ for molecules of the first gas can be divided into the parts $\Delta_1\bar{\phi}_1, \Delta_2\bar{\phi}_1, \ldots$, due respectively to encounters with molecules of the first, second, ..., gases. Thus

$$\Delta\bar{\phi}_1 = \Delta_1\bar{\phi}_1 + \Delta_2\bar{\phi}_1 + \ldots. \tag{3.51, 1}$$

In an encounter of a molecule of the first gas, the value of ϕ_1 for the molecule, which when written in full (as in 2.22) is $\phi_1(\boldsymbol{c}_1, \boldsymbol{r}, t)$, is changed to ϕ'_1, signifying $\phi_1(\boldsymbol{c}'_1, \boldsymbol{r}, t)$. Thus the ϕ for this molecule is altered by the amount $\phi'_1 - \phi_1$. The change in $\Sigma\phi_1$ due to all encounters of the special type considered in 3.5, between a molecule of the first gas and one of the second gas, is therefore

$$(\phi'_1 - \phi_1) f_1 f_2 g \alpha_{12} d\mathbf{e}' d\boldsymbol{c}_1 d\boldsymbol{c}_2 d\boldsymbol{r} \, dt. \tag{3.51, 2}$$

* L. Waldmann, *Handbuch der Physik*, vol. 12, 295–514 (1958).

† In some cases this transformation needs careful treatment, because $\partial b/\partial\chi$ is not always of the same sign for molecular pairs of particular types, so that b is not a one-valued function of χ. Where such difficulties arise, the symbol $\alpha_{12} d\mathbf{e}'$ may be regarded as merely a convenient brief notation for $b \, db \, d\epsilon$.

Integration, first over all permissible values of $\mathbf{e}'$, and then over all values of $\boldsymbol{c}_1$ and $\boldsymbol{c}_2$, gives the total change during dt in $\Sigma\phi_1$, summed over all molecules of the first gas in $d\boldsymbol{r}$, due to their encounters with molecules of the second gas. Since the number of molecules m_1 in $d\boldsymbol{r}$ is $n_1 d\boldsymbol{r}$, this integral must equal $n_1 d\boldsymbol{r}\, \Delta_2\overline{\phi}_1\, dt$. Dividing by $d\boldsymbol{r}\, dt$, we get

$$n_1\Delta_2\overline{\phi}_1 = \iiint(\phi_1' - \phi_1) f_1 f_2\, g\alpha_{12}\, d\mathbf{e}'\, d\boldsymbol{c}_1\, d\boldsymbol{c}_2. \tag{3.51, 3}$$

The variable $\boldsymbol{c}_1'$ in ϕ_1' is a function of $\boldsymbol{c}_1$, $\boldsymbol{c}_2$, and $\mathbf{e}'$, given (cf. (3.41, 6, 7)) by

$$\boldsymbol{c}_1' - \boldsymbol{c}_1 = -M_2(\boldsymbol{g}_{21}' - \boldsymbol{g}_{21}), \tag{3.51, 4}$$

where $\boldsymbol{g}_{21}' = g\mathbf{e}'$.

The value of $n_1\Delta_1\overline{\phi}_1$ can be obtained from (3.51, 3) as a special case by replacing m_2 by m_1 in the relation between $\boldsymbol{c}_1'$ and $\boldsymbol{c}_1$, $\boldsymbol{c}_2$, $\mathbf{e}'$, and using the relation between α_{12} and $\mathbf{e}'$ (or between χ and b, g) appropriate to the law of interaction between two like molecules m_1 instead of that between the unlike molecules m_1 and m_2; the symbol α_{12} is then replaced by α_1. To distinguish between the initial velocities of the two encountering molecules, one velocity is denoted by $\boldsymbol{c}_1$, as before, and the other is written without suffix, as $\boldsymbol{c}$. Similarly the two functions f_1, f_2 are written as f_1, f, being now identical except that in the former the variables are $\boldsymbol{c}_1$, $\boldsymbol{r}$, t and in the latter they are $\boldsymbol{c}$, $\boldsymbol{r}$, t. Thus

$$n_1\Delta_1\overline{\phi}_1 = \iiint(\phi_1' - \phi_1) f f_1\, g\alpha_1\, d\mathbf{e}'\, d\boldsymbol{c}\, d\boldsymbol{c}_1. \tag{3.51, 5}$$

When the gas is simple, that is, when molecules of one kind only are present, the suffix 1 in the symbol $n_1\Delta_1\overline{\phi}_1$ may be omitted; but it must be retained in the integral, in order to distinguish between the initial velocities of two molecules (now of equal mass) involved in an encounter, since the two velocities are separate variables of integration.

3.52. The calculation of $\partial_e f/\partial t$

Like $\Delta\overline{\phi}_1$, $\partial_e f/\partial t$ may be divided into the parts $(\partial_e f_1/\partial t)_1$, $(\partial_e f_1/\partial t)_2$, ..., due to the encounters of molecules m_1 with molecules m_1, m_2, ... respectively; thus

$$\frac{\partial_e f_1}{\partial t} = \left(\frac{\partial_e f_1}{\partial t}\right)_1 + \left(\frac{\partial_e f_1}{\partial t}\right)_2 + \dots. \tag{3.52, 1}$$

When an expression for $(\partial_e f_1/\partial t)_2$ has been obtained, the values of the other parts of $\partial_e f_1/\partial t$ can be derived by changes of suffix.

Consider the set of molecules of the first kind, situated within $d\boldsymbol{r}$, which have velocities within the range $\boldsymbol{c}_1$, $d\boldsymbol{c}_1$; the expression

$$\left(\frac{\partial_e f_1}{\partial t}\right)_2 d\boldsymbol{c}_1\, d\boldsymbol{r}\, dt$$

signifies the net increase, during dt, in the number of molecules of this set, due to encounters with molecules of the second kind (without restriction as to the velocity of these latter molecules). This net increase is the difference between the numbers of molecules of the first kind, within $d\mathbf{r}$, which during dt *enter* and *leave* the set, owing to encounters with molecules of the second kind.

Every encounter of a molecule of the set results in a change of velocity, and so involves the loss of the molecule to the set. Thus the number of molecules lost to the set during time dt, owing to the particular group of encounters with molecules m_2 such that $\mathbf{c}_2$ and $\mathbf{e}'$ lie in ranges $d\mathbf{c}_2$ and $d\mathbf{e}'$, is by (3.5, 5) equal to

$$f_1 f_2 g\alpha_{12} d\mathbf{e}' d\mathbf{c}_1 d\mathbf{c}_2 d\mathbf{r}\, dt.$$

The total loss for all values of $\mathbf{c}_2$ and $\mathbf{e}'$ is found by integrating with respect to these variables; this gives

$$d\mathbf{c}_1 d\mathbf{r}\, dt \iint f_1 f_2 g\alpha_{12} d\mathbf{e}' d\mathbf{c}_2. \tag{3.52, 2}$$

The number of molecules m_1 entering the set $\mathbf{c}_1$, $d\mathbf{c}_1$ owing to encounters with molecules m_2 during dt may be found in like manner. We must, for this purpose, consider encounters such that the velocity of a molecule m_1 *after* encounter lies in the range $\mathbf{c}_1$, $d\mathbf{c}_1$. Such encounters will be termed *inverse* encounters; those in which the *initial* velocity of the molecule m_1 lies in the range, $\mathbf{c}_1$, $d\mathbf{c}_1$ may be styled *direct* encounters. Corresponding to any direct encounter with initial velocities $\mathbf{c}_1$, $\mathbf{c}_2$ and final velocities $\mathbf{c}_1'$, $\mathbf{c}_2'$, there is an associated inverse encounter in which the initial velocities are $\mathbf{c}_1'$, $\mathbf{c}_2'$, the final velocities are $\mathbf{c}_1$, $\mathbf{c}_2$, and the apse-line has the direction $-\mathbf{k}$. The correspondence is illustrated in Figs. 3*a*, *b* (p. 56), which refer to the encounter of molecules behaving like centres of repulsive force, and show the motion of one molecule relative to the other.

The number of inverse encounters with initial velocities $\mathbf{c}_1'$, $\mathbf{c}_2'$ in the ranges $d\mathbf{c}_1'$, $d\mathbf{c}_2'$, and such that the direction $\mathbf{e}$ of the final relative velocity lies in the solid angle $d\mathbf{e}$, is

$$f_1' f_2' g\alpha_{12}(g, \chi)\, d\mathbf{e}\, d\mathbf{c}_1' d\mathbf{c}_2' d\mathbf{r}\, dt; \tag{3.52, 3}$$

the factor $\alpha_{12} \equiv \alpha_{12}(g, \chi)$ is the same in (3.52, 2, 3) because g and χ have the same values in the direct encounter and the corresponding inverse encounter.

Now let

$$\mathrm{J} \equiv \frac{\partial(\mathbf{G}, \mathbf{g}_{21})}{\partial(\mathbf{c}_1, \mathbf{c}_2)}$$

denote a Jacobian similar to those of 1.411. Then (cf. (3.41, 6, 7) and (1.411, 1, 2))

$$\mathrm{J} = \frac{\partial(\mathbf{c}_1 + M_2\mathbf{g}_{21}, \mathbf{g}_{21})}{\partial(\mathbf{c}_1, \mathbf{c}_2)} = \frac{\partial(\mathbf{c}_1, \mathbf{g}_{21})}{\partial(\mathbf{c}_1, \mathbf{c}_2)}$$

$$= \frac{\partial(\mathbf{c}_1, \mathbf{c}_2 - \mathbf{c}_1)}{\partial(\mathbf{c}_1, \mathbf{c}_2)} = \frac{\partial(\mathbf{c}_1, \mathbf{c}_2)}{\partial(\mathbf{c}_1, \mathbf{c}_2)} = 1 \tag{3.52, 4}$$

and similarly

$$\mathrm{J}' \equiv \frac{\partial(\mathbf{G}, \mathbf{g}_{21}')}{\partial(\mathbf{c}_1', \mathbf{c}_2')} = 1.$$

Hence if $\mathbf{c}_1$, $\mathbf{c}_2$ lie in ranges $d\mathbf{c}_1$, $d\mathbf{c}_2$, by the theory of Jacobians $\mathbf{G}$, $\mathbf{g}_{21}$ lie in ranges $d\mathbf{G}$, $d\mathbf{g}_{21}$,* where the sixfold (positive) differential elements $d\mathbf{c}_1 d\mathbf{c}_2$ and $d\mathbf{G}\,d\mathbf{g}_{21}$ are connected by the relation

$$d\mathbf{G}\,d\mathbf{g}_{21} = |\mathrm{J}|\,d\mathbf{c}_1 d\mathbf{c}_2 = d\mathbf{c}_1 d\mathbf{c}_2; \tag{3.52, 5}$$

and similarly

$$d\mathbf{G}\,d\mathbf{g}'_{21} = d\mathbf{c}'_1 d\mathbf{c}'_2.$$

Again, since $d\mathbf{e}$, $d\mathbf{e}'$ are elements of solid angle, and $\mathbf{g}_{21} = g\mathbf{e}$, $\mathbf{g}'_{21} = g\mathbf{e}'$,

$$d\mathbf{g}_{21} = g^2 dg\,d\mathbf{e}, \quad d\mathbf{g}'_{21} = g^2 dg\,d\mathbf{e}'.$$

On combining these results,

$$\begin{aligned} d\mathbf{e}\,d\mathbf{c}'_1 d\mathbf{c}'_2 &= d\mathbf{e}\,d\mathbf{G}\,d\mathbf{g}'_{21} = g^2 dg\,d\mathbf{G}\,d\mathbf{e}\,d\mathbf{e}' \\ &= d\mathbf{e}'\,d\mathbf{c}_1 d\mathbf{c}_2 \end{aligned} \tag{3.52, 6}$$

by symmetry. Hence (3.52, 3) can be expressed in the form

$$f'_1 f'_2 g\alpha_{12} d\mathbf{e}'\,d\mathbf{c}_1 d\mathbf{c}_2 d\mathbf{r}\,dt. \tag{3.52, 7}$$

This represents the number of encounters such that the final velocities lie in the ranges $d\mathbf{c}_1$, $d\mathbf{c}_2$, and the direction $\mathbf{e}'$ of the initial relative velocity lies in the range $d\mathbf{e}'$. On integrating over all possible values of $\mathbf{e}'$ and $\mathbf{c}_2$, the total gain by encounters during dt to the set of molecules m_1, $\mathbf{c}_1$, $d\mathbf{c}_1$ in the volume $d\mathbf{r}$ is found to be

$$d\mathbf{c}_1 d\mathbf{r}\,dt \iint f'_1 f'_2 g\alpha_{12} d\mathbf{e}'\,d\mathbf{c}_2. \tag{3.52, 8}$$

Combining (3.52, 2 and 8), we get

$$d\mathbf{c}_1 d\mathbf{r}\,dt \iint (f'_1 f'_2 - f_1 f_2) g\alpha_{12} d\mathbf{e}'\,d\mathbf{c}_2$$

for the net gain to this set.

This net gain is denoted by $(\partial_e f_1/\partial t)_2 d\mathbf{c}_1 d\mathbf{r}\,dt$. Hence, dividing by $d\mathbf{c}_1 d\mathbf{r}\,dt$, we find that

$$\left(\frac{\partial_e f_1}{\partial t}\right)_2 = \iint (f'_1 f'_2 - f_1 f_2) g\alpha_{12} d\mathbf{e}'\,d\mathbf{c}_2. \tag{3.52, 9}$$

This is the required expression for $(\partial_e f_1/\partial t)_2$.

* This statement is not strictly accurate. To a volume-element $d\mathbf{c}_1 d\mathbf{c}_2$ in the six-dimensional space in which the coordinates of a point are the components of $\mathbf{c}_1$ and $\mathbf{c}_2$ there corresponds a volume element δ in the six-dimensional space in which the coordinates are the components of $\mathbf{G}$ and $\mathbf{g}_{21}$: but δ cannot in general be put into the form $d\mathbf{G}\,d\mathbf{g}_{21}$, any more than the element of area between the pairs of curves $\xi = \phi(x,y)$, $\xi + d\xi = \phi(x,y)$, and $\eta = \psi(x,y)$, $\eta + d\eta = \psi(x,y)$, can in general be expressed as equal to an elementary rectangle $dx\,dy$. However, just as the small area between these pairs of curves can be divided up into a large number of still smaller *rectangles* $dx\,dy$, so δ can be divided up into elementary volumes $d\mathbf{G}\,d\mathbf{g}_{21}$. It is necessary in equations like (3.52, 5) to regard the notation $d\mathbf{G}\,d\mathbf{g}_{21}$ as referring to the general volume-element δ, formed from a sum of products of three-dimensional elements in spaces in which the coordinates are the components of $\mathbf{G}$ and $\mathbf{g}_{21}$ respectively. With this understanding the expression (3.52, 7) is readily derived, as in the text.

From this and (3.11, 2) we obtain a second expression for $\Delta_2\bar{\phi}_1$, namely

$$n_1\Delta_2\bar{\phi}_1 = \int \phi_1\left(\frac{\partial_e f_1}{\partial t}\right)_2 dc_1$$

$$= \iiint \phi_1(f_1' f_2' - f_1 f_2) g\alpha_{12} de' \, dc_1 dc_2. \qquad (3.52, 10)$$

The corresponding formulae for $(\partial_e f_1/\partial t)_1$ and $\Delta_1\bar{\phi}_1$ are

$$\left(\frac{\partial_e f_1}{\partial t}\right)_1 = \iint (f' f_1' - f f_1) g\alpha_1 de' \, dc, \qquad (3.52, 11)$$

$$n_1\Delta_1\bar{\phi}_1 = \iiint \phi_1(f' f_1' - f f_1) g\alpha_1 de' \, dc dc_1. \qquad (3.52, 12)$$

3.53. Alternative expressions for $n\Delta\bar{\phi}$; proof of equality

The equality of the expressions for $n_1\Delta_2\bar{\phi}_1$ given by (3.51, 3) and (3.52, 10) can readily be established. In the integral

$$\iiint \phi_1 f_1' f_2' g\alpha_{12} de' \, dc_1 dc_2 \qquad (3.53, 1)$$

let the variables of integration $\mathbf{c}_1$, $\mathbf{c}_2$, $\mathbf{e}'$ be changed to $\mathbf{c}_1'$, $\mathbf{c}_2'$, $\mathbf{e}$, these being functions of $\mathbf{c}_1$, $\mathbf{c}_2$, $\mathbf{e}'$ such that $d\mathbf{c}_1 d\mathbf{c}_2 d\mathbf{e}' = d\mathbf{c}_1' d\mathbf{c}_2' d\mathbf{e}$ (cf. (3.52, 6)). Then the integral becomes

$$\iiint \phi_1 f_1' f_2' g\alpha_{12}(g, \chi) \, de \, dc_1' dc_2'.$$

Now $\mathbf{c}_1'$, $\mathbf{c}_2'$, $\mathbf{e}$ are variables specifying a certain encounter, namely, the encounter inverse to that specified by $\mathbf{c}_1$, $\mathbf{c}_2$, $\mathbf{e}'$: and integration over all possible values of $\mathbf{c}_1'$, $\mathbf{c}_2'$, $\mathbf{e}$ is equivalent to a summation over all possible inverse encounters, or, since every encounter is inverse to another encounter, over all possible encounters. Since $\mathbf{c}_1'$, $\mathbf{c}_2'$, $\mathbf{e}$ are variables specifying an encounter, $\mathbf{c}_1$, $\mathbf{c}_2$, $\mathbf{e}'$ may be written in their stead; the variables $\mathbf{c}_1$, $\mathbf{c}_2$, $\mathbf{e}'$ that specify the inverse encounter must then be replaced by $\mathbf{c}_1'$, $\mathbf{c}_2'$, $\mathbf{e}$. Hence the integral becomes equal to

$$\iiint \phi_1(\mathbf{c}_1', \mathbf{r}, t) f_1(\mathbf{c}_1, \mathbf{r}, t) f_2(\mathbf{c}_2, \mathbf{r}, t) g\alpha_{12}(g, \chi) \, de' \, dc_1 dc_2$$

or, in brief, to

$$\iiint \phi_1' f_1 f_2 g\alpha_{12} de' \, dc_1 dc_2. \qquad (3.53, 2)$$

From the equality of (3.53, 1) and (3.53, 2) that of the two expressions for $\Delta_2\bar{\phi}_1$ at once follows.

3.54. Transformations of some integrals

The proof in 3.53 is independent of the nature of the functions ϕ, f, and similar arguments serve to establish a number of analogous analytical results which we quote here for later reference.

First, if F, G, and ϕ are any functions of velocity, position, and time, such arguments show that

$$\iiint \phi_1 F_1' G_2' g\alpha_{12} de' \, dc_1 dc_2 = \iiint \phi_1' F_1 G_2 g\alpha_{12} de' \, dc_1 dc_2. \qquad (3.54, 1)$$

In this equation replace ϕ_1 by unity and F_1 by $\phi_1 F_1$; then it becomes

$$\iiint \phi_1' F_1' G_2' g\alpha_{12} d\mathbf{e}' d\mathbf{c}_1 d\mathbf{c}_2 = \iiint \phi_1 F_1 G_2 g\alpha_{12} d\mathbf{e}' d\mathbf{c}_1 d\mathbf{c}_2.$$

From this equation and (3.54, 1) it follows that

$$\iiint \phi_1(F_1 G_2 - F_1' G_2') g\alpha_{12} d\mathbf{e}' d\mathbf{c}_1 d\mathbf{c}_2 = -\iiint \phi_1'(F_1 G_2 - F_1' G_2') g\alpha_{12} d\mathbf{e}' d\mathbf{c}_1 d\mathbf{c}_2$$
$$= \tfrac{1}{2}\iiint (\phi_1 - \phi_1')(F_1 G_2 - F_1' G_2') g\alpha_{12} d\mathbf{e}' d\mathbf{c}_1 d\mathbf{c}_2. \quad (3.54, 2)$$

From (3.54, 2), omitting the suffix 2, the corresponding equation for encounters between pairs of molecules m_1 is obtained; this is

$$\iiint \phi_1(F_1 G - F_1' G') g\alpha_1 d\mathbf{e}' d\mathbf{c}_1 d\mathbf{c}$$
$$= \tfrac{1}{2}\iiint (\phi_1 - \phi_1')(F_1 G - F_1' G') g\alpha_1 d\mathbf{e}' d\mathbf{c}_1 d\mathbf{c}. \quad (3.54, 3)$$

Since $\mathbf{c}_1$ and $\mathbf{c}$ both refer to molecules m_1, an interchange of $\mathbf{c}_1$ and $\mathbf{c}$ does not alter the value of either integral. Thus, making this interchange on the right-hand side of (3.54, 3), we have

$$\iiint \phi_1(F_1 G - F_1' G') g\alpha_1 d\mathbf{e}' d\mathbf{c}_1 d\mathbf{c} = \tfrac{1}{2}\iiint (\phi - \phi')(F G_1 - F' G_1') g\alpha_1 d\mathbf{e}' d\mathbf{c}_1 d\mathbf{c}.$$

Adding to (3.54, 3) this equation and two similar equations, in which $F_1 G - F_1' G'$ on the left-hand side is replaced by $FG_1 - F'G_1'$, we get

$$\iiint \phi_1(F_1 G + F G_1 - F_1' G' - F' G_1') g\alpha_1 d\mathbf{e}' d\mathbf{c}_1 d\mathbf{c}$$
$$= \tfrac{1}{4}\iiint (\phi + \phi_1 - \phi' - \phi_1')(F_1 G + F G_1 - F_1' G' - F' G_1') g\alpha_1 d\mathbf{e}' d\mathbf{c}_1 d\mathbf{c}. \quad (3.54, 4)$$

In this equation put $F = G$. Then

$$\iiint \phi_1(F F_1 - F' F_1') g\alpha_1 d\mathbf{e}' d\mathbf{c}_1 d\mathbf{c}$$
$$= \tfrac{1}{4}\iiint (\phi + \phi_1 - \phi' - \phi_1')(F F_1 - F' F_1') g\alpha_1 d\mathbf{e}' d\mathbf{c}_1 d\mathbf{c}. \quad (3.54, 5)$$

3.6. The limiting range of molecular influence

The integration with respect to $\mathbf{e}'$ in 3.51–3.54 was tacitly supposed to be taken over all permissible values of $\mathbf{e}'$; that is, since $\alpha_{12} d\mathbf{e}' = b\,db\,d\epsilon$, over all values of b from 0 to ∞, and all values of ϵ from 0 to 2π. But this statement requires some qualification when applied to integrals such as (3.52, 2), since such integrals become infinite if the range of integration with respect to b is infinite. Some upper limit for b, representing the limiting range of molecular influence, should really be taken; this distance might vary to some extent with g, but with gases of moderate density it is usually considerably less than the mean distance between neighbouring molecules. To make the distance definite we may, for example, choose to ignore all 'grazing' encounters during the whole course of which the deflection χ of the relative velocity is less than a very small angle δ.

The value of the integral (3.52, 2) will depend entirely on the upper limit chosen for b, but in such integrals as (3.51, 3, 5), (3.52, 9–12), where the integral contains a factor such as $\phi_1 - \phi_1'$ or $f_1' f_2' - f_1 f_2$, the case is different; as b tends to infinity, $\mathbf{c}_1'$, $\mathbf{c}_2'$ approach the initial values $\mathbf{c}_1$, $\mathbf{c}_2$, so that ϕ_1' becomes equal to ϕ_1, and $f_1' f_2'$ to $f_1 f_2$. It is found that, when the relation connecting χ, b and g corresponds to the laws of force that hold good in the case of most actual gases, the larger values of b contribute very little to the integral, and the result is not appreciably affected if, for analytical convenience, the integration is extended up to $b = \infty$.

Where this procedure is illegitimate (as, for example—cf. 10.34—when the molecules repel or attract according to the inverse square law) b must be restricted, and a discussion of the effect of encounters in which more than two molecules participate is required. In this case, moreover, it is not true (as was assumed in the derivation of Boltzmann's equation) that molecules are appreciably influenced by forces of interaction only along a small portion of their paths; hence the results derived from Boltzmann's equation cannot be expected to give more than the correct *order of magnitude* of the quantities concerned.

4

BOLTZMANN'S H-THEOREM AND THE MAXWELLIAN VELOCITY-DISTRIBUTION

4.1. Boltzmann's H-theorem: the uniform steady state

Consider a simple gas whose molecules are spherical, possess only energy of translation, and are subject to no external forces. If its state is uniform, so that the velocity-distribution function f is independent of $\boldsymbol{r}$, Boltzmann's equation (3.1, 1) reduces to

$$\frac{\partial f}{\partial t} = \iint (f'f_1' - ff_1) g\alpha_1 d\mathbf{e}' d\mathbf{c}_1, \qquad (4.1, 1)$$

after substituting for $\partial_e f/\partial t$ from (3.52, 11).

Let H be the complete integral (that is, the integral over all values of the velocities) defined by the equation

$$H = \int f \ln f d\mathbf{c}. \qquad (4.1, 2)$$

Then H is a number, independent of $\boldsymbol{r}$, but a function of t, depending only on the mode of distribution of the molecular velocities. Also

$$\frac{\partial H}{\partial t} = \int \frac{\partial}{\partial t}(f \ln f) d\mathbf{c} = \int (1 + \ln f) \frac{\partial f}{\partial t} d\mathbf{c}$$

$$= \iiint (1 + \ln f)(f'f_1' - ff_1) g\alpha_1 d\mathbf{e}' d\mathbf{c} d\mathbf{c}_1 \qquad (4.1, 3)$$

by (4.1, 1). Hence, using (3.54, 5),

$$\frac{\partial H}{\partial t} = \frac{1}{4}\iiint (1 + \ln f + 1 + \ln f_1 - 1 - \ln f' - 1 - \ln f_1') \times (f'f_1' - ff_1) g\alpha_1 d\mathbf{e}' d\mathbf{c} d\mathbf{c}_1$$

$$= \frac{1}{4}\iiint \ln (ff_1/f'f_1')(f'f_1' - ff_1) g\alpha_1 d\mathbf{e}' d\mathbf{c} d\mathbf{c}_1. \qquad (4.1, 4)$$

Now $\ln (ff_1/f'f_1')$ is positive or negative according as ff_1 is greater or less than $f'f_1'$, and is therefore always opposite in sign to $f'f_1' - ff_1$. Thus the integral on the right-hand side of (4.1, 4) is either negative or zero, and so H can never increase. This is known as Boltzmann's H-theorem.

Since H is bounded below,* it cannot decrease indefinitely, but must tend

* This is so because $H = -\infty$ only if $\int f \ln f d\mathbf{c}$ diverges. The integral $\int f . \frac{1}{2}mc^2 d\mathbf{c}$ certainly converges, since it represents the (finite) total energy of translation of the molecules; hence if $\int f \ln f d\mathbf{c}$ were to diverge, $-\ln f$ would have to tend to infinity more rapidly than c^2 as $c \to \infty$. But this would imply that f tends to zero more rapidly than e^{-c^2}, in which case $\int f \ln f d\mathbf{c}$ certainly converges.

to a limit, corresponding to a state of the gas in which $\partial H/\partial t = 0$. By (4.1, 4), this can occur only if, for all values of $\boldsymbol{c}$, $\boldsymbol{c}_1$,

$$f'f_1' = ff_1, \tag{4.1, 5}^*$$

or, what is equivalent, $$\ln f' + \ln f_1' = \ln f + \ln f_1. \tag{4.1, 6}$$

Comparing (4.1, 1) and (4.1, 5), we see that, if $\partial H/\partial t = 0$, then $\partial f/\partial t = 0$ also, so that the state of the gas is steady as well as uniform. Conversely, if the gas is in a uniform steady state, not only must $\partial f/\partial t = 0$, but also, since H depends only on f, $\partial H/\partial t = 0$, for which (4.1, 5) is necessary. That is to say, the solution of

$$\iint (f'f_1' - ff_1) g\alpha_1 d\boldsymbol{e}' d\boldsymbol{c}_1 = 0 \tag{4.1, 7}$$

is (4.1, 5) or (4.1, 6).

Equation (4.1, 5) implies that the number of encounters between molecules with velocities in the ranges $d\boldsymbol{c}$, $d\boldsymbol{c}_1$, such that the direction $\boldsymbol{e}'$ of the final relative velocity lies in the range $d\boldsymbol{e}'$, is equal to the number of inverse encounters of a similar type that result in molecules entering these velocity-ranges (cf. 3.52). This is expressed analytically by the equation

$$f'f_1'\alpha_1(g,\chi)\, d\boldsymbol{e}\, d\boldsymbol{c}'\, d\boldsymbol{c}_1' = ff_1\alpha_1(g,\chi)\, d\boldsymbol{e}'\, d\boldsymbol{c}\, d\boldsymbol{c}_1,$$

which is satisfied in virtue of (4.1, 5) and the relation $d\boldsymbol{e}\, d\boldsymbol{c}'\, d\boldsymbol{c}_1' = d\boldsymbol{e}'\, d\boldsymbol{c}\, d\boldsymbol{c}_1$ (cf. (3.52, 6)). Thus not only is the state of the gas steady, so that encounters as a whole produce no effect, but also the effect of every type of encounter is exactly balanced by the effect of the inverse process. This is an example of *detailed balancing*, recognized as a general principle in statistical mechanics.†

Equation (4.1, 6) shows that $\ln f$ is a summational invariant for encounters (3.2). Thus it must be a linear combination of the three summational invariants $\psi^{(i)}$ of (3.2, 2), so that

$$\ln f = \Sigma \alpha^{(i)}\psi^{(i)} = \alpha^{(1)} + \boldsymbol{\alpha}^{(2)}.m\boldsymbol{c} - \alpha^{(3)}.\tfrac{1}{2}mc^2, \tag{4.1, 8}$$

where, since $\ln f$ is a scalar, $\alpha^{(1)}$ and $\alpha^{(3)}$ are scalars and $\boldsymbol{\alpha}^{(2)}$ is a vector; all three must be independent of $\boldsymbol{r}$, t, since the state of the gas is uniform and steady. This equation is equivalent to

$$\begin{aligned}\ln f &= \alpha^{(1)} + m(\alpha_x^{(2)}u + \alpha_y^{(2)}v + \alpha_z^{(2)}w) - \tfrac{1}{2}\alpha^{(3)}m(u^2+v^2+w^2)\\ &= \ln \alpha^{(0)} - \alpha^{(3)}.\tfrac{1}{2}m\{(u-\alpha_x^{(2)}/\alpha^{(3)})^2 + (v-\alpha_y^{(2)}/\alpha^{(3)})^2 + (w-\alpha_z^{(2)}/\alpha^{(3)})^2\},\end{aligned}$$

where $\alpha^{(0)}$ is a new constant. Thus, if $\boldsymbol{C}' = \boldsymbol{c} - \boldsymbol{\alpha}^{(2)}/\alpha^{(3)}$,

$$f = \alpha^{(0)} e^{-\alpha^{(3)}.\frac{1}{2}mC'^2}. \tag{4.1, 9}$$

This result was first obtained by Maxwell, and a gas in the state defined by (4.1, 9) may be said to be in the Maxwellian state.‡

* Equation (4.1, 5) first occurs in J. C. Maxwell, *Phil. Trans. Roy. Soc.* **157**, 49 (1867); *Collected Papers*, **2**, 45.

† R. H. Fowler, *Statistical Mechanics*, p. 417 (1929), or p. 660 (1936).

‡ The variable $\boldsymbol{C}'$ of (4.1, 9) should not be confused with the peculiar velocity of a molecule after collision, for which the same symbol $\boldsymbol{C}'$ will later be used.

The constants $\alpha^{(0)}$, $\boldsymbol{\alpha}^{(2)}$ and $\alpha^{(3)}$ can be evaluated in terms of the number-density n, the mean velocity $\boldsymbol{c}_0$, and the temperature T. First we have

$$n = \int f d\boldsymbol{c} = \alpha^{(0)} \int e^{-\alpha^{(3)} \cdot \frac{1}{2} m C'^2} d\boldsymbol{C}',$$

whence, on expressing $\boldsymbol{C}'$ in terms of polar coordinates C', θ, ϕ,

$$n = \alpha^{(0)} \int_0^\infty C'^2 e^{-\alpha^{(3)} \cdot \frac{1}{2} m C'^2} dC' \int_0^\pi \sin\theta \, d\theta \int_0^{2\pi} d\phi$$

$$= \alpha^{(0)} \left(\frac{2\pi}{m\alpha^{(3)}}\right)^{\frac{3}{2}},$$

using (1.4, 2). Again,

$$n\boldsymbol{c}_0 = \int \boldsymbol{c} f d\boldsymbol{c}$$

$$= \int (\boldsymbol{\alpha}^{(2)}/\alpha^{(3)} + \boldsymbol{C}') f d\boldsymbol{C}'$$

$$= n\boldsymbol{\alpha}^{(2)}/\alpha^{(3)} + \alpha^{(0)} \int e^{-\alpha^{(3)} \cdot \frac{1}{2} m C'^2} \boldsymbol{C}' d\boldsymbol{C}'.$$

The second term vanishes because the integrand is an odd function of the components of $\boldsymbol{C}'$; hence $\boldsymbol{c}_0 = \boldsymbol{\alpha}^{(2)}/\alpha^{(3)}$ and $\boldsymbol{C}'$ is identical with $\boldsymbol{C}$, the peculiar velocity (2.2), so that (4.1, 9) takes the form

$$f = \alpha^{(0)} e^{-\alpha^{(3)} \cdot \frac{1}{2} m C^2} = n \left(\frac{m\alpha^{(3)}}{2\pi}\right)^{\frac{3}{2}} e^{-\alpha^{(3)} \cdot \frac{1}{2} m C^2}.$$

Finally, from the definition of temperature in the uniform steady state,

$$\tfrac{3}{2} kT = \tfrac{1}{2} m \overline{C^2}$$

$$= \frac{m}{2n} \int C^2 f d\boldsymbol{c}$$

$$= \frac{m}{2} \left(\frac{m\alpha^{(3)}}{2\pi}\right)^{\frac{3}{2}} \int C^2 e^{-\alpha^{(3)} \cdot \frac{1}{2} m C^2} d\boldsymbol{C}$$

$$= \frac{3}{2\alpha^{(3)}}$$

by (1.4, 2). Hence $$\alpha^{(3)} = 1/kT.$$

Consequently (4.1, 9) is equivalent to

$$f = n \left(\frac{m}{2\pi kT}\right)^{\frac{3}{2}} e^{-mC^2/2kT}, \qquad (4.1, 10)$$

which is the usual form of Maxwell's velocity-distribution function.

It therefore appears that when the density, mean velocity, and temperature of a uniform gas are assigned, there is only one possible permanent mode of distribution of the molecular velocities, and that the actual mode, if different, will tend to approach this mode.

4.11. Properties of the Maxwellian state

The number of molecules per unit volume with velocities in the range $\boldsymbol{c}$, $d\boldsymbol{c}$ is $f d\boldsymbol{c}$ or $f\,du\,dv\,dw$. Hence the number of molecules per unit volume, in the Maxwellian state, whose component velocities lie between the limits u and $u+du$, v and $v+dv$, and w and $w+dw$, may be written

$$n(m/2\pi kT)^{\frac{3}{2}}\, e^{-\frac{1}{2}m(u-u_0)^2/kT}\, du \,.\, e^{-\frac{1}{2}m(v-v_0)^2/kT}\, dv \,.\, e^{-\frac{1}{2}m(w-w_0)^2/kT}\, dw.$$

This indicates that the distribution of u is independent of the values of v, w; thus the probability that the x-component of the velocity of a molecule lies between given limits is independent of the value of the component perpendicular to Ox; the x-component is distributed about its mean value u_0 proportionately to the Gaussian function

$$e^{-\frac{1}{2}m(u-u_0)^2/kT}$$

or e^{-s^2}, where $s^2 = \frac{1}{2}m(u-u_0)^2/kT$.

Writing $d\boldsymbol{c} = d\boldsymbol{C} = C^2 \sin\theta\, dC\, d\theta\, d\varphi$, and integrating with respect to θ and φ, the number of molecules per unit volume whose peculiar speeds lie between C and $C+dC$ is found to be

$$\left(\frac{2}{\pi}\right)^{\frac{1}{2}} n \left(\frac{m}{kT}\right)^{\frac{3}{2}} C^2 e^{-mC^2/2kT}\, dC, \qquad (4.11, 1)$$

which is proportional to $s^2 e^{-s^2}$, where $s^2 = mC^2/2kT$. Graphs of the two functions e^{-s^2}, $s^2 e^{-s^2}$ are given in Fig. 7. The first illustrates the distribution of any component of the peculiar velocity, and the second shows that of the peculiar speed C.

The mean value of any function of the molecular velocity for a gas in the Maxwellian state can be found from the equation

$$n\bar{\phi} = \int \phi f\, d\boldsymbol{c} = n\left(\frac{m}{2\pi kT}\right)^{\frac{3}{2}} \int \phi\, e^{-mC^2/2kT}\, d\boldsymbol{C}.$$

If the function is of odd degree in any component U, V, or W of the peculiar velocity, its mean value vanishes.

The mean value of the peculiar speed C is given by

$$\begin{aligned}\bar{C} &= \left(\frac{m}{2\pi kT}\right)^{\frac{3}{2}} \int C e^{-mC^2/2kT}\, d\boldsymbol{C} \\ &= 4\pi \left(\frac{m}{2\pi kT}\right)^{\frac{3}{2}} \int_0^\infty C^3 e^{-mC^2/2kT}\, dC \\ &= \left(\frac{8kT}{\pi m}\right)^{\frac{1}{2}} \qquad (4.11, 2)\end{aligned}$$

by (1.4, 3).

The mean value of C^2 is given by

$$\overline{C^2} = \frac{3kT}{m} \tag{4.11, 3}$$

(cf. the definition of temperature in (2.41, 1)). The root-mean-square of the peculiar speed is defined as equal to $\sqrt{(\overline{C^2})}$. It is thus not equal to the mean speed; in fact

$$\sqrt{(\overline{C^2})} = \bar{C}\sqrt{(3\pi/8)} = 1{\cdot}086\bar{C}. \tag{4.11, 4}$$

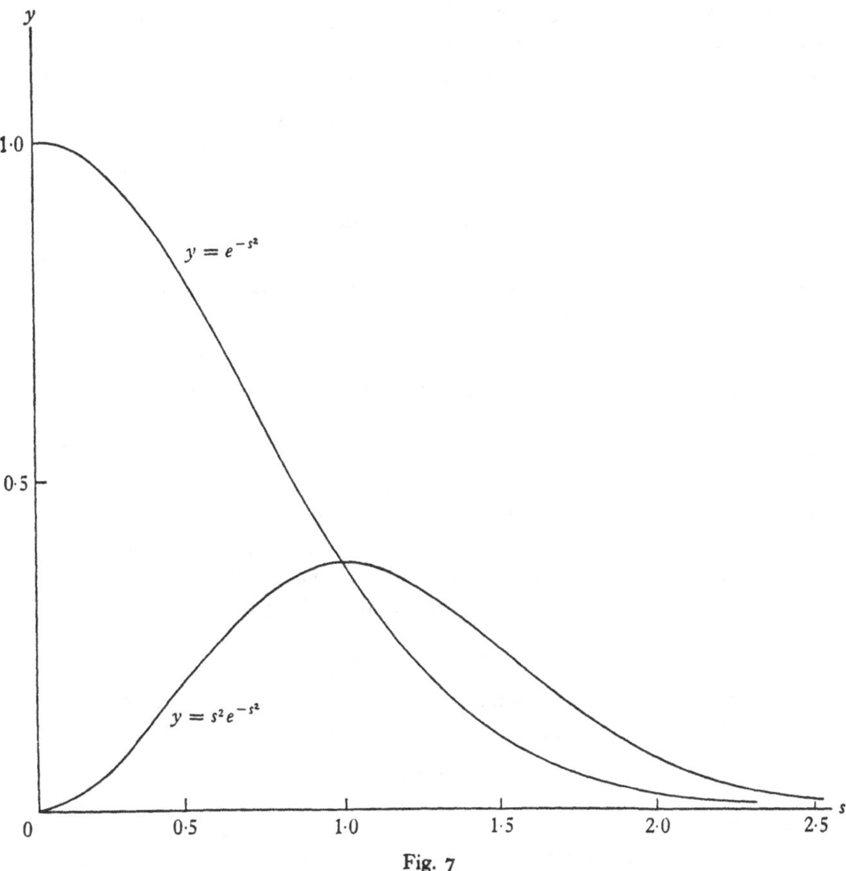

Fig. 7

Another mean value needed later is that of the z-component of the peculiar velocity, averaged over those molecules at a given point for which this component is positive; it will be denoted by $\overline{W}_+$. Since the number-density of such molecules is $\frac{1}{2}n$,

$$\tfrac{1}{2}n\overline{W}_+ = \int_+ fW d\mathbf{C}, \tag{4.11, 5}$$

where the integration on the right extends over all values of $\mathbf{C}$ for which

$W > 0$. Hence, using (1.4, 2, 3),

$$\overline{W}_+ = 2\left(\frac{m}{2\pi kT}\right)^{\frac{3}{2}} \int_{-\infty}^{+\infty} e^{-mU^2/2kT}\, dU \int_{-\infty}^{\infty} e^{-mV^2/2kT}\, dV \int_0^{\infty} We^{-mW^2/2kT}\, dW$$
$$= \left(\frac{2kT}{\pi m}\right)^{\frac{1}{2}}$$
$$= \tfrac{1}{2}\bar{C}. \qquad (4.11, 6)$$

The components of the pressure tensor (2.31, 3) can readily be found. Since the mean value of any function of the velocity odd in U, V, or W is zero, the non-diagonal terms of the tensor vanish, while by symmetry

$$p_{xx} = p_{yy} = p_{zz} = \tfrac{1}{3}(p_{xx}+p_{yy}+p_{zz})$$
$$= p = knT, \quad \text{and} \quad \mathsf{p} = knT\mathsf{U}. \qquad (4.11, 7)$$

Hence in this case the pressure system is hydrostatic (2.32).

4.12. Maxwell's original treatment of velocity-distribution

The above law of distribution of velocities, (4.1, 10), was first given by Maxwell* for the case of a gas at rest. His original argument is of historical interest, though not mathematically rigorous. He assumed that, as the component velocities u, v, and w of a molecule are perpendicular to each other, the distribution of one of these components among the molecules will be independent of the values of the other components. Assume, then, that $F(u)\,du$ is the probability that a molecule should possess an x-component of velocity between u and $u+du$, and that $F(u)$ is independent of v, w. Then the probabilities that its y- and z-components of velocity should have values between v and $v+dv$, w and $w+dw$, are similarly $F(v)\,dv$, $F(w)\,dw$; hence if

$$f(u,v,w)\,du\,dv\,dw$$

denotes the number of molecules per unit volume whose component velocities lie in the ranges du, dv, dw,

$$f(u,v,w)\,du\,dv\,dw = n\,F(u)\,du\,F(v)\,dv\,F(w)\,dw.$$

Now in a gas at rest there is nothing to distinguish one direction from another; thus $f(u,v,w)$ can depend on u, v, w only through the invariant $u^2+v^2+w^2$. Thus

$$n\,F(u)\,F(v)\,F(w) = f(u,v,w) = \phi(u^2+v^2+w^2),$$

say. The solution of this functional equation is given by

$$F(u) = \mathrm{X}e^{\mathrm{Y}u^2},$$

$$f(u,v,w) = \phi(u^2+v^2+w^2) = n\mathrm{X}^3 e^{\mathrm{Y}(u^2+v^2+w^2)},$$

* J. C. Maxwell, *Collected Papers* 1, 377; *Phil. Mag.* (4), 19, 22 (1860).

where X, Y are arbitrary constants; this agrees with the form derived above for f, taking $n\mathrm{X}^3 = \alpha^{(0)}$, $\mathrm{Y} = -\frac{1}{2}m\alpha^{(3)}$.

The unsatisfactory feature of this proof is the assumption that the distribution of each of the three velocity-components among the molecules is independent of the values of the others. As these three components do not enter independently into the equations governing a collision, it would be natural to suppose that the distributions of the components are not independent.

On account of this defect, Maxwell* attempted a second proof, which was also imperfect; he showed only that *if* in a gas the Maxwellian distribution of velocities was once attained, it would not alter thereafter (since $f'f_1' = ff_1$, so that $\partial f/\partial t = 0$). Boltzmann first showed, by his H-theorem,† that the gas would tend to the Maxwellian state. The demonstration was improved later by Lorentz,‡ substantially to the form given above in 4.1. This proof also is open to some objection, because of the assumption in 3.5 that there is no correlation between the velocity and the position of a molecule. In very dense gases it is, in fact, probable that the velocity of one molecule is related to the velocities of other neighbouring molecules, in whose close proximity it remains for some time by reason of the close packing of the molecules; but for gases under ordinary conditions the assumption appears to be valid.§

4.13. The steady state in a smooth vessel

Maxwell's form for f also applies to the steady state of a gas at rest in a smooth-walled vessel under no forces.

Consider H_0, defined by

$$H_0 \equiv \int H\,d\boldsymbol{r} = \iint f \ln f\,d\boldsymbol{c}\,d\boldsymbol{r} \tag{4.13, 1}$$

(cf. (4.1, 2)), the space-integration extending throughout the volume of the vessel. In the time-derivative of this equation,

$$\frac{\partial H_0}{\partial t} = \iint (\ln f + 1)\frac{\partial f}{\partial t}\,d\boldsymbol{c}\,d\boldsymbol{r}, \tag{4.13, 2}$$

we substitute for $\partial f/\partial t$ from Boltzmann's equation (3.1, 1), omitting the term containing $\boldsymbol{F}$, since $\boldsymbol{F} = 0$. We thus have

$$\begin{aligned}\frac{\partial H_0}{\partial t} &= \iint (\ln f + 1)\left(\frac{\partial_e f}{\partial t} - \boldsymbol{c}\cdot\frac{\partial f}{\partial \boldsymbol{r}}\right)d\boldsymbol{c}\,d\boldsymbol{r}\\ &= \iint (\ln f + 1)\frac{\partial_e f}{\partial t}\,d\boldsymbol{c}\,d\boldsymbol{r} - \iint \boldsymbol{c}\cdot\frac{\partial f \ln f}{\partial \boldsymbol{r}}\,d\boldsymbol{c}\,d\boldsymbol{r}.\end{aligned} \tag{4.13, 3}$$

* J. C. Maxwell, *Collected Papers*, **2**, 43; *Phil. Trans. R. Soc.*, **157**, 49 (1867).
† L. Boltzmann, *Wien. Sitz.* **66**, 275 (1872).
‡ H. A. Lorentz, *Wien. Sitz.* **95** (2), 127 (1887).
§ See J. H. Jeans, *Dynamical Theory of Gases* (4th ed.), pp. 59–64 (1925), and H. Grad, *Handbuch der Physik*, vol. 12, 218–33 (1958).

The second term becomes, on transformation by Green's theorem,

$$-\iint c_n f \ln f \, d\mathbf{c} \, dS$$

or

$$-\int n\overline{c_n \ln f} \, dS,$$

where c_n is the component of $\mathbf{c}$ along the outward normal to the element dS of the surface of the vessel. Consider the contribution to this integral from any element dS, which we may without loss of generality take perpendicular to Ox, since the directions of the axes of reference are arbitrary. Since the vessel is smooth, the molecules that leave are the same as those that strike it; their x-components of velocity are exactly reversed, and their y- and z-components are unaltered. Hence near dS

$$f(-u, v, w) = f(u, v, w),$$

and so

$$\overline{c_n \ln f} = \overline{u \ln f} = 0.$$

Thus the contribution to the integral from this element vanishes. Since the element dS is arbitrary, the integral as a whole vanishes.

It follows that

$$\frac{\partial H_0}{\partial t} = \iint (\ln f + 1) \frac{\partial_e f}{\partial t} \, d\mathbf{c} \, d\mathbf{r}. \tag{4.13, 4}$$

After substituting for $\partial_e f/\partial t$ from (3.52, 11), we may show, as in 4.1, that

$$\frac{\partial H_0}{\partial t} \leqslant 0,$$

and hence that, when the state of the gas is steady (so that $\partial f/\partial t = 0$ and $\partial H_0/\partial t = 0$), $\ln f$ must be a summational invariant, as in (4.1, 8). Consequently $\partial_e f/\partial t = 0$, and

$$f = n\left(\frac{m}{2\pi kT}\right)^{\frac{3}{2}} e^{-mc^2/2kT},$$

the variable c taking the place of C because the gas is supposed to be at rest.

The quantities n, T in this equation might conceivably be functions of $\mathbf{r}$, as the state has not been assumed uniform. But Boltzmann's equation for f, which now reduces to

$$\mathbf{c} \cdot \frac{\partial f}{\partial \mathbf{r}} = 0,$$

must be satisfied for all values of $\mathbf{c}$; this implies that n and T are independent of $\mathbf{r}$.

When the walls of the vessel are not smooth, no such simple proof that Maxwell's formula remains valid seems available.*

* On this point see R. H. Fowler, *Statistical Mechanics*, pp. 697–99 (1936).

4.14. The steady state in the presence of external forces

We next consider the case when an external force $m\boldsymbol{F}$ acts on each molecule ($\boldsymbol{F}$ being, as before, independent of the velocity $\boldsymbol{c}$ of the molecule). As in 4.1

$$\begin{aligned}\frac{\partial H}{\partial t} &= \int (1+\ln f)\frac{\partial f}{\partial t}d\boldsymbol{c} \\ &= \int (1+\ln f)\left(\frac{\partial_e f}{\partial t} - \boldsymbol{c}.\frac{\partial f}{\partial \boldsymbol{r}} - \boldsymbol{F}.\frac{\partial f}{\partial \boldsymbol{c}}\right)d\boldsymbol{c} \\ &= \int\left(\frac{\partial_e(f\ln f)}{\partial t} - \boldsymbol{c}.\frac{\partial(f\ln f)}{\partial \boldsymbol{r}} - \boldsymbol{F}.\frac{\partial(f\ln f)}{\partial \boldsymbol{c}}\right)d\boldsymbol{c}. \qquad (4.14, 1)\end{aligned}$$

In the last term we have three components of the integral

$$\int \frac{\partial(f\ln f)}{\partial \boldsymbol{c}}\,d\boldsymbol{c} \qquad (4.14, 2)$$

to consider. In the component

$$\iiint \frac{\partial(f\ln f)}{\partial u}\,du\,dv\,dw$$

an integration with respect to u gives

$$\iint \Big[f\ln f\Big]_{u=-\infty}^{u=\infty} dv\,dw,$$

which vanishes since $f\ln f$ must vanish when $\boldsymbol{c}$ or any of its components tends to $\pm\infty$. Hence the integral (4.14, 2) vanishes.

If the gas is in a smooth-walled vessel at rest, or if its density tends to zero in all directions, it may be shown, as in 4.13, that the second term on the right of (4.14, 1) contributes nothing to H_0, defined as before by

$$H_0 = \int H\,d\boldsymbol{r}.$$

Thus again it follows that $\partial H_0/\partial t \leqslant 0$, and in the steady state $\partial_e f/\partial t = 0$ and

$$f = n\left(\frac{m}{2\pi kT}\right)^{\frac{3}{2}} e^{-mC^2/2kT}, \qquad (4.14, 3)$$

where $\boldsymbol{C} = \boldsymbol{c} - \boldsymbol{c}_0$ and n, $\boldsymbol{c}_0$, T are independent of $\boldsymbol{c}$ and t, but may now depend on $\boldsymbol{r}$.

To examine this dependence, we substitute from (4.14, 3) into Boltzmann's equation (3.1, 2), which then reduces to $\mathscr{D}f = 0$. We use the form for $\mathscr{D}f$ given in (3.12, 2), since in (4.14, 3) f is expressed in terms of the peculiar velocity $\boldsymbol{C}$. Since the state is steady, D/Dt in (3.13, 2) may be replaced by $\boldsymbol{c}_0.\partial/\partial\boldsymbol{r}$. Hence, on dividing by f, Boltzmann's equation becomes

$$\boldsymbol{c}_0.\frac{\partial \ln f}{\partial \boldsymbol{r}} + \boldsymbol{C}.\frac{\partial \ln f}{\partial \boldsymbol{r}} + \left\{\boldsymbol{F} - \left(\boldsymbol{c}_0.\frac{\partial}{\partial \boldsymbol{r}}\right)\boldsymbol{c}_0\right\}.\frac{\partial \ln f}{\partial \boldsymbol{C}} - \frac{\partial \ln f}{\partial \boldsymbol{C}}\boldsymbol{C}:\frac{\partial}{\partial \boldsymbol{r}}\boldsymbol{c}_0 = 0. \qquad (4.14, 4)$$

Since $$\ln f = \ln(n/T^{\frac{3}{2}}) - mC^2/2kT + \text{constant},$$

we have $$\frac{\partial \ln f}{\partial \boldsymbol{r}} = \frac{\partial \ln(n/T^{\frac{3}{2}})}{\partial \boldsymbol{r}} + \frac{mC^2}{2kT^2}\frac{\partial T}{\partial \boldsymbol{r}}, \quad \frac{\partial \ln f}{\partial \boldsymbol{C}} = -\frac{m\boldsymbol{C}}{kT}.$$

Using these values, we can express the left-hand side of (4.14, 4) as the sum of a part independent of $\boldsymbol{C}$, and parts involving $\boldsymbol{C}$ to the first, second, and third powers of its components; these must vanish separately, since the equation is an identity in $\boldsymbol{C}$.

The part of the third degree in $\boldsymbol{C}$ is

$$\frac{mC^2}{2kT^2}\boldsymbol{C}\cdot\frac{\partial T}{\partial \boldsymbol{r}} = 0,$$

whence it follows that $\partial T/\partial \boldsymbol{r} = 0$, that is, the temperature must be uniform throughout the gas. Allowing for this, the part which is of the second degree in $\boldsymbol{C}$ becomes

$$\frac{m}{kT}\boldsymbol{C}\boldsymbol{C}:\frac{\partial}{\partial \boldsymbol{r}}\boldsymbol{c}_0 = 0,$$

whence it follows that $$\overline{\overline{\frac{\partial}{\partial \boldsymbol{r}}\boldsymbol{c}_0}} = 0,$$

or $\mathsf{e} = 0$ in the notation of 1.33. This implies that

$$\frac{\partial u_0}{\partial x} = \frac{\partial v_0}{\partial y} = \frac{\partial w_0}{\partial z} = 0, \quad \frac{\partial v_0}{\partial z} + \frac{\partial w_0}{\partial y} = \frac{\partial w_0}{\partial x} + \frac{\partial u_0}{\partial z} = \frac{\partial u_0}{\partial y} + \frac{\partial v_0}{\partial x} = 0.$$

The solution of these equations* is

$$\boldsymbol{c}_0 = \boldsymbol{c}' + \boldsymbol{\omega} \wedge \boldsymbol{r}, \tag{4.14, 5}$$

where $\boldsymbol{c}'$ and $\boldsymbol{\omega}$ are arbitrary constants. Thus the mean velocity of the gas at any point is the same as the velocity of a rigid body moving with a screw motion.

First consider the special case when $\boldsymbol{c}_0 = 0$. This ensures the vanishing of the part of (4.14, 4) which is independent of $\boldsymbol{C}$; using the conditions $\boldsymbol{c}_0 = 0$ and $\partial T/\partial \boldsymbol{r} = 0$, the remaining part, of the first degree in $\boldsymbol{C}$, becomes

$$\boldsymbol{C}\cdot\left(\frac{\partial \ln n}{\partial \boldsymbol{r}} - \frac{m\boldsymbol{F}}{kT}\right).$$

Since this vanishes for all values of $\boldsymbol{C}$,

$$\frac{\partial \ln n}{\partial \boldsymbol{r}} = \frac{m}{kT}\boldsymbol{F}. \tag{4.14, 6}$$

The steady state is therefore possible only if $\boldsymbol{F}$ is the gradient of the scalar function $(kT/m)\ln n$, so that the field of force must possess a potential Ψ, satisfying the equation

$$\Psi = -\frac{kT}{m}\ln n + \text{const.}$$

* Cf. A. E. H. Love, *The Mathematical Theory of Elasticity*, § 18 (1927).

The density distribution is given in terms of Ψ by

$$n = n_0 e^{-m\Psi/kT}, \tag{4.14, 7}$$

where n_0 is a constant, being, in fact, the number-density at points at which $\Psi = 0$. Hence the complete expression for f is

$$f = n_0 \left(\frac{m}{2\pi kT}\right)^{\frac{3}{2}} e^{-m(2\Psi+c^2)/2kT}. \tag{4.14, 8}$$

This result was first given by Maxwell* as a deduction from his equation (4.1, 5). Boltzmann† later gave the same result (apparently unaware that Maxwell had already published it); his proof was based on the H-theorem, and supplied a needed foundation for Maxwell's deduction.

Consider next the case when $\boldsymbol{c}_0$ does not vanish. Let Oz be taken as the axis of the screw motion; then the components of $\boldsymbol{c}_0$ are $(-\omega y, \omega x, c')$, and we find

$$\left(\boldsymbol{c}_0 \cdot \frac{\partial}{\partial \boldsymbol{r}}\right) \boldsymbol{c}_0 = \frac{\partial \Psi_0}{\partial \boldsymbol{r}},$$

where
$$\Psi_0 = -\tfrac{1}{2}\omega^2(x^2+y^2). \tag{4.14, 9}$$

The equations obtained on equating the terms of first and zero degrees in $\boldsymbol{C}$ to zero are now

$$\boldsymbol{C}.\left(\frac{\partial \ln n}{\partial \boldsymbol{r}} - \frac{m\boldsymbol{F}}{kT} + \frac{m}{kT}\frac{\partial \Psi_0}{\partial \boldsymbol{r}}\right) = 0, \quad \boldsymbol{c}_0 . \frac{\partial \ln n}{\partial \boldsymbol{r}} = 0. \tag{4.14, 10}$$

The first of these implies that $\boldsymbol{F}$ is again derivable from a potential Ψ, and that the number-density n is given in terms of Ψ by

$$n = n_0 e^{-m(\Psi+\Psi_0)/kT}. \tag{4.14, 11}$$

Comparing this with (4.14, 7), we see that the effect of the motion on the density-distribution is the same as if a field of centrifugal force, of potential Ψ_0, acted on the gas.

Using this expression for n, and remembering that $\boldsymbol{c}_0 . \partial\Psi_0/\partial\boldsymbol{r} = 0$, we obtain from the second of equations (4.14, 10) the condition

$$\boldsymbol{c}_0 . \frac{\partial \Psi}{\partial \boldsymbol{r}} = 0,$$

expressing that the motion of the gas must at every point be along an equipotential surface Ψ = constant. Thus if $\omega = 0$ and $c' \neq 0$, Ψ does not depend on z, and if $c' = 0$ and $\omega \neq 0$, Ψ must be symmetric about Oz: if both c' and ω do not vanish, Ψ is constant along spiral curves with Oz as axis.

If the gas is enclosed in a smooth stationary vessel, the motions must be consistent with the shape of the vessel: that is, the gas must in general

* J. C. Maxwell, *Nature, Lond.* **8**, 537 (1873); *Collected Papers*, **2**, 351.

† L. Boltzmann, *Wien. Ber.* **72**, 427 (1875).

be at rest, but if the vessel possesses symmetry about an axis a rotation about this axis is possible.

When $c' = 0$ and $\boldsymbol{F} = 0$ the form of f is

$$f = n_0 \left(\frac{m}{2\pi kT}\right)^{\frac{3}{2}} \exp\left[-(m/2kT)\{u^2+v^2+w^2+2\omega(uy-vx)\}\right].$$

This form of the velocity-distribution function for a rotating gas was first indicated by Maxwell.*

4.2. The H-theorem and entropy

For a gas in the uniform steady state the quantity H, defined by the equation

$$H = \int f \ln f \, d\boldsymbol{c} = n \overline{\ln f},$$

can be expressed in terms of n and T. For in this case

$$\ln f = \ln n + \tfrac{3}{2} \ln (m/2\pi kT) - mC^2/2kT,$$

whence
$$H = n\{\ln n + \tfrac{3}{2} \ln (m/2\pi kT) - \tfrac{3}{2}\}$$

(cf. (2.41, 1)). If the total mass of gas present is M, the volume occupied by the gas is M/ρ, or M/mn. On integrating H through this volume, it follows that

$$H_0 \equiv \int H \, d\boldsymbol{r} = (M/m)\{\ln n + \tfrac{3}{2} \ln (m/2\pi kT) - \tfrac{3}{2}\}.$$

The entropy S of the gas (cf. 2.431) is such that

$$\delta \mathrm{S} = M\left\{c_v \frac{\delta T}{T} + \frac{k}{m} \frac{\delta \mathrm{V}}{\mathrm{V}}\right\}.$$

Since $c_v = 3k/2m$ in this case (there being no *communicable* internal energy), and $n\mathrm{V}$ is the total number of molecules, which is constant,

$$\delta \mathrm{S} = \frac{Mk}{m}\left(\frac{3}{2}\frac{\delta T}{T} - \frac{\delta n}{n}\right),$$

whence, on integration, we find

$$\mathrm{S} = \frac{Mk}{m} \ln (T^{\frac{3}{2}}/n) + \text{const.};$$

thus
$$\mathrm{S} + kH_0 = -\frac{3M}{2m}\{\ln (2\pi k/m) + 1\} + \text{const.}$$

The right-hand side of this equation is independent of the state of the gas; hence, except for an additive constant,

$$\mathrm{S} = -kH_0. \qquad (4.2, 1)$$

* J. C. Maxwell, *Nature, Lond.* **16**, 244 (1877).

This relation connects H_0 with the entropy* when the gas is in a uniform steady state. For non-uniform or non-steady states there is no thermodynamic definition of entropy, but the concept of entropy may be generalized to such states, taking equation (4.2, 1) as the definition. Boltzmann's H-theorem, which shows that for a gas not in a steady state H_0 must decrease, is a generalization of the thermodynamic law that entropy cannot diminish. This association of H with S was indicated by Boltzmann in 1872.

4.21. The *H*-theorem and reversibility

Suppose that at a certain instant the velocity of every molecule in a mass of gas in a uniform state under no forces is reversed; the value of H or $n\overline{\ln f}$ is unaltered by this process. The molecules will now retrace their previous paths. Since, in general, $\partial H/\partial t < 0$ before the change, one infers that $\partial H/\partial t > 0$ after the change, which contradicts the H-theorem. Thus a paradox arises.

The H-theorem is, of course, a probability theorem. Its proof depends on probability concepts, e.g. in the definition of the velocity-distribution function and the calculation of the number of encounters of a given type. Thus the theorem is to be interpreted as implying, not that H for a given mass of gas must necessarily decrease during a given short interval, but that it is far more likely to decrease than to increase. Even so, the theorem appears inconsistent with reversibility, which implies that to every state of the gas for which H is decreasing there is one for which H is increasing equally rapidly.

The following considerations help towards resolving the paradox. The argument leading to the expression for $\partial_e f/\partial t$ assumes molecular chaos before encounters, i.e. that the velocities of two molecules just before they collide are uncorrelated. This is reasonable, because in gases at moderate densities the two molecules will have come from regions far enough apart to rule out the possibility of their having recently influenced each other. The velocities of the molecules immediately after encounter are not, however, in the same sense uncorrelated. If the velocities of all the molecules are suddenly reversed, encounters are described in the reverse sense, and one is no longer considering a state with molecular chaos before encounters. Thus the value found earlier for $\partial H/\partial t$ no longer applies. This in part resolves the difficulty; on physical grounds, states characterized by molecular chaos before collisions are far more probable than states deviating widely from such chaos.

* In Boltzmann's papers of 1872 and 1875, in which the H-theorem was introduced, he used the symbol E (presumably because it is the initial letter of *entropy*) for what is now denoted by H; S. H. Burbury (*Phil. Mag.* **30**, 301 (1890)) seems to have introduced the symbol H, though later he used B to denote an almost identical function. Boltzmann used the symbol E as late as 1893, but in 1895 adopted the letter H; cf. *Nature, Lond.* **139**, 931 (1937).

But the mathematical difficulty remains that to each state characterized by molecular chaos there corresponds one (that with velocities reversed) not so characterized.

The reversibility paradox can be related to another occurring in the kinetic theory. Consider, for example, the following paradox relating to an atmosphere in a steady state under gravity. Any one molecule has a constant downward acceleration due to gravity, and since with the velocity-distribution of (4.14, 8) velocities in all directions are equally likely, collisions with other molecules may impede, but will not completely destroy, the downward motion of the molecule.* Hence the gas as a whole must descend; that is, it cannot be in a steady state.

This second paradox is easily resolved. If the atmosphere is held up against gravity, it must be held up by a surface, and collisions with this surface interrupt the steady descent. To see how the steady state is maintained at a level well above this surface, consider two neighbouring horizontal planes A and B, of which A is above B. Because of the action of gravity it is more probable that a molecule which at the beginning of a short interval is at A will sink to B during that interval, than that a molecule initially at B will rise to A. As, however, the density of molecules is greater at B than at A, the smaller proportion of the molecules initially at B which rise to A can exactly balance the larger proportion of molecules initially at A which sink to B. Hence each molecule can tend to diffuse downward with a certain velocity, and yet the mean velocity of molecules at a given point can vanish.

A comparison with the second paradox helps further to clarify the reversibility paradox. The value of $\partial H/\partial t$ found earlier is, as it were, a velocity of diffusion, analogous to that of a molecule at A in the preceding paragraph, with which H tends to approach its minimum value as the actual state of the gas varies among the different possible states. It may be negative, even though the mean value of the 'velocity' $\partial H/\partial t$ for all possible states with a given H is zero, provided that the possible states with a smaller value of H are more numerous than those with a larger value, i.e. provided that smaller values of H are intrinsically more probable than larger. This agrees with the result of statistical mechanics that the Maxwellian function gives the most probable distribution of the molecular velocities.†

4.3. The *H*-theorem for gas-mixtures; equipartition of kinetic energy of peculiar motion

The velocity-distribution functions for a mixture of gases in a uniform steady state under no forces may be obtained by a generalization of the argument of 4.1. For simplicity, we consider only a binary mixture. By analogy with (4.1, 1), the equations satisfied by the velocity-distribution

* Cf. 5.5, on the persistence of velocity after collision.

† J. H. Jeans, *Dynamical Theory of Gases* (4th ed.), chapter 3 (1925).

functions f_1, f_2 are

$$0 = \frac{\partial f_1}{\partial t} = \iint (f_1'f' - f_1 f) g\alpha_1 d\mathbf{e}'\, d\mathbf{c} + \iint (f_1' F_2' - f_1 F_2) g\alpha_{12} d\mathbf{e}'\, d\mathbf{c}_2,$$

$$0 = \frac{\partial F_2}{\partial t} = \iint (F_2'F' - F_2 F) g\alpha_2 d\mathbf{e}'\, d\mathbf{c} + \iint (f_1' F_2' - f_1 F_2) g\alpha_{12} d\mathbf{e}'\, d\mathbf{c}_1, \tag{4.3, 1}$$

where, for the present, f_2 has been replaced by F_2, in order to emphasize the distinction between the functions $f_1(\mathbf{c}_1, \mathbf{r}, t)$ and $f_1(\mathbf{c}, \mathbf{r}, t)$ for molecules of the first kind, and the functions $f_2(\mathbf{c}_2, \mathbf{r}, t)$ and $f_2(\mathbf{c}, \mathbf{r}, t)$ for molecules of the second kind: these are represented in the above integrals by f_1, f and F_2, F.

Multiply the first of equations (4.3, 1) by $\ln f_1 d\mathbf{c}_1$, and the second by $\ln F_2 d\mathbf{c}_2$, and integrate over all values of $\mathbf{c}_1$, $\mathbf{c}_2$ respectively; then, on transformation by (3.54, 2) and (3.54, 5) they become

$$\tfrac{1}{4}\iiint (\ln f_1 + \ln f - \ln f_1' - \ln f')(f'f_1' - ff_1) g\alpha_1 d\mathbf{e}'\, d\mathbf{c}\, d\mathbf{c}_1$$
$$+ \tfrac{1}{2}\iiint (\ln f_1 - \ln f_1')(f_1'F_2' - f_1 F_2) g\alpha_{12} d\mathbf{e}'\, d\mathbf{c}_1 d\mathbf{c}_2 = 0,$$

$$\tfrac{1}{4}\iiint (\ln F_2 + \ln F - \ln F_2' - \ln F')(F'F_2' - FF_2) g\alpha_2 d\mathbf{e}'\, d\mathbf{c}\, d\mathbf{c}_2$$
$$+ \tfrac{1}{2}\iiint (\ln F_2 - \ln F_2')(f_1'F_2' - f_1 F_2) g\alpha_{12} d\mathbf{e}'\, d\mathbf{c}_1 d\mathbf{c}_2 = 0,$$

whence, on addition, we obtain

$$\tfrac{1}{4}\iiint \ln (ff_1/f'f_1')(f'f_1' - ff_1) g\alpha_1 d\mathbf{e}'\, d\mathbf{c}\, d\mathbf{c}_1$$
$$+ \tfrac{1}{2}\iiint \ln (f_1 F_2/f_1'F_2')(f_1'F_2' - f_1 F_2) g\alpha_{12} d\mathbf{e}'\, d\mathbf{c}_1 d\mathbf{c}_2$$
$$+ \tfrac{1}{4}\iiint \ln (FF_2/F'F_2')(F'F_2' - FF_2) g\alpha_2 d\mathbf{e}'\, d\mathbf{c}\, d\mathbf{c}_2 = 0.$$

In none of these three integrals can the integrand be positive, so that their sum can be zero only if, for all values of the variables, the integrands vanish. Hence for all types of encounter, between like or unlike molecules,

$$f_1 f = f_1' f', \quad f_1 F_2 = f_1' F_2', \quad F_2 F = F_2' F', \tag{4.3, 2}$$

and therefore $\ln f_1$, $\ln F_2$ are solutions of the three equations

$$\psi_1 + \psi = \psi_1' + \psi', \quad \psi_1 + \Psi_2 = \psi_1' + \Psi_2', \quad \Psi_2 + \Psi = \Psi_2' + \Psi'. \tag{4.3, 3}$$

The first and third of these equations show that ψ_1, Ψ_2 are of the forms

$$\psi_1 = \alpha_1^{(1)} + \boldsymbol{\alpha}_1^{(2)}.m_1\mathbf{c}_1 + \alpha_1^{(3)}.\tfrac{1}{2}m_1 c_1^2, \quad \Psi_2 = \alpha_2^{(1)} + \boldsymbol{\alpha}_2^{(2)}.m_2\mathbf{c}_2 + \alpha_2^{(3)}.\tfrac{1}{2}m_2 c_2^2;$$

the middle equation then requires that $\boldsymbol{\alpha}_1^{(2)} = \boldsymbol{\alpha}_2^{(2)} = \boldsymbol{\alpha}^{(2)}$ (say); $\alpha_1^{(3)} = \alpha_2^{(3)} = \alpha^{(3)}$ (say), to satisfy the equations of conservation of momentum and energy at encounters of unlike molecules. Thus

$$\ln f_1 = \alpha_1^{(1)} + \boldsymbol{\alpha}^{(2)}.m_1\mathbf{c}_1 + \alpha^{(3)}.\tfrac{1}{2}m_1 c_1^2 = \ln A_1 + \alpha^{(3)}.\tfrac{1}{2}m_1\Sigma(u_1 - u_0)^2,$$
$$\ln f_2 = \alpha_2^{(1)} + \boldsymbol{\alpha}^{(2)}.m_2\mathbf{c}_2 + \alpha^{(3)}.\tfrac{1}{2}m_2 c_2^2 = \ln A_2 + \alpha^{(3)}.\tfrac{1}{2}m_2\Sigma(u_2 - u_0)^2,$$

where A_1, A_2, u_0, v_0, w_0 are new constants. It may be shown, as in 4.1, that u_0, v_0, w_0 are the components of the mean velocity of either constituent, and so of the mixture: and that the mean kinetic energies of peculiar motion of molecules of the two constituents are the same, and equal to $-3/2\alpha^{(3)}$. Hence, if T is the temperature of the gas, $kT = -1/\alpha^{(3)}$, and the velocity-distribution functions f_1, f_2 may be expressed in the forms

$$\left.\begin{aligned} f_1 &= n_1\left(\frac{m_1}{2\pi kT}\right)^{\frac{3}{2}} \exp(-m_1 C_1^2/2kT), \\ f_2 &= n_2\left(\frac{m_2}{2\pi kT}\right)^{\frac{3}{2}} \exp(-m_2 C_2^2/2kT), \end{aligned}\right\} \tag{4.3, 4}$$

The result that the mean kinetic energies of peculiar motion of molecules of the different constituents are equal is a special case of the statistical-mechanical theorem of equipartition of energy, referred to in 2.431.

The results of 4.13, 4.14 may also readily be generalized to apply to a mixture of gases. Thus, for example, if the molecules of the two gases are subject to fields of force of potentials Ψ_1, Ψ_2, the velocity-distributions will be given by equations of the same form as (4.3, 4) where n_1, n_2 are given by

$$n_1 = N_1 e^{-m_1\Psi_1/kT}, \quad n_2 = N_2 e^{-m_2\Psi_2/kT}, \tag{4.3, 5}$$

N_1, N_2 being constants.

4.4. Integral theorems; $I(F)$, $[F, G]$, $\{F, G\}$

To conclude this chapter certain integral theorems, similar to those of 3.54, will be proved. Only a binary gas-mixture will be considered; the corresponding results for a simple gas, also required later, are merely special cases of these.

The velocity-spaces of the different sets of molecules will be regarded as distinct domains; thus functions of the velocity may be differently defined in the two domains. Let $f^{(0)}$ denote Maxwell's velocity-distribution function

$$f^{(0)} = n\left(\frac{m}{2\pi kT}\right)^{\frac{3}{2}} e^{-mC^2/2kT}, \tag{4.4, 1}$$

where the suffix 1 or 2 must be appended to $f^{(0)}$, n, m and C, while an accent, as in $f^{(0)\prime}$, will indicate that the variable C' replaces C. Then, by (4.3, 2),

$$f_1^{(0)\prime} f^{(0)\prime} = f_1^{(0)} f^{(0)}, \quad f_1^{(0)\prime} f_2^{(0)\prime} = f_1^{(0)} f_2^{(0)}, \quad f_2^{(0)\prime} f^{(0)\prime} = f_2^{(0)} f^{(0)}. \tag{4.4, 2}$$

Let F be a function of the velocity defined in the first velocity-domain; we write

$$n_1^2 I_1(F) \equiv \iint f_1^{(0)} f^{(0)} (F_1 + F - F_1' - F') g\alpha_1 \, de' \, dc. \tag{4.4, 3}$$

The quantity $I_2(F)$ is similarly defined when F is defined in the second velocity-domain. Again, if K is any function of $\mathbf{c}_1$ and $\mathbf{c}_2$, and K' is the same function of c_1' and c_2', we write

$$n_1 n_2 I_{12}(K) \equiv \iint f_1^{(0)} f_2^{(0)} (K - K') g\alpha_{12} d\mathbf{e}'\, d\mathbf{c}_2, \tag{4.4, 4}$$

$$n_1 n_2 I_{21}(K) \equiv \iint f_1^{(0)} f_2^{(0)} (K - K') g\alpha_{12} d\mathbf{e}'\, d\mathbf{c}_1. \tag{4.4, 5}$$

Since F and K appear linearly in the above expressions,

$$I(\phi+\psi) = I(\phi) + I(\psi), \quad I(a\phi) = aI(\phi), \tag{4.4, 6}$$

where a is any constant, and I may have any of the above suffixes. The functions I possess a certain similarity to $\partial_e f/\partial t$ in being only partly integrated, so that $I_1(F)$, $I_{12}(K)$ are functions of $\mathbf{c}_1$, while $I_{21}(K)$, $I_2(F)$ are functions of $\mathbf{c}_2$.

Complete integrals related to these functions are defined as follows. First, if F and G are functions defined in the first velocity-domain, we write

$$[F, G]_1 \equiv \int G_1 I_1(F) d\mathbf{c}_1. \tag{4.4, 7}$$

Then, by (4.4, 2) and (3.54, 4),

$$[F, G]_1 = \frac{1}{4n_1^2} \iiint f^{(0)} f_1^{(0)} (F + F_1 - F' - F_1')(G + G_1 - G' - G_1') g\alpha_1 d\mathbf{e}'\, d\mathbf{c}\, d\mathbf{c}_1, \tag{4.4, 8}$$

whence, by symmetry, $\qquad [F, G]_1 = [G, F]_1.$

For functions F, G defined in the second domain, $[F, G]_2$ may be defined in like manner.

Again, when F, H are defined in the first domain, and G, K in the second, let

$$[F_1 + G_2, H_1 + K_2]_{12} \equiv \int F_1 I_{12}(H_1 + K_2) d\mathbf{c}_1 + \int G_2 I_{21}(H_1 + K_2) d\mathbf{c}_2. \tag{4.4, 9}$$

Then, by (4.4, 2) and (3.54, 3),

$$\begin{aligned}[F_1 + G_2, H_1 + K_2]_{12} &= \frac{1}{2n_1 n_2} \iiint f_1^{(0)} f_2^{(0)} (F_1 + G_2 - F_1' - G_2') \\ &\qquad \times (H_1 + K_2 - H_1' - K_2') g\alpha_{12} d\mathbf{e}'\, d\mathbf{c}_1\, d\mathbf{c}_2 \\ &= [H_1 + K_2, F_1 + G_2]_{12}. \end{aligned} \tag{4.4, 10}$$

If n_1, n_2 both refer to the same gas, so that $\alpha_{12} = \alpha_1$, it follows from (4.4, 9) that

$$[F_1, G_1 + G_2]_{12} = [F, G]_1. \tag{4.4, 11}$$

These complete integrals bear a certain resemblance to the expressions for $\Delta_1 \overline{\phi}_1$, $\Delta_2 \overline{\phi}_1$. The following compound of them is likewise analogous to $\Delta(\overline{\phi}_1 + \overline{\phi}_2)$:

$$n^2\{F, G\} \equiv n_1^2 [F, G]_1 + n_1 n_2 [F_1 + F_2, G_1 + G_2]_{12} + n_2^2 [F, G]_2, \tag{4.4, 12}$$

where F and G are each defined in both velocity-domains, and $n = n_1 + n_2$.

On account of the linearity of these complete integrals in the functions F, G, etc., relations typified by the following hold good for each of the functions defined in (4.4, 7, 9, 12).

$$\{F,G\} = \{G,F\}, \quad \{F,G+H\} = \{F,G\}+\{F,H\}, \quad \{F,aG\} = a\{F,G\}, \tag{4.4, 13}$$

where a is any constant.

In $\{F,G\}$ and $[F,G]$ the functions F, G may be either scalars or vectors or tensors, so long as they are of like kind. The incomplete integrals $I(F)$ are of the same nature as F; the integrands in $[F,G]$ and $\{F,G\}$ are understood to contain the scalar products of G and $I(F)$.

If the functions F, G do not involve the number-densities n_1 or n_2 explicitly, the functions I_1 and I_{12}, and the square-bracket expressions $[F,G]_1$, $[F_1+G_2, H_1+K_2]_{12}, \ldots$, are absolutely independent of the number-densities; the curly bracket expressions $\{F,G\}$, on the other hand, in general depend on the concentrations n_1/n, n_2/n, though they also are independent of the total number-density n.

4.41. Inequalities concerning the bracket expressions $[F,G]$, $\{F,G\}$

It follows from (4.4, 8) that

$$[F,F]_1 = \frac{1}{4n_1^2}\iiint f^{(0)}f_1^{(0)}(F+F_1-F'-F_1')^2 g\alpha_1 d\mathbf{e}'\, d\mathbf{c}\, d\mathbf{c}_1 \geqslant 0, \tag{4.41, 1}$$

since the integrand is essentially positive. Similarly, by (4.4, 10)

$$[F_1+G_2, F_1+G_2]_{12} \geqslant 0,$$

and so

$$\{F,F\} \geqslant 0. \tag{4.41, 2}$$

The sign of equality in (4.41, 1) is valid only if

$$F+F_1 = F'+F_1', \tag{4.41, 3}$$

i.e. if F is one of the summational invariants $\psi^{(i)}$ of 3.2, or a linear combination of them. Thus, if F is a scalar quantity, the complete solution of the equation $I_1(F) = 0$, which, by our definitions, has $[F,F]_1 = 0$ as an immediate consequence, is

$$F_1 = \alpha_1^{(1)} + \boldsymbol{\alpha}_1^{(2)}.m_1\mathbf{C}_1 + \alpha_1^{(3)}E_1, \tag{4.41, 4}$$

where $\alpha_1^{(1)}$, $\boldsymbol{\alpha}_1^{(2)}$, $\alpha_1^{(3)}$ are arbitrary magnitudes independent of $\mathbf{c}_1$, which may, however, be functions of $\mathbf{r}$, t. Similarly

$$[F_1+G_2, F_1+G_2]_{12} = 0$$

implies that

$$F_1+G_2 = F_1'+G_2',$$

and

$$\{F,F\} = 0 \tag{4.41, 5}$$

requires that

$$[F,F]_1 = 0, \quad [F_1+F_2, F_1+F_2]_{12} = 0, \quad [F,F]_2 = 0,$$

and hence that

$$F+F_1 = F'+F_1', \quad F_1+F_2 = F_1'+F_2', \quad F_2+F = F_2'+F'. \qquad (4.41, 6)$$

Thus, as in 4.3, if F is a scalar, the solution of (4.41, 5) is

$$F_1 = \alpha_1^{(1)} + \boldsymbol{\alpha}^{(2)}.m_1\mathbf{C}_1 + \alpha^{(3)}E_1, \quad F_2 = \alpha_2^{(1)} + \boldsymbol{\alpha}^{(2)}.m_2\mathbf{C}_2 + \alpha^{(3)}E_2. \qquad (4.41, 7)$$

The results (4.41, 1, 2, 3, 6) apply equally to scalar, vector and tensor functions of the velocity.

5

THE FREE PATH, THE COLLISION-FREQUENCY AND PERSISTENCE OF VELOCITIES

5.1. Smooth rigid elastic spherical molecules

The work of this and the following chapter refers to molecules that are smooth rigid elastic spheres not surrounded by fields of force. In this case the molecules affect each other's motion only at collisions. The path of a rigid molecule between two successive collisions is called a *free path*. For non-rigid molecules an encounter has no definite beginning and end; the concept of a free path then involves difficulties, and will therefore not be employed.

Consider the collision of two molecules of diameters σ_1, σ_2; let

$$\sigma_{12} = \tfrac{1}{2}(\sigma_1 + \sigma_2).$$

The angle ψ made by the relative velocity $\boldsymbol{g}_{21}$ with the direction $\mathbf{k}$ of the line of centres of the molecules at collision can take any value between 0 and $\pi/2$. The deflection χ of the relative velocity in the collision (see Fig. 5, p. 58) is given by

$$\chi = \pi - 2\psi,$$

and so χ takes all values from 0 to π. Also, as in (3.44, 2), the encounter variable b satisfies the equation

$$b = \sigma_{12} \cos \tfrac{1}{2}\chi,$$

whence, by (3.5, 4)

$$\alpha_{12} = \tfrac{1}{4}\sigma_{12}^2. \qquad (5.1, 1)$$

Again, the angles χ, ϵ are polar angles giving the orientation of $\mathbf{e}'$ about an axis parallel to $\boldsymbol{g}_{21}$; thus

$$d\mathbf{e}' = \sin\chi \, d\chi \, d\epsilon. \qquad (5.1, 2)$$

5.2. The frequency of collisions

Consider collisions occurring between pairs of molecules m_1, m_2 in a gas-mixture at rest in a uniform steady state. The number of collisions per unit volume and time such that $\mathbf{e}'$ lies in $d\mathbf{e}'$ and the velocities of the colliding molecules lie in ranges $\boldsymbol{c}_1, d\boldsymbol{c}_1$, and $\boldsymbol{c}_2, d\boldsymbol{c}_2$, is, by (3.5, 5),

$$f_1 f_2 g \alpha_{12} \, d\mathbf{e}' \, d\boldsymbol{c}_1 \, d\boldsymbol{c}_2$$

or, using the value of $\alpha_{12} d\mathbf{e}'$ found in 5.1,

$$\tfrac{1}{4} f_1 f_2 g \sigma_{12}^2 \sin\chi \, d\chi \, d\epsilon \, d\boldsymbol{c}_1 \, d\boldsymbol{c}_2.$$

The total number of collisions occurring per unit volume and time between pairs of molecules m_1, m_2 is obtained by integrating over all values of $\mathbf{e}', \mathbf{c}_1, \mathbf{c}_2$; thus it is N_{12}, where

$$N_{12} \equiv \tfrac{1}{4}\iiiint f_1 f_2 g \sigma_{12}^2 \sin\chi \, d\chi \, d\epsilon \, d\mathbf{c}_1 \, d\mathbf{c}_2.$$

The integrations with respect to χ and ϵ offer no difficulty. The limits of integration are 0 and π for χ, and 0 and 2π for ϵ; hence, integrating and substituting the forms for f_1, f_2 appropriate to the uniform steady state,

$$N_{12} = \frac{\pi n_1 n_2 (m_1 m_2)^{\frac{3}{2}} \sigma_{12}^2}{(2\pi kT)^3} \iint \exp\{-(m_1 c_1^2 + m_2 c_2^2)/2kT\} g \, d\mathbf{c}_1 \, d\mathbf{c}_2. \tag{5.2, 1}$$

To evaluate this expression, the variables of integration are changed from $\mathbf{c}_1, \mathbf{c}_2$ to the variables $\mathbf{G}, \mathbf{g}_{21}$ introduced in 3.41. Then, by (3.41, 8),

$$m_1 c_1^2 + m_2 c_2^2 = m_0(G^2 + M_1 M_2 g^2).$$

Also, by (3.52, 5) the element $d\mathbf{c}_1 d\mathbf{c}_2$ may be replaced by $d\mathbf{G} \, d\mathbf{g}_{21}$, and so

$$N_{12} = \frac{\pi n_1 n_2 (m_1 m_2)^{\frac{3}{2}} \sigma_{12}^2}{(2\pi kT)^3} \iint \exp\{-m_0(G^2 + M_1 M_2 g^2)/2kT\} g \, d\mathbf{G} \, d\mathbf{g}_{21}. \tag{5.2, 2}$$

On integrating over all directions of $\mathbf{g}_{21}$ and $\mathbf{G}$, we get

$$N_{12} = \frac{2 n_1 n_2 (m_1 m_2)^{\frac{3}{2}} \sigma_{12}^2}{(kT)^3} \int_0^\infty \int_0^\infty \exp\{-m_0(G^2 + M_1 M_2 g^2)/2kT\} g^3 G^2 \, dG \, dg.$$

Thus, using (1.4, 2) and (1.4, 3) to effect the integrations with respect to G and g,

$$N_{12} = \frac{\pi^{\frac{1}{2}} n_1 n_2 \sigma_{12}^2}{2} \left(\frac{2 m_1 m_2}{m_0 kT}\right)^{\frac{3}{2}} \int_0^\infty \exp\{-m_0 M_1 M_2 g^2/2kT\} g^3 \, dg \tag{5.2, 3}$$

$$= 2 n_1 n_2 \sigma_{12}^2 \left(\frac{2\pi kT m_0}{m_1 m_2}\right)^{\frac{1}{2}}. \tag{5.2, 4}$$

5.21. The mean free path

Changing the suffix 2 to 1 in (5.2, 4), we get

$$N_{11} = 4 n_1^2 \sigma_1^2 (\pi kT/m_1)^{\frac{1}{2}}. \tag{5.21, 1}$$

The number of collisions between pairs of molecules m_1, per unit volume and time, is $\frac{1}{2}N_{11}$, because N_{11} counts each collision between a pair of molecules A, B (say) twice over, once regarding A as the molecule with the velocity $\mathbf{c}_1$, and once as that with the velocity $\mathbf{c}_2$. On the other hand, the average number of collisions of any one molecule of the first constituent, per unit time, with similar molecules, is N_{11}/n_1, not $N_{11}/2n_1$, since each collision affects two molecules at once.

The average number of collisions undergone by each molecule per unit time is called the *collision frequency*. Thus the frequency for a molecule m_1, for collisions with like molecules, is N_{11}/n_1; for collisions with molecules m_2 it is N_{12}/n_1, etc. The frequency for collisions of all kinds is

$$(N_{11}+N_{12}+\ldots)/n_1,$$

the number of terms in the bracket being equal to the number of constituents in the mixture. The *collision interval*, or mean time between successive collisions, is therefore τ_1, where

$$\tau_1 = n_1/(N_{11}+N_{12}+\ldots). \qquad (5.21, 2)$$

The mean distance l_1 travelled by a molecule m_1 between successive collisions in a given time t is called its *mean free path*. This is found by dividing the total distance $n_1\bar{c}_1 t$ travelled by molecules m_1 in this time by the total number $n_1 t/\tau_1$ of their collisions; thus

$$l_1 = \bar{c}_1\tau_1 = n_1\bar{c}_1/(N_{11}+N_{12}+\ldots).$$

From (4.11, 2) $\bar{c}_1 = \bar{C}_1 = 2(2kT/\pi m_1)^{\frac{1}{2}}$. Using this and the values found above for N_{11}, N_{12}, etc.,

$$l_1 = 1/\pi\{n_1\sigma_1^2\sqrt{2}+n_2\sigma_{12}^2\sqrt{(1+m_1/m_2)}+\ldots\}. \qquad (5.21, 3)$$

In particular, if only one gas is present,

$$l = 1/\pi n\sigma^2\sqrt{2} = 0{\cdot}707/\pi n\sigma^2. \qquad (5.21, 4)$$

Another kind of mean free path was used by Tait,* who defined it as the mean distance moved by a molecule between a given instant and its next collision. The calculation of Tait's mean free path involves the evaluation of an integral by quadrature; for a simple gas its value is

$$0{\cdot}677/\pi n\sigma^2.$$

5.22. Numerical values

For a gas at S.T.P. the number of molecules in a cubic centimetre is approximately $2{\cdot}687\times 10^{19}$. For hydrogen the molecular radius, found by comparison of the experimental values of the coefficient of viscosity with the formula deduced on the assumption that the molecules are rigid elastic spheres, is about $1{\cdot}372\times 10^{-8}$ cm.; the radii found similarly for other molecules are of the same order, though in general somewhat larger. As the mass of a hydrogen molecule is $3{\cdot}347\times 10^{-24}$ g., the number of collisions ($\frac{1}{2}N_{11}$) occurring per second between hydrogen molecules in 1 c.c. at S.T.P. is found to be $2{\cdot}05\times 10^{29}$, and the collision frequency is $1{\cdot}5\times 10^{10}$ sec.$^{-1}$; also the mean free path of a hydrogen molecule is $1{\cdot}112\times 10^{-5}$ cm., and the mean time between two successive collisions is $6{\cdot}6\times 10^{-11}$ sec.

* P. G. Tait, *Trans. R. Soc. Edinb.* **33**, 74 (1886).

The length of the mean free path does not depend on the mass of the molecule, nor on the temperature (unless the diameters of the molecules are supposed to vary with the temperature; see 12.3). Thus, as the molecular diameters for different gases are of the same order, the mean free path in any gas at S.T.P. is of order 10^{-5} cm., and so is several hundred times the diameter of the molecule. This is the justification for the assumption of molecular chaos made in 3.5: at the beginning of a free path which terminates in the collision of two molecules, these are at a relatively large distance apart, so that an appreciable correlation between their velocities is improbable.

The mean free path is inversely proportional to the density; thus in a highly rarefied gas at, say, 0·01 mm. pressure, the free path is of order 1 cm., and so may be comparable with the dimensions of the containing vessel. For a gas at a pressure of, say, 100 atm., on the other hand, the free path is comparable with the dimensions of a molecule; in this case the assumption of molecular chaos may be invalid.

5.3. The distribution of relative velocity, and of energy, in collisions

The total number of collisions occurring in a gas-mixture, per unit volume and time, between molecules m_1 and m_2 is, by (5.2, 3),

$$N_{12} = n_1 n_2 \sigma_{12}^2 (2\pi)^{\frac{1}{2}} \left(\frac{m_1 m_2}{m_0 kT}\right)^{\frac{3}{2}} \int_0^\infty \exp(-m_0 M_1 M_2 g^2/2kT) g^3 dg.$$

The element

$$n_1 n_2 \sigma_{12}^2 (2\pi)^{\frac{1}{2}} \left(\frac{m_1 m_2}{m_0 kT}\right)^{\frac{3}{2}} \exp(-m_0 M_1 M_2 g^2/2kT) g^3 dg \qquad (5.3, 1)$$

of this expression represents the number of such collisions in which the relative velocity g lies in the range dg. Hence the number in which g exceeds an assigned value g_0 is

$$n_1 n_2 \sigma_{12}^2 (2\pi)^{\frac{1}{2}} \left(\frac{m_1 m_2}{m_0 kT}\right)^{\frac{3}{2}} \int_{g_0}^\infty \exp(-m_0 M_1 M_2 g^2/2kT) g^3 dg.$$

If we write $x \equiv g\sqrt{(m_0 M_1 M_2/2kT)}$, $x_0 \equiv g_0\sqrt{(m_0 M_1 M_2/2kT)}$, this integral reduces to

$$4 n_1 n_2 \sigma_{12}^2 \left(\frac{2\pi m_0 kT}{m_1 m_2}\right)^{\frac{1}{2}} \int_{x_0}^\infty e^{-x^2} x^3 dx,$$

i.e. to

$$2 n_1 n_2 \sigma_{12}^2 \left(\frac{2\pi m_0 kT}{m_1 m_2}\right)^{\frac{1}{2}} e^{-x_0^2}(x_0^2 + 1). \qquad (5.3, 2)$$

In the theory of activation of gas reactions it is sometimes assumed that a reaction between two unlike molecules occurs at collision in a certain proportion of those cases in which the kinetic energy relative to their mass-centre exceeds a critical value E_0. As the kinetic energy in question is $\frac{1}{2} m_0 M_1 M_2 g^2$ (cf. (3.41, 8)), the number of encounters per unit volume and time for which it exceeds E_0 is given by (5.3, 2), provided that

$$E_0 = \tfrac{1}{2} m_0 M_1 M_2 g_0^2 = kT x_0^2. \qquad (5.3, 3)$$

We can also use (5.3, 1) to find the mean value of any function $\phi(g)$ of the relative velocity g averaged over all collisions. This will be $\overline{\phi}$, where

$$N_{12}\overline{\phi}(g) = n_1 n_2 \sigma_{12}^2 (2\pi)^{\frac{1}{2}} \left(\frac{m_1 m_2}{m_0 kT}\right)^{\frac{3}{2}} \int_0^\infty \exp(-m_0 M_1 M_2 g^2/2kT) g^3 \phi \, dg. \tag{5.3, 4}$$

In particular, if $\phi(g) = \frac{1}{2}m_0 M_1 M_2 g^2 \equiv E'$, the initial or final kinetic energy of a pair of colliding molecules *relative to axes moving with their mass-centre*,

$$N_{12}\overline{E'} = n_1 n_2 \sigma_{12}^2 (2\pi)^{\frac{1}{2}} \left(\frac{m_1 m_2}{m_0 kT}\right)^{\frac{3}{2}} \int_0^\infty \exp(-m_0 M_1 M_2 g^2/2kT) \tfrac{1}{2} m_0 M_1 M_2 g^5 \, dg$$

$$= 4 n_1 n_2 \sigma_{12}^2 \left(\frac{2\pi m_0}{m_1 m_2}\right)^{\frac{1}{2}} (kT)^{\frac{3}{2}},$$

whence, substituting for N_{12} from (5.2, 4),

$$\overline{E'} = \tfrac{1}{2} m_0 M_1 M_2 \overline{g^2} = 2kT. \tag{5.3, 5}$$

This may be compared with the average kinetic energy ($3kT$) for such pairs of colliding molecules, relative to axes with respect to which the gas is at rest.

5.4. Dependence of collision frequency and mean free path on speed

The number of collisions occurring during dt between pairs of molecules m_1, m_2, such that $\mathbf{c}_1$, $\mathbf{c}_2$, χ, ϵ lie in ranges $d\mathbf{c}_1$, $d\mathbf{c}_2$, $d\chi$, $d\epsilon$ is

$$dt \, . \, \tfrac{1}{4} f_1 f_2 g \sigma_{12}^2 \sin \chi \, d\chi \, d\epsilon \, d\mathbf{c}_1 \, d\mathbf{c}_2.$$

The total number of such collisions during dt such that $\mathbf{c}_1$ lies in $d\mathbf{c}_1$ is found by integrating over all values of $\mathbf{c}_2$, χ, ϵ; it will be proportional to dt, and to the number $f_1(\mathbf{c}_1)\,d\mathbf{c}_1$ of molecules m_1 within the given velocity-range. As is shown below, it is independent of the *direction* of $\mathbf{c}_1$. Hence this number may be denoted by $P_{12}(c_1) f_1 \, d\mathbf{c}_1 \, dt$; the function $P_{12}(c_1)$ thus signifies the average number of collisions per unit time *per molecule* of speed c_1, with molecules of the second kind. It is called the collision frequency for a molecule m_1 having the speed c_1, with molecules m_2.

Dividing by $f_1 \, d\mathbf{c}_1 \, dt$, we obtain

$$P_{12}(c_1) = \tfrac{1}{4} \iiint f_2 g \sigma_{12}^2 \sin \chi \, d\chi \, d\epsilon \, d\mathbf{c}_2.$$

The integration with respect to χ and ϵ is simple, the limits being 0 and π for χ, and 0 and 2π for ϵ; thus

$$P_{12}(c_1) = \pi \sigma_{12}^2 \int f_2 g \, d\mathbf{c}_2. \tag{5.4, 1}$$

Let $\mathbf{c}_2$ be expressed in terms of polar coordinates c_2, θ, φ about $\mathbf{c}_1$ as axis; then $d\mathbf{c}_2 = c_2^2 \sin \theta \, dc_2 \, d\theta \, d\varphi$, and

$$g^2 = c_1^2 + c_2^2 - 2 c_1 c_2 \cos \theta. \tag{5.4, 2}$$

Hence, on integrating with respect to ϕ between the limits 0 and 2π, we get

$$P_{12}(c_1) = 2\pi^2\sigma_{12}^2 \iint f_2 g c_2^2 \sin\theta \, dc_2 \, d\theta. \qquad (5.4, 3)$$

In this we change the variable of integration from θ to g; then, by (5.4, 2)

$$g\,dg = c_1 c_2 \sin\theta \, d\theta,$$

and the limits of g are $c_1 \sim c_2$ and $c_1 + c_2$. Hence

$$\begin{aligned}\int g \sin\theta \, d\theta &= \frac{1}{c_1 c_2}\int g^2 \, dg \\ &= \frac{1}{3c_1 c_2}\{(c_1+c_2)^3 - (c_1 \sim c_2)^3\} \\ &= \frac{2}{3c_1}(3c_1^2 + c_2^2) \quad \text{if} \quad c_1 > c_2,\end{aligned}$$

or

$$= \frac{2}{3c_2}(3c_2^2 + c_1^2) \quad \text{if} \quad c_2 > c_1.$$

Using this result in (5.4, 3), and substituting the value of f_2, we obtain

$$\begin{aligned}P_{12}(c_1) = \tfrac{4}{3}\pi^{\frac{1}{2}} n_2 \sigma_{12}^2 \left(\frac{m_2}{2kT}\right)^{\frac{3}{2}} &\times \left\{\frac{1}{c_1}\int_0^{c_1} \exp(-m_2 c_2^2/2kT)\, c_2^2(c_2^2 + 3c_1^2)\, dc_2 \right. \\ &\left. + \int_{c_1}^{\infty} \exp(-m_2 c_2^2/2kT)\, c_2(3c_2^2 + c_1^2)\, dc_2\right\}.\end{aligned}$$

The second integral is equal to

$$\exp(-m_2 c_1^2/2kT)\left(\frac{m_2 c_1^2}{kT} + \frac{3}{2}\right)\left(\frac{2kT}{m_2}\right)^2;$$

the first, after two integrations by parts, reduces to

$$-\exp(-m_2 c_1^2/2kT)\left(\frac{2kT}{m_2}\right)^2\left(\frac{m_2 c_1^2}{kT} + \frac{3}{4}\right) + \frac{3}{4c_1}\left(\frac{2kT}{m_2}\right)^2\left(\frac{m_2 c_1^2}{kT} + 1\right) \times \int_0^{c_1} \exp(-m_2 c_2^2/2kT)\, dc_2.$$

Thus

$$P_{12}(c_1) = n_2 \sigma_{12}^2 \left(\frac{2\pi kT}{m_2}\right)^{\frac{1}{2}} \left\{\exp(-m_2 c_1^2/2kT) + \frac{1}{c_1}\left(\frac{m_2 c_1^2}{kT} + 1\right) \times \int_0^{c_1} \exp(-m_2 c_2^2/2kT)\, dc_2\right\}.$$

In terms of the error function $\text{Erf}(x)$* defined by the equation

$$\text{Erf}(x) = \int_0^x e^{-y^2}\, dy, \qquad (5.4, 4)$$

* The notation is that of E. T. Whittaker and G. N. Watson, *Modern Analysis*, 4th ed., footnote to p. 341.

we have $$P_{12}(c_1) = n_2\sigma_{12}^2\left(\frac{2\pi kT}{m_2}\right)^{\frac{1}{2}}\{e^{-x^2}+(2x+1/x)\,\mathrm{Erf}\,(x)\},\qquad (5\cdot4,\ 5)$$

where $$x = c_1\surd(m_2/2kT).\qquad (5\cdot4,\ 6)$$

The frequency $P_{11}(c_1)$ of collisions of the molecule with molecules of the same kind is given by a similar expression; and the total collision frequency of the molecule is

$$P_1(c_1) = P_{11}(c_1)+P_{12}(c_1)+\ldots.$$

Hence the collision-interval $\tau_1(c_1)$ for molecules of speed c_1 is given by

$$1/\tau_1(c_1) = P_1(c_1) = P_{11}(c_1)+P_{12}(c_1)+\ldots,\qquad (5\cdot4,\ 7)$$

and the mean length $l_1(c_1)$ of their free paths is equal to $c_1\tau_1$. In particular, when only a single gas is present

$$l_1(c_1) = c_1/P_{11}$$
$$= x^2/\pi^{\frac{1}{2}}n_1\sigma_1^2E(x),\qquad (5\cdot4,\ 8)$$

where $E(x)$ denotes the function

$$E(x) = xe^{-x^2}+(2x^2+1)\,\mathrm{Erf}\,(x),\qquad (5\cdot4,\ 9)$$

and x is now given by $$x = c_1\surd(m_1/2kT).\qquad (5\cdot4,\ 10)$$

Values of $E(x)$ have been tabulated by Tait.* Using his table, a table giving the ratio $l(c)/l$ is obtained which is given below.†

Table 2

$c/\bar{c}$	x^2	$l(c)/l$
0	0	0
0·25	0·080	0·3445
0·5	0·318	0·6411
0·627	0·5	0·7647
0·886	1	0·9611
1·0	1·273	1·0257
1·253	2	1·1340
1·535	3	1·2127
1·772	4	1·2572
2	5·093	1·2878
3	11·459	1·3551
4	—	1·3803
5	—	1·3923
6	—	1·3989
∞	∞	1·4142

* P. G. Tait, *Trans. R. Soc. Edinb.* 33, 74 (1886).

† This table is taken from O. E. Meyer's *Kinetic Theory of Gases* (Longman, 1899), p. 429.

5.41. Probability of a free path of a given length

Let $p(l', c_1)$ denote the probability that a molecule moving with a speed c_1 shall describe a free path at least equal to an assigned value l'. The probability that it should undergo a collision while travelling a further distance dl' is

$$(dl'/c_1)P_1(c_1) = dl'/l_1(c_1).$$

Hence the probability $p(l'+dl', c_1)$ that the molecule describes a free path at least equal to $l'+dl'$ is

$$p(l', c_1)\{1 - dl'/l_1(c_1)\},$$

and so
$$p(l', c_1)\{1 - dl'/l_1(c_1)\} = p(l', c_1) + dl'\frac{\partial p(l', c_1)}{\partial l'},$$

whence
$$\frac{\partial \ln p(l', c_1)}{\partial l'} = -\frac{1}{l_1(c_1)}.$$

Integrating, and using the fact that $p(0, c_1) = 1$, we find that

$$p(l', c_1) = e^{-l'/l_1(c_1)}. \tag{5.41, 1}$$

The probability $p(l')$ that a molecule of *any* speed should describe a free path at least equal to l' is not e^{-l'/l_1}, because of the variation of the collision frequency with speed. Jeans* has found by quadrature that for a simple gas, over the range of l' for which $p(l')$ is appreciable, it never differs by more than about 1% from $e^{-1.04l'/l}$, which is the value for molecules moving with speed $(\sqrt{\pi})\bar{c}/2$.

It is clear from (5.41, 1) only a very small proportion of molecules have free paths many times as long as the mean free path.

5.5. The persistence of velocities after collision

After a collision with another molecule the velocity of a given molecule will, on the average, still retain a component in the direction of its original motion. This phenomenon is known as the persistence of velocities after collision. As a consequence, the average distance traversed by a molecule in the direction of its velocity at a given instant, before (on the average) it loses its component motion in that direction, is somewhat greater than its (Tait) mean free path.

The collision frequency for a molecule m_1, having the speed c_1, in collisions with molecules m_2, is $P_{12}(c_1)$; for the particular set of such collisions in which the velocities of the latter molecules lie in the range $\boldsymbol{c}_2$, $d\boldsymbol{c}_2$, and χ, ϵ lie in ranges $d\chi$, $d\epsilon$, the collision frequency is

$$\tfrac{1}{4}f_2 g\sigma_{12}^2 \sin\chi\, d\chi\, d\epsilon\, d\boldsymbol{c}_2.$$

Let $\bar{\boldsymbol{c}}_1'(\boldsymbol{c}_1)$ denote the mean velocity, after collision, of molecules m_1 which before collision have the velocity $\boldsymbol{c}_1$, the average being taken over collisions

* J. H. Jeans, *Dynamical Theory of Gases* (4th ed.), p. 258 (1925).

with molecules m_2 for all possible values of $\mathbf{c}_2$, χ, ϵ; then

$$\bar{c}_1'(c_1) = \frac{1}{4P_{12}}\iiint f_2 \mathbf{c}_1' g\sigma_{12}^2 \sin\chi \, d\chi \, d\epsilon \, d\mathbf{c}_2$$

or, using (3.51, 4),

$$\bar{c}_1'(c_1) = \frac{1}{4P_{12}}\iiint f_2\{\mathbf{c}_1 + M_2(\mathbf{g}_{21} - \mathbf{g}_{21}')\} g\sigma_{12}^2 \sin\chi \, d\chi \, d\epsilon \, d\mathbf{c}_2.$$

The components of $\mathbf{g}_{21}'$ parallel and perpendicular to $\mathbf{g}_{21}$ are $g\cos\chi$, $g\sin\chi$. As ϵ varies, the second component rotates about $\mathbf{g}_{21}$; hence its integral with respect to ϵ is zero, so that

$$\bar{c}_1'(c_1) = \frac{\pi}{2P_{12}}\iint f_2\{\mathbf{c}_1 + M_2 \mathbf{g}_{21}(1 - \cos\chi)\} g\sigma_{12}^2 \sin\chi \, d\chi \, d\mathbf{c}_2.$$

The integration with respect to χ between the limits 0 and π is elementary, and yields the result

$$\bar{c}_1'(c_1) = \frac{\pi}{P_{12}}\int f_2(\mathbf{c}_1 + M_2 \mathbf{g}_{21}) g\sigma_{12}^2 \, d\mathbf{c}_2 = \frac{\pi}{P_{12}}\int f_2(M_1\mathbf{c}_1 + M_2\mathbf{c}_2) g\sigma_{12}^2 \, d\mathbf{c}_2,$$

or, using (5.4, 1),

$$\bar{c}_1'(c_1) = M_1\mathbf{c}_1 + \frac{M_2\pi\sigma_{12}^2}{P_{12}}\int f_2 \mathbf{c}_2 g \, d\mathbf{c}_2.$$

The integral may be evaluated by the method used in 5.4, expressing $\mathbf{c}_2$ in terms of polar coordinates c_2, θ, ϕ about $\mathbf{c}_1$ as axis. On integration over all values of ϕ, the component of $\mathbf{c}_2$ perpendicular to $\mathbf{c}_1$ vanishes; as the component in the direction of $\mathbf{c}_1$ is $c_2\cos\theta$, the mean value of $\mathbf{c}_1'$ is $\varpi_{12}(c_1)\,\mathbf{c}_1$, where

$$\varpi_{12}(c_1) = M_1 + \frac{2M_2\pi^2\sigma_{12}^2}{P_{12}}\iint f_2 \frac{c_2}{c_1}\cos\theta \, g c_2^2 \sin\theta \, dc_2 \, d\theta.$$

The quantity $\varpi_{12}(c_1)$, which is the ratio of the mean value of the velocity of a molecule after collision to the velocity before collision when the latter velocity is of magnitude c_1, may be termed the *persistence-ratio* for molecules of speed c_1.

On changing the variable of integration from θ to g, as in 5.4, it follows that

$$\int g\cos\theta\sin\theta\,d\theta = \int_{c_1\sim c_2}^{c_1+c_2} g\frac{c_1^2+c_2^2-g^2}{2c_1c_2}\frac{g\,dg}{c_1c_2}$$

$$= \frac{2c_2}{15c_1^2}(c_2^2 - 5c_1^2) \quad \text{if} \quad c_1 > c_2$$

$$= \frac{2c_1}{15c_2^2}(c_1^2 - 5c_2^2) \quad \text{if} \quad c_2 > c_1.$$

The expression for $\varpi_{12}(c_1)$ accordingly reduces to

$$\varpi_{12}(c_1) = M_1 + \frac{4M_2\pi^2\sigma_{12}^2}{15P_{12}}\left\{\int_0^{c_1} f_2\frac{c_2^4}{c_1^3}(c_2^2-5c_1^2)\,dc_2 + \int_{c_1}^{\infty} f_2c_2(c_1^2-5c_2^2)\,dc_2\right\}. \tag{5.5, 1}$$

The second of the integrals in the bracket can be evaluated in finite terms, and the first integral can be expressed in terms of the error function Erf (x) of (5.4, 4); on so doing, the value of $\varpi_{12}(c_1)$ is found to be

$$\varpi_{12}(c_1) = M_1 + \frac{n_2M_2\sigma_{12}^2}{2P_{12}}\left(\frac{2\pi kT}{m_2}\right)^{\frac{1}{2}}\left\{-\frac{1}{x^2}e^{-x^2}+\frac{1-2x^2}{x^3}\operatorname{Erf}(x)\right\}, \tag{5.5, 2}$$

where x is given by (5.4, 6).

The value of $\varpi_{12}(c_1)$ given by (5.5, 2) varies between $M_1 - M_2/3$ and M_1 as x varies between 0 and ∞, i.e. as c_1 ranges from 0 to ∞. For collisions between like particles, for which $M_1 = M_2 = \frac{1}{2}$, $\varpi_{12}(c_1)$ lies between $\frac{1}{3}$ and $\frac{1}{2}$.

5.51. The mean persistence-ratio

An expression for ϖ_{12}, the mean value of $\varpi_{12}(c_1)$ averaged over all values of c_1, may be derived as follows. The number of molecules m_1 per unit volume whose speeds lie between c_1 and c_1+dc_1 is $4\pi f_1c_1^2dc_1$; the number of collisions per unit time with molecules of the second gas in which these molecules participate is $4\pi f_1c_1^2dc_1.P_{12}$. Hence, by (5.5, 1), as the total number of collisions between molecules m_1 and m_2 per unit time is N_{12}, the mean value of $\varpi_{12}(c_1)$ averaged over all possible collisions is given by

$$N_{12}\varpi_{12} = M_1\int_0^{\infty} 4\pi f_1c_1^2P_{12}dc_1 + \frac{16M_2\pi^3\sigma_{12}^2}{15}$$
$$\times\left\{\int_0^{\infty} f_1c_1^2\left[\int_0^{c_1} f_2\frac{c_2^4}{c_1^3}(c_2^2-5c_1^2)\,dc_2+\int_{c_1}^{\infty} f_2c_2(c_1^2-5c_2^2)\,dc_2\right]dc_1\right\},$$

or, since
$$N_{12} = \int_0^{\infty} 4\pi f_1c_1^2P_{12}dc_1,$$

by
$$\varpi_{12} = M_1 + \frac{16M_2\pi^3\sigma_{12}^2}{15N_{12}}$$
$$\times\left\{\int_0^{\infty} f_1c_1^2\left[\int_0^{c_1} f_2\frac{c_2^4}{c_1^3}(c_2^2-5c_1^2)\,dc_2+\int_{c_1}^{\infty} f_2c_2(c_1^2-5c_2^2)\,dc_2\right]dc_1\right\}.$$

On inserting the known expressions for f_1, f_2, and N_{12} (cf. (5.2, 4)), this becomes

$$\varpi_{12} = M_1 + \frac{8m_0^{\frac{7}{2}}M_1^2M_2^3}{15\pi^{\frac{1}{2}}(2kT)^{\frac{7}{2}}}\left\{\int_0^{\infty}\int_0^{c_1}\exp\{-(m_1c_1^2+m_2c_2^2)/2kT\}\frac{c_2^4}{c_1}(c_2^2-5c_1^2)\,dc_2dc_1\right.$$
$$\left.+\int_0^{\infty}\int_{c_1}^{\infty}\exp\{-(m_1c_1^2+m_2c_2^2)/2kT\}c_1^2c_2(c_1^2-5c_2^2)\,dc_2dc_1\right\}.$$

In this write $c_2 = \theta c_1$; then

$$\varpi_{12} = M_1 + \frac{8m_0^{\frac{7}{2}} M_1^2 M_2^3}{15\pi^{\frac{1}{2}}(2kT)^{\frac{7}{2}}} \left\{ \int_0^1 I(\theta)\,\theta^4(\theta^2 - 5)\,d\theta + \int_1^\infty I(\theta)\,\theta(1 - 5\theta^2)\,d\theta \right\},$$

where

$$I(\theta) = \int_0^\infty \exp\{-c_1^2(m_1 + m_2\theta^2)/2kT\}\,c_1^6\,dc_1$$

$$= \frac{15\sqrt{\pi}}{16}\left(\frac{2kT}{m_1 + m_2\theta^2}\right)^{\frac{7}{2}}.$$

Thus $$\varpi_{12} = M_1 + \tfrac{1}{2}M_1^2 M_2^3 \left\{ \int_0^1 \frac{\theta^4(\theta^2 - 5)\,d\theta}{(M_1 + M_2\theta^2)^{\frac{7}{2}}} + \int_1^\infty \frac{\theta(1 - 5\theta^2)\,d\theta}{(M_1 + M_2\theta^2)^{\frac{7}{2}}} \right\},$$

whence, on evaluating the integrals,

$$\varpi_{12} = \tfrac{1}{2}M_1 + \tfrac{1}{2}M_1^2 M_2^{-\frac{1}{2}} \ln\left[(M_2^{\frac{1}{2}} + 1)/M_1^{\frac{1}{2}}\right]. \qquad (5.51, 1)$$

The value of ϖ_{12} given by this equation increases from zero to unity as m_1/m_2 increases from zero to infinity; its value for the collisions of like molecules, when $M_1 = M_2 = \frac{1}{2}$, is $\frac{1}{4} + \frac{1}{4\sqrt{2}}\ln(1 + \sqrt{2})$, or 0·406. For the encounter of heavy molecules with light molecules, the persistence ratio for the heavy molecules is nearly unity, and that for the light molecules nearly zero: this implies that, as might be expected, the heavy molecule continues its path nearly undisturbed, while the light molecule bounces off in a direction unrelated to that of its previous motion.

6

ELEMENTARY THEORIES OF THE TRANSPORT PHENOMENA

6.1. The transport phenomena

The phenomena of viscosity, thermal conduction, and diffusion in non-uniform gases represent tendencies towards uniformity of mass-velocity, temperature and composition. The kinetic theory attributes these tendencies to the motion of molecules from point to point. This tends to equalize conditions at the two ends of each free path, by transporting to the further end an average amount of momentum and energy that is characteristic of the starting-point. Hence we may speak of the phenomena in question as the transport phenomena, or the free-path phenomena.

It was shown in Chapter 4 that a gas in a uniform steady state has a Maxwellian velocity-distribution function. When the gas departs slightly from a uniform steady state, Maxwell's function gives a first approximation to the actual distribution of velocities. Thus the results of Chapter 5 are still approximately true; they may therefore be used in obtaining approximate expressions for the coefficients of viscosity, thermal conduction and diffusion in a gas composed of rigid elastic spheres of diameter σ.

6.2. Viscosity

Consider a simple gas, uniform in temperature and density, moving parallel to Ox with a mass-velocity u_0 which is a function of z alone. Thus the gas is in laminar motion parallel to $z = 0$, and $w_0 = 0$, $w = W$.

Consider the rate of transport of x-momentum across unit area of the plane $z = 0$. The number of molecules crossing this area in unit time from the negative side (that on which z is negative) to the positive side is

$$\int_{+} wf\,d\mathbf{c} = \int_{+} Wf\,d\mathbf{C},$$

the integration extending over all values of $\mathbf{C}$ for which W is positive. Similarly, the number crossing from the positive side to the negative is

$$\int_{-} (-w)f\,d\mathbf{c} = \int_{-} (-W)f\,d\mathbf{C},$$

the integration extending over values of $\mathbf{C}$ for which W is negative. Since the gas has no mass-velocity parallel to Oz, these two numbers are equal. Their value, to a first approximation, is $\frac{1}{4}n\bar{C}$, the same as for a gas in a uniform steady state (cf. (4.11, 5, 6)).

The mean x-velocity of molecules crossing the area from the negative side to the positive is not the value of u_0 appropriate to the plane $z = 0$. It corresponds to the mean layer in which their free paths began, or, if allowance is made for the persistence of velocities after collision, to a somewhat more distant layer. Thus their mean x-velocity is that of a layer $z = -\mathrm{u}l$, where l is the mean free path of the molecules, and u a numerical factor of order unity. Their total x-momentum is

$$\tfrac{1}{4}n\bar{C}.m(u_0)_{z=-\mathrm{u}l},$$

or, as l is usually very small compared with the scale of variation of the mass-properties of the gas,

$$\tfrac{1}{4}\rho\bar{C}\left(u_0 - \mathrm{u}l\frac{\partial u_0}{\partial z}\right);$$

the values of u_0 and $\partial u_0/\partial z$ here refer to the plane $z = 0$.

Similarly, the mean x-velocity of molecules crossing the area per unit time from the positive side to the negative is that corresponding to $z = +\mathrm{u}l$, and their total x-momentum is

$$\tfrac{1}{4}\rho\bar{C}(u_0)_{z=\mathrm{u}l} = \tfrac{1}{4}\rho\bar{C}\left(u_0 + \mathrm{u}l\frac{\partial u_0}{\partial z}\right),$$

u_0 and $\partial u_0/\partial z$ in the last expression again referring to $z = 0$. Thus the net rate of transport of x-momentum across unit area of the plane $z = 0$ from the negative side to the positive is

$$\tfrac{1}{4}\rho\bar{C}\left(u_0 - \mathrm{u}l\frac{\partial u_0}{\partial z}\right) - \tfrac{1}{4}\rho\bar{C}\left(u_0 + \mathrm{u}l\frac{\partial u_0}{\partial z}\right) = -\tfrac{1}{2}\mathrm{u}\rho\bar{C}l\frac{\partial u_0}{\partial z}.$$

This momentum transport is equivalent to a force of this amount per unit area, exerted parallel to Ox, by the gas on the negative side of $z = 0$, upon the gas on the positive side. According to the usual definition of the coefficient of viscosity μ, this force is $-\mu\,\partial u_0/\partial z$. Hence

$$\mu = \tfrac{1}{2}\mathrm{u}\rho\bar{C}l = \frac{\mathrm{u}}{\pi^{\frac{3}{2}}}\frac{\sqrt{(kmT)}}{\sigma^2}, \qquad (6.2, 1)$$

by (4.11, 2) and (5.21, 4). The value of u is found (cf. (12.1, 6)), by the exact methods described later in the book, to be $0{\cdot}1792\pi^{\frac{3}{2}}$ or $0{\cdot}998$.

6.21. Viscosity at low pressures

In a gas at a sufficiently low pressure the mean free path is comparable with the dimensions of the containing vessel; this produces an apparent diminution in the viscosity of the gas. Consider as a typical example of this phenomenon the motion of gas between two infinite parallel walls $z = 0$ and $z = d$, the first of which is at rest, while the other moves parallel to Ox with speed q, the motion of the gas being supposed due solely to that of the walls, and not to imposed pressure-gradients. Let q_1, q_2 denote the mean x-velocities of molecules just before striking and just after leaving the second wall; then the mean x-velocity of the gas at this wall is $\frac{1}{2}(q_1+q_2)$.

Experiment indicates that some of the molecules striking the moving wall enter the material, and later leave it with the temperature of the wall, and a mean x-velocity equal to its velocity q. The remainder are reflected elastically. The proportion θ of those that enter depends on the nature of the wall and the gas, and on the condition of the surface of the wall. The mean x-velocity of molecules leaving the wall is a weighted mean of q_1 (for the elastically reflected molecules) and q (for those that enter the wall) in the proportion $1-\theta$ to θ; that is,

$$q_2 = (1-\theta)\,q_1+\theta q. \qquad (6.21,\ 1)$$

Now, as in 6.2, the value of q_1 is equal to the mass-velocity of the gas at a distance ul from the wall, which differs from that at the wall by u$l(\partial u_0/\partial z)$, where $\partial u_0/\partial z$ denotes the (constant) gradient of the mass-velocity of the gas. Thus

$$q_1 = \tfrac{1}{2}(q_1+q_2)-\mathrm{u}l\frac{\partial u_0}{\partial z}. \qquad (6.21,\ 2)$$

On combining (6.21, 1) and (6.21, 2) we find

$$\tfrac{1}{2}(q_1+q_2) = q-\frac{2-\theta}{\theta}\,\mathrm{u}l\frac{\partial u_0}{\partial z},$$

so that the gas near the wall is slipping along the wall with a speed

$$\frac{2-\theta}{\theta}\,\mathrm{u}l\frac{\partial u_0}{\partial z}. \qquad (6.21,\ 3)$$

There is a similar speed of slip at the other wall; hence the difference between the mean velocities of the gas near the two walls is

$$q-2.\frac{2-\theta}{\theta}\,\mathrm{u}l\frac{\partial u_0}{\partial z},$$

and since this is equal to $(\partial u_0/\partial z)\,d$,

$$\frac{\partial u_0}{\partial z} = q\bigg/\left(d+2\mathrm{u}l\frac{2-\theta}{\theta}\right).$$

The viscous stress transmitted by the gas is $\mu\,\partial u_0/\partial z$ per unit area. In the absence of slipping at the walls this would be equal to $\mu q/d$; if it be denoted by $\mu' q/d$,

$$\mu' = \mu d\bigg/\left(d+2\mathrm{u}l\frac{2-\theta}{\theta}\right). \qquad (6.21,\ 4)$$

The effect of slipping at the walls is that the apparent viscosity, μ', of the gas, is smaller than the true viscosity μ. The reduction is very small for ordinary pressures, but for pressures so small that l becomes comparable with d it is very pronounced.

The above discussion fails when l exceeds d, since molecules pass direct from one wall to the other, and the transport of momentum ceases to be a free-path phenomenon. In this case experiment shows that μ' decreases to zero with the pressure.*

6.3. Thermal conduction

The simple theory of thermal conduction is similar to that of viscosity. We consider now a simple gas at rest, whose temperature T is a function of z; it is required to find the rate of flow of heat across unit area of the plane $z = 0$.

Let E denote the total thermal energy of a molecule; then $\bar{E}$, the mean thermal energy of molecules at a given point, is a function of T, and so of z; also, as in 2.43, the specific heat c_v of the gas is given by the relation

$$c_v = \frac{d}{dT}\left(\frac{\bar{E}}{m}\right). \qquad (6.3, 1)$$

The number of molecules crossing unit area of the plane $z = 0$ from the negative side to the positive in unit time is $\frac{1}{4}n\bar{C}$, as in 6.2. Each of these carries with it thermal energy equal, on the average, to the value of $\bar{E}$, not at the plane $z = 0$, but at $z = -\mathrm{u}'l$, where l is the length of the free path, and u' is another numerical constant which, like u (cf. 6.2), is of order unity. Thus the total thermal energy which they carry across $z = 0$ is

$$\tfrac{1}{4}n\bar{C}(\bar{E})_{z=-\mathrm{u}'l} = \tfrac{1}{4}n\bar{C}\left(\bar{E} - \mathrm{u}'l\frac{\partial \bar{E}}{\partial z}\right),$$

where, in the last expression, $\bar{E}$ and $\partial\bar{E}/\partial z$ refer to $z = 0$. Similarly, the total thermal energy carried by molecules that cross unit area of the plane $z = 0$ from the positive side to the negative is

$$\tfrac{1}{4}n\bar{C}(\bar{E})_{z=\mathrm{u}'l} = \tfrac{1}{4}n\bar{C}\left(\bar{E} + \mathrm{u}'l\frac{\partial \bar{E}}{\partial z}\right),$$

where again $\bar{E}$ and $\partial\bar{E}/\partial z$ refer to $z = 0$. Thus the net rate of flow of thermal energy across unit area of the plane $z = 0$ from the negative side to the positive is

$$\tfrac{1}{4}n\bar{C}\left(\bar{E} - \mathrm{u}'l\frac{\partial \bar{E}}{\partial z}\right) - \tfrac{1}{4}n\bar{C}\left(\bar{E} + \mathrm{u}'l\frac{\partial \bar{E}}{\partial z}\right)$$
$$= -\tfrac{1}{2}n\bar{C}\mathrm{u}'l\frac{\partial \bar{E}}{\partial z}$$
$$= -\tfrac{1}{2}mn\bar{C}\mathrm{u}'lc_v\frac{\partial T}{\partial z},$$

by (6.3, 1). In the theory of the conduction of heat this rate of flow is written $-\lambda\,\partial T/\partial z$, where λ is the coefficient of thermal conduction of the material. Accordingly

$$\lambda = \tfrac{1}{2}\rho\mathrm{u}'l\bar{C}c_v, \qquad (6.3, 2)$$

* W. Crookes, *Phil. Trans.* 172, 387 (1882). See also M. H. C. Knudsen, *Kinetic Theory of Gases* (Methuen, 1946).

whence, by (6.2, 1), $$\lambda = \mathrm{f}\mu c_v, \qquad (6.3, 3)$$

where f is a new numerical constant, equal to u′/u.

The constant u′ of this section is not equal to the constant u of 6.2, because of the correlation between the energies of molecules and their velocities; molecules possessing most energy are in general the most rapid, and therefore possess the longest free paths. In consequence, u′ is in general greater than u, and f is greater than unity. For the transport of internal energy, however, which is but slightly correlated with the molecular velocity, an equation similar to (6.3, 3), with f roughly equal to unity, may be expected to hold; the symbols λ and c_v now refer to the coefficient of conduction, and the specific heat, of the internal energy. In general, the larger the ratio of internal to translatory energy, the smaller is f.

It is immaterial, in the formulae of this section, whether thermal energy is measured in thermal or mechanical units; the effect of a change from one set of units to the other is that E, c_v, and λ are all multiplied by the same factor.

6.31. Temperature-drop at a wall

Just as the mean velocity of gas near a moving wall differs from that of the wall, so there is a difference between the temperature of a hot body and the gas that conducts heat away from it. This temperature-difference is, by analogy with (6.21, 3), equal to

$$\frac{2-\theta}{\theta}\,\mathrm{u}' l \frac{\partial T}{\partial z}, \qquad (6.31, 1)$$

where $\partial T/\partial z$ denotes the temperature gradient near the body.

One method of measuring the thermal conductivity of a gas is to determine the loss of heat through the gas from a hot wire of small diameter. In such experiments the temperature-drop at the surface is usually very important.

6.4. Diffusion

Consider a binary gas-mixture, uniform in temperature and pressure, whose composition varies with z. If n_1, n_2 are the number-densities of the two gases, then, as the pressure is uniform,

$$0 = \frac{\partial p}{\partial z} = kT\frac{\partial(n_1+n_2)}{\partial z},$$

whence $$\frac{\partial n_1}{\partial z} = -\frac{\partial n_2}{\partial z}. \qquad (6.4, 1)$$

We assume that the pressure remains uniform as diffusion proceeds in the mixture; this is very nearly true.

The number of molecules m_1 crossing unit area of the plane $z = 0$ from the negative side to the positive in unit time is $\frac{1}{4}n_1\bar{C}_1$, as before: in this expression

n_1 no longer refers to $z = 0$, as the gas is non-uniform, but to $z = -\mathrm{u}_1 l_1$, say. Since collisions between pairs of molecules m_1 do not directly affect the mean diffusion velocity $\overline{w}_1$, l_1 is taken to be the mean free path of molecules m_1 between successive collisions with molecules m_2; u_1 is a number of order unity. The number of molecules in question is thus

$$\tfrac{1}{4}\bar{C}_1(n_1)_{z=-\mathrm{u}_1 l_1} = \tfrac{1}{4}\bar{C}_1\left(n_1 - \mathrm{u}_1 l_1 \frac{\partial n_1}{\partial z}\right).$$

The number of molecules m_1 crossing unit area of $z = 0$ in the opposite direction per unit time is similarly

$$\tfrac{1}{4}\bar{C}_1(n_1)_{z=\mathrm{u}_1 l_1} = \tfrac{1}{4}\bar{C}_1\left(n_1 + \mathrm{u}_1 l_1 \frac{\partial n_1}{\partial z}\right).$$

Hence the net number-flow of molecules m_1 per unit area and time across $z = 0$ from the negative side to the positive is

$$\tfrac{1}{4}\bar{C}_1\left(n_1 - \mathrm{u}_1 l_1 \frac{\partial n_1}{\partial z}\right) - \tfrac{1}{4}\bar{C}_1\left(n_1 + \mathrm{u}_1 l_1 \frac{\partial n_1}{\partial z}\right) = -\tfrac{1}{2}\mathrm{u}_1 l_1 \bar{C}_1 \frac{\partial n_1}{\partial z}.$$

Similarly the net number-flow of molecules m_2 per unit area and time across $z = 0$ in the opposite direction is, by (6.4, 1),

$$-\tfrac{1}{2}\mathrm{u}_2 l_2 \bar{C}_2 \frac{\partial n_1}{\partial z},$$

where l_2 is the mean free path of molecules m_2 between successive collisions with molecules m_1, and u_2 is another number of order unity. This does not balance the number-flow of molecules m_1; however, to ensure that the pressure remains uniform there must be no resultant flow of molecules. It is therefore necessary to assume that the first trace of a pressure difference in the gas immediately produces a mass flow of the gas, with a velocity

$$\frac{1}{n_1 + n_2}(\tfrac{1}{2}\mathrm{u}_1 l_1 \bar{C}_1 - \tfrac{1}{2}\mathrm{u}_2 l_2 \bar{C}_2)\frac{\partial n_1}{\partial z}$$

just sufficient to counter the resultant number-flow. When this mass-flow is taken into account, the number-flow of molecules m_1 becomes

$$-\frac{\tfrac{1}{2}n_2 \mathrm{u}_1 l_1 \bar{C}_1 + \tfrac{1}{2}n_1 \mathrm{u}_2 l_2 \bar{C}_2}{n_1 + n_2}\frac{\partial n_1}{\partial z};$$

the number-flow of molecules m_2 is equal in magnitude, but opposite in direction. We write the two as $-D_{12}\,\partial n_1/\partial z$ and $-D_{12}\,\partial n_2/\partial z$, where D_{12} is called the coefficient of mutual diffusion of the two constituents in the mixture. Then

$$D_{12} = \frac{\tfrac{1}{2}(n_2 \mathrm{u}_1 l_1 \bar{C}_1 + n_1 \mathrm{u}_2 l_2 \bar{C}_2)}{n_1 + n_2}. \tag{6.4, 2}$$

In (6.4, 2), l_1 and l_2 are respectively inversely proportional to n_2 and n_1; thus (6.4, 2) implies that D_{12} is inversely proportional to the total pressure, whereas the coefficients of viscosity and thermal conduction are independent of the pressure. This result agrees with the exact theory, which shows, however, that D_{12} varies somewhat with the proportions of the two gases in the mixture (cf. 14.3). Thus u_1, u_2 must be supposed also to depend somewhat on the proportions.

When the two gases in the mixture are identical, the process under consideration is the diffusion of certain selected molecules of the gas relative to the rest. In this case D_{12} is replaced by D_{11}, which may be termed the coefficient of self-diffusion of the gas; since now $n_2 l_1 = n_1 l_2 = (n_1+n_2) l$, where l is the free path when all the collisions are taken into account, we may write

$$D_{11} = \tfrac{1}{2}u_{11} l\bar{C}. \tag{6.4, 3}$$

Hence, by (6.2, 1),

$$D_{11} = u'_{11}\mu/\rho, \tag{6.4, 4}$$

where u'_{11} denotes a new numerical constant of order unity.

In a gas-mixture, diffusion can also arise as a consequence of a temperature gradient. No really satisfactory simple theory of this 'thermal diffusion' can be given, employing only the simple free-path ideas of this chapter. The reason is that thermal diffusion is an interaction phenomenon, arising from the processing by collisions of the heat-flow which is the immediate consequence of the temperature gradient. Similar remarks apply to the inverse 'diffusion thermo-effect', in which diffusion due to inhomogeneity of composition is found to be accompanied by a heat flux in a gas-mixture initially at uniform temperature.

6.5. Defects of free-path theories

The free-path formulae derived above depend on the undetermined numbers u, u', u_1, u_2, f, u_{11}, u'_{11}. Elementary arguments are often used to derive rough values for these numbers. Implicit reliance on the formulae so obtained may, however, lead to serious difficulties, especially in the theory of diffusion.*

Better estimates of the numbers can be obtained by taking successively into account such factors as the difference between the lengths of the free paths of different molecules, the persistence of velocities after collision, and so on. This is the plan adopted by Meyer and others.† Even in its most refined form, however, this mode of attack leads to somewhat inaccurate results, and the precise magnitude of the error has to be found by other methods. Also the free-path methods apply only to rigid spherical molecules. Similar objections apply to other approximate theories, such as those described later in this chapter. To obtain reliable results, it is better to proceed by evaluating the velocity-distribution function, as in the later chapters of this book.

* See S. Chapman, 'On approximate theories of diffusion phenomena', *Phil. Mag.* **5**, 630 (1928).

† Cf. O. E. Meyer, *Kinetic Theory of Gases*, or J. H. Jeans, *Dynamical Theory of Gases*, chapters 11–13.

6.6. Collision interval theories

A rather different kind of approximate theory is based on the collision interval τ instead of the free path.* Consider a simple gas under no forces. As was shown in Chapter 4, collisions cause the velocity-distribution function f to approach the Maxwellian value $f^{(0)}$ corresponding to the local values of n, $\mathbf{c}_0$ and T. The collision interval theory assumes, as its basic approximation, that during a time dt a fraction dt/τ of the molecules in a given small volume undergo collisions, which alter their velocity-distribution function from f to $f^{(0)}$. This is equivalent to assuming that the rate of change $\partial_e f/\partial t$ of f due to collisions is $-(f-f^{(0)})/\tau$, so that the Boltzmann equation becomes

$$\frac{\partial f}{\partial t}+\mathbf{c}.\frac{\partial f}{\partial \mathbf{r}} = -\frac{f-f^{(0)}}{\tau}. \tag{6.6, 1}$$

Because of the smallness of τ this equation implies that, in a gas whose state is not varying rapidly with time, $f-f^{(0)}$ must be small. Replacing f by $f^{(0)}$ on the left of (6.6, 1) we find†

$$f = f^{(0)} - \tau\frac{\partial f^{(0)}}{\partial t} - \mathbf{c}\tau.\frac{\partial f^{(0)}}{\partial \mathbf{r}}. \tag{6.6, 2}$$

For a gas of uniform density and temperature, streaming parallel to Ox with a mass-velocity u_0 which is a function of z alone, equation (6.6, 2) becomes

$$f = f^{(0)} - w\tau\frac{\partial u_0}{\partial z}\frac{\partial f^{(0)}}{\partial u_0}. \tag{6.6, 3}$$

The viscous stress in the x-direction across a plane z = const. is

$$p_{xz} = \int m(u-u_0)\,wf\,d\mathbf{c}. \tag{6.6, 4}$$

Substitute for f from (6.6, 3); then, since $f^{(0)}$ is an even function of $u-u_0$, (6.6, 4) reduces to $p_{xz} = -\mu\,\partial u_0/\partial z$, where

$$\begin{aligned}\mu &= \tau\int m(u-u_0)\,w^2\frac{\partial f^{(0)}}{\partial u_0}\,d\mathbf{c}\\ &= \frac{\partial}{\partial u_0}\left\{\tau\int m(u-u_0)\,w^2 f^{(0)}\,d\mathbf{c}\right\}+\tau\int mw^2 f^{(0)}\,d\mathbf{c}.\end{aligned}$$

The integral in the brackets vanishes, the integrand being an odd function of

* Such theories, used in the theory of ionized gases for a number of years, have attained wide vogue through the work of P. L. Bhatragar, E. P. Gross and M. Krook, *Phys. Rev.* **94**, 511 (1954).

† The exact solution of (6.6, 1) is

$$f = \int_0^\infty e^{-t'/\tau} f^{(0)}(\mathbf{c}, \mathbf{r}-\mathbf{c}t', t-t')\,\tau^{-1}\,dt';$$

this can also be derived directly by classifying the molecules in the range $\mathbf{c}$, $d\mathbf{c}$ that reach the volume $\mathbf{r}$, $d\mathbf{r}$ at time t according to the times $t-t'$ at which they began the free paths that bring them into the volume. Equation (6.6, 2) follows from this integral on expanding $f^{(0)}$ in powers of t' and neglecting t'^2 and higher powers.

$u-u_0$; the second integral is equal to p, by (1.42, 1) and 2.32. Hence the viscosity μ is given by

$$\mu = p\tau. \tag{6.6, 5}$$

Similarly for a gas at rest, uniform in pressure but with a temperature which is a function of z, equation (6.6, 2) becomes

$$f = f^{(0)} - w\tau \frac{\partial T}{\partial z} \frac{\partial f^{(0)}}{\partial T}. \tag{6.6, 6}$$

If the molecules have no internal energy, the corresponding heat flow $-\lambda\, \partial T/\partial z$ in the z-direction is $\int \frac{1}{2}mc^2 wf\, dc$. On evaluating the integral by (1.42, 1) and (1.4, 2) the thermal conductivity λ is found to be

$$\lambda = \tau \frac{\partial}{\partial T} \int \tfrac{1}{2}mc^2 w^2 f^{(0)}\, dc = \tau \frac{\partial}{\partial T}\left(\frac{5}{2}\frac{pkT}{m}\right)$$

$$= \tfrac{5}{3} p\tau c_v, \tag{6.6, 7}$$

since p is constant in the gas, and $c_v = 3k/2m$.

6.61. Relaxation times

The collision interval τ as used in 6.6 takes no account of persistence of velocities after collisions, or of the variation of collision frequency with molecular speed. Thus it must be regarded as an ideal collision interval, which may be expected to differ from the collision interval of Chapter 5 by a numerical factor analogous to the numbers u, u′ of 6.2 and 6.3. For example, to make (6.6, 5) and (6.2, 1) agree we must take

$$\tau = 3\mathrm{u}l\bar{C}/2\overline{C^2} = 4\mathrm{u}l/\pi\bar{C},$$

whereas the τ of Chapter 5 was $l/\bar{C}$.

To elucidate further the significance of the present τ, consider a uniform gas with a non-Maxwellian velocity distribution. For such a gas (6.6, 1) becomes

$$\frac{\partial f}{\partial t} = -\frac{f - f^{(0)}}{\tau}, \tag{6.61, 1}$$

so that $f - f^{(0)}$ decays like the exponential $e^{-t/\tau}$. Thus τ is the time in which a disturbance to the Maxwellian state decreases to $1/e$ of its original value; it is called the time of relaxation of the disturbance.

Exact theory shows that an equation of form (6.61, 1) holds good only when $f - f^{(0)}$ is small, and then only for a specific set of 'eigenfunctions' $f - f^{(0)}$, the relaxation time differing from one eigenfunction to another (cf. 10.321). If $f_{(1)}, f_{(2)}, \ldots,$ are the successive eigenfunctions, and $\tau_1, \tau_2, \ldots,$ the corresponding relaxation times, any small deviation of f from $f^{(0)}$ can be expressed by a series

$$f - f^{(0)} = \sum_m A_m f_{(m)};$$

corresponding to this

$$\frac{\partial f}{\partial t} = -\sum_m \frac{A_m f_{(m)}}{\tau_m}. \tag{6.61, 2}$$

Equation (6.61, 1) is to be regarded as an approximation to (6.61, 2), the τ in (6.61, 1) representing an average value of $\tau_1, \tau_2, \ldots$, which varies from one function $f - f^{(0)}$ to another. In particular, τ in equations (6.6, 5) and (6.6, 7) may be expected to stand for two distinct average relaxation times, unequal but similar in order of magnitude, characteristic respectively of viscosity and heat conduction.*

6.62. Relaxation and diffusion

An approximate theory of diffusion can be given in terms of a relaxation time without explicitly determining the velocity-distribution function. Suppose that diffusion in a binary mixture leads to a force between the two gases proportional to their relative velocity $\bar{\mathbf{c}}_1 - \bar{\mathbf{c}}_2$, tending to destroy this velocity. Let τ_{12} be a time such that the forces per unit volume on the two gases are $\pm(\rho_1\rho_2/\rho\tau_{12})(\bar{\mathbf{c}}_1 - \bar{\mathbf{c}}_2)$. Then, if the factors leading to diffusion were suddenly to cease to operate, the two gases would experience accelerations

$$-\rho_2(\bar{\mathbf{c}}_1 - \bar{\mathbf{c}}_2)/\rho\tau_{12}, \quad \rho_1(\bar{\mathbf{c}}_1 - \bar{\mathbf{c}}_2)/\rho\tau_{12}$$

and their relative acceleration would be $-(\bar{\mathbf{c}}_1 - \bar{\mathbf{c}}_2)/\tau_{12}$. Thus the velocity of diffusion $\bar{\mathbf{c}}_1 - \bar{\mathbf{c}}_2$ would die away exponentially like $e^{-t/\tau_{12}}$; that is, τ_{12} is the relaxation time for diffusion.

In steady diffusion, the two diffusing gases are assumed to move independently, each under the action of its own partial pressure $p_s \equiv n_s kT$, together with the drag $\pm(\rho_1\rho_2/\rho\tau_{12})(\bar{\mathbf{c}}_1 - \bar{\mathbf{c}}_2)$ exerted by the other gas. Since the gases are unaccelerated,

$$-\nabla p_1 = \rho_1\rho_2(\bar{\mathbf{c}}_1 - \bar{\mathbf{c}}_2)/\rho\tau_{12} = \nabla p_2, \tag{6.62, 1}$$

where $\nabla \equiv \partial/\partial \mathbf{r}$ (cf. 1.2). In the case of diffusion at constant temperature, this gives

$$\bar{\mathbf{c}}_1 - \bar{\mathbf{c}}_2 = -\rho kT\tau_{12}(\nabla n_1)/\rho_1\rho_2. \tag{6.62, 2}$$

According to 6.4, in this case

$$n_1\bar{\mathbf{c}}_1 = -D_{12}\nabla n_1 = -n_2\bar{\mathbf{c}}_2.$$

These equations are consistent if

$$D_{12} = \rho\tau_{12}kT/nm_1m_2. \tag{6.62, 3}$$

Provided that this equation is satisfied, the results of the relaxation theory duplicate those of free-path theory. In terms of D_{12}, the drag forces which the two gases exert on each other per unit volume are

$$\pm p_1p_2(\bar{\mathbf{c}}_1 - \bar{\mathbf{c}}_2)/pD_{12}. \tag{6.62, 4}$$

We may relate τ_{12} to the collision processes by assuming that there are N_{12} collisions between unlike molecules per unit volume and time, and that these

* For a more detailed discussion of the relation of relaxation times to the transport phenomena, see E. P. Gross and E. A. Jackson, *Phys. Fluids*, **2**, 432 (1959).

reduce the mean velocities $\bar{c}_1$ and $\bar{c}_2$ of the colliding molecules to a common mean velocity $\boldsymbol{c}'$. By conservation of momentum

$$m_1\bar{c}_1+m_2\bar{c}_2=(m_1+m_2)\boldsymbol{c}'; \tag{6.62, 5}$$

hence the mean momentum change of a molecule m_1 at a collision is $m_1(\boldsymbol{c}'-\bar{c}_1)$, or $-m_1m_2(\bar{c}_1-\bar{c}_2)/(m_1+m_2)$. The drag $-\rho_1\rho_2(\bar{c}_1-\bar{c}_2)/\rho\tau_{12}$ on the first gas is N_{12} times this change of momentum, and so

$$\tau_{12}=n_1n_2(m_1+m_2)/\rho N_{12}=(\rho_2n_1+\rho_1n_2)/\rho N_{12}. \tag{6.62, 6}$$

Thus τ_{12} is a weighted mean of the collision intervals n_1/N_{12}, n_2/N_{12} for molecules m_1, m_2 in mutual collisions, with weighting factors ρ_2, ρ_1. If m_2/m_1 and ρ_2/ρ_1 are small, τ_{12} approximates to n_2/N_{12}, the collision interval for molecules m_2.

By (6.62, 3, 6),

$$D_{12}=\frac{(m_1+m_2)p_1p_2}{m_1m_2pN_{12}}. \tag{6.62, 7}$$

Since $N_{12}\propto n_1n_2$, this formula indicates that D_{12} is, at a given temperature, inversely proportional to the total pressure, and independent of the proportions of the mixture. As noted in 6.4, exact theory indicates a small dependence of D_{12} on the proportions of the mixture, so that to this extent the relaxation approach is inadequate. Also the relaxation theory, like the simple free-path theory, is not able to explain thermal diffusion. However, because of its simplicity a relaxation argument has often been employed in complex problems, particularly in relation to ionized gases.

Consider, for example, a moving binary mixture in which, in addition to the drags (6.62, 4) and their partial pressures, the gases are subject to applied forces $m_s\boldsymbol{F}_s$ per molecule. Since the velocities of the two gases differ only by the small velocity of diffusion, the acceleration of each gas is, to a first approximation, equal to the acceleration $D\boldsymbol{c}_0/Dt$ of the mass as a whole. Thus the equations of motion of the two gases are

$$\rho_1\frac{D\boldsymbol{c}_0}{Dt}=\rho_1\boldsymbol{F}_1-\nabla p_1-\frac{\rho_1\rho_2}{\rho\tau_{12}}(\bar{c}_1-\bar{c}_2), \tag{6.62, 8}$$

$$\rho_2\frac{D\boldsymbol{c}_0}{Dt}=\rho_2\boldsymbol{F}_2-\nabla p_2+\frac{\rho_1\rho_2}{\rho\tau_{12}}(\bar{c}_1-\bar{c}_2). \tag{6.62, 9}$$

Adding these equations, we get the approximate equation of motion of the gas as a whole,

$$\rho\frac{D\boldsymbol{c}_0}{Dt}=\rho_1\boldsymbol{F}_1+\rho_2\boldsymbol{F}_2-\nabla p.$$

On the other hand, eliminating $D\boldsymbol{c}_0/Dt$,

$$\bar{c}_1-\bar{c}_2=\tau_{12}\{\boldsymbol{F}_1-\rho_1^{-1}\nabla p_1-(\boldsymbol{F}_2-\rho_2^{-1}\nabla p_2)\}. \tag{6.62, 10}$$

This is a general equation of diffusion. It implies that relative diffusion occurs whenever the accelerations that would be given to the two gases by the forces $\rho_s \boldsymbol{F}_s - \nabla p_s$ acting on them are unequal; the diffusion velocity is τ_{12} times the difference of those accelerations.

The diffusion equation can also be derived from the Boltzmann equations of the two gases, written in the approximate forms

$$\frac{\partial f_1}{\partial t} + \boldsymbol{c}_1 \cdot \frac{\partial f_1}{\partial \boldsymbol{r}} + \boldsymbol{F}_1 \cdot \frac{\partial f_1}{\partial \boldsymbol{c}_1} = -\frac{f_1 - f_1^{(0)}}{\tau_1}, \tag{6.62, 11}$$

$$\frac{\partial f_2}{\partial t} + \boldsymbol{c}_2 \cdot \frac{\partial f_2}{\partial \boldsymbol{r}} + \boldsymbol{F}_2 \cdot \frac{\partial f_2}{\partial \boldsymbol{c}_2} = -\frac{f_2 - f_2^{(0)}}{\tau_2}, \tag{6.62, 12}$$

where $\tau_1 = n_1/N_{12}$, $\tau_2 = n_2/N_{12}$, and $f_1^{(0)}$, $f_2^{(0)}$ denote Maxwellian velocity-distributions about the mean velocity $\boldsymbol{c}'$ given by (6.62, 5). Correct to terms of first order in τ_1, τ_2 this gives results agreeing with (6.62, 8, 9). However, modified forms of (6.62, 11, 12) may be used in certain more general diffusion problems for which the simple relaxation method is inadequate.*

6.63. Gas mixtures

An approximate theory of the viscosity of a binary or multiple mixture of gases can be given as follows. Each gas is assumed to have its own relaxation time; that of the sth gas is τ_s, where τ_s^{-1} is the collision-frequency $\sum_t N_{st}/n_s$. Write $N_{st} = \mu_{st}^{-1} n_s n_t kT$; then

$$\tau_s^{-1} = kT \sum_t n_t \mu_{st}^{-1}. \tag{6.63, 1}$$

It is further assumed that, since the viscosity is a measure of the ease with which momentum is transmitted by the molecular motions, the viscosity of the gas as a whole is the sum of contributions from the separate gases, calculated as in (6.6, 5), i.e.

$$\mu = \sum_s p_s \tau_s = \sum_s \frac{n_s}{\mu_{s1}^{-1} n_1 + \mu_{s2}^{-1} n_2 + \dots}. \tag{6.63, 2}$$

By taking all save one of the quantities n_s equal to zero, μ_{ss} can be seen to be the viscosity of the sth gas when pure; μ_{st} can similarly be interpreted as a 'mutual viscosity' of the sth and tth gases. More correctly $\mu_{st}^{-1} n_t$, or $N_{st}/n_s kT$, is to be interpreted as a measure of the extent to which collisions with molecules m_t impede the transfer of momentum by a molecule m_s. Thus (6.63, 2) implies that the contribution of the sth gas to the viscosity is proportional directly to the number-density n_s, and inversely to the sum of terms $\mu_{s1}^{-1} n_1$, $\mu_{s2}^{-1} n_2$, ..., measuring the extent to which collisions with molecules $m_1, m_2, \dots$ impede the flux of momentum.

Equation (6.63, 2) is similar to one derived by Sutherland† by free-path methods. His derivation, like that above, gave $\mu_{st} = \mu_{ts}$. However, he then

* P. L. Bhatnagar, E. P. Gross and M. Krook, *Phys. Rev.* **94**, 511 (1954).

† W. Sutherland, *Phil. Mag.* **40**, 421 (1895).

introduced a semi-empirical correction to allow for differences of velocity-persistence of the two colliding molecules, equivalent to taking

$$\mu_{st} = \mu_{ts}(m_s/m_t)^{\frac{1}{4}}.$$

Even if the possibility $\mu_{st} \neq \mu_{ts}$ is admitted, a formula of the form (6.63, 2) can only be approximate; collisions do not simply obstruct the flux of momentum, but transfer it from one gas to another.

An approximate formula similar to (6.63, 2) was given by Wassiljewa* for the thermal conductivity; it is

$$\lambda = \sum_s \frac{n_s}{\lambda_{s1}^{-1} n_1 + \lambda_{s2}^{-1} n_2 + \ldots}, \tag{6.63, 3}$$

where λ_{ss} is the thermal conductivity of the sth gas when pure, and λ_{st} a 'mutual conductivity'. A formula of this type, with $\lambda_{st} \neq \lambda_{ts}$, is found to give a fair degree of agreement with experiment for the variation of λ with composition, even for gases whose molecules possess internal energy.

Finally, the methods of 6.62 can be extended to apply to diffusion in a multiple gas-mixture.† In such a mixture, collisions with molecules of the tth gas lead to a drag force

$$\frac{p_s p_t}{p D_{st}} (\bar{\boldsymbol{c}}_t - \bar{\boldsymbol{c}}_s)$$

per unit volume acting on the sth gas (cf. (6.62, 4)). Here D_{st} is the coefficient of mutual diffusion of the sth and tth gases, at the pressure of the actual mixture (it should be noted that pD_{st} is independent of the pressure). Hence equations (6.62, 8 and 9) can be generalized to read

$$\rho_s \frac{D\boldsymbol{c}_0}{Dt} = \rho_s \boldsymbol{F}_s - \frac{\partial p_s}{\partial \boldsymbol{r}} + \sum_t \frac{p_s p_t}{p D_{st}} (\bar{\boldsymbol{c}}_t - \bar{\boldsymbol{c}}_s). \tag{6.63, 4}$$

These equations can be solved to give $D\boldsymbol{c}_0/Dt$ and the velocities of diffusion, relative to an arbitrarily chosen gas of the mixture, of all the remaining gases.

In certain problems, notably those referring to ionized gases in a rapidly alternating electric field, it is inadequate to equate the accelerations of the different gases to $D\boldsymbol{c}_0/Dt$. In such problems, $D\boldsymbol{c}_0/Dt$ is to be replaced on the left of (6.63, 4) by $D_s \bar{\boldsymbol{c}}_s/Dt$, where D_s/Dt is the derivative following the motion $\bar{\boldsymbol{c}}_s$ of the sth gas.

* A. Wassiljewa, *Phys. Z.* **5**, 737 (1904).

† See M. H. Johnson and E. O. Hulburt, *Phys. Rev.* **79**, 802 (1950); M. H. Johnson, *Phys. Rev.* **82**, 298 (1951); A. Schlüter, *Z. Naturf.* **5a**, 72 (1950), and **6a**, 73 (1951).

7

THE NON-UNIFORM STATE FOR A SIMPLE GAS

7.1. The method of solution of Boltzmann's equation

The present chapter deals with the method by which Enskog solved Boltzmann's equation in the general case. We consider here a simple gas, and in Chapters 8 and 18 binary and multiple gas-mixtures, whose molecules possess translatory energy only. The corresponding theory for a gas whose molecules possess internal energy is given in Chapter 11.

Enskog's method is one of successive approximation. Boltzmann's equation can be expressed in the general form $\xi(f) = 0$, where $\xi(f)$ denotes the result of certain operations performed on the unknown function f. Suppose that the solution is expressible in the form of an infinite series*

$$f = f^{(0)} + f^{(1)} + f^{(2)} + \ldots. \tag{7.1, 1}$$

Suppose also that when ξ operates on this series, the result can be expressed as a series in which the rth term involves only the first r terms of the series in (7.1, 1); that is,

$$\xi(f) = \xi(f^{(0)} + f^{(1)} + f^{(2)} + \ldots) = \xi^{(0)}(f^{(0)}) + \xi^{(1)}(f^{(0)}, f^{(1)}) + \xi^{(2)}(f^{(0)}, f^{(1)}, f^{(2)}) + \ldots. \tag{7.1, 2}$$

The functions $f^{(r)}$, which as yet are subject only to the condition that their sum is a solution of $\xi(f) = 0$, are now assumed to satisfy the separate equations

$$\xi^{(0)}(f^{(0)}) = 0, \tag{7.1, 3}$$

$$\xi^{(1)}(f^{(0)}, f^{(1)}) = 0, \tag{7.1, 4}$$

$$\xi^{(2)}(f^{(0)}, f^{(1)}, f^{(2)}) = 0, \tag{7.1, 5}$$

$$\ldots\ldots\ldots\ldots\ldots\ldots\ldots\ldots$$

which together ensure that $\xi(f) = 0$. The function $f^{(0)}$ is determined from (7.1, 3), and $f^{(1)}, f^{(2)}, \ldots$ are then found successively from (7.1, 4, 5, ...); each equation contains only one unknown when all the preceding equations have been solved.

The division of $\xi(f)$ into the parts $\xi^{(0)}, \xi^{(1)}, \ldots$, if possible at all, is not unique, because any term $\xi^{(r)}$ can be divided into any number of parts, any of which can be transferred to any subsequent term without altering the general form of the expression. The division must be made in such a way

* All the series introduced are supposed to converge uniformly.

that the equations (7.1, 3, 4, ...) are all soluble, and regard must also be had to the convenience of solution of these equations.

Arbitrary elements may enter into the solution of the various equations: they must be so chosen or grouped that the number of arbitrary elements in the final result does not exceed that appropriate to $\xi(f) = 0$. The expressions $f^{(0)}, f^{(0)}+f^{(1)}, f^{(0)}+f^{(1)}+f^{(2)}, \ldots$ will be successive approximations to f.

7.11. The subdivision of $\xi(f)$; the first approximation $f^{(0)}$

In the case of Boltzmann's equation for a simple gas,

$$\xi(f) = J(ff_1) + \mathscr{D}f, \tag{7.11, 1}$$

where

$$J(FG_1) \equiv \iint (FG_1 - F'G_1')g\alpha_1 \, d\mathbf{e}' d\mathbf{c}_1 \tag{7.11, 2}$$

so that $J(ff_1) = -\partial_e f/\partial t$, by (3.52, 11); also, as in (3.1, 3),

$$\mathscr{D}f = \frac{\partial f}{\partial t} + \mathbf{c}\cdot\frac{\partial f}{\partial \mathbf{r}} + \mathbf{F}\cdot\frac{\partial f}{\partial \mathbf{c}}. \tag{7.11, 3}$$

Substituting from (7.1, 1) into (7.11, 1), we have

$$\begin{aligned} \xi(f) &= J\{(\Sigma f^{(r)})(\Sigma f_1^{(s)})\} + \mathscr{D}\Sigma f^{(r)} \\ &= \Sigma\Sigma J(f^{(r)}f_1^{(s)}) + \Sigma\mathscr{D}f^{(r)}. \end{aligned}$$

Enskog wrote

$$\begin{aligned} J^{(r)} &\equiv J^{(r)}(f^{(0)}, f^{(1)}, \ldots, f^{(r)}) \\ &= J(f^{(0)}f_1^{(r)}) + J(f^{(1)}f_1^{(r-1)}) + \ldots + J(f^{(r)}f_1^{(0)}), \end{aligned} \tag{7.11, 4}$$

corresponding to the mode of grouping of the terms in the expression for the product of two infinite series Σx_r and Σy_r, as a third series

$$\Sigma(x_0 y_r + x_1 y_{r-1} + \ldots + x_r y_0).$$

Enskog also divided $\Sigma\mathscr{D}f^{(r)}$ into a series of parts $\mathscr{D}^{(r)}$, but not by the obvious method of writing $\mathscr{D}^{(r)} = \mathscr{D}f^{(r)}$. His method of division will be explained later (7.14); it suffices here to say that he took

$$\mathscr{D}^{(0)} = 0, \tag{7.11, 5}$$

and for $r > 0$ he made $\mathscr{D}^{(r)}$ depend only on $f^{(0)}, \ldots, f^{(r-1)}$. Thus, writing

$$\xi^{(r)} = J^{(r)} + \mathscr{D}^{(r)} \tag{7.11, 6}$$

we have

$$\xi^{(0)} = J^{(0)} = 0, \tag{7.11, 7}$$

$$\xi^{(r)} = J^{(r)} + \mathscr{D}^{(r)} = 0 \quad (r > 0). \tag{7.11, 8}$$

Now (7.11, 7), or

$$J(f^{(0)}f_1^{(0)}) = 0,$$

is identical in form with the equation (4.1, 7), which determines the velocity-distribution function f in the uniform steady state. The general solution is therefore of the form found in 4.1; that is, $\ln f^{(0)}$ is a linear combination of the summational invariants $\psi^{(i)}$ of (3.2, 2), say

$$\ln f^{(0)} = \alpha^{(1)} + \boldsymbol{\alpha}^{(2)}.m\boldsymbol{c} + \alpha^{(3)}.\tfrac{1}{2}mc^2,$$

where $\alpha^{(1)}$, $\boldsymbol{\alpha}^{(2)}$ and $\alpha^{(3)}$ are arbitrary quantities, which are independent of $\boldsymbol{c}$, but may depend on $\boldsymbol{r}$ and t. By a simple transformation as in 4.1, we get

$$f^{(0)} = n\left(\frac{m}{2\pi kT}\right)^{\frac{3}{2}} e^{-mC^2/2kT}, \tag{7.11, 9}$$

where $\boldsymbol{C} = \boldsymbol{c} - \boldsymbol{c}_0$; n, $\boldsymbol{c}_0$, and T here denote arbitrary quantities related to $\alpha^{(1)}$, $\boldsymbol{\alpha}^{(2)}$, and $\alpha^{(3)}$ by simple relations. So far as (7.11, 7) is concerned they are not necessarily identical with the number-density, mass-velocity and temperature of the gas. Their values are, however, entirely at our disposal, and it is convenient to identify them with these quantities. This amounts to the choice, as a valid first approximation to f at $\boldsymbol{r}$, t, of the Maxwellian function corresponding to the number-density, mass-velocity and temperature at $\boldsymbol{r}$, t. Later, in 7.15, it is shown that this choice is the only one that leads to a properly ordered form for the whole solution f.

In consequence of our identifications of n, $\boldsymbol{c}_0$, and T,

$$\int f^{(0)} d\boldsymbol{c} = n = \int f d\boldsymbol{c},$$

and similarly

$$\int f^{(0)} m\boldsymbol{C}\, d\boldsymbol{c} = \int f m\boldsymbol{C}\, d\boldsymbol{c}, \quad \int f^{(0)} \tfrac{1}{2}mC^2 d\boldsymbol{c} = \int f \tfrac{1}{2}mC^2 d\boldsymbol{c}.$$

These relations may be written

$$\int (f - f^{(0)})\, \psi^{(i)} d\boldsymbol{c} = 0,$$

where now $\psi^{(i)} = 1, m\boldsymbol{C}, \tfrac{1}{2}mC^2$. By (7.1, 1), this is identical with

$$\int \sum_{r=1}^{\infty} f^{(r)} \psi^{(i)} d\boldsymbol{c} = 0. \tag{7.11, 10}$$

It follows from (7.11, 9) that

$$f^{(0)} f_1^{(0)} = f^{(0)\prime} f_1^{(0)\prime}. \tag{7.11, 11}$$

7.12. The complete formal solution

Equation (7.11, 8) may be written in the form

$$J(f^{(0)} f_1^{(r)}) + J(f^{(r)} f_1^{(0)}) = -\mathscr{D}^{(r)} - J(f^{(1)} f_1^{(r-1)}) - \ldots - J(f^{(r-1)} f_1^{(1)}), \tag{7.12, 1}$$

there being r terms on the right (when $r = 1$ the single term on the right is $-\mathscr{D}^{(1)}$). The right-hand side involves only $f^{(0)}, f^{(1)}, \ldots, f^{(r-1)}$, and these are known from the solutions of the previous equations $\xi^{(0)} = 0, \ldots, \xi^{(r-1)} = 0$; the unknown function $f^{(r)}$ appears only on the left, and occurs there linearly. Hence if $F^{(r)}$ is any one solution, any other solution will differ from $F^{(r)}$ by a quantity $\chi^{(r)}$ which is a solution of

$$J(f^{(0)}\chi_1^{(r)}) + J(\chi^{(r)}f_1^{(0)}) = 0; \tag{7.12, 2}$$

$F^{(r)}$ and $\chi^{(r)}$ correspond to the particular integral and complementary function of a linear differential equation. The most general solution of (7.12, 1) is the sum of $F^{(r)}$ and the most general solution of (7.12, 2).

To obtain the general solution of (7.12, 2), write $\chi^{(r)} = \phi^{(r)}f^{(0)}$, so that $\phi^{(r)}$ becomes the function to be determined. Now, if $\phi^{(r)}$ is any function of $\boldsymbol{c}$, by (7.11, 2) and (4.4, 3),

$$\begin{aligned} J(f^{(0)}f_1^{(0)}\phi_1^{(r)}) + J(f^{(0)}\phi^{(r)}f_1^{(0)}) &= \iint f^{(0)}f_1^{(0)}(\phi^{(r)} + \phi_1^{(r)} - \phi^{(r)\prime} - \phi_1^{(r)\prime})g\alpha_1\,d\boldsymbol{e}'\,d\boldsymbol{c}_1 \\ &= n^2 I(\phi^{(r)}), \end{aligned} \tag{7.12, 3}$$

where the suffix has been omitted from $I(\phi)$, since the gas contains molecules of one kind only. Thus (7.12, 2) takes the form $I(\phi^{(r)}) = 0$, of which the solution was shown in 4.41 to be

$$\phi^{(r)} = \alpha^{(1,r)} + \boldsymbol{\alpha}^{(2,r)}.m\boldsymbol{C} + \alpha^{(3,r)}.\tfrac{1}{2}mC^2,$$

where $\alpha^{(1,r)}$, $\boldsymbol{\alpha}^{(2,r)}$, $\alpha^{(3,r)}$ are arbitrary functions of $\boldsymbol{r}$, t. Hence

$$\chi^{(r)} = (\alpha^{(1,r)} + \boldsymbol{\alpha}^{(2,r)}.m\boldsymbol{C} + \alpha^{(3,r)}.\tfrac{1}{2}mC^2)f^{(0)}.$$

Moreover, as any solution $f^{(r)}$ of (7.12, 1) is of the form $F^{(r)} + \chi^{(r)}$, by a suitable choice of $\alpha^{(1,r)}$, $\boldsymbol{\alpha}^{(2,r)}$, $\alpha^{(3,r)}$ the quantities

$$\int f^{(r)}\psi^{(i)}\,d\boldsymbol{c} \quad (i = 1, 2, 3)$$

can be made to take any arbitrary values. It is convenient to choose $\alpha^{(1,r)}$, $\boldsymbol{\alpha}^{(2,r)}$, $\alpha^{(3,r)}$ such that, for all values of r greater than zero,

$$\int f^{(r)}\psi^{(i)}\,d\boldsymbol{c} = 0 \quad (i = 1, 2, 3). \tag{7.12, 4}$$

This choice ensures the satisfaction of (7.11, 10), which is the only restriction as yet placed on the functions $f^{(r)}$, beyond the fact that they satisfy (7.12, 1): it also implies that at every stage of the approximation to f the new constants introduced depend on no parameters other than n, $\boldsymbol{c}_0$, and T, and their space- and time-derivatives of all orders, at $\boldsymbol{r}$, t. This choice of $\alpha^{(1,r)}$, $\boldsymbol{\alpha}^{(2,r)}$, and $\alpha^{(3,r)}$ may seem unduly restrictive, since it substitutes, for the one condition (7.11, 10), the infinity of relations (7.12, 4); but actually the same value is found for the *sum* $\Sigma f^{(r)}$, whether this is made to satisfy the single relation (7.11, 10), or its parts are made to satisfy the set of relations (7.12, 4) (see 7.2).

7.13. The conditions of solubility

Equation (7.11, 8) is as yet indefinite because $\mathscr{D}f$ has not so far been divided into its component parts $\mathscr{D}^{(r)}$. The division cannot be performed arbitrarily, for, according to the theory of integral equations, the equation is soluble only if $\mathscr{D}^{(r)}$ satisfies certain conditions. For, if the equation is satisfied,

$$\int (J^{(r)} + \mathscr{D}^{(r)})\,\psi^{(i)}\,d\boldsymbol{c} = 0,$$

where $\psi^{(i)}$ denotes any one of the summational invariants of (3.2, 2). Now $J^{(r)}$ is divisible into pairs of terms such as $J(f^{(p)}f_1^{(q)})+J(f^{(q)}f_1^{(p)})$, together with a term $J(f^{(p)}f_1^{(p)})$ if r is even. When these are multiplied by $\psi^{(i)}d\boldsymbol{c}$ and integrated over the whole range of $\boldsymbol{c}$, integrals similar to those appearing in (3.54, 4, 5) are obtained; these vanish since

$$\psi+\psi_1-\psi'-\psi_1'=0.$$

Hence a necessary condition for (7.11, 8) to be soluble is that

$$\int \mathscr{D}^{(r)}\psi^{(i)}d\boldsymbol{c}=0. \tag{7.13, 1}$$

This is, moreover, a sufficient condition; for if we write

$$f^{(r)}=f^{(0)}\Phi^{(r)}, \tag{7.13, 2}$$

then, using (7.12, 3), equation (7.11, 8) becomes

$$n^2I(\Phi^{(r)})=-\mathscr{D}^{(r)}-J(f^{(1)}f_1^{(r-1)})-\ldots-J(f^{(r-1)}f_1^{(1)}). \tag{7.13, 3}$$

It can be shown* that $I(\Phi^{(r)})$, which is a function of $\boldsymbol{c}$, can be expressed in the form

$$K_0(\boldsymbol{c})\Phi^{(r)}(\boldsymbol{c})+\int K(\boldsymbol{c},\boldsymbol{c}_1)\Phi^{(r)}(\boldsymbol{c}_1)\,d\boldsymbol{c}_1,$$

where $K(\boldsymbol{c},\boldsymbol{c}_1)$ is a symmetric function of $\boldsymbol{c}$, $\boldsymbol{c}_1$. Consequently (7.13, 3) is a linear orthogonal non-homogeneous integral equation of the second kind; the associated linear orthogonal homogeneous integral equation of the second kind is $n^2I(\Phi^{(r)})=0$, whose independent solutions are $\Phi^{(r)}=\psi^{(i)}$ $(i=1,2,3)$, as was shown in 4.41. From the theory of integral equations† it follows that (7.13, 3) possesses a solution if, and only if, the associated 'conditions of orthogonality' are satisfied, namely

$$\int \psi^{(i)}\{-\mathscr{D}^{(r)}-J(f^{(1)}f_1^{(r-1)})-\ldots-J(f^{(r-1)}f_1^{(1)})\}\,d\boldsymbol{c}=0.$$

On omitting vanishing terms from this equation, it reduces to (7.13, 1). Hence this is also a sufficient condition for the solubility of (7.11, 8).

7.14. The subdivision of $\mathscr{D}f$

The equations (3.21, 3, 5), giving the time-derivatives of $\boldsymbol{c}_0$ and T, involve mean-value functions, p and $\boldsymbol{q}$ (cf. 2.31 and 2.45), which can be evaluated only when f is known. This causes difficulty in the determination of f by successive approximation, because the time-derivatives of n, $\boldsymbol{c}_0$ and

* This was first proved by D. Hilbert (*Integralgleichungen*, p. 267; *Math. Annalen*, **72**, 567 (1912)) for rigid elastic spherical molecules; the extension to general spherical molecules was made by Enskog, whose proof was included as 7.6 in earlier editions of this book.

† See, e.g., R. Courant and D. Hilbert, *Methoden der Math. Physik*, vol. 1 (2nd ed.), 99, 129.

T determine that of $f^{(0)}$, which affects the equation from which $f^{(1)}$ is determined. At this stage, however, as f is only incompletely known, p and $\boldsymbol{q}$ also cannot be completely evaluated. This difficulty was overcome by Enskog by an appropriate method of division of the part $\mathscr{D}f$ of $\xi(f)$.

Let ϕ denote any function of $\boldsymbol{c}$, $\boldsymbol{r}$, t; then

$$\bar{\phi} = \frac{1}{n}\int f\phi \, d\boldsymbol{c}$$

$$= \frac{1}{n}\int \sum_0^\infty f^{(r)}\phi \, d\boldsymbol{c}$$

$$= \Sigma\bar{\phi}^{(r)}, \tag{7.14, 1}$$

where
$$\bar{\phi}^{(r)} = \frac{1}{n}\int \phi f^{(r)} d\boldsymbol{c}. \tag{7.14, 2}$$

In particular, taking in turn $\phi = m\boldsymbol{CC}$, $\phi = E\boldsymbol{C}$, we have

$$\mathsf{p} = \sum_0^\infty \mathsf{p}^{(r)}, \quad \boldsymbol{q} = \sum_0^\infty \boldsymbol{q}^{(r)}, \tag{7.14, 3}$$

where
$$\mathsf{p}^{(r)} = \int m\boldsymbol{CC}f^{(r)} d\boldsymbol{c}, \quad \boldsymbol{q}^{(r)} = \int E\boldsymbol{C}f^{(r)} d\boldsymbol{c}; \tag{7.14, 4}$$

in the present chapter, of course, $E = \frac{1}{2}mC^2$ and (cf. 2.44) $N = 3$.

Owing to the form of $f^{(0)}$,
$$\boldsymbol{q}^{(0)} = 0; \tag{7.14, 5}$$

also the non-diagonal elements of $\mathsf{p}^{(0)}$ vanish, and each diagonal element is equal to knT. Consequently

$$\mathsf{p}^{(0)} = knT\mathsf{U} = \mathsf{U}p, \tag{7.14, 6}$$

where U is the unit tensor (1.3), and p is the hydrostatic pressure.

The expressions (3.21, 1, 3, 5) giving $\partial n/\partial t$, $\partial \boldsymbol{c}_0/\partial t$ and $\partial T/\partial t$ may now be written

$$\frac{\partial n}{\partial t} = -\frac{\partial}{\partial \boldsymbol{r}}\cdot(n\boldsymbol{c}_0),$$

$$\frac{\partial \boldsymbol{c}_0}{\partial t} = -\left(\boldsymbol{c}_0\cdot\frac{\partial}{\partial \boldsymbol{r}}\right)\boldsymbol{c}_0 + \boldsymbol{F} - \frac{1}{\rho}\frac{\partial}{\partial \boldsymbol{r}}\cdot\Sigma\mathsf{p}^{(r)},$$

$$\frac{\partial T}{\partial t} = -\boldsymbol{c}_0\cdot\frac{\partial T}{\partial \boldsymbol{r}} - \frac{2}{3kn}\left\{(\Sigma\mathsf{p}^{(r)}):\frac{\partial}{\partial \boldsymbol{r}}\boldsymbol{c}_0 + \frac{\partial}{\partial \boldsymbol{r}}\cdot\Sigma\boldsymbol{q}^{(r)}\right\}.$$

These time-derivatives are divided into parts, as follows,

$$\frac{\partial n}{\partial t} = \Sigma\frac{\partial_r n}{\partial t}, \quad \frac{\partial \boldsymbol{c}_0}{\partial t} = \Sigma\frac{\partial_r \boldsymbol{c}_0}{\partial t}, \quad \frac{\partial T}{\partial t} = \Sigma\frac{\partial_r T}{\partial t}, \tag{7.14, 7}$$

where the quantities on the right are not themselves time-derivatives, but are defined by

$$\frac{\partial_0 n}{\partial t} = -\frac{\partial}{\partial \boldsymbol{r}}\cdot(n\boldsymbol{c}_0), \tag{7.14, 8}$$

$$\frac{\partial_r n}{\partial t} = 0 \quad (r>0), \tag{7.14, 9}$$

$$\frac{\partial_0 \boldsymbol{c}_0}{\partial t} = -\left(\boldsymbol{c}_0\cdot\frac{\partial}{\partial \boldsymbol{r}}\right)\boldsymbol{c}_0 + \boldsymbol{F} - \frac{1}{\rho}\frac{\partial}{\partial \boldsymbol{r}}\cdot\mathsf{p}^{(0)}$$

$$= -\left(\boldsymbol{c}_0\cdot\frac{\partial}{\partial \boldsymbol{r}}\right)\boldsymbol{c}_0 + \boldsymbol{F} - \frac{1}{\rho}\frac{\partial p}{\partial \boldsymbol{r}}, \tag{7.14, 10}$$

$$\frac{\partial_r \boldsymbol{c}_0}{\partial t} = -\frac{1}{\rho}\frac{\partial}{\partial \boldsymbol{r}}\cdot\mathsf{p}^{(r)} \quad (r>0), \tag{7.14, 11}$$

$$\frac{\partial_0 T}{\partial t} = -\boldsymbol{c}_0\cdot\frac{\partial T}{\partial \boldsymbol{r}} - \frac{2}{3kn}\left\{\mathsf{p}^{(0)}:\frac{\partial}{\partial \boldsymbol{r}}\boldsymbol{c}_0 + \frac{\partial}{\partial \boldsymbol{r}}\cdot\boldsymbol{q}^{(0)}\right\}$$

$$= -\boldsymbol{c}_0\cdot\frac{\partial T}{\partial \boldsymbol{r}} - \frac{2T}{3}\frac{\partial}{\partial \boldsymbol{r}}\cdot\boldsymbol{c}_0, \tag{7.14, 12}$$

$$\frac{\partial_r T}{\partial t} = -\frac{2}{3kn}\left\{\mathsf{p}^{(r)}:\frac{\partial}{\partial \boldsymbol{r}}\boldsymbol{c}_0 + \frac{\partial}{\partial \boldsymbol{r}}\cdot\boldsymbol{q}^{(r)}\right\} \quad (r>0). \tag{7.14, 13}$$

It is convenient also to write

$$\frac{D_0}{Dt} \equiv \frac{\partial_0}{\partial t} + \boldsymbol{c}_0\cdot\frac{\partial}{\partial \boldsymbol{r}}, \tag{7.14, 14}$$

so that D_0/Dt denotes a first approximation to D/Dt, and

$$\frac{D_0 n}{Dt} = -n\frac{\partial}{\partial \boldsymbol{r}}\cdot\boldsymbol{c}_0, \quad \frac{D_0 \boldsymbol{c}_0}{Dt} = \boldsymbol{F} - \frac{1}{\rho}\frac{\partial p}{\partial \boldsymbol{r}}, \tag{7.14, 15}$$

$$\frac{D_0 T}{Dt} = -\frac{2T}{3}\left(\frac{\partial}{\partial \boldsymbol{r}}\cdot\boldsymbol{c}_0\right). \tag{7.14, 16}$$

The notation $\partial_r/\partial t$ can also be applied to functions involving n, $\boldsymbol{c}_0$ and T and their space-derivatives, $\partial_r/\partial t$ being assumed to obey the same laws of differentiation as $\partial/\partial t$. Thus, for example, from (7.14, 15, 16) it follows that

$$\frac{D_0}{Dt}(n/T^{\frac{3}{2}}) = 0, \tag{7.14, 17}$$

so that, to this first approximation, the variation of temperature during the motion of the gas follows the adiabatic law

$$\rho \propto T^{\frac{3}{2}}.$$

Again, when $\partial_r/\partial t$ is followed by a space-derivative of n, $\boldsymbol{c}_0$, or T, it is understood that

$$\frac{\partial_r}{\partial t}\frac{\partial}{\partial \boldsymbol{r}} = \frac{\partial}{\partial \boldsymbol{r}}\frac{\partial_r}{\partial t}.$$

Also, if F is a function which depends on t only through parameters λ_s (these being n, $\boldsymbol{c}_0$, T and their space-derivatives of any order),

$$\frac{\partial_r F}{\partial t} = \sum_s \frac{\partial F}{\partial \lambda_s}\frac{\partial_r \lambda_s}{\partial t}. \qquad (7.14, 18)$$

We now assume that each $f^{(r)}$ in (7.11, 1) depends on t only through n, $\boldsymbol{c}_0$, T and their space-derivatives. Then

$$\mathscr{D}f = \left(\Sigma\frac{\partial_r}{\partial t} + \boldsymbol{c}.\frac{\partial}{\partial \boldsymbol{r}} + \boldsymbol{F}.\frac{\partial}{\partial \boldsymbol{c}}\right)\Sigma f^{(s)}$$

$$= \Sigma\Sigma\frac{\partial_r f^{(s)}}{\partial t} + \Sigma\left(\boldsymbol{c}.\frac{\partial f^{(s)}}{\partial \boldsymbol{r}} + \boldsymbol{F}.\frac{\partial f^{(s)}}{\partial \boldsymbol{c}}\right).$$

Enskog's division of $\mathscr{D}f$ can now be indicated. He took $\mathscr{D}^{(0)} = 0$, and, for $r > 0$,

$$\mathscr{D}^{(r)} \equiv \mathscr{D}^{(r)}(f^{(0)}, f^{(1)}, \ldots, f^{(r-1)})$$

$$\equiv \frac{\partial_0 f^{(r-1)}}{\partial t} + \frac{\partial_1 f^{(r-2)}}{\partial t} + \ldots + \frac{\partial_{r-1} f^{(0)}}{\partial t} + \boldsymbol{c}.\frac{\partial f^{(r-1)}}{\partial \boldsymbol{r}} + \boldsymbol{F}.\frac{\partial f^{(r-1)}}{\partial \boldsymbol{c}}, \qquad (7.14, 19)$$

each term in which is determinate when $f^{(0)}, \ldots, f^{(r-1)}$ are known. This procedure overcomes the difficulty mentioned at the beginning of this section. It can be shown that with this mode of division the conditions of solubility

$$\int \mathscr{D}^{(r)}\psi^{(i)}d\boldsymbol{c} = 0 \quad (i = 1, 2, 3) \qquad (7.14, 20)$$

are satisfied. This is so because $\mathscr{D}^{(0)} = 0$, while for $r > 0$ the conditions (7.14, 20) are equivalent to (7.14, 8–13), the equations for the subdivision of $\partial n/\partial t$, $\partial \boldsymbol{c}_0/\partial t$, $\partial T/\partial t$.

To see the truth of the last assertion, we note that the values of $\int \psi^{(i)}\mathscr{D}f d\boldsymbol{c}$ given in (3.21, 1, 3, 4) can be analysed at sight to give the parts corresponding to the separate expressions $\mathscr{D}^{(r)}$. In these parts, p, $\boldsymbol{q}$ are replaced by their parts $\mathsf{p}^{(r-1)}$, $\boldsymbol{q}^{(r-1)}$; also, whenever the time-derivative $\partial\bar{\phi}/\partial t$ of the mean value of a function ϕ occurs, it is replaced by

$$\frac{\partial_0 \bar{\phi}^{(r-1)}}{\partial t} + \frac{\partial_1 \bar{\phi}^{(r-2)}}{\partial t} + \ldots + \frac{\partial_{r-1} \bar{\phi}^{(0)}}{\partial t},$$

corresponding to the subdivision of $\partial f/\partial t$ among the expressions $\mathscr{D}^{(r)}$. But in the equations of 3.21 the only time-derivatives present are those of mean values of the functions $\psi^{(i)}$, which, by (7.12, 4), are such that

$$\overline{\psi^{(i)}}^{(r)} = 0 \quad (r > 0).$$

Hence the only time-derivatives appearing in $\int \mathscr{D}^{(r)}\psi^{(i)} d\mathbf{c}$ are $\partial_{r-1}n/\partial t$, $\partial_{r-1}\mathbf{c}_0/\partial t$, $\partial_{r-1}T/\partial t$. The equivalence of (7.14, 20) with (7.14, 8–15) follows directly. The rules for the subdivision of $\partial n/\partial t$, $\partial \mathbf{c}_0/\partial t$, $\partial T/\partial t$ could, in fact, have been derived from the integrability condition (7.14, 20).

Since the conditions of solubility are satisfied, f can be determined to any desired degree of approximation. The time derivatives of n, $\mathbf{c}_0$, and T are then known to the same degree: they do not appear as arbitrary adjustable parameters, but are uniquely determinable in terms of n, $\mathbf{c}_0$ and T and their space-derivatives at $\mathbf{r}$, t.

7.15. The parametric expression of Enskog's method of solution

The preceding description of Enskog's method will now be summarized.

Boltzmann's equation is expressed in the form $\xi(f) = 0$, and a solution $f = \Sigma f^{(r)}$ is assumed; $\xi(f)$ is expressed in the form $\Sigma\xi^{(r)}$, where $\xi^{(r)} = J^{(r)} + \mathscr{D}^{(r)}$, and $J^{(r)}$, $\mathscr{D}^{(r)}$ are given by (7.11, 4) and (7.14, 19). The first term ($f^{(0)}$) of f is the same function of the number-density n, the mass-velocity $\mathbf{c}_0$, and the temperature T at the point in question as would represent the velocity-distribution function in a uniform gas in the steady state in which everywhere the number-density, mass-velocity and temperature are n, $\mathbf{c}_0$, and T. The later terms, $f^{(r)}$, are the solutions of the equations $\xi^{(r)} = 0$ such that

$$\int f^{(r)}\psi^{(i)} d\mathbf{c} = 0 \quad (i = 1, 2, 3);$$

the solutions thus defined are unique.

The division of ξ into $\Sigma\xi^{(r)}$ was expressed concisely by Enskog in the following way. He introduced a parameter θ into the series for f, writing

$$f = \frac{1}{\theta} f^{(0)} + f^{(1)} + \theta f^{(2)} + \theta^2 f^{(3)} + \dots \qquad (7.15, 1)$$

(if we put $\theta = 1$ this agrees with (7.1, 1)). Then

$$J(ff_1) = \frac{1}{\theta^2} J^{(0)} + \frac{1}{\theta} J^{(1)} + J^{(2)} + \theta J^{(3)} + \dots,$$

where $J^{(r)}$ is the same as in (7.11, 4). Similarly the equations of 7.14 may be modified as follows:

$$\frac{1}{\theta}\overline{\phi} = \frac{1}{n}\int \phi \left(\frac{1}{\theta} f^{(0)} + f^{(1)} + \theta f^{(2)} + \dots\right) d\mathbf{c}$$

$$= \frac{1}{\theta}\overline{\phi}^{(0)} + \overline{\phi}^{(1)} + \theta\overline{\phi}^{(2)} + \dots,$$

or

$$\overline{\phi} = \sum_0^\infty \theta^r \overline{\phi}^{(r)}, \qquad (7.15, 2)$$

and $$\frac{\partial}{\partial t} = \sum_0^\infty \theta^r \frac{\partial_r}{\partial t},$$

$$\mathscr{D}f = \left(\sum_0^\infty \theta^r \frac{\partial_r}{\partial t} + \mathbf{c}.\frac{\partial}{\partial \mathbf{r}} + \mathbf{F}.\frac{\partial}{\partial \mathbf{c}}\right)\left(\frac{1}{\theta}f^{(0)} + f^{(1)} + \ldots\right)$$

$$= \frac{1}{\theta}\mathscr{D}^{(1)} + \mathscr{D}^{(2)} + \theta\mathscr{D}^{(3)} + \ldots,$$

where $\mathscr{D}^{(r)}$ is defined as in 7.14. It follows that

$$\xi(f) = \frac{1}{\theta^2}J^{(0)} + \frac{1}{\theta}(J^{(1)} + \mathscr{D}^{(1)}) + \ldots$$

$$= \frac{1}{\theta^2}\xi^{(0)} + \frac{1}{\theta}\xi^{(1)} + \xi^{(2)} + \theta\xi^{(3)} + \ldots,$$

where $\xi^{(r)} = J^{(r)} + \mathscr{D}^{(r)}$, as before. If $\xi^{(r)} = 0$ for every r, then $\xi(f)$ vanishes for all values of θ. The preceding description of Enskog's method corresponds formally to the case $\theta = 1$.

On evaluating $f^{(0)}, f^{(1)}, \ldots$, with the above choice of arbitrary parameters, it is found that $f^{(0)}$ is proportional to n (or ρ), $f^{(1)}$ is independent of n, $f^{(2)}$ is proportional to $1/n$, and so on. Hence the power of $1/\theta$ appearing in any term of the equations is the same as that of n appearing in the same term. If a different choice of arbitrary parameters were made, the solution would not be thus simply ordered according to powers of the density, though the value of the complete solution would be unaltered (7.2).

When the density of a gas is comparable with that of the atmosphere near the ground, and the non-uniformities in n, $\mathbf{c}_0$, T are such as ordinarily occur in laboratory experiments, the terms $f^{(r)}$ in f decrease rapidly as r increases,* and $f^{(0)} + f^{(1)}$ is a sufficiently good approximation to f for most purposes. In rarer gases, naturally, the later terms are relatively more important, so that for some purposes it is worth while to determine $f^{(2)}$. Unfortunately the complexity of the successive terms $f^{(r)}$ increases rapidly with r.

Hilbert expressed f by an equation of the form (7.15, 1), but his discussion of Boltzmann's integral equation did not afford a convenient determination of f beyond the first term $f^{(0)}$, because he did not introduce (7.15, 2) before proceeding to the division of $\mathscr{D}f$.

7.2. The arbitrary parameters in f

The solution f of Boltzmann's equation which is obtained by Enskog's method depends on no parameters other than the values of n, $\mathbf{c}_0$, and T throughout the gas. The conclusion that the physical state of the gas depends

* The ratio of $f^{(r)}$ to $f^{(r-1)}$ is in general comparable with l/L for rigid spherical molecules, where l is the mean free path, and L a scale-length characteristic of the spatial variations in density, mass-velocity and temperature; for more general molecules the ratio is comparable with $\bar{C}\tau/L$, where τ is the relaxation time (cf. 6.61).

on these parameters, and on no others, appears to be in accordance with experiment. However, Enskog's solution of Boltzmann's equation is not the most general one: it is possible to assign an arbitrary value to f at an initial instant, and Boltzmann's equation merely determines the way in which f subsequently varies.

It might be supposed that Enskog's solution lacks the arbitrariness of the general solution because of the special values which have been adopted for n, $\boldsymbol{c}_0$, and T in the expression (7.11, 9) for $f^{(0)}$, and the special values adopted for the arbitrary constants $\alpha^{(1,r)}$, $\boldsymbol{\alpha}^{(2,r)}$ and $\alpha^{(3,r)}$, appearing in the expression for $f^{(r)}$ in 7.12. This, however, is not so: it can be proved (though the proof is too long to be given here) that, subject to certain conditions of convergence, every solution given by Enskog's method of subdivision, corresponding to a given distribution of n, $\boldsymbol{c}_0$, and T, is identical with the one obtained with these special values of the arbitrary elements. The lack of arbitrariness is due to the fact that $\partial f^{(r)}/\partial t$ does not contribute to the equation determining $f^{(r)}$: if it did, $f^{(r)}$ could be arbitrary, and the equation would merely determine $\partial f^{(r)}/\partial t$.

To see the relation of Enskog's solution to the general one, suppose that the initial velocity-distribution function f corresponds to values of n, $\boldsymbol{c}_0$, and T, whose scale-lengths of spatial variation are large compared with a free path, but that f is otherwise arbitrary. Then in general f at each point in the gas begins by varying rapidly on account of molecular encounters, with a time-scale of variation comparable with the mean collision-interval. However, by analogy with the results for a gas in a uniform state, f can be expected quickly to approach a limiting form in which molecular encounters no longer produce such rapid time-variations. The limiting value of f is called a 'normal solution' of Boltzmann's equation, and the corresponding state of the gas is a 'normal state'. The time of relaxation in which the gas effectively attains a normal state is of the same order as the collision-interval (cf. 6.61).

If the gas is uniform, the normal solution is the Maxwellian distribution function. In a slightly non-uniform gas the normal solution is a function of $\boldsymbol{r}$ and t, though at any time it approximates locally to the Maxwellian form. The normal state is characterized by a balance between the effects of non-uniformity, which tends to produce departures from a locally Maxwellian state, and those of molecular encounters, which tend to destroy such departures.

Normal solutions of Boltzmann's equation are distinguished from non-normal solutions by having no 'irregularities' which can be rapidly smoothed out by molecular encounters. In consequence, they depend on fewer parameters than the non-normal solutions. Since the mass, momentum and energy per unit volume are unaffected by encounters, the values of n, $\boldsymbol{c}_0$, and T throughout the gas are among the parameters on which a normal solution must depend; conversely, since there are only three summational invariants, we can expect that n, $\boldsymbol{c}_0$ and T (determined by these invariants) are the only local properties on which a normal solution can depend. Actually, since the state at the point P is appreciably affected only by the gas immediately adjacent (i.e. within a distance of a few mean free paths) a normal solution at P can be taken to depend on the values at P of n, $\boldsymbol{c}_0$, T and their space-derivatives; it may also depend on the force $m\boldsymbol{F}$ acting on the molecules. Enskog's method of approximation provides a solution depending on just these parameters, and on no others. Since a normal solution depends on a minimum number of parameters, Enskog's solution is such a solution, and so the sequence 7.1, if it converges, defines a normal state of the gas.*

It must be stressed that Enskog's method is justified only because the collision-interval is normally small compared with the time-scale of variation of the macroscopic properties. This justifies our regarding $\partial f^{(r)}/\partial t$ as smaller in order of magnitude than the contribution of $f^{(r)}$ to $\partial_e f^{(r)}/\partial t$ (cf. (7.11, 4) and (7.14, 19)). It also has the consequence that deviations from uniformity affect f only slightly; the first approximation $f^{(0)}$ depends on n, $\boldsymbol{c}_0$, and T, but neglects their space-derivatives; the second approximation $f^{(0)}+f^{(1)}$ involves their first derivatives linearly; the third, $f^{(0)}+f^{(1)}+f^{(2)}$, includes terms involving second derivatives linearly, or quadratic in first derivatives; and so on. When conditions are such that the time-scale of variation of any macroscopic variables is not large compared with the collision-interval, the whole method of approximation breaks down.

7.3. The second approximation to f

The remainder of this chapter is devoted to the detailed evaluation of the second approximation to f (involving the determination of $\Phi^{(1)}$); this leads to the determination of $\boldsymbol{q}^{(1)}$ and $\mathsf{p}^{(1)}$, and hence to expressions for the coefficients of thermal conductivity (λ) and viscosity (μ).

The equation from which $f^{(1)}$ or $f^{(0)}\Phi^{(1)}$ is to be determined is

$$\xi^{(1)} \equiv \mathscr{D}^{(1)} + J^{(1)} = 0. \qquad (7.3, 1)$$

The differential part $\mathscr{D}^{(1)}$ of $\xi^{(1)}$ depends only on $f^{(0)}$. As $f^{(0)}$ is a function of $\boldsymbol{C}$, it is convenient to express $\mathscr{D}^{(1)}$ in a form analogous to that of $\mathscr{D}f$ given in (3.12, 2), namely

$$\mathscr{D}^{(1)} = \frac{D_0 f^{(0)}}{Dt} + \boldsymbol{C}.\frac{\partial f^{(0)}}{\partial \boldsymbol{r}} + \left(\boldsymbol{F} - \frac{D_0 \boldsymbol{c}_0}{Dt}\right).\frac{\partial f^{(0)}}{\partial \boldsymbol{C}} - \frac{\partial f^{(0)}}{\partial \boldsymbol{C}}\boldsymbol{C}:\frac{\partial}{\partial \boldsymbol{r}}\boldsymbol{c}_0,$$

or, on substitution from (7.14, 15),

$$\mathscr{D}^{(1)} = f^{(0)}\left\{\frac{D_0 \ln f^{(0)}}{Dt} + \boldsymbol{C}.\frac{\partial \ln f^{(0)}}{\partial \boldsymbol{r}} + \frac{1}{\rho}\frac{\partial p}{\partial \boldsymbol{r}}.\frac{\partial \ln f^{(0)}}{\partial \boldsymbol{C}} - \frac{\partial \ln f^{(0)}}{\partial \boldsymbol{C}}\boldsymbol{C}:\frac{\partial}{\partial \boldsymbol{r}}\boldsymbol{c}_0\right\}. \qquad (7.3, 2)$$

* A very full discussion of normal and non-normal solutions of Boltzmann's equation has been given by H. Grad, *Handbuch der Physik*, vol. 12, 241–51, 266–93 (Springer, 1958); *Phys. Fluids*, 6, 147 (1963).

Now, by (7.11, 9)

$$\ln f^{(0)} = \text{const.} + \ln(n/T^{\frac{3}{2}}) - mC^2/2kT,$$

and so

$$\frac{\partial \ln f^{(0)}}{\partial \boldsymbol{C}} = -\frac{m\boldsymbol{C}}{kT},$$

while, by (7.14, 16, 17),

$$\frac{D_0 \ln f^{(0)}}{Dt} = \frac{mC^2}{2kT^2}\frac{D_0 T}{Dt} = -\frac{mC^2}{3kT}\frac{\partial}{\partial \boldsymbol{r}}.\boldsymbol{c}_0.$$

Thus the sum of the first and last terms in the bracket on the right-hand side of (7.3, 2) is

$$\frac{m}{kT}\overset{\circ}{\boldsymbol{C}\boldsymbol{C}}:\frac{\partial}{\partial \boldsymbol{r}}\boldsymbol{c}_0,$$

where $\overset{\circ}{\boldsymbol{C}\boldsymbol{C}}$ is given by (1.32, 2). Also the sum of the two middle terms is

$$\boldsymbol{C}.\left(\frac{\partial \ln f^{(0)}}{\partial \boldsymbol{r}} - \frac{m}{\rho kT}\frac{\partial p}{\partial \boldsymbol{r}}\right) = \boldsymbol{C}.\frac{\partial \ln (f^{(0)}/nkT)}{\partial \boldsymbol{r}}$$

$$= \boldsymbol{C}.\left(\frac{\partial \ln T^{-\frac{5}{2}}}{\partial \boldsymbol{r}} + \frac{mC^2}{2kT^2}\frac{\partial T}{\partial \boldsymbol{r}}\right)$$

$$= \left(\frac{mC^2}{2kT} - \frac{5}{2}\right)\boldsymbol{C}.\frac{\partial \ln T}{\partial \boldsymbol{r}}.$$

Accordingly (7.3, 2) may be written in the form

$$\mathscr{D}^{(1)} = f^{(0)}\left\{\left(\frac{mC^2}{2kT} - \frac{5}{2}\right)\boldsymbol{C}.\frac{\partial \ln T}{\partial \boldsymbol{r}} + \frac{m}{kT}\overset{\circ}{\boldsymbol{C}\boldsymbol{C}}:\frac{\partial}{\partial \boldsymbol{r}}\boldsymbol{c}_0\right\}$$

$$= f^{(0)}\left\{(\mathscr{C}^2 - \tfrac{5}{2})\boldsymbol{C}.\frac{\partial \ln T}{\partial \boldsymbol{r}} + 2\overset{\circ}{\boldsymbol{\mathscr{C}}\boldsymbol{\mathscr{C}}}:\frac{\partial}{\partial \boldsymbol{r}}\boldsymbol{c}_0\right\}, \tag{7.3, 3}$$

where $\boldsymbol{\mathscr{C}}$ is a dimensionless variable defined by the equation

$$\boldsymbol{\mathscr{C}} \equiv \left(\frac{m}{2kT}\right)^{\frac{1}{2}}\boldsymbol{C}, \tag{7.3, 4}$$

with components $\mathscr{U}, \mathscr{V}, \mathscr{W}$ and magnitude $\mathscr{C}$. In terms of this variable

$$f^{(0)} = n\left(\frac{m}{2\pi kT}\right)^{\frac{3}{2}} e^{-\mathscr{C}^2}, \quad f^{(0)}d\boldsymbol{c} = \frac{n}{\pi^{\frac{3}{2}}}e^{-\mathscr{C}^2}d\boldsymbol{\mathscr{C}}. \tag{7.3, 5}$$

The integral part $J^{(1)}$ of $\xi^{(1)}$ is given by the equation

$$J^{(1)} = J(f^{(0)}f_1^{(1)}) + J(f^{(1)}f_1^{(0)}).$$

Substitute $f^{(1)} = f^{(0)}\Phi^{(1)}$: then, using (7.12, 3), we get

$$J^{(1)} = n^2 I(\Phi^{(1)}). \tag{7.3, 6}$$

The equation $\xi^{(1)} = 0$ is therefore equivalent to

$$n^2 I(\Phi^{(1)}) = -f^{(0)}\{(\mathscr{C}^2 - \tfrac{5}{2})\boldsymbol{C}.\nabla \ln T + 2\overset{\circ}{\mathscr{C}\mathscr{C}} : \nabla \boldsymbol{c}_0\}, \tag{7.3, 7}$$

introducing the notation ∇ in place of $\partial/\partial\boldsymbol{r}$ (cf. 1.2).

7.31. The function $\Phi^{(1)}$

Since $\Phi^{(1)}$, like $f^{(1)}$ itself, is a scalar, it suffices to consider only the scalar solutions of (7.3, 7). Now $I(\Phi^{(1)})$ is linear in $\Phi^{(1)}$, and the right-hand side of this equation is linear in the space-derivatives of T and of u_0, v_0, w_0. Hence the most general scalar solution $\Phi^{(1)}$ is the sum of three parts: (i) a linear combination of the components of ∇T: for this to be a scalar, it must be given by the scalar product of ∇T and another vector; (ii) a linear combination of the components of $\nabla \boldsymbol{c}_0$: this must similarly be the scalar product of $\nabla \boldsymbol{c}_0$ and another tensor; (iii) the most general scalar solution of the equation $I(\Phi^{(1)}) = 0$. Of these (i) and (ii) correspond to the particular integral of a differential equation, and (iii) to the complementary function. Thus we can write

$$\Phi^{(1)} = -\frac{1}{n}\left(\frac{2kT}{m}\right)^{\frac{1}{2}} \boldsymbol{A}.\nabla \ln T - \frac{2}{n}\mathsf{B} : \nabla \boldsymbol{c}_0 + \alpha^{(1,1)} + \boldsymbol{\alpha}^{(2,1)}.m\boldsymbol{C} + \alpha^{(3,1)}.\tfrac{1}{2}mC^2, \tag{7.31, 1}$$

where $\boldsymbol{A}$ and $\boldsymbol{\alpha}^{(2,1)}$ are vectors, and B is a tensor: $\boldsymbol{A}$, B are functions of $\boldsymbol{C}$, while $\alpha^{(1,1)}$, $\boldsymbol{\alpha}^{(2,1)}$, $\alpha^{(3,1)}$ are constants.

Substituting from (7.31, 1) into (7.3, 7), and equating the coefficients of the different components of ∇T, $\nabla \boldsymbol{c}_0$, we find that $\boldsymbol{A}$, B are special solutions of the equations

$$nI(\boldsymbol{A}) = f^{(0)}(\mathscr{C}^2 - \tfrac{5}{2})\boldsymbol{\mathscr{C}}, \tag{7.31, 2}$$

$$nI(\mathsf{B}) = f^{(0)}\overset{\circ}{\mathscr{C}\mathscr{C}}. \tag{7.31, 3}$$

It is easily verified that the conditions of solubility of these integral equations are satisfied, i.e. (cf. 7.13) that

$$\int f^{(0)}(\mathscr{C}^2 - \tfrac{5}{2})\boldsymbol{\mathscr{C}}\psi^{(i)} d\boldsymbol{c} = 0, \quad \int f^{(0)}\overset{\circ}{\mathscr{C}\mathscr{C}}\psi^{(i)} d\boldsymbol{c} = 0 \quad (i = 1, 2, 3).$$

The only variables involved in (7.31, 2, 3) are $\boldsymbol{C}$, or $\boldsymbol{\mathscr{C}}$, and the values of n, T at the point in question; $\boldsymbol{c}_0$ does not appear explicitly, but only as involved in $\boldsymbol{\mathscr{C}}$. Hence $\boldsymbol{A}$ and B must be functions of n, T and $\boldsymbol{\mathscr{C}}$. The only vector which can be formed from these elements is $\boldsymbol{\mathscr{C}}$ itself, multiplied by some function of n, T and $\mathscr{C}$, the latter being the one independent scalar connected with $\boldsymbol{\mathscr{C}}$. Hence we can write

$$\boldsymbol{A} = A(\mathscr{C})\boldsymbol{\mathscr{C}}, \tag{7.31, 4}$$

$A(\mathscr{C})$ being a function of $\mathscr{C}$ (and of n, T).

Equation (7.31, 3) can be separated into nine component equations; typical among these equations are

$$nI(B_{xx}) = f^{(0)}(\mathscr{U}^2 - \tfrac{1}{3}\mathscr{C}^2), \quad nI(B_{xy}) = f^{(0)}\mathscr{U}\mathscr{V}.$$

Addition of the three equations of the first type shows that

$$I(B_{xx}+B_{yy}+B_{zz}) = 0;$$

also, from those of the second type,

$$I(B_{xy}-B_{yx}) = 0.$$

Thus the nine equations possess solutions such that

$$B_{xx}+B_{yy}+B_{zz} = 0, \quad B_{xy} = B_{yx}, \quad \text{etc.},$$

i.e. such that B is a symmetrical and non-divergent tensor. Now B depends only on n, T and $\mathscr{C}$, and the only symmetrical non-divergent tensors that can be formed from these elements are multiples of $\overset{\circ}{\mathscr{C}\mathscr{C}}$ by factors that are functions of n, T and $\mathscr{C}$. Hence for the solution of (7.31, 3) we can write

$$\mathsf{B} = \overset{\circ}{\mathscr{C}\mathscr{C}}\, B(\mathscr{C}), \tag{7.31, 5}$$

where $B(\mathscr{C})$ is a function of $\mathscr{C}$, n and T.

The constants $\alpha^{(1,1)}$, $\boldsymbol{\alpha}^{(2,1)}$, $\alpha^{(3,1)}$ in (7.31, 1) are to be chosen such that the corresponding form for $f^{(1)}$ satisfies equations (7.12, 4), i.e. that

$$\begin{aligned} 0 &= \int \psi^{(i)} f^{(1)}\, d\mathbf{c} = \int \psi^{(i)} f^{(0)} \Phi^{(1)}\, d\mathbf{c} \\ &= \int \psi^{(i)} f^{(0)} \left\{ -\frac{1}{n}\left(\frac{2kT}{m}\right)^{\frac{1}{2}} A(\mathscr{C})\boldsymbol{\mathscr{C}}.\nabla \ln T - \frac{2}{n} B(\mathscr{C}) \overset{\circ}{\boldsymbol{\mathscr{C}}\boldsymbol{\mathscr{C}}} : \nabla \mathbf{c}_0 \right. \\ &\qquad \left. + \alpha^{(1,1)} + \boldsymbol{\alpha}^{(2,1)}.m\mathbf{C} + \alpha^{(3,1)}.\tfrac{1}{2}mC^2 \right\} d\mathbf{c}. \end{aligned}$$

Neglecting vanishing integrals, and simplifying by (1.42, 4), these equations, for $i = 1, 2, 3$, become

$$\int f^{(0)}(\alpha^{(1,1)} + \alpha^{(3,1)}.\tfrac{1}{2}mC^2)\, d\mathbf{c} = 0,$$

$$\int f^{(0)} \left(-\frac{1}{n} A(\mathscr{C}) \nabla \ln T + m\boldsymbol{\alpha}^{(2,1)} \right) mC^2\, d\mathbf{c} = 0,$$

$$\int f^{(0)}(\alpha^{(1,1)} + \alpha^{(3,1)}.\tfrac{1}{2}mC^2)\, \tfrac{1}{2}mC^2\, d\mathbf{c} = 0.$$

The first and third of these relations show that $\alpha^{(1,1)} = 0$, $\alpha^{(3,1)} = 0$; the second shows that $\boldsymbol{\alpha}^{(2,1)}$ is proportional to ∇T, so that in (7.31, 1) the term $\boldsymbol{\alpha}^{(2,1)}.m\mathbf{C}$ may be absorbed into the first term on the right: hence we can also write $\boldsymbol{\alpha}^{(2,1)} = 0$, when the second of these relations becomes

$$\int f^{(0)} A(\mathscr{C}) \mathscr{C}^2\, d\mathbf{c} = 0. \tag{7.31, 6}$$

Also (7.31, 1) reduces to

$$\Phi^{(1)} = -\frac{1}{n}\left(\frac{2kT}{m}\right)^{\frac{1}{2}} \mathbf{A}.\nabla \ln T - \frac{2}{n}\mathsf{B} : \nabla \mathbf{c}_0. \tag{7.31, 7}$$

7.4. Thermal conductivity

Before considering the details of the method by which $\boldsymbol{A}$, B are determined, we shall derive general expressions for the second approximation to the thermal flux $\boldsymbol{q}$ (this is $\boldsymbol{q}^{(1)}$, since $\boldsymbol{q}^{(0)} = 0$; cf. (7.14, 5)) and to the deviation of the pressure system from the hydrostatic pressure given by $\mathsf{p}^{(0)}$ or $p\mathsf{U}$; this deviation is, to the approximation considered, given by $\mathsf{p}^{(1)}$. The equation giving $\boldsymbol{q}^{(1)}$ is

$$\boldsymbol{q}^{(1)} = \tfrac{1}{2}m\int f^{(1)}C^2\boldsymbol{C}\,d\boldsymbol{c} = \tfrac{1}{2}m\int f^{(0)}\Phi^{(1)}C^2\boldsymbol{C}\,d\boldsymbol{c}.$$

On substituting for $\Phi^{(1)}$ from (7.31, 7), and neglecting integrals of odd functions of the components of $\mathscr{C}$ or $\boldsymbol{C}$, this becomes

$$\boldsymbol{q}^{(1)} = -\frac{m}{2n}\int f^{(0)}\boldsymbol{C}C^2\left(\frac{2kT}{m}\right)^{\frac{1}{2}}A(\mathscr{C})\,\mathscr{C}.\nabla\ln T\,d\boldsymbol{c}$$

$$= -\frac{m}{2n}\left(\frac{2kT}{m}\right)^{2}\int f^{(0)}\mathscr{C}\mathscr{C}^2A(\mathscr{C})\,\mathscr{C}.\nabla\ln T\,d\boldsymbol{c}$$

$$= -\frac{1}{3}\frac{2k^2T^2}{mn}\nabla\ln T\int f^{(0)}\mathscr{C}^4A(\mathscr{C})\,d\boldsymbol{c}$$

by (1.42, 4). Using (7.31, 6), we get

$$\boldsymbol{q}^{(1)} = -\frac{2k^2T}{3mn}\nabla T\int f^{(0)}(\mathscr{C}^4 - \tfrac{5}{2}\mathscr{C}^2)\,A(\mathscr{C})\,d\boldsymbol{c}$$

$$= -\frac{2k^2T}{3mn}\nabla T\int \mathscr{C}A(\mathscr{C}).f^{(0)}(\mathscr{C}^2 - \tfrac{5}{2})\,\mathscr{C}\,d\boldsymbol{c},$$

whence, by (7.31, 2) and (4.4, 7),

$$\boldsymbol{q}^{(1)} = -\frac{2k^2T}{3m}\nabla T\int \boldsymbol{A}.I(\boldsymbol{A})\,d\boldsymbol{c}$$

$$= -\frac{2k^2T}{3m}[\boldsymbol{A},\boldsymbol{A}]\,\nabla T.$$

Thus, if

$$\lambda = \frac{2k^2T}{3m}[\boldsymbol{A},\boldsymbol{A}], \tag{7.4, 1}$$

so that, by (4.41, 1), λ is essentially positive, the value of $\boldsymbol{q}^{(1)}$ is given by

$$\boldsymbol{q}^{(1)} = -\lambda\nabla T. \tag{7.4, 2}$$

The thermal flow is therefore, to the present order of approximation, opposite in direction and proportional in magnitude to the temperature gradient; this agrees with the approximate theory of Chapter 6. The quantity λ is the coefficient of thermal conduction of the gas.

7.41. Viscosity

The equation for $\mathsf{p}^{(1)}$ is

$$\mathsf{p}^{(1)} = m\int f^{(1)}\mathbf{CC}\,d\mathbf{c} = m\int f^{(0)}\Phi^{(1)}\mathbf{CC}\,d\mathbf{c}.$$

In this we substitute for $\Phi^{(1)}$ from (7.31, 7): then, on omitting integrals of odd functions of the components of $\boldsymbol{C}$,

$$\mathsf{p}^{(1)} = -\frac{2m}{n}\int f^{(0)} B(\mathscr{C})(\overset{\circ}{\mathscr{C}\mathscr{C}} : \nabla \mathbf{c}_0)\,\mathbf{CC}\,d\mathbf{c}$$

$$= -\frac{4kT}{n}\int f^{(0)} B(\mathscr{C})(\overset{\circ}{\mathscr{C}\mathscr{C}} : \nabla \mathbf{c}_0)\,\mathscr{C}\mathscr{C}\,d\mathbf{c}$$

whence, using the theorem of 1.421 and (7.31, 3), and writing e for $\overline{\overline{\nabla \mathbf{c}_0}}$ (cf. 1.33),

$$\mathsf{p}^{(1)} = -\frac{4kT}{5n}\overset{\circ}{\mathsf{e}}\int f^{(0)} B(\mathscr{C})(\overset{\circ}{\mathscr{C}\mathscr{C}} : \overset{\circ}{\mathscr{C}\mathscr{C}})\,d\mathbf{c}$$

$$= -\frac{4kT}{5n}\overset{\circ}{\mathsf{e}}\int f^{(0)}\overset{\circ}{\mathscr{C}\mathscr{C}} : \mathsf{B}\,d\mathbf{c}$$

$$= -\tfrac{4}{5}kT\overset{\circ}{\mathsf{e}}\int \mathsf{B} : I(\mathsf{B})\,d\mathbf{c}$$

$$= -\tfrac{4}{5}kT[\mathsf{B}, \mathsf{B}]\,\overset{\circ}{\mathsf{e}}.$$

Thus, if
$$\mu = \tfrac{2}{5}kT[\mathsf{B}, \mathsf{B}], \qquad (7.41, 1)$$
so that μ, like λ, is an essentially positive quantity, the above expression for $\mathsf{p}^{(1)}$ becomes

$$\mathsf{p}^{(1)} = -2\mu\overset{\circ}{\mathsf{e}} \equiv -2\mu\overset{\circ}{\overline{\overline{\nabla \mathbf{c}_0}}}. \qquad (7.41, 2)$$

By (7.14, 3, 6), the second approximation to the complete pressure tensor p is $p\mathsf{U}+\mathsf{p}^{(1)}$, or

$$p\mathsf{U} - 2\mu\overset{\circ}{\overline{\overline{\nabla \mathbf{c}_0}}},$$

by (7.41, 2). Thus the values of the second approximations to the typical elements p_{xx}, p_{yz} of the pressure tensor p are

$$p_{xx} = p - \tfrac{2}{3}\mu\left(\frac{2\partial u_0}{\partial x} - \frac{\partial v_0}{\partial y} - \frac{\partial w_0}{\partial z}\right),$$

$$p_{yz} = -\mu\left(\frac{\partial v_0}{\partial z} + \frac{\partial w_0}{\partial y}\right).$$

These are identical with the expressions for the corresponding stress-components in a medium of viscosity μ. They are generalizations of the result of 6.2, which corresponds to the case when $v_0 = w_0 = 0$, and u_0 is a function of z only; the above approximations to the six components of stress then reduce to

$$p_{xx} = p_{yy} = p_{zz} = p, \quad p_{xy} = p_{yx} = -\mu\frac{\partial u_0}{\partial z}, \quad p_{yz} = p_{zy} = 0 = p_{zx} = p_{xz}.$$

7.5. Sonine polynomials

The quantities $\boldsymbol{A}$ and B are determined by expressing them in terms of certain polynomials $S_m^{(n)}(x)$,* defined as follows.

Let s be a positive number less than unity, and x, m be any real numbers; also let

$$S = s/(1-s). \tag{7.5, 1}$$

Then the polynomial $S_m^{(n)}(x)$ is defined by the expansion

$$(S/s)^{m+1}e^{-xS} \equiv (1-s)^{-m-1}e^{-xs/(1-s)} = \sum_{n=0}^{\infty} s^n S_m^{(n)}(x). \tag{7.5, 2}$$

Now

$$(1-s)^{-m-1}e^{-xs/(1-s)} = \sum_p (-xs)^p (1-s)^{-p-m-1}/p!$$

$$= \sum_p \sum_q (-xs)^p s^q (m+p+q)_q/p!\,q!,$$

where r_q denotes the product of the q factors $r, r-1, \ldots, r-q+1$; selection of the terms such that $p+q=n$ gives

$$S_m^{(n)}(x) = \sum_{p=0}^{n} (-x)^p (m+n)_{n-p}/p!\,(n-p)!. \tag{7.5, 3}$$

In particular

$$S_m^{(0)}(x) = 1, \quad S_m^{(1)} = m+1-x. \tag{7.5, 4}$$

The polynomials $S_m^{(n)}(x)$ are numerical multiples of Sonine's polynomials,† which arise in the study of Bessel functions.

Since, writing $T = t/(1-t)$,

$$(1-s)^{-m-1}(1-t)^{-m-1}\int_0^\infty e^{-x(1+S+T)}x^m\,dx$$

$$= (1-s)^{-m-1}(1-t)^{-m-1}\int_0^\infty e^{-x(1-st)/(1-s)(1-t)}x^m\,dx$$

$$= (1-st)^{-m-1}\Gamma(m+1),$$

on equating coefficients of s^pt^q on the two sides of this equation, we find that

$$\left.\begin{aligned}\int_0^\infty e^{-x}S_m^{(p)}(x)\,S_m^{(q)}(x)\,x^m\,dx &= 0 \qquad (p \neq q)\\ &= \Gamma(m+p+1)/p! \quad (p=q).\end{aligned}\right\} \tag{7.5, 5}$$

* These polynomials were first used in the kinetic theory of gases by D. Burnett, *Proc. Lond. Math. Soc.* **39**, 385 (1935). The polynomial called $S_m^{(n)}(x)$ by Burnett differs by a constant factor $1/\Gamma(m+n+1)$ from that defined above. An alternative discussion in terms of generalized Hermite polynomials was given by H. Grad (*Comm. Pure Appl. Maths.* **2**, 331 (1949)). These polynomials are associated with an expansion in Cartesian coordinates in the velocity space, whereas the Sonine polynomials are associated with an expansion in polar coordinates.

† N. J. Sonine, *Math. Ann.* **16**, 41 (1880). The typical Sonine polynomial is equal to

$$\Gamma(m+n+1)\,(-1)^n\,S_m^{(n)}(x).$$

It is related to the generalized Laguerre polynomials L used by E. Schrödinger (*Ann. Phys.* **80**, 483 (1926)), which are such that

$$\sum_{k=0}^{\infty} L_{n+k}^{n}(x)\,\frac{t^k}{\Gamma(n+k+1)} = (-1)^n\,(1-t)^{-n-1}e^{-xt/(1-t)}.$$

7.51. The formal evaluation of A and λ

Suppose that the function $A(\mathscr{C})$ can be expanded in a convergent series of the form*

$$A(\mathscr{C}) = \sum_{p=0}^{\infty} a_p S_{\frac{3}{2}}^{(p)}(\mathscr{C}^2), \tag{7.51, 1}$$

where the coefficients a_p are independent of $\mathscr{C}$. The relation (7.31, 6) is now (cf. (7.3, 5)),

$$n\pi^{-\frac{3}{2}} \sum_{p=0}^{\infty} a_p \int e^{-\mathscr{C}^2} \mathscr{C}^2 S_{\frac{3}{2}}^{(p)}(\mathscr{C}^2)\, d\mathscr{C} = 0,$$

or, on integrating over all directions of $\mathscr{C}$, and noting that $S_{\frac{3}{2}}^{(0)}(\mathscr{C}^2) = 1$,

$$0 = 2n\pi^{-\frac{1}{2}} \sum_{p=0}^{\infty} a_p \int_0^{\infty} e^{-\mathscr{C}^2} \mathscr{C}^3 S_{\frac{3}{2}}^{(0)}(\mathscr{C}^2)\, S_{\frac{3}{2}}^{(p)}(\mathscr{C}^2)\, d\mathscr{C}^2$$

$$= 2n\pi^{-\frac{1}{2}} \Gamma(\tfrac{5}{2})\, a_0,$$

by (7.5, 5). Hence $a_0 = 0$, and

$$\boldsymbol{A} \equiv A(\mathscr{C})\boldsymbol{\mathscr{C}} = \sum_{p=1}^{\infty} a_p \boldsymbol{a}^{(p)}, \tag{7.51, 2}$$

where

$$\boldsymbol{a}^{(p)} = S_{\frac{3}{2}}^{(p)}(\mathscr{C}^2)\boldsymbol{\mathscr{C}}. \tag{7.51, 3}$$

The values of $a_1, a_2, \ldots$ are to be determined from (7.31, 2), which is equivalent to

$$nI(\boldsymbol{A}) = -f^{(0)} S_{\frac{3}{2}}^{(1)}(\mathscr{C}^2)\boldsymbol{\mathscr{C}}. \tag{7.51, 4}$$

We multiply this by $\boldsymbol{a}^{(q)} = S_{\frac{3}{2}}^{(q)}(\mathscr{C}^2)\boldsymbol{\mathscr{C}}$ and integrate with respect to $\boldsymbol{c}$. Then, using (4.4, 7) and (7.3, 5), we obtain

$$[\boldsymbol{a}^{(q)}, \boldsymbol{A}] = \alpha_q, \tag{7.51, 5}$$

where

$$\alpha_q = -2\pi^{-\frac{1}{2}} \int_0^{\infty} e^{-\mathscr{C}^2} \mathscr{C}^3 S_{\frac{3}{2}}^{(1)}(\mathscr{C}^2)\, S_{\frac{3}{2}}^{(q)}(\mathscr{C}^2)\, d(\mathscr{C}^2)$$

$$\left.\begin{aligned} &= -\tfrac{15}{4} \quad (q=1) \\ &= 0 \quad\;\; (q \neq 1) \end{aligned}\right\} \tag{7.51, 6}$$

by (7.5, 5). Let

$$a_{pq} \equiv [\boldsymbol{a}^{(p)}, \boldsymbol{a}^{(q)}]. \tag{7.51, 7}$$

Then, in virtue of (7.51, 2), equation (7.51, 5) becomes

$$\sum_{p=1}^{\infty} a_p a_{pq} = \alpha_q \quad (q = 1, 2, \ldots, \infty). \tag{7.51, 8}$$

As the functions $\boldsymbol{a}^{(p)}$ are known, a_{pq} may be regarded as known. Hence (7.51, 8) is an infinite set of equations to determine the infinite set of coefficients a_p. We need not go into the theory of infinite sets of linear equations: it is sufficient to state that when they have a unique solution (and

* Problems of convergence were considered for certain molecular models by D. Burnett, *loc. cit.* For the general theory of the expressibility of a function by a series of polynomials, see R. Courant and D. Hilbert, *Meth. der Math. Phys.* vol. 1, chapter 2.

physical considerations make it clear that this is so in the present problem) the solution can be found by a method of successive approximation. Enskog assumed that $\boldsymbol{A}^{(m)}$, given by

$$\boldsymbol{A}^{(m)} = \sum_{p=1}^{m} a_p^{(m)} \boldsymbol{a}^{(p)}, \tag{7.51, 9}$$

where
$$\sum_{p=1}^{m} a_p^{(m)} a_{pq} = \alpha_q \quad (q=1, 2, \ldots, m) \tag{7.51, 10}$$

can be taken to be an mth approximation to $\boldsymbol{A}$. This approximation amounts to neglecting the terms in (7.51, 2) and (7.51, 8) depending on $\boldsymbol{a}^{(r)}$, where $r > m$. It is assumed that, as m tends to infinity, $a_p^{(m)}$ tends to a_p, and $\boldsymbol{A}^{(m)}$ to $\boldsymbol{A}$. Using the values of α_q given in (7.51, 6), we obtain

$$a_q^{(m)} = -\tfrac{15}{4}\mathscr{A}_{1q}^{(m)}/\mathscr{A}^{(m)} \quad (q=1, 2, \ldots, m) \tag{7.51, 11}$$

as the solution of equations (7.51, 10).* Here $\mathscr{A}^{(m)}$ is the symmetric determinant with elements a_{pq} $(p, q = 1, 2, \ldots, m)$ and $\mathscr{A}_{1q}^{(m)}$ is the cofactor of a_{1q} in $\mathscr{A}^{(m)}$.

The thermal conductivity λ depends only on the coefficient a_1; for, from (7.4, 1),

$$\begin{aligned}\lambda &= \frac{2k^2T}{3m}[\boldsymbol{A}, \boldsymbol{A}] \\ &= \frac{2k^2T}{3m}\sum_{p=1}^{\infty} a_p[\boldsymbol{a}^{(p)}, \boldsymbol{A}] \\ &= -\frac{5k^2T}{2m}a_1\end{aligned}$$

using (7.51, 5 and 6). Since the molecules of the gas are supposed to possess only translatory energy, the specific heat c_v of the gas is $3k/2m$. Thus, using (7.51, 11), we may write

$$\lambda = \tfrac{25}{4}c_v kT \lim_{m\to\infty} \mathscr{A}_{11}^{(m)}/\mathscr{A}^{(m)}. \tag{7.51, 12}$$

The limit on the right can be replaced by an infinite series. By Jacobi's theorem on the minors of a determinant and its reciprocal (adjugate) determinant,†

$$\mathscr{A}_{11}^{(m)}\mathscr{A}^{(m-1)} - (\mathscr{A}_{1m}^{(m)})^2 = \mathscr{A}_{11}^{(m-1)}\mathscr{A}^{(m)},$$

and so
$$\frac{\mathscr{A}_{11}^{(m)}}{\mathscr{A}^{(m)}} - \frac{\mathscr{A}_{11}^{(m-1)}}{\mathscr{A}^{(m-1)}} = \frac{(\mathscr{A}_{1m}^{(m)})^2}{\mathscr{A}^{(m)}\mathscr{A}^{(m-1)}}. \tag{7.51, 13}$$

Also $\mathscr{A}_{11}^{(1)}/\mathscr{A}^{(1)} = 1/a_{11}$; hence, by repeated application of (7.51, 13),

$$\frac{\mathscr{A}_{11}^{(m)}}{\mathscr{A}^{(m)}} = \frac{1}{a_{11}} + \frac{(\mathscr{A}_{12}^{(2)})^2}{\mathscr{A}^{(1)}\mathscr{A}^{(2)}} + \ldots + \frac{(\mathscr{A}_{1m}^{(m)})^2}{\mathscr{A}^{(m-1)}\mathscr{A}^{(m)}}.$$

* In practice, when numerical values of the a_{pq} are known, the most convenient way of finding $a_q^{(m)}$ is by one of the standard numerical techniques (e.g. that of Gauss) for the solution of a system of linear equations.

† See, e.g., L. Mirsky, *Introduction to Linear Algebra* (Oxford, 1961), p. 25.

Substituting from this into (7.51, 12), we get

$$\lambda = \tfrac{25}{4} c_v kT \left[\frac{1}{a_{11}} + \frac{(\mathscr{A}_{12}^{(2)})^2}{\mathscr{A}^{(1)}\mathscr{A}^{(2)}} + \frac{(\mathscr{A}_{13}^{(3)})^2}{\mathscr{A}^{(2)}\mathscr{A}^{(3)}} + \dots \right]. \qquad (7.51, 14)$$

By (7.51, 7), the quantities a_{pq} do not depend on the number-density n; hence the same is true of the series in (7.51, 14), and so of λ.

Every term of the series in (7.51, 14) is essentially positive; for if $x_1, x_2, \dots, x_m$ are arbitrary non-zero parameters,

$$\left[\sum_{p=1}^{m} x_p \boldsymbol{a}_p, \sum_{p=1}^{m} x_p \boldsymbol{a}_p \right] = \sum_{p=1}^{m} \sum_{q=1}^{m} a_{pq} x_p x_q.$$

The expression on the left is essentially positive; it cannot be zero, since $\sum_{p=1}^{\infty} x_p \boldsymbol{a}_p$ is not a summational invariant. Thus the quadratic form on the right is positive definite, which can be true only if the determinant $\mathscr{A}^{(m)}$ of the coefficients is positive. This holds for all m; since $a_{11} = \mathscr{A}^{(1)}$, the series in (7.51, 14) is one of positive terms.

When the expressions a_{pq} are known, the mth approximation to λ is found by taking m terms of the series in (7.51, 14). The calculation of a_{pq} is effected in Chapters 9 and 10.

7.52. The formal evaluation of B and μ

The tensor B and the coefficient of viscosity μ are evaluated by methods similar to those used in 7.51. Since B is of the form $\overset{\circ}{\mathscr{C}\mathscr{C}} B(\mathscr{C})$, we express it as a series of the form

$$\mathsf{B} = \sum_{p=1}^{\infty} b_p \mathsf{b}^{(p)}, \qquad (7.52, 1)$$

where

$$\mathsf{b}^{(p)} = \overset{\circ}{\mathscr{C}\mathscr{C}} S_{\frac{5}{2}}^{(p-1)}(\mathscr{C}^2) \qquad (7.52, 2)$$

and the coefficients b_p are constants to be determined. Multiplying (7.31, 3) by $\mathsf{b}^{(q)}$, and integrating with respect to $\boldsymbol{c}$, we obtain

$$[\mathsf{b}^{(q)}, \mathsf{B}] = \beta_q \qquad (7.52, 3)$$

or, by (7.52, 1),

$$\sum_{p=1}^{\infty} b_p b_{pq} = \beta_q, \qquad (7.52, 4)$$

where

$$b_{pq} = [\mathsf{b}^{(p)}, \mathsf{b}^{(q)}], \qquad (7.52, 5)$$

$$\beta_q = \frac{1}{n} \int f^{(0)} \overset{\circ}{\mathscr{C}\mathscr{C}} : S_{\frac{5}{2}}^{(q-1)}(\mathscr{C}^2) \overset{\circ}{\mathscr{C}\mathscr{C}} \, d\boldsymbol{c}.$$

On using (1.32, 9), (7.3, 5) and (7.5, 4, 5) this gives

$$\begin{aligned} \beta_q &= \frac{4}{3\pi^{\frac{1}{2}}} \int_0^\infty e^{-\mathscr{C}^2} \mathscr{C}^5 S_{\frac{5}{2}}^{(q-1)}(\mathscr{C}^2)\, d(\mathscr{C}^2) \\ &= \tfrac{5}{2} \quad (q=1), \\ &= 0 \quad (q>1). \end{aligned} \qquad (7.52, 6)$$

The viscosity μ depends only on b_1, for (cf. (7.41, 1))

$$\mu = \tfrac{2}{5}kT[\mathsf{B},\mathsf{B}] = \tfrac{2}{5}kT\sum_p b_p[\mathsf{b}^{(p)},\mathsf{B}]$$
$$= kTb_1$$

by (7.52, 3) and (7.52, 6). Thus, solving (7.52, 4) by the same method of approximation as that used in 7.51, we find

$$\mu = \tfrac{5}{2}kT\lim_{m\to\infty}\mathscr{B}_{11}^{(m)}/\mathscr{B}^{(m)} \qquad (7.52, 7)$$

$$= \tfrac{5}{2}kT\left[\frac{1}{b_{11}}+\frac{(\mathscr{B}_{12}^{(2)})^2}{\mathscr{B}^{(1)}\mathscr{B}^{(2)}}+\frac{(\mathscr{B}_{13}^{(3)})^2}{\mathscr{B}^{(2)}\mathscr{B}^{(3)}}+\dots\right], \qquad (7.52, 8)$$

where $\mathscr{B}^{(m)}$ is the determinant with elements b_{pq} ($p, q = 1, 2, \dots, m$), and $\mathscr{B}_{1q}^{(m)}$ is the cofactor of b_{1q} in $\mathscr{B}^{(m)}$. The expression for μ, like that for λ, is independent of the number-density; also every term in the series in (7.52, 8) is positive, so that the successive approximations to μ form a monotonic increasing sequence.

Series similar to (7.52, 8) and (7.51, 14), were given by Enskog.*

Historical note

Enskog's results appeared in his Uppsala Dissertation in 1917.† A little earlier, Chapman‡ had independently given formulae for λ and μ, and for the coefficient of diffusion in a gas mixture, essentially equivalent to those of Enskog. Chapman, following Maxwell, considered equations of change of molecular properties rather than attempting explicitly to solve the Boltzmann equation. Equations (7.51, 8) and (7.52, 4) are readily identifiable as equations of change for the molecular properties $a^{(p)}$, $\mathsf{b}^{(p)}$, and similarly for the corresponding equations in Chapter 8. Essentially, Chapman used such equations in approximate forms like (7.51, 10), whence the agreement of his results with those of Enskog.§

* Inaugural Dissertation, Uppsala (1917). Enskog also showed that of all functions $A^{(m)}$ given by an equation of the form (7.5, 9), the one in which the coefficients $a^{(m)}$ are given by (7.5, 10), is the one that makes the essentially positive quantity $[A-A_p^{(m)}, A-A^{(m)}]$ a minimum. This implies that the coefficients $a_p^{(m)}$ are chosen such as to make $A^{(m)}$, in a certain sense, the best possible approximation to A of the form (7.5, 9). A similar argument applies to B. The minimum principle can conversely be used to establish (7.5, 10) and the corresponding equations for $b_p^{(m)}$: see E. J. Hellund and E. A. Uehling, *Phys. Rev.* **56**, 818 (1939), and M. Kohler, *Z. Phys.* **124**, 772 (1948).

† See also D. Enskog, *Arkiv. Mat. Astr. Fys.* **16**, 1 (1921).

‡ S. Chapman, *Phil. Trans. R. Soc.* A, **216**, 279 (1916), and **217**, 115 (1917); *Proc. R. Soc.* A, **93**, 1 (1916).

§ See also the Historical Summary (p. 407).

8

THE NON-UNIFORM STATE FOR A BINARY GAS-MIXTURE

8.1. Boltzmann's equation, and the equation of transfer, for a binary mixture

The general problem of evaluating the velocity-distribution function in a gas-mixture, when the composition, mass-velocity, and temperature of the gas vary from point to point, is solved in a manner analogous to that employed for a simple gas. In this chapter we consider a mixture of two constituents; the corresponding solution for a multiple gas-mixture is given in Chapter 18.

The definitions of mass-velocity, partial pressures, densities, and so forth, for a gas-mixture, are given in 2.5, p. 44.

The functions f_1, f_2 that specify the distribution of velocities of molecules of the two gases satisfy equations of the form

$$\frac{\partial f_1}{\partial t}+\boldsymbol{c}_1.\frac{\partial f_1}{\partial \boldsymbol{r}}+\boldsymbol{F}_1.\frac{\partial f_1}{\partial \boldsymbol{c}_1}=\frac{\partial_e f_1}{\partial t}, \tag{8.1, 1}$$

$$\frac{\partial f_2}{\partial t}+\boldsymbol{c}_2.\frac{\partial f_2}{\partial \boldsymbol{r}}+\boldsymbol{F}_2.\frac{\partial f_2}{\partial \boldsymbol{c}_2}=\frac{\partial_e f_2}{\partial t}, \tag{8.1, 2}$$

the values of $\partial_e f_1/\partial t$, $\partial_e f_2/\partial t$ being given by (3.52, 1, 9, 11) and similar equations. If f_1 is expressed as a function of $\boldsymbol{C}_1$, $\boldsymbol{r}$ and t instead of $\boldsymbol{c}_1$, $\boldsymbol{r}$ and t, the left-hand side of (8.1, 1) becomes

$$\frac{Df_1}{Dt}+\boldsymbol{C}_1.\frac{\partial f_1}{\partial \boldsymbol{r}}+\left(\boldsymbol{F}_1-\frac{D\boldsymbol{c}_0}{Dt}\right).\frac{\partial f_1}{\partial \boldsymbol{C}_1}-\frac{\partial f_1}{\partial \boldsymbol{C}_1}\boldsymbol{C}_1:\frac{\partial}{\partial \boldsymbol{r}}\boldsymbol{c}_0;$$

there is a similar expression for the left-hand side of (8.1, 2).

The equation of change of a function $\phi_1(\boldsymbol{C}_1, \boldsymbol{r}, t)$ of the velocity of molecules of the first gas is (cf. (3.13, 2))

$$\frac{D(n_1\bar{\phi}_1)}{Dt}+n_1\bar{\phi}_1\frac{\partial}{\partial \boldsymbol{r}}.\boldsymbol{c}_0+\frac{\partial}{\partial \boldsymbol{r}}.n_1\overline{\phi_1\boldsymbol{C}_1}$$

$$-n_1\left\{\frac{D\bar{\phi}_1}{Dt}+\overline{\boldsymbol{C}_1.\frac{\partial\phi_1}{\partial \boldsymbol{r}}}+\left(\boldsymbol{F}_1-\frac{D\boldsymbol{c}_0}{Dt}\right).\overline{\frac{\partial\phi_1}{\partial \boldsymbol{C}_1}}-\overline{\frac{\partial\phi_1}{\partial \boldsymbol{C}_1}\boldsymbol{C}_1}:\frac{\partial}{\partial \boldsymbol{r}}\boldsymbol{c}_0\right\}=n_1\Delta\bar{\phi}_1; \tag{8.1, 3}$$

a similar equation holds for a function $\phi_2(\boldsymbol{C}_2, \boldsymbol{r}, t)$ of the velocities of the molecules of the second gas.

As in 3.21, $\Delta\bar{\phi}$ vanishes for certain values of ϕ; the corresponding equations of change are of special importance.

Case 1. Let $\phi_1 = 1$; then $n_1\Delta\bar{\phi}_1$ is the rate at which the number-density of the first gas is being altered by encounters; this rate is, of course, zero. Thus the equation becomes

$$\frac{Dn_1}{Dt}+n_1\frac{\partial}{\partial \boldsymbol{r}}\cdot\boldsymbol{c}_0+\frac{\partial}{\partial \boldsymbol{r}}\cdot(n_1\overline{\boldsymbol{C}}_1) = 0, \qquad (8.1, 4)$$

which is the equation of conservation of molecules of the first gas. On adding this to the corresponding equation for the second gas, the equation of conservation of molecules of the mixture is obtained; this is

$$\frac{Dn}{Dt}+n\frac{\partial}{\partial \boldsymbol{r}}\cdot\boldsymbol{c}_0+\frac{\partial}{\partial \boldsymbol{r}}\cdot(n_1\overline{\boldsymbol{C}}_1+n_2\overline{\boldsymbol{C}}_2) = 0. \qquad (8.1, 5)$$

Again, multiply the equations for the two gases by m_1 and m_2 and add: then, since $\rho_1\overline{\boldsymbol{C}}_1+\rho_2\overline{\boldsymbol{C}}_2 = 0$ (cf. (2.5, 9)), the resulting equation is

$$\frac{D\rho}{Dt}+\rho\frac{\partial}{\partial \boldsymbol{r}}\cdot\boldsymbol{c}_0 = 0, \qquad (8.1, 6)$$

which is the equation of conservation of mass for the mixture.

Case 2. Let $\phi_1 = m_1\boldsymbol{C}_1$; then the equation of change is

$$\frac{D}{Dt}(\rho_1\overline{\boldsymbol{C}}_1)+\rho_1\overline{\boldsymbol{C}}_1\left(\frac{\partial}{\partial \boldsymbol{r}}\cdot\boldsymbol{c}_0\right)+\frac{\partial}{\partial \boldsymbol{r}}\cdot\rho_1\overline{\boldsymbol{C}_1\boldsymbol{C}_1}-\rho_1\left(\boldsymbol{F}_1-\frac{D\boldsymbol{c}_0}{Dt}\right)$$

$$+\rho_1\overline{\boldsymbol{C}}_1\cdot\frac{\partial}{\partial \boldsymbol{r}}\boldsymbol{c}_0 = n_1\Delta m_1\overline{\boldsymbol{C}}_1. \qquad (8.1, 7)$$

On adding this to the corresponding equation for the second gas, and using the relation

$$n_1\Delta m_1\overline{\boldsymbol{C}}_1+n_2\Delta m_2\overline{\boldsymbol{C}}_2 = 0,$$

expressing the fact that the total momentum of the molecules is unaltered by encounters, the equation of motion of the mixture is obtained, namely

$$\frac{\partial}{\partial \boldsymbol{r}}\cdot\mathsf{p} = \rho_1\boldsymbol{F}_1+\rho_2\boldsymbol{F}_2-\rho\frac{D\boldsymbol{c}_0}{Dt}. \qquad (8.1, 8)$$

Case 3. Let $\phi_1 = E_1$. Then, since the translational motion of a molecule contributes a term $\frac{1}{2}m_1C_1^2$ to E_1, $\partial\phi_1/\partial\boldsymbol{C}_1 = m_1\boldsymbol{C}_1$. The corresponding equation of change is thus

$$\frac{D(n_1\bar{E}_1)}{Dt}+n_1\bar{E}_1\left(\frac{\partial}{\partial \boldsymbol{r}}\cdot\boldsymbol{c}_0\right)+\frac{\partial}{\partial \boldsymbol{r}}\cdot\boldsymbol{q}_1-\rho_1\overline{\boldsymbol{C}}_1\cdot\left(\boldsymbol{F}_1-\frac{D\boldsymbol{c}_0}{Dt}\right)$$

$$+\rho_1\overline{\boldsymbol{C}_1\boldsymbol{C}_1}:\frac{\partial}{\partial \boldsymbol{r}}\boldsymbol{c}_0 = n_1\Delta\bar{E}_1. \qquad (8.1, 9)$$

On adding this to the corresponding equation for the second gas, and using the relation

$$n_1\Delta\bar{E}_1+n_2\Delta\bar{E}_2 = 0,$$

expressing the conservation of energy at encounters, the equation of energy of the whole gas is found to be

$$\frac{D(n\bar{E})}{Dt}+n\bar{E}\left(\frac{\partial}{\partial \boldsymbol{r}}\cdot\boldsymbol{c}_0\right)+\frac{\partial}{\partial \boldsymbol{r}}\cdot\boldsymbol{q}=\rho_1\overline{\boldsymbol{C}}_1\cdot\boldsymbol{F}_1+\rho_2\overline{\boldsymbol{C}}_2\cdot\boldsymbol{F}_2-\mathsf{p}:\frac{\partial}{\partial \boldsymbol{r}}\boldsymbol{c}_0,$$

which, like (3.21, 4), may be interpreted term by term. Writing $d\bar{E}/dT=\frac{1}{2}Nk$, as in (2.44, 1), and using (8.1, 5), this equation reduces to the form

$$\tfrac{1}{2}Nkn\frac{DT}{Dt}+\frac{\partial}{\partial \boldsymbol{r}}\cdot\boldsymbol{q}=\bar{E}\frac{\partial}{\partial \boldsymbol{r}}\cdot(n_1\overline{\boldsymbol{C}}_1+n_2\overline{\boldsymbol{C}}_2)$$
$$+\rho_1\overline{\boldsymbol{C}}_1\cdot\boldsymbol{F}_1+\rho_2\overline{\boldsymbol{C}}_2\cdot\boldsymbol{F}_2-\mathsf{p}:\frac{\partial}{\partial \boldsymbol{r}}\boldsymbol{c}_0. \quad (8.1,\ 10)$$

8.2. The method of solution

The equations (8.1, 1, 2) for f_1, f_2 are solved, as in Chapter 7, by a method of successive approximation. They may be written in the forms

$$\mathscr{D}_1 f_1+J_1(f_1f)+J_{12}(f_1f_2)=0,\quad \mathscr{D}_2 f_2+J_2(f_2f)+J_{21}(f_2f_1)=0, \quad (8.2,\ 1)$$

where $\mathscr{D}_1 f_1$, $\mathscr{D}_2 f_2$ denote the expressions on the left-hand sides of (8.1, 1, 2) and

$$J_1(f_1f)\equiv\iint(f_1f-f_1'f')g\alpha_1\,d\boldsymbol{e}'\,d\boldsymbol{c}, \quad (8.2,\ 2)$$

$$J_{12}(f_1f_2)\equiv\iint(f_1f_2-f_1'f_2')g\alpha_{12}\,d\boldsymbol{e}'\,d\boldsymbol{c}_2, \quad (8.2,\ 3)$$

with similar definitions for the expressions $J_2(f_2f)$, $J_{21}(f_2f_1)$.

The expansions for f_1, f_2 corresponding to (7.1, 1) are

$$f_1=f_1^{(0)}+f_1^{(1)}+f_1^{(2)}+\ldots,\quad f_2=f_2^{(0)}+f_2^{(1)}+f_2^{(2)}+\ldots, \quad (8.2,\ 4)$$

and the equations (8.2, 1) are similarly subdivided into the sets of equations

$$\mathscr{D}_1^{(r)}+J_1^{(r)}=0,\quad \mathscr{D}_2^{(r)}+J_2^{(r)}=0, \quad (8.2,\ 5)$$

where $J_1^{(0)}\equiv J_1(f_1^{(0)}f^{(0)})+J_{12}(f_1^{(0)}f_2^{(0)})$, (8.2, 6)

$$J_1^{(r)}\equiv J_1(f_1^{(0)}f^{(r)})+J_1(f_1^{(1)}f^{(r-1)})+\ldots+J_1(f_1^{(r)}f^{(0)})$$
$$+J_{12}(f_1^{(0)}f_2^{(r)})+\ldots+J_{12}(f_1^{(r)}f_2^{(0)}), \quad (8.2,\ 7)$$

and $\mathscr{D}_1^{(r)}$ is a function of $f_1^{(0)}, f_1^{(1)}, \ldots, f_1^{(r-1)}$, but not of $f_1^{(r)}$, such that $\mathscr{D}_1^{(0)}=0$, and

$$\sum_{r=1}^{\infty}\mathscr{D}_1^{(r)}\equiv\mathscr{D}_1(f_1^{(0)}+f_1^{(1)}+f_1^{(2)}+\ldots)=\mathscr{D}_1 f_1. \quad (8.2,\ 8)$$

Similar remarks apply to $\mathscr{D}_2^{(r)}$ and $J_2^{(r)}$.

The first approximations to f_1, f_2 are given by the equations

$$J_1^{(0)}\equiv J_1(f_1^{(0)}f^{(0)})+J_{12}(f_1^{(0)}f_2^{(0)})=0,\quad J_2^{(0)}\equiv J_2(f_2^{(0)}f^{(0)})+J_{21}(f_2^{(0)}f_1^{(0)})=0.$$

These are identical in form with (4.3, 1); hence, as in 4.3, they possess the solutions

$$f_1^{(0)} = n_1\left(\frac{m_1}{2\pi kT}\right)^{\frac{3}{2}} \exp\left\{-\frac{m_1}{2kT}[(u_1-u_0)^2+(v_1-v_0)^2+(w_1-w_0)^2]\right\}, \tag{8.2, 9}$$

$$f_2^{(0)} = n_2\left(\frac{m_2}{2\pi kT}\right)^{\frac{3}{2}} \exp\left\{-\frac{m_2}{2kT}[(u_2-u_0)^2+(v_2-v_0)^2+(w_2-w_0)^2]\right\}, \tag{8.2, 10}$$

where $n_1, n_2, \boldsymbol{c}_0, T$ are arbitrary functions of $\boldsymbol{r}, t$. These are chosen, as in 7.11, such that n_1, n_2 are the number-densities of the two gases in the mixture, and $\boldsymbol{c}_0$ and T are the mass-velocity and temperature of the mixture: consequently for a given gas at a given time t they are known functions of $\boldsymbol{r}$. This choice implies that

$$\int f_1 d\boldsymbol{c}_1 = \int f_1^{(0)} d\boldsymbol{c}_1, \quad \int f_2 d\boldsymbol{c}_2 = \int f_2^{(0)} d\boldsymbol{c}_2,$$

$$\int f_1 m_1 \boldsymbol{C}_1 d\boldsymbol{c}_1 + \int f_2 m_2 \boldsymbol{C}_2 d\boldsymbol{c}_2 = \int f_1^{(0)} m_1 \boldsymbol{C}_1 d\boldsymbol{c}_1 + \int f_2^{(0)} m_2 \boldsymbol{C}_2 d\boldsymbol{c}_2,$$

$$\int f_1 E_1 d\boldsymbol{c}_1 + \int f_2 E_2 d\boldsymbol{c}_2 = \int f_1^{(0)} E_1 d\boldsymbol{c}_1 + \int f_2^{(0)} E_2 d\boldsymbol{c}_2,$$

i.e. that

$$\int \sum_{r=1}^{\infty} f_1^{(r)} d\boldsymbol{c}_1 = 0, \qquad \int \sum_{r=1}^{\infty} f_2^{(r)} d\boldsymbol{c}_2 = 0, \tag{8.2, 11}$$

$$\int \sum_{r=1}^{\infty} f_1^{(r)} m_1 \boldsymbol{C}_1 d\boldsymbol{c}_1 + \int \sum_{r=1}^{\infty} f_2^{(r)} m_2 \boldsymbol{C}_2 d\boldsymbol{c}_2 = 0, \tag{8.2, 12}$$

$$\int \sum_{r=1}^{\infty} f_1^{(r)} E_1 d\boldsymbol{c}_1 + \int \sum_{r=1}^{\infty} f_2^{(r)} E_2 d\boldsymbol{c}_2 = 0. \tag{8.2, 13}$$

The equations from which $f_1^{(r)}, f_2^{(r)}$ $(r>0)$ are determined are of the form (8.2, 5). By an argument similar to that used in 7.12 it may be shown that, if $F_1^{(r)}$, $F_2^{(r)}$ are any pair of solutions of these equations, any other pair will be of the form $F_1^{(r)}+\chi_1$, $F_2^{(r)}+\chi_2$, where χ_1, χ_2 are solutions of the equations

$$J_1(f_1^{(0)}\chi) + J_1(\chi_1 f^{(0)}) + J_{12}(f_1^{(0)}\chi_2) + J_{12}(\chi_1 f_2^{(0)}) = 0,$$

$$J_2(f_2^{(0)}\chi) + J_2(\chi_2 f^{(0)}) + J_{21}(f_2^{(0)}\chi_1) + J_{21}(\chi_2 f_1^{(0)}) = 0,$$

which, on writing $\chi_1 = f_1^{(0)}\phi_1, \chi_2 = f_2^{(0)}\phi_2$, reduce to

$$n_1^2 I_1(\phi_1) + n_1 n_2 I_{12}(\phi_1+\phi_2) = 0,$$

$$n_2^2 I_2(\phi_2) + n_1 n_2 I_{21}(\phi_2+\phi_1) = 0,$$

in the notation of 4.4. We multiply these equations by $\phi_1 d\boldsymbol{c}_1$, $\phi_2 d\boldsymbol{c}_2$, integrate over all values of $\boldsymbol{c}_1$, $\boldsymbol{c}_2$, and add. The resulting equation is

$$\{\phi, \phi\} = 0,$$

the solution of which was found in 4.41 to be

$$\phi_1 = \alpha_1^{(1)} + \boldsymbol{\alpha}^{(2)}.m_1\boldsymbol{C}_1 + \alpha^{(3)}E_1, \quad \phi_2 = \alpha_2^{(1)} + \boldsymbol{\alpha}^{(2)}.m_2\boldsymbol{C}_2 + \alpha^{(3)}E_2,$$

where $\alpha_1^{(1)}$, $\alpha_2^{(1)}$, $\boldsymbol{\alpha}^{(2)}$, $\alpha^{(3)}$ are arbitrary functions of $\boldsymbol{r}$, t.

Any pair of solutions of (8.2, 5) can be expressed in the forms $F_1^{(r)} + f_1^{(0)}\phi_1$, $F_2^{(r)} + f_2^{(0)}\phi_2$. By a suitable choice of $\alpha_1^{(1)}$, $\alpha_2^{(1)}$, $\boldsymbol{\alpha}^{(2)}$ and $\alpha^{(3)}$, solutions $f_1^{(r)}, f_2^{(r)}$ may be constructed such that

$$\int f_1^{(r)} d\boldsymbol{c}_1 = 0, \quad \int f_2^{(r)} d\boldsymbol{c}_2 = 0, \tag{8.2, 14}$$

$$\int f_1^{(r)} m_1 \boldsymbol{C}_1 d\boldsymbol{c}_1 + \int f_2^{(r)} m_2 \boldsymbol{C}_2 d\boldsymbol{c}_2 = 0, \tag{8.2, 15}$$

$$\int f_1^{(r)} E_1 d\boldsymbol{c}_1 + \int f_2^{(r)} E_2 d\boldsymbol{c}_2 = 0. \tag{8.2, 16}$$

We adopt these as the values of $f_1^{(r)}$ and $f_2^{(r)}$ in the series (8.2, 4) when $r > 0$. Equations (8.2, 11, 12, 13) are then satisfied.

8.21. The subdivision of $\mathscr{D}f$

The subdivision adopted for $\mathscr{D}_1 f_1$ and $\mathscr{D}_2 f_2$ is similar to that for $\mathscr{D}f$ in 7.14. We define $\mathsf{p}^{(r)}$ and $\boldsymbol{q}^{(r)}$ by

$$\mathsf{p}^{(r)} = \int f_1^{(r)} m_1 \boldsymbol{C}_1 \boldsymbol{C}_1 d\boldsymbol{c}_1 + \int f_2^{(r)} m_2 \boldsymbol{C}_2 \boldsymbol{C}_2 d\boldsymbol{c}_2,$$

$$\boldsymbol{q}^{(r)} = \int f_1^{(r)} E_1 \boldsymbol{C}_1 d\boldsymbol{c}_1 + \int f_2^{(r)} E_2 \boldsymbol{C}_2 d\boldsymbol{c}_2.$$

In the same way

$$\overline{\boldsymbol{C}}_1^{(r)} = \frac{1}{n_1}\int f_1^{(r)} \boldsymbol{C}_1 d\boldsymbol{c}_1, \quad \overline{\boldsymbol{C}}_2^{(r)} = \frac{1}{n_2}\int f_2^{(r)} \boldsymbol{C}_2 d\boldsymbol{c}_2,$$

so that

$$\overline{\boldsymbol{C}}_1 = \sum_{r=0}^{\infty} \overline{\boldsymbol{C}}_1^{(r)}, \quad \overline{\boldsymbol{C}}_2 = \sum_{r=0}^{\infty} \overline{\boldsymbol{C}}_2^{(r)},$$

and (8.2, 15) becomes

$$\rho_1 \overline{\boldsymbol{C}}_1^{(r)} + \rho_2 \overline{\boldsymbol{C}}_2^{(r)} = 0.$$

The time-derivatives in equations (8.1, 4, 8, 10) are expanded in accordance with the formal scheme

$$\frac{\partial}{\partial t} = \frac{\partial_0}{\partial t} + \frac{\partial_1}{\partial t} + \frac{\partial_2}{\partial t} + \ldots, \tag{8.21, 1}$$

the subdivision being made as follows, for each of the variables n_1, $\boldsymbol{c}_0$, and T on which $\partial/\partial t$ operates in those equations; in each the operations $\partial_1/\partial t$, $\partial_2/\partial t$, ... are directly specified, while the operation $\partial_0/\partial t$ is specified in terms of D_0/Dt, signifying $\partial_0/\partial t + \boldsymbol{c}_0 . \partial/\partial \boldsymbol{r}$:

$$\frac{D_0 n_1}{Dt} \equiv \frac{\partial_0 n_1}{\partial t} + \boldsymbol{c}_0 . \frac{\partial n_1}{\partial \boldsymbol{r}} = -n_1 \frac{\partial}{\partial \boldsymbol{r}} . \boldsymbol{c}_0, \tag{8.21, 2}$$

$$\frac{\partial_r n_1}{\partial t} \equiv -\frac{\partial}{\partial \boldsymbol{r}} . (n_1 \overline{\boldsymbol{C}}_1^{(r)}) \quad (r > 0), \tag{8.21, 3}$$

$$\rho \frac{D_0 \boldsymbol{c}_0}{Dt} \equiv \rho \left\{ \frac{\partial_0 \boldsymbol{c}_0}{\partial t} + \left(\boldsymbol{c}_0 \cdot \frac{\partial}{\partial \boldsymbol{r}} \right) \boldsymbol{c}_0 \right\} = \rho_1 \boldsymbol{F}_1 + \rho_2 \boldsymbol{F}_2 - \frac{\partial}{\partial \boldsymbol{r}} \cdot \mathsf{p}^{(0)}$$

$$= \rho_1 \boldsymbol{F}_1 + \rho_2 \boldsymbol{F}_2 - \frac{\partial p}{\partial \boldsymbol{r}}, \qquad (8.21, 4)$$

$$\rho \frac{\partial_r \boldsymbol{c}_0}{\partial t} \equiv -\frac{\partial}{\partial \boldsymbol{r}} \cdot \mathsf{p}^{(r} \quad (r > 0), \qquad (8.21, 5)$$

$$\tfrac{3}{2} nk \frac{D_0 T}{Dt} \equiv \tfrac{3}{2} nk \left\{ \frac{\partial_0 T}{\partial t} + \boldsymbol{c}_0 \cdot \frac{\partial T}{\partial \boldsymbol{r}} \right\} = -p \frac{\partial}{\partial \boldsymbol{r}} \cdot \boldsymbol{c}_0, \qquad (8.21, 6)$$

$$\tfrac{3}{2} nk \frac{\partial_r T}{\partial t} \equiv \tfrac{3}{2} kT \frac{\partial}{\partial \boldsymbol{r}} \cdot (n_1 \overline{\boldsymbol{C}}_1^{(r)} + n_2 \overline{\boldsymbol{C}}_2^{(r)}) + \rho_1 \overline{\boldsymbol{C}}_1^{(r)} . \boldsymbol{F}_1 + \rho_2 \overline{\boldsymbol{C}}_2^{(r)} . \boldsymbol{F}_2$$

$$- \frac{\partial}{\partial \boldsymbol{r}} \cdot \boldsymbol{q}^{(r)} - \mathsf{p}^{(r)} : \frac{\partial}{\partial \boldsymbol{r}} \boldsymbol{c}_0 \quad (r > 0). \quad (8.21, 7)$$

In (8.21, 6, 7) it has been assumed that the molecules have only translational energy, so that $N = 3$. Where $\partial_r / \partial t$ operates on any function of n_1, n_2, $\boldsymbol{c}_0$, and T and their space-derivatives, it obeys the ordinary rules of differentiation, as in 7.14.

With this subdivision of the time-derivatives, $\mathscr{D}_1^{(r)}$ may be expressed in the form

$$\mathscr{D}_1^{(r)} = \frac{\partial_{r-1} f_1^{(0)}}{\partial t} + \frac{\partial_{r-2} f_1^{(1)}}{\partial t} + \ldots + \frac{\partial_0 f_1^{(r-1)}}{\partial t} + \left(\boldsymbol{c}_1 \cdot \frac{\partial}{\partial \boldsymbol{r}} + \boldsymbol{F}_1 \cdot \frac{\partial}{\partial \boldsymbol{c}_1} \right) f_1^{(r-1)}, \qquad (8.21, 8)$$

analogous to (7.14, 19).

The equations

$$\mathscr{D}_1^{(r)} + J_1^{(r)} = 0, \quad \mathscr{D}_2^{(r)} + J_2^{(r)} = 0$$

are soluble if and only if certain relations analogous to (7.13, 1) are satisfied by $\mathscr{D}_1^{(r)}$, $\mathscr{D}_2^{(r)}$. Using (3.54, 2, 4, 5) and the definitions of $J_1^{(r)}$, $J_2^{(r)}$, it can be shown that

$$\int J_1^{(r)} d\boldsymbol{c}_1 = 0, \quad \int J_2^{(r)} d\boldsymbol{c}_2 = 0,$$

$$\int J_1^{(r)} m_1 \boldsymbol{C}_1 d\boldsymbol{C}_1 + \int J_2^{(r)} m_2 \boldsymbol{C}_2 d\boldsymbol{c}_2 = 0,$$

$$\int J_1^{(r)} \tfrac{1}{2} m_1 C_1^2 d\boldsymbol{c}_1 + \int J_2^{(r)} \tfrac{1}{2} m_2 C_2^2 d\boldsymbol{c}_2 = 0.$$

Hence for the equations to be soluble it is necessary and sufficient that

$$\int \mathscr{D}_1^{(r)} d\boldsymbol{c}_1 = 0, \quad \int \mathscr{D}_2^{(r)} d\boldsymbol{c}_2 = 0,$$

$$\int \mathscr{D}_1^{(r)} m_1 \boldsymbol{C}_1 d\boldsymbol{c}_1 + \int \mathscr{D}_2^{(r)} m_2 \boldsymbol{C}_2 d\boldsymbol{c}_2 = 0,$$

$$\int \mathscr{D}_1^{(r)} \tfrac{1}{2} m_1 C_1^2 d\boldsymbol{c}_1 + \int \mathscr{D}_2^{(r)} \tfrac{1}{2} m_2 C_2^2 d\boldsymbol{c}_2 = 0.$$

It may readily be verified that with the above choice of $\mathscr{D}_1^{(r)}$, $\mathscr{D}_2^{(r)}$ these conditions are, in fact, satisfied.

8.3. The second approximation to f

Attention will be confined to a study of the second approximations to f_1, f_2. If $f_1^{(1)}, f_2^{(1)}$ are written in the form $f_1^{(0)}\Phi_1^{(1)}, f_2^{(0)}\Phi_2^{(1)}$, then

$$J_1^{(1)} = J_1(f_1^{(0)}f^{(0)}\Phi^{(1)}) + J_1(f_1^{(0)}\Phi_1^{(1)}f^{(0)}) + J_{12}(f_1^{(0)}f_2^{(0)}\Phi_2^{(1)}) + J_{12}(f_1^{(0)}\Phi_1^{(1)}f_2^{(0)})$$
$$= n_1^2 I_1(\Phi_1^{(1)}) + n_1 n_2 I_{12}(\Phi_1^{(1)} + \Phi_2^{(1)}),$$

in the notation of 4.4. Thus the equations satisfied by $\Phi_1^{(1)}$, $\Phi_2^{(1)}$ are

$$\mathscr{D}_1^{(1)} = -n_1^2 I_1(\Phi_1^{(1)}) - n_1 n_2 I_{12}(\Phi_1^{(1)} + \Phi_2^{(1)}), \tag{8.3, 1}$$

$$\mathscr{D}_2^{(1)} = -n_2^2 I_2(\Phi_2^{(1)}) - n_1 n_2 I_{21}(\Phi_1^{(1)} + \Phi_2^{(1)}). \tag{8.3, 2}$$

If f_1 is regarded as a function of $\boldsymbol{C}_1$, $\boldsymbol{r}$, t,

$$\mathscr{D}_1^{(1)} = f_1^{(0)}\left\{\frac{D_0 \ln f_1^{(0)}}{Dt} + \boldsymbol{C}_1 . \frac{\partial \ln f_1^{(0)}}{\partial \boldsymbol{r}}\right.$$
$$\left. + \left(\boldsymbol{F}_1 - \frac{D_0 \boldsymbol{c}_0}{Dt}\right) . \frac{\partial \ln f_1^{(0)}}{\partial \boldsymbol{C}_1} - \frac{\partial \ln f_1^{(0)}}{\partial \boldsymbol{C}_1} \boldsymbol{C}_1 : \frac{\partial}{\partial \boldsymbol{r}} \boldsymbol{c}_0\right\}, \tag{8.3, 3}$$

as in 7.3; also, as in that section,

$$\frac{D_0 \ln f_1^{(0)}}{Dt} = -\frac{m_1 C_1^2}{3kT} \frac{\partial}{\partial \boldsymbol{r}} . \boldsymbol{c}_0, \quad \frac{\partial \ln f_1^{(0)}}{\partial \boldsymbol{C}_1} = -\frac{m_1 \boldsymbol{C}_1}{kT}.$$

Thus the first and last terms in the bracket in (8.3, 3) reduce to

$$\frac{m_1}{kT} \overset{\circ}{\boldsymbol{C}_1 \boldsymbol{C}_1} : \frac{\partial}{\partial \boldsymbol{r}} \boldsymbol{c}_0.$$

Since (8.21, 4) is equivalent to

$$\boldsymbol{F}_1 - \frac{D_0 \boldsymbol{c}_0}{Dt} = \frac{1}{\rho}\left\{\frac{\partial p}{\partial \boldsymbol{r}} + \rho_2(\boldsymbol{F}_1 - \boldsymbol{F}_2)\right\}, \tag{8.3, 4}$$

the middle two terms in the bracket in (8.3, 3) may be written in the form

$$\boldsymbol{C}_1 . \left[\frac{\partial \ln (n_1 T^{-\frac{5}{2}})}{\partial \boldsymbol{r}} + \frac{m_1 C_1^2}{2kT} \frac{\partial \ln T}{\partial \boldsymbol{r}} - \frac{m_1}{\rho kT}\left\{\rho_2(\boldsymbol{F}_1 - \boldsymbol{F}_2) + \frac{\partial p}{\partial \boldsymbol{r}}\right\}\right]. \tag{8.3, 5}$$

Let p_1, p_2 be the partial pressures $n_1 kT$, $n_2 kT$ of the two gases; also let

$$\mathscr{C}_1 \equiv (m_1/2kT)^{\frac{1}{2}} \boldsymbol{C}_1, \quad \mathscr{C}_2 \equiv (m_2/2kT)^{\frac{1}{2}} \boldsymbol{C}_2, \tag{8.3, 6}$$

$$x_1 \equiv n_1/n = p_1/p, \qquad x_2 \equiv n_2/n = p_2/p, \tag{8.3, 7}$$

so that x_1, x_2 denote the proportions by volume of the two gases in the mixture. Then (8.3, 5) can be written in the form

$$\boldsymbol{C}_1 . \{(\mathscr{C}_1^2 - \tfrac{5}{2}) \nabla \ln T + x_1^{-1} \boldsymbol{d}_{12}\},$$

where

$$\boldsymbol{d}_{12} \equiv x_1 \nabla \ln p_1 - \frac{\rho_1 \rho_2}{\rho p}(\boldsymbol{F}_1 - \boldsymbol{F}_2) - \frac{\rho_1}{\rho p} \nabla p. \tag{8.3, 8}$$

Hence (8.3, 3) becomes

$$\mathscr{D}_1^{(1)} = f_1^{(0)}\{(\mathscr{C}_1^2 - \tfrac{5}{2}) \boldsymbol{C}_1 . \nabla \ln T + x_1^{-1} \boldsymbol{d}_{12} . \boldsymbol{C}_1 + 2\overset{\circ}{\mathscr{C}_1 \mathscr{C}_1} : \nabla \boldsymbol{c}_0\}. \tag{8.3, 9}$$

If $\boldsymbol{d}_{21}$ is defined by an equation similar to (8.3, 8) we find in the same way

$$\mathscr{D}_2^{(1)} = f_2^{(0)}\{(\mathscr{C}_2^2-\tfrac{5}{2})\,\boldsymbol{C}_2.\nabla \ln T + x_2^{-1}\boldsymbol{d}_{21}.\boldsymbol{C}_2 + 2\overset{\circ}{\mathscr{C}_2\mathscr{C}_2}:\nabla\boldsymbol{c}_0\}. \quad (8.3,\ 10)$$

Using the equations $x_1 = p_1/p$, $p = p_1+p_2$, two alternative expressions can be derived for $\boldsymbol{d}_{12}$, i.e.

$$\boldsymbol{d}_{12} = \frac{\rho_1\rho_2}{\rho p}\left\{\boldsymbol{F}_2 - \frac{1}{\rho_2}\nabla p_2 - \left(\boldsymbol{F}_1 - \frac{1}{\rho_1}\nabla p_1\right)\right\}, \quad (8.3,\ 11)$$

$$\boldsymbol{d}_{12} = \nabla x_1 + \frac{n_1 n_2(m_2-m_1)}{n\rho}\nabla \ln p - \frac{\rho_1\rho_2}{\rho p}(\boldsymbol{F}_1-\boldsymbol{F}_2). \quad (8.3,\ 12)$$

From either of these (since $\nabla x_2 = -\nabla x_1$) it follows that $\boldsymbol{d}_{21} = -\boldsymbol{d}_{12}$.

The expression for $\boldsymbol{d}_{12}$ given by (8.3, 8) is much simplified if the gas is at rest. Taking $D_0\boldsymbol{c}_0/Dt = 0$ in (8.21, 4) we find

$$\nabla p = \rho_1\boldsymbol{F}_1 + \rho_2\boldsymbol{F}_2$$

and (8.3, 8) becomes

$$\boldsymbol{d}_{12} = x_1\nabla \ln p_1 - \rho_1\boldsymbol{F}_1/p = (\nabla p_1 - \rho_1\boldsymbol{F}_1)/p. \quad (8.3,\ 13)$$

8.31. The functions $\Phi^{(1)}$, $\boldsymbol{A}$, $\boldsymbol{D}$, B

Since, by (8.3, 9, 10), the independent expressions $\nabla \ln T$, $\boldsymbol{d}_{12}$ (or $-\boldsymbol{d}_{21}$) and $\nabla\boldsymbol{c}_0$ occur *linearly* in the left-hand sides of equations (8.3, 1, 2), we can prove, as in 7.31, that $\Phi_1^{(1)}$, $\Phi_2^{(1)}$ are expressible in the forms

$$\Phi_1^{(1)} = -\boldsymbol{A}_1.\frac{\partial \ln T}{\partial \boldsymbol{r}} - \boldsymbol{D}_1.\boldsymbol{d}_{12} - 2\mathsf{B}_1:\frac{\partial}{\partial \boldsymbol{r}}\boldsymbol{c}_0, \quad (8.31,\ 1)$$

$$\Phi_2^{(1)} = -\boldsymbol{A}_2.\frac{\partial \ln T}{\partial \boldsymbol{r}} - \boldsymbol{D}_2.\boldsymbol{d}_{12} - 2\mathsf{B}_2:\frac{\partial}{\partial \boldsymbol{r}}\boldsymbol{c}_0, \quad (8.31,\ 2)$$

the functions $\boldsymbol{A}$, $\boldsymbol{D}$ being vectors and the functions B being non-divergent tensors, such that

$$\boldsymbol{A} = \boldsymbol{C}A(C), \quad \boldsymbol{D} = \boldsymbol{C}D(C), \quad \mathsf{B} = \overset{\circ}{\boldsymbol{C}\boldsymbol{C}}B(C), \quad (8.31,\ 3)$$

with the suffix 1 or 2 throughout. The functions $\boldsymbol{A}$, $\boldsymbol{D}$, B must satisfy the equations

$$\left.\begin{aligned} f_1^{(0)}(\mathscr{C}_1^2-\tfrac{5}{2})\,\boldsymbol{C}_1 &= n_1^2 I_1(\boldsymbol{A}_1) + n_1 n_2 I_{12}(\boldsymbol{A}_1+\boldsymbol{A}_2),\\ f_2^{(0)}(\mathscr{C}_2^2-\tfrac{5}{2})\,\boldsymbol{C}_2 &= n_2^2 I_2(\boldsymbol{A}_2) + n_1 n_2 I_{21}(\boldsymbol{A}_1+\boldsymbol{A}_2),\end{aligned}\right\} \quad (8.31,\ 4)$$

$$\left.\begin{aligned} x_1^{-1}f_1^{(0)}\boldsymbol{C}_1 &= n_1^2 I_1(\boldsymbol{D}_1) + n_1 n_2 I_{12}(\boldsymbol{D}_1+\boldsymbol{D}_2),\\ -x_2^{-1}f_2^{(0)}\boldsymbol{C}_2 &= n_2^2 I_2(\boldsymbol{D}_2) + n_1 n_2 I_{21}(\boldsymbol{D}_1+\boldsymbol{D}_2),\end{aligned}\right\} \quad (8.31,\ 5)$$

$$\left.\begin{aligned} f_1^{(0)}\overset{\circ}{\mathscr{C}_1\mathscr{C}_1} &= n_1^2 I_1(\mathsf{B}_1) + n_1 n_2 I_{12}(\mathsf{B}_1+\mathsf{B}_2),\\ f_2^{(0)}\overset{\circ}{\mathscr{C}_2\mathscr{C}_2} &= n_2^2 I_2(\mathsf{B}_2) + n_1 n_2 I_{21}(\mathsf{B}_1+\mathsf{B}_2).\end{aligned}\right\} \quad (8.31,\ 6)$$

These equations may be shown to satisfy the conditions of solubility, which are similar to those of the original equations (8.3, 1, 2). For (8.2, 14–16) to be satisfied, $\boldsymbol{A}$, $\boldsymbol{D}$ must also be chosen such that

$$\int f_1^{(0)} m_1 \boldsymbol{C}_1 . \boldsymbol{A}_1 d\boldsymbol{c}_1 + \int f_2^{(0)} m_2 \boldsymbol{C}_2 . \boldsymbol{A}_2 d\boldsymbol{c}_2 = 0, \tag{8.31, 7}$$

$$\int f_1^{(0)} m_1 \boldsymbol{C}_1 . \boldsymbol{D}_1 d\boldsymbol{c}_1 + \int f_2^{(0)} m_2 \boldsymbol{C}_2 . \boldsymbol{D}_2 d\boldsymbol{c}_2 = 0. \tag{8.31, 8}$$

Thus, to a second approximation, f_1 and f_2 are given by

$$f_1 = f_1^{(0)}\{1 - A_1(C_1)\boldsymbol{C}_1 . \nabla \ln T - D_1(C_1)\boldsymbol{C}_1 . \boldsymbol{d}_{12} - 2B_1(C_1)\overset{\circ}{\boldsymbol{C}_1\boldsymbol{C}_1} : \nabla \boldsymbol{c}_0\}, \tag{8.31, 9}$$

$$f_2 = f_2^{(0)}\{1 - A_2(C_2)\boldsymbol{C}_2 . \nabla \ln T - D_2(C_2)\boldsymbol{C}_2 . \boldsymbol{d}_{12} - 2B_2(C_2)\overset{\circ}{\boldsymbol{C}_2\boldsymbol{C}_2} : \nabla \boldsymbol{c}_0\}. \tag{8.31, 10}$$

It may readily be verified with the aid of these expressions that, correct to the second approximation, the mean kinetic energy of the peculiar motion of the molecules is the same for each of the two constituent gases.

In virtue of (8.31, 4–6), if $\boldsymbol{a}$ is any vector-function, and b any tensor-function, each defined in both velocity-domains, then

$$n^2\{\boldsymbol{A}, \boldsymbol{a}\} = \int f_1^{(0)}(\mathscr{C}_1^2 - \tfrac{5}{2})\boldsymbol{C}_1 . \boldsymbol{a}_1 d\boldsymbol{c}_1 + \int f_2^{(0)}(\mathscr{C}_2^2 - \tfrac{5}{2})\boldsymbol{C}_2 . \boldsymbol{a}_2 d\boldsymbol{c}_2, \tag{8.31, 11}$$

$$n^2\{\boldsymbol{D}, \boldsymbol{a}\} = x_1^{-1}\int f_1^{(0)}\boldsymbol{C}_1 . \boldsymbol{a}_1 d\boldsymbol{c}_1 - x_2^{-1}\int f_2^{(0)}\boldsymbol{C}_2 . \boldsymbol{a}_2 d\boldsymbol{c}_2, \tag{8.31, 12}$$

$$n^2\{\mathsf{B}, \mathsf{b}\} = \int f_1^{(0)}\overset{\circ}{\boldsymbol{\mathscr{C}}_1\boldsymbol{\mathscr{C}}_1} : \mathsf{b}_1 d\boldsymbol{c}_1 + \int f_2^{(0)}\overset{\circ}{\boldsymbol{\mathscr{C}}_2\boldsymbol{\mathscr{C}}_2} : \mathsf{b}_2 d\boldsymbol{c}_2. \tag{8.31, 13}$$

8.4. Diffusion and thermal diffusion

The two constituents of the gas-mixture are said to be diffusing relative to one another if the mean velocities of the two sets of molecules at $\boldsymbol{r}$, t are not the same, that is, if $\bar{\boldsymbol{c}}_1 - \bar{\boldsymbol{c}}_2$ or (what is equivalent) $\overline{\boldsymbol{C}}_1 - \overline{\boldsymbol{C}}_2$ is not equal to zero. Now

$$\overline{\boldsymbol{C}}_1 - \overline{\boldsymbol{C}}_2 = \frac{1}{n_1}\int f_1 \boldsymbol{C}_1 d\boldsymbol{c}_1 - \frac{1}{n_2}\int f_2 \boldsymbol{C}_2 d\boldsymbol{c}_2.$$

On substituting from (8.31, 9, 10), the first and last terms in the expressions for f_1 and f_2 give integrals of odd functions of the components of $\boldsymbol{C}$; hence these integrals vanish. The remaining terms readily yield the expression

$$\begin{aligned}
\overline{\boldsymbol{C}}_1 - \overline{\boldsymbol{C}}_2 &= -\frac{1}{3}\left[\left\{\frac{1}{n_1}\int f_1^{(0)} C_1^2 D_1(C_1)\, d\boldsymbol{c}_1 - \frac{1}{n_2}\int f_2^{(0)} C_2^2 D_2(C_2)\, d\boldsymbol{c}_2\right\}\boldsymbol{d}_{12}\right.\\
&\qquad \left. + \left\{\frac{1}{n_1}\int f_1^{(0)} C_1^2 A_1(C_1)\, d\boldsymbol{c}_1 - \frac{1}{n_2}\int f_2^{(0)} C_2^2 A_2(C_2)\, d\boldsymbol{c}_2\right\}\nabla \ln T\right]\\
&= -\frac{1}{3}\left[\left\{\frac{1}{n_1}\int f_1^{(0)}\boldsymbol{C}_1 . \boldsymbol{D}_1 d\boldsymbol{c}_1 - \frac{1}{n_2}\int f_2^{(0)}\boldsymbol{C}_2 . \boldsymbol{D}_2 d\boldsymbol{c}_2\right\}\boldsymbol{d}_{12}\right.\\
&\qquad \left. + \left\{\frac{1}{n_1}\int f_1^{(0)}\boldsymbol{C}_1 . \boldsymbol{A}_1 d\boldsymbol{c}_1 - \frac{1}{n_2}\int f_2^{(0)}\boldsymbol{C}_2 . \boldsymbol{A}_2 d\boldsymbol{c}_2\right\}\nabla \ln T\right]\\
&= -\tfrac{1}{3}n[\{\boldsymbol{D}, \boldsymbol{D}\}\boldsymbol{d}_{12} + \{\boldsymbol{D}, \boldsymbol{A}\}\nabla \ln T]
\end{aligned} \tag{8.4, 1}$$

by (8.31, 12).

The velocity of diffusion $\overline{C}_1 - \overline{C}_2$ accordingly has one component in the direction of $-\boldsymbol{d}_{12}$, since the factor $\{\boldsymbol{D}, \boldsymbol{D}\}$ is positive, by (4.41, 2). Since $-\boldsymbol{d}_{12}$ itself (cf. (8.3, 12)) has components proportional, with different (positive) factors, to $-\nabla x_1$, to $\boldsymbol{F}_1 - \boldsymbol{F}_2$, and to $-(m_2 - m_1)\nabla \ln p$, the diffusive motion of molecules of the first gas relative to the second has components in the directions of these vectors. The first of these three components corresponds to diffusion tending to reduce the inhomogeneity of a gas whose composition is not uniform; the second component indicates that diffusion also occurs, as one would expect, when the accelerative effects of the forces acting on the molecules of the two gases are unequal; the third component shows that, when the pressure is non-uniform, the heavier molecules tend to diffuse towards the regions of greater pressure.

The velocity of diffusion also possesses a component in the direction of the temperature gradient; at this stage no general statement can be made about the sign of the coefficient $\{\boldsymbol{D}, \boldsymbol{A}\}$. This thermal diffusion tends to produce a non-uniform steady state in a gas enclosed in a vessel, different parts of which are maintained at different steady temperatures.

The definition of the coefficient of diffusion D_{12} given in Chapter 6 refers to a state of the gas in which no forces act on the molecules, and the pressure and temperature of the gas are uniform. In this case n does not depend on $\boldsymbol{r}$, and so (8.3, 12) reduces to

$$\boldsymbol{d}_{12} = \nabla x_1 = n^{-1}\nabla n_1.$$

Thus

$$\overline{c}_1 - \overline{c}_2 = \overline{C}_1 - \overline{C}_2 = -\tfrac{1}{3}\{\boldsymbol{D}, \boldsymbol{D}\}\nabla n_1. \tag{8.4, 2}$$

The vector $n_1\overline{c}_1$ giving the flow of molecules of the first gas is equal, as in 6.4, to $-D_{12}\nabla n_1$, and similarly for molecules of the second gas. Hence

$$\begin{aligned}\overline{c}_1 - \overline{c}_2 &= -D_{12}\left(\frac{1}{n_1}\nabla n_1 - \frac{1}{n_2}\nabla n_2\right)\\ &= -D_{12}\left(\frac{1}{n_1} + \frac{1}{n_2}\right)\nabla n_1\\ &= -D_{12}\frac{n}{n_1 n_2}\nabla n_1.\end{aligned} \tag{8.4, 3}$$

Comparing (8.4, 2 and 3), we see that

$$D_{12} = (n_1 n_2/3n)\{\boldsymbol{D}, \boldsymbol{D}\}. \tag{8.4, 4}$$

We also write

$$D_T \equiv (n_1 n_2/3n)\{\boldsymbol{D}, \boldsymbol{A}\} \tag{8.4, 5}$$

and

$$k_T \equiv D_T/D_{12} = \{\boldsymbol{D}, \boldsymbol{A}\}/\{\boldsymbol{D}, \boldsymbol{D}\}. \tag{8.4, 6}$$

The coefficient D_T is called the coefficient of *thermal diffusion*, and k_T will be termed the *thermal-diffusion ratio*. In terms of D_{12}, D_T and k_T, (8.4, 1) may be written as

$$\overline{\boldsymbol{C}}_1-\overline{\boldsymbol{C}}_2 = -\frac{n^2}{n_1 n_2}\{D_{12}\boldsymbol{d}_{12}+D_T\nabla\ln T\}$$

$$= -\frac{n^2}{n_1 n_2}D_{12}\{\boldsymbol{d}_{12}+\mathrm{k}_T\nabla\ln T\}, \qquad (8.4, 7)$$

which is the general equation of diffusion (to the present order of approximation). It is also convenient* to introduce a *thermal-diffusion factor* α_{12} defined by

$$\mathrm{k}_T = \mathrm{x}_1\mathrm{x}_2\alpha_{12}. \qquad (8.4, 8)$$

Then the term in $\overline{\boldsymbol{C}}_1-\overline{\boldsymbol{C}}_2$ due to the temperature gradient becomes $-\alpha_{12}D_{12}\nabla\ln T$.

As can most simply be seen by interchanging the suffices 1 and 2 in (8.4, 7) and using the relation $\boldsymbol{d}_{21} = -\boldsymbol{d}_{12}$, the quantities D_{12}, α_{12} satisfy the relations $D_{21} = D_{12}$, $\alpha_{21} = -\alpha_{12}$.

8.41. Thermal conduction

Since we are at present considering a gas-mixture in which all the molecules possess only translatory communicable energy, the thermal flux is given (cf. (2.5, 13)) by

$$\boldsymbol{q} = \int f_1 \tfrac{1}{2}m_1 C_1^2 \boldsymbol{C}_1 d\boldsymbol{c}_1 + \int f_2 \tfrac{1}{2}m_2 C_2^2 \boldsymbol{C}_2 d\boldsymbol{c}_2. \qquad (8.41, 1)$$

Thus

$$(\boldsymbol{q}/kT)-\tfrac{5}{2}(n_1\overline{\boldsymbol{C}}_1+n_2\overline{\boldsymbol{C}}_2) = \int f_1(\mathscr{C}_1^2-\tfrac{5}{2})\boldsymbol{C}_1 d\boldsymbol{c}_1+\int f_2(\mathscr{C}_2^2-\tfrac{5}{2})\boldsymbol{C}_2 d\boldsymbol{c}_2. \quad (8.41, 2)$$

After substituting for f_1, f_2 from (8.31, 9, 10), and omitting terms that do not contribute to the final result, the right-hand side of this equation becomes (to the present order of approximation)

$$-\tfrac{1}{3}\int f_1^{(0)}(\mathscr{C}_1^2-\tfrac{5}{2})\{(\boldsymbol{C}_1.\boldsymbol{D}_1)\boldsymbol{d}_{12}+(\boldsymbol{C}_1.\boldsymbol{A}_1)\nabla\ln T\}d\boldsymbol{c}_1$$

$$-\tfrac{1}{3}\int f_2^{(0)}(\mathscr{C}_2^2-\tfrac{5}{2})\{(\boldsymbol{C}_2.\boldsymbol{D}_2)\boldsymbol{d}_{12}+(\boldsymbol{C}_2.\boldsymbol{A}_2)\nabla\ln T\}d\boldsymbol{c}_2,$$

or, using (8.31, 11), $\quad -\tfrac{1}{3}n^2[\{\boldsymbol{A},\boldsymbol{D}\}\boldsymbol{d}_{12}+\{\boldsymbol{A},\boldsymbol{A}\}\nabla\ln T]$.

Hence our approximation to $\boldsymbol{q}$ is

$$\boldsymbol{q} = \tfrac{5}{2}kT(n_1\overline{\boldsymbol{C}}_1+n_2\overline{\boldsymbol{C}}_2)-\tfrac{1}{3}kn^2T[\{\boldsymbol{A},\boldsymbol{D}\}\boldsymbol{d}_{12}+\{\boldsymbol{A},\boldsymbol{A}\}\nabla\ln T],$$

or, eliminating $\boldsymbol{d}_{12}$ between this equation and (8.4, 1),

$$\boldsymbol{q} = \tfrac{5}{2}kT(n_1\overline{\boldsymbol{C}}_1+n_2\overline{\boldsymbol{C}}_2)+knT(\overline{\boldsymbol{C}}_1-\overline{\boldsymbol{C}}_2)(\{\boldsymbol{A},\boldsymbol{D}\}/\{\boldsymbol{D},\boldsymbol{D}\})-\lambda\nabla T$$

$$= -\lambda\nabla T+\tfrac{5}{2}kT(n_1\overline{\boldsymbol{C}}_1+n_2\overline{\boldsymbol{C}}_2)+knT\mathrm{k}_T(\overline{\boldsymbol{C}}_1-\overline{\boldsymbol{C}}_2), \qquad (8.41, 3)$$

where $\qquad \lambda \equiv \tfrac{1}{3}kn^2[\{\boldsymbol{A},\boldsymbol{A}\}-\{\boldsymbol{A},\boldsymbol{D}\}^2/\{\boldsymbol{D},\boldsymbol{D}\}]. \qquad (8.41, 4)$

* W. H. Furry, R. Clark Jones and L. Onsager, *Phys. Rev.* 55, 1083 (1939).

We write

$$\bar{A}_1 \equiv A_1 - \frac{\{A, D\}}{\{D, D\}} D_1 = A_1 - \mathrm{k}_T D_1;\ \bar{A}_2 \equiv A_2 - \mathrm{k}_T D_2, \qquad (8.41, 5)$$

so that

$$\{\bar{A}, \bar{A}\} = \{A, A\} - 2\mathrm{k}_T\{A, D) + \mathrm{k}_T^2\{D, D\}$$
$$= \{A, A\} - \{A, D\}^2/\{D, D\}.$$

Then (8.41, 4) becomes $\lambda = \frac{1}{3}kn^2\{\bar{A}, \bar{A}\}$, (8.41, 6)

showing that λ is essentially positive. If there is no mutual diffusion of the gases in the mixture, so that $\overline{C}_1$, $\overline{C}_2$ both vanish, the thermal flux becomes equal to $-\lambda\nabla T$; this is the case when the composition has attained the steady state corresponding to the given temperature distribution. Clearly λ is identical with the coefficient of thermal conduction, as usually defined.

Equation (8.41, 3) indicates that the thermal flux is in general made up of three parts. First, there is the ordinary flow of heat resulting from inequalities of temperature in the gas. Next, when diffusion is proceeding, there is a heat flow resulting from the flux of $n_1\overline{C}_1 + n_2\overline{C}_2$ molecules per unit time relative to the mass-velocity: each molecule carries, on an average, a quantity $\frac{5}{2}kT$* of heat energy. This flow appears because the thermal flux is measured relative to the mass-velocity c_0 of the gas, not relative to the mean velocity $\bar{c}$ of the molecules; if it were measured relative to $\bar{c}$, this term in the flow would disappear. Lastly, even if the flow of heat were measured relative to $\bar{c}$, diffusion would contribute a term $knT\mathrm{k}_T(\overline{C}_1 - \overline{C}_2)$ to the heat flow. Only the first of these three parts of the thermal flux is usually measured in laboratory experiments.

The flow $knT\mathrm{k}_T(\overline{C}_1 - \overline{C}_2)$ is, however, important in certain circumstances. It indicates a heat flow due to diffusion, which may be regarded as an effect inverse to thermal diffusion; it is called the *diffusion thermo-effect*. The direction of the flow is such that, if it generates a temperature gradient,

* The factor $\frac{5}{2}$ occurs instead of the usual $\frac{3}{2}$ because the kinetic energy of the molecules, as well as the thermal flow, is that measured relative to the mass-velocity c_0, not the mean velocity $\bar{c}$. The following analysis is due to Enskog. By analogy with (8.41, 1), the flow of energy relative to the mean velocity $\bar{c}$ of the molecules is

$$\tfrac{1}{2}(\rho_1\overline{(c_1-\bar{c}).(c_1-\bar{c})\,(c_1-\bar{c})} + \rho_2\overline{(c_2-\bar{c}).(c_2-\bar{c})\,(c_2-\bar{c})})$$
$$= \tfrac{1}{2}(\rho_1\overline{(C_1-\overline{C}).(C_1-\overline{C})\,(C_1-\overline{C})} + \rho_2\overline{(C_2-\overline{C}).(C_2-\overline{C})\,(C_2-\overline{C})}).$$

Expanding this, and using the fact that $\rho_1\overline{C}_1 + \rho_2\overline{C}_2 = 0$, we obtain the expression

$$\tfrac{1}{2}(\rho_1\overline{C_1^2C_1} + \rho_2\overline{C_2^2C_2}) - \tfrac{1}{2}\rho\overline{C}.\overline{C}\,\overline{C} - \tfrac{1}{2}(\rho_1\overline{C_1^2} + \rho_2\overline{C_2^2})\,\overline{C} - \rho_1\overline{C_1C_1}.\overline{C} - \rho_2\overline{C_2C_2}.\overline{C}.$$

Now $\overline{C}$ is a small quantity: thus we may neglect its square, and its products with other small quantities. Hence we can neglect $\frac{1}{2}\rho\overline{C}.\overline{C}\,\overline{C}$, and in the product $\rho_1\overline{C_1C_1}.\overline{C}$ we can replace $\overline{C_1C_1}$ by its first approximation $\overline{C_1C_1}^{(0)} = (kT/m_1)\,\mathsf{U}$, where U is the unit tensor. Then the expression becomes

$$q - \tfrac{5}{2}knT\overline{C} = q - \tfrac{5}{2}kT(n_1\overline{C}_1 + n_2\overline{C}_2).$$

Thus the term $\frac{5}{2}kT(n_1\overline{C}_1 + n_2\overline{C}_2)$ in (8.41, 3) arises because the thermal energy and thermal flow are both relative to the mass-velocity c_0, not to $\bar{c}$.

thermal diffusion due to this gradient decreases the primary diffusion. In other words, the heat flow is in the direction of diffusion of molecules which, in thermal diffusion, would seek the cold side. The equality of the coefficient k_T in the expression for the heat-flow with the thermal diffusion ratio is a special case of a general reciprocal theorem in the thermodynamics of irreversible processes.*

In a gas in which there is no diffusion, $\boldsymbol{d}_{12} = -k_T \nabla \ln T$ (cf. (8.4, 7)). Hence for such a gas (8.31, 1, 2) take the forms

$$\left.\begin{aligned} \Phi_1^{(1)} &= -\boldsymbol{A}_1 . \nabla \ln T - 2\mathsf{B}_1 : \nabla \boldsymbol{c}_0, \\ \Phi_2^{(1)} &= -\boldsymbol{A}_2 . \nabla \ln T - 2\mathsf{B}_2 : \nabla \boldsymbol{c}_0. \end{aligned}\right\} \tag{8.41, 7}$$

These relations give the vectors $\boldsymbol{A}_1$, $\boldsymbol{A}_2$ an independent interest. The conditions that (8.41, 7) should correspond to zero diffusion are

$$\int f_1^{(0)} \boldsymbol{A}_1 . \boldsymbol{C}_1 d\boldsymbol{c}_1 = 0; \quad \int f_2^{(0)} \boldsymbol{A}_2 . \boldsymbol{C}_2 d\boldsymbol{c}_2 = 0. \tag{8.41, 8}$$

8.42. Viscosity

As in the case of a simple gas, the first approximation to the pressure system reduces to the hydrostatic pressure. The second approximation adds to this the pressure system given by $\mathsf{p}^{(1)}$, where

$$\mathsf{p}^{(1)} \equiv n_1 m_1 (\overline{\boldsymbol{C}_1 \boldsymbol{C}_1})^{(1)} + n_2 m_2 (\overline{\boldsymbol{C}_2 \boldsymbol{C}_2})^{(1)},$$

which represents, to this degree of approximation, the deviation of the pressure system from the hydrostatic pressure p.

Only the terms in (8.31, 9, 10) that contain $\nabla \boldsymbol{c}_0$ contribute to the value of $\mathsf{p}^{(1)}$; thus, as in 7.41,

$$\begin{aligned} \mathsf{p}^{(1)} &= -2m_1 \int f_1^{(0)} \boldsymbol{C}_1 \boldsymbol{C}_1 (\mathsf{B}_1 : \nabla \boldsymbol{c}_0) \, d\boldsymbol{c}_1 - 2m_2 \int f_2^{(0)} \boldsymbol{C}_2 \boldsymbol{C}_2 (\mathsf{B}_2 : \nabla \boldsymbol{c}_0) \, d\boldsymbol{c}_2 \\ &= -\tfrac{2}{5} \{ m_1 \int f_1^{(0)} \overset{\circ}{\boldsymbol{C}_1 \boldsymbol{C}_1} : \mathsf{B}_1 d\boldsymbol{c}_1 + m_2 \int f_2^{(0)} \overset{\circ}{\boldsymbol{C}_2 \boldsymbol{C}_2} : \mathsf{B}_2 d\boldsymbol{c}_2 \} \, \mathsf{e} \\ &= -\tfrac{4}{5} k n^2 T \{\mathsf{B}, \mathsf{B}\} \, \mathsf{e} \end{aligned}$$

by (8.31, 3), the theorem of 1.421, and (8.31, 13). Hence, if we write

$$\mu \equiv \tfrac{2}{5} k n^2 T \{\mathsf{B}, \mathsf{B}\}, \tag{8.42, 1}$$

the last equation becomes

$$\mathsf{p}^{(1)} = -2\mu \mathsf{e} \equiv -2\mu \overline{\overline{\frac{\partial}{\partial \boldsymbol{r}} \boldsymbol{c}_0}}^{\circ}, \tag{8.42, 2}$$

which is identical with the expression for the viscous stress system in any medium (cf. 7.41). Hence μ can be identified with the coefficient of viscosity. As in the case of a simple gas, μ is essentially positive.

* L. Onsager, *Phys. Rev.* **37**, 405 (1931); **38**, 2265 (1931).

8.5. The four first gas-coefficients

Thus, when the velocity-distribution functions for a gas-mixture are determined to a second approximation, it appears that there are four coefficients concerned in the phenomena of the non-uniform state, namely D_{12}, D_T, λ and μ. These we call the 'first' gas-coefficients for a non-uniform gas. They and the associated ratios f and k_T (cf. (6.3, 3) and (8.4, 6)) depend on the four integral expressions

$$\{\bar{\boldsymbol{A}}, \bar{\boldsymbol{A}}\}, \quad \{\boldsymbol{A}, \boldsymbol{D}\}, \quad \{\boldsymbol{D}, \boldsymbol{D}\}, \quad \{\mathsf{B}, \mathsf{B}\},$$

which may be evaluated by methods similar to those of 7.51, 7.52.

8.51. The coefficients of conduction, diffusion, and thermal diffusion

It is assumed that $\bar{\boldsymbol{A}}_1$, $\bar{\boldsymbol{A}}_2$, $\boldsymbol{D}_1$, $\boldsymbol{D}_2$ may be expanded in series, which, for convenience in later work, we write in the forms

$$\bar{\boldsymbol{A}}_1 = \sum_{-\infty}^{+\infty}{}' a_p \boldsymbol{a}_1^{(p)}, \quad \bar{\boldsymbol{A}}_2 = \sum_{-\infty}^{+\infty}{}' a_p \boldsymbol{a}_2^{(p)}, \tag{8.51, 1}$$

$$\boldsymbol{D}_1 = \sum_{-\infty}^{+\infty} d_p \boldsymbol{a}_1^{(p)}, \quad \boldsymbol{D}_2 = \sum_{-\infty}^{+\infty} d_p \boldsymbol{a}_2^{(p)}. \tag{8.51, 2}$$

Here
$$\boldsymbol{a}_1^{(0)} \equiv M_1^{\frac{1}{2}} \rho_2 \mathscr{C}_1/\rho, \quad \boldsymbol{a}_2^{(0)} \equiv -M_2^{\frac{1}{2}} \rho_1 \mathscr{C}_2/\rho \tag{8.51, 3}$$

(where $M_1 = m_1/m_0$, $M_2 = m_2/m_0$, $m_0 = m_1 + m_2$ as before), and for values of p greater than zero

$$\left.\begin{aligned} \boldsymbol{a}_1^{(p)} &\equiv S_{\frac{3}{2}}^{(p)}(\mathscr{C}_1^2)\mathscr{C}_1, \quad \boldsymbol{a}_1^{(-p)} \equiv 0, \\ \boldsymbol{a}_2^{(p)} &\equiv 0, \quad \boldsymbol{a}_2^{(-p)} \equiv S_{\frac{3}{2}}^{(p)}(\mathscr{C}_2^2)\mathscr{C}_2. \end{aligned}\right\} \tag{8.51, 4}$$

The expansions (8.51, 1), (8.51, 2) resemble those of 7.51. It will be noted that in the two equations (8.51, 1) the same coefficients a_p are written, and likewise the coefficients in (8.51, 2) are the same: but except when $p = 0$, one or other of $\boldsymbol{a}_1^{(p)}$ and $\boldsymbol{a}_2^{(p)}$ is always zero, so that the equality of their coefficients has only formal significance. The notation Σ' implies that the summations in (8.51, 1) do not include a term corresponding to $p = 0$; this is shown to be the case by an argument similar to that leading to (7.51, 2), the equations (8.41, 8) taking the place of (7.31, 6). The equality of the coefficients of $\boldsymbol{a}_1^{(0)}$ and $\boldsymbol{a}_2^{(0)}$ in (8.51, 2) follows from equation (8.31, 8). On substituting from (8.51, 2) into the latter equation, the parts of the integrals involving the functions $\boldsymbol{a}^{(p)}$ vanish (as in 7.51) when $p \neq 0$. Hence, since the same coefficient has been chosen for $\boldsymbol{a}_1^{(0)}$ and $\boldsymbol{a}_2^{(0)}$, the equation reduces to

$$\int f_1^{(0)} m_1 \boldsymbol{C}_1 . \boldsymbol{a}_1^{(0)} d\boldsymbol{c}_1 + \int f_2^{(0)} m_2 \boldsymbol{C}_2 . \boldsymbol{a}_2^{(0)} d\boldsymbol{c}_2 = 0.$$

With the values of $\boldsymbol{a}_1^{(0)}$ and $\boldsymbol{a}_2^{(0)}$ given by (8.51, 3), this is equivalent to

$$\int \exp(-\mathscr{C}_1^2)\mathscr{C}_1^2 d\mathscr{C}_1 - \int \exp(-\mathscr{C}_2^2)\mathscr{C}_2^2 d\mathscr{C}_2 = 0,$$

which is obviously satisfied. Thus (8.31, 8) is satisfied by expressions for $\boldsymbol{D}_1, \boldsymbol{D}_2$ of the form (8.51, 2) if the coefficients of $\boldsymbol{a}_1^{(0)}$, $\boldsymbol{a}_2^{(0)}$ are equal, but not otherwise.

To determine the coefficients a_p, d_p we note that, by (8.31, 11, 12),

$$\{\boldsymbol{A}, \boldsymbol{a}^{(q)}\} = \alpha_q, \quad \{\boldsymbol{D}, \boldsymbol{a}^{(q)}\} = \delta_q, \tag{8.51, 5}$$

where $$n^2\alpha_q \equiv \int f_1^{(0)}(\mathscr{C}_1^2 - \tfrac{5}{2})\, \mathbf{C}_1 . \boldsymbol{a}_1^{(q)}\, d\boldsymbol{c}_1 + \int f_2^{(0)}(\mathscr{C}_2^2 - \tfrac{5}{2})\, \mathbf{C}_2 . \boldsymbol{a}_2^{(q)}\, d\boldsymbol{c}_2, \tag{8.51, 6}$$

$$n^2\delta_q \equiv \mathrm{x}_1^{-1}\int f_1^{(0)} \boldsymbol{C}_1 . \boldsymbol{a}_1^{(q)}\, d\boldsymbol{c}_1 - \mathrm{x}_2^{-1}\int f_2^{(0)} \boldsymbol{C}_2 . \boldsymbol{a}_2^{(q)}\, d\boldsymbol{c}_2. \tag{8.51, 7}$$

Evaluating these integrals by the methods of 7.51, using (7.5, 4, 5), we find

$$\alpha_1 = -\frac{15 n_1}{4n^2}\left(\frac{2kT}{m_1}\right)^{\frac{1}{2}}, \quad \alpha_{-1} = -\frac{15 n_2}{4n^2}\left(\frac{2kT}{m_2}\right)^{\frac{1}{2}}, \quad \delta_0 = \frac{3}{2n}\left(\frac{2kT}{m_0}\right)^{\frac{1}{2}}, \tag{8.51, 8}$$

whereas for all other values of q

$$\alpha_q = 0, \quad \delta_q = 0. \tag{8.51, 9}$$

On substituting from (8.51, 2) into (8.51, 5) we find

$$\sum_{p=-\infty}^{\infty} d_p a_{pq} = \delta_q, \tag{8.51, 10}$$

where $$a_{pq} \equiv \{\boldsymbol{a}^{(p)}, \boldsymbol{a}^{(q)}\} \equiv a_{qp}. \tag{8.51, 11}$$

These equations determine the coefficients d_p. Again

$$\{\bar{\boldsymbol{A}}, \boldsymbol{a}^{(q)}\} = \{\boldsymbol{A}, \boldsymbol{a}^{(q)}\} - \mathrm{k}_T\{\boldsymbol{D}, \boldsymbol{a}^{(q)}\} = \alpha_q - \mathrm{k}_T\delta_q. \tag{8.51, 12}$$

Now $\delta_q = 0$ if $q \neq 0$; hence, on substituting from (8.51, 1) and using (8.51, 11) we have

$$\sum_{p=-\infty}^{\infty}{}' a_p a_{pq} = \alpha_q \quad (q \neq 0). \tag{8.51, 13}$$

This provides one equation less than (8.51, 10), corresponding to the fact that there is no coefficient a_0 to be determined.

Equations (8.51, 10) and (8.51, 13) can be solved by successive approximation in the same way as (7.51, 8). The solutions are

$$d_p = \delta_0 \lim_{m\to\infty} \mathscr{A}_{0p}^{(m)}/\mathscr{A}^{(m)} \tag{8.51, 14}$$

and $$a_p = \lim_{m\to\infty} (\alpha_1 \mathscr{A}'^{(m)}_{1p} + \alpha_{-1}\mathscr{A}'^{(m)}_{-1,p})/\mathscr{A}'^{(m)} \quad (p \neq 0). \tag{8.51, 15}$$

Here $\mathscr{A}^{(m)}$ is the determinant of $2m+1$ rows and columns with elements $a_{pq}\,(-m \leqslant p, q \leqslant m)$, and $\mathscr{A}_{0p}^{(m)}$ is the cofactor of a_{0p} in the expansion of $\mathscr{A}^{(m)}$; also $\mathscr{A}'^{(m)}$ denotes $\mathscr{A}_{00}^{(m)}$, and $\mathscr{A}'^{(m)}_{qp}$ is the cofactor of a_{qp} in the expansion of $\mathscr{A}'^{(m)}$.

By (8.51, 1, 2, 5, 9 and 12),

$$\{\boldsymbol{D},\boldsymbol{D}\} = \Sigma d_p\{\boldsymbol{a}^{(p)},\boldsymbol{D}\} = d_0\delta_0,$$

$$\{\boldsymbol{D},\boldsymbol{A}\} = \Sigma d_p\{\boldsymbol{a}^{(p)},\boldsymbol{A}\} = d_1\alpha_1 + d_{-1}\alpha_{-1},$$

$$\{\bar{\boldsymbol{A}},\bar{\boldsymbol{A}}\} = \Sigma' a_p\{\boldsymbol{a}^{(p)},\bar{\boldsymbol{A}}\} = a_1\alpha_1 + a_{-1}\alpha_{-1}.$$

Hence, using (8.51, 8, 14, 15), the formulae for the transport coefficients given in 8.4 and 8.41 become

$$D_{12} = \frac{3}{2}\frac{x_1 x_2 kT}{nm_0}\lim_{m\to\infty}\mathscr{A}'^{(m)}/\mathscr{A}^{(m)}, \tag{8.51, 16}$$

$$k_T = -\tfrac{5}{2}\lim_{m\to\infty}\{x_1 M_1^{-\frac{1}{2}}\mathscr{A}_{01}^{(m)} + x_2 M_2^{-\frac{1}{2}}\mathscr{A}_{0,-1}^{(m)}\}/\mathscr{A}'^{(m)}, \tag{8.51, 17}$$

$$\lambda = \tfrac{75}{8}k^2 T\lim_{m\to\infty}\{x_1^2 m_1^{-1}\mathscr{A}'^{(m)}_{11} + 2x_1x_2(m_1m_2)^{-\frac{1}{2}}\mathscr{A}'^{(m)}_{1-1} + x_2^2 m_2^{-1}\mathscr{A}'^{(m)}_{-1-1}\}/\mathscr{A}'^{(m)}. \tag{8.51, 18}$$

As in the latter part of 7.51, it is possible to prove various theorems about the determinants introduced above, and in particular that the successive approximations to D_{12} and λ form monotonic increasing sequences. On account of the complexity of the results for gas-mixtures, however, these results are not given here.

From (8.51, 3, 4 and 11) the quantities a_{pq} depend on the composition, but are independent of the total density for a mixture of given composition. Thus the same is true of λ and k_T; however, D_{12} is inversely proportional to the total density.

8.52. The coefficient of viscosity

As in 8.51, it is assumed that B_1, B_2 can be expressed in series, of the form

$$\mathsf{B}_1 = \sum_{p=-\infty}^{+\infty}{}' b_p \mathsf{b}_1^{(p)}, \quad \mathsf{B}_2 = \sum_{p=-\infty}^{+\infty}{}' b_p \mathsf{b}_2^{(p)}, \tag{8.52, 1}$$

with the same coefficients in the two series: the functions $\mathsf{b}^{(p)}$ are defined in the two velocity-domains by the equations

$$\left.\begin{aligned} &\mathsf{b}_1^{(p)} \equiv 0, && \mathsf{b}_2^{(-p)} \equiv 0 && (p<0),\\ &\mathsf{b}_1^{(p)} \equiv S_{\frac{5}{2}}^{(p-1)}(\mathscr{C}_1^2)\,\mathscr{C}_1^{\circ}\mathscr{C}_1, && \mathsf{b}_2^{(-p)} \equiv S_{\frac{5}{2}}^{(p-1)}(\mathscr{C}_2^2)\,\mathscr{C}_2^{\circ}\mathscr{C}_2 && (p>0).\end{aligned}\right\} \tag{8.52, 2}$$

As in 8.51, the notation Σ' signifies that the value $p = 0$ does not appear in the summation.

From (8.31, 13)
$$\{\mathsf{B},\mathsf{b}^{(q)}\} = \beta_q, \tag{8.52, 3}$$
where
$$n^2\beta_q \equiv \int f_1^{(0)}\mathscr{C}_1\mathscr{C}_1 : \mathsf{b}_1^{(q)}\,d\boldsymbol{c}_1 + \int f_2^{(0)}\mathscr{C}_2\mathscr{C}_2 : \mathsf{b}_2^{(q)}\,d\boldsymbol{c}_2. \tag{8.52, 4}$$

Integrating as in 7.52, we find

$$\beta_1 = \frac{5}{2}\frac{n_1}{n^2}, \quad \beta_{-1} = \frac{5}{2}\frac{n_2}{n^2}, \quad \beta_q = 0 \quad (q \neq \pm 1). \tag{8.52, 5}$$

On combining (8.52, 1, 3) we obtain

$$\sum_{p=-\infty}^{\infty}{}' \, b_p b_{pq} = \beta_q, \tag{8.52, 6}$$

where

$$b_{pq} = \{\mathsf{b}^{(p)}, \mathsf{b}^{(q)}\}. \tag{8.52, 7}$$

These equations are similar to (8.51, 11, 13). Their solution is

$$b_p = \lim_{m\to\infty} \{\beta_1 \mathscr{B}_{1p}^{(m)} + \beta_{-1} \mathscr{B}_{-1p}^{(m)}\}/\mathscr{B}^{(m)}. \tag{8.52, 8}$$

Here $\mathscr{B}^{(m)}$ is the symmetric determinant, having $2m$ rows and columns, in which the elements are b_{pq}, p and q taking all values other than zero between $-m$ and m; $\mathscr{B}_{qp}^{(m)}$ is the cofactor of b_{qp} in the expansion of $\mathscr{B}^{(m)}$.

From (8.52, 1, 3, 5)

$$\{\mathsf{B}, \mathsf{B}\} = \sum_p b_p \{\mathsf{B}, \mathsf{b}^{(p)}\} = \beta_1 b_1 + \beta_{-1} b_{-1}. \tag{8.52, 9}$$

Hence (8.42, 1) becomes, using (8.52, 5, 8),

$$\mu = \tfrac{5}{2}kT \lim_{m\to\infty} \{\mathrm{x}_1^2 \mathscr{B}_{11}^{(m)} + 2\mathrm{x}_1\mathrm{x}_2 \mathscr{B}_{1-1}^{(m)} + \mathrm{x}_2^2 \mathscr{B}_{-1-1}^{(m)}\}/\mathscr{B}^{(m)}. \tag{8.52, 10}$$

As before, it can be shown that the successive approximations to μ form a monotonic increasing sequence.

Like λ, the viscosity μ depends on the composition, but is independent of the total density for a mixture of given composition.

9

VISCOSITY, THERMAL CONDUCTION, AND DIFFUSION: GENERAL EXPRESSIONS

9.1. The evaluation of $[\boldsymbol{a}^{(p)}, \boldsymbol{a}^{(q)}]$ and $[\mathsf{b}^{(p)}, \mathsf{b}^{(q)}]$

In order to determine the coefficients of viscosity, thermal conduction, and diffusion of a gas by the methods of the last two chapters, it is necessary first to evaluate $\{\boldsymbol{a}^{(p)}, \boldsymbol{a}^{(q)}\}$ and $\{\mathsf{b}^{(p)}, \mathsf{b}^{(q)}\}$, and therefore, by (4.4, 12),

$$[\boldsymbol{a}_1^{(p)}, \boldsymbol{a}_1^{(q)}]_1,\quad [\boldsymbol{a}_1^{(p)}, \boldsymbol{a}_1^{(q)}]_{12},\quad [\boldsymbol{a}_1^{(p)}, \boldsymbol{a}_2^{(q)}]_{12},\quad [\mathsf{b}_1^{(p)}, \mathsf{b}_1^{(q)}]_1,\quad [\mathsf{b}_1^{(p)}, \mathsf{b}_1^{(q)}]_{12},\quad [\mathsf{b}_1^{(p)}, \mathsf{b}_2^{(q)}]_{12}.$$

This involves integration over all the variables specifying an encounter between pairs of like or unlike molecules; such integration can be completely effected only when the nature of the interaction between the molecules is specified. As noted in 3.42, the special law of interaction affects only the relation between the deflection (χ) of the relative velocity at encounter, and the variables g and b. The integration over the variables other than g and b can be performed without a knowledge of the law of interaction, and is effected in this chapter. In the next chapter various special laws of interaction are considered, and the corresponding values of the above expressions are determined.

9.2. Velocity transformations

Certain general relations are grouped here for convenience of reference. The velocities $\boldsymbol{c}_1, \boldsymbol{c}_2$ of two molecules of masses m_1, m_2 before encounter are given in terms of the variables $\boldsymbol{G}, \boldsymbol{g}_{21}$ by the equations (3.41, 6), i.e.

$$\boldsymbol{c}_1 = \boldsymbol{G} - M_2\boldsymbol{g}_{21},\quad \boldsymbol{c}_2 = \boldsymbol{G} + M_1\boldsymbol{g}_{21},$$

where, as in (3.41, 1),

$$M_1 \equiv m_1/m_0,\quad M_2 \equiv m_2/m_0,\quad m_0 \equiv m_1 + m_2,$$

so that

$$M_1 + M_2 = 1.$$

Similar relations connect the velocities after encounter, $\boldsymbol{c}_1', \boldsymbol{c}_2'$, with the variables $\boldsymbol{G}, \boldsymbol{g}_{21}'$. The magnitudes of $\boldsymbol{g}_{21}, \boldsymbol{g}_{21}'$ are equal, and are denoted by g.

Let $\boldsymbol{G}_0$ denote the velocity of the mass-centre of the pair of molecules, relative to axes moving with the mass-velocity of the gas, so that

$$\boldsymbol{G}_0 = \boldsymbol{G} - \boldsymbol{c}_0. \tag{9.2, 1}$$

Then the peculiar velocities $\boldsymbol{C}_1, \boldsymbol{C}_2$ of the molecules are given by the equations

$$\boldsymbol{C}_1 = \boldsymbol{G}_0 - M_2\boldsymbol{g}_{21},\quad \boldsymbol{C}_2 = \boldsymbol{G}_0 + M_1\boldsymbol{g}_{21}; \tag{9.2, 2}$$

similar equations give $\boldsymbol{C}_1'$, $\boldsymbol{C}_2'$. From (9.2, 2) it follows that

$$\tfrac{1}{2}m_1 C_1^2 + \tfrac{1}{2}m_2 C_2^2 = \tfrac{1}{2}m_0(G_0^2 + M_1 M_2 g^2) \tag{9.2, 3}$$

and (cf. (3.52, 4))

$$\frac{\partial(\boldsymbol{G}_0, \boldsymbol{g}_{21})}{\partial(\boldsymbol{C}_1, \boldsymbol{C}_2)} = 1. \tag{9.2, 4}$$

By analogy with the equations (8.3, 6), namely

$$\mathscr{C}_1 \equiv (m_1/2kT)^{\frac{1}{2}} \boldsymbol{C}_1, \quad \mathscr{C}_2 \equiv (m_2/2kT)^{\frac{1}{2}} \boldsymbol{C}_2, \tag{9.2, 5}$$

we define new variables $\mathscr{G}_0$, $\mathscr{g}$, $\mathscr{g}'$ by the equations

$$\mathscr{G}_0 \equiv (m_0/2kT)^{\frac{1}{2}} \boldsymbol{G}_0, \quad \mathscr{g} \equiv (m_0 M_1 M_2/2kT)^{\frac{1}{2}} \boldsymbol{g}_{21}, \quad \mathscr{g}' \equiv (m_0 M_1 M_2/2kT)^{\frac{1}{2}} \boldsymbol{g}_{21}', \tag{9.2, 6}$$

so that $\mathscr{g} = \mathscr{g}'$.

From these definitions and (9.2, 2–4) it follows that

$$\mathscr{C}_1 = M_1^{\frac{1}{2}} \mathscr{G}_0 - M_2^{\frac{1}{2}} \mathscr{g}, \quad \mathscr{C}_2 = M_2^{\frac{1}{2}} \mathscr{G}_0 + M_1^{\frac{1}{2}} \mathscr{g}, \quad \mathscr{C}_1' = M_1^{\frac{1}{2}} \mathscr{G}_0 - M_2^{\frac{1}{2}} \mathscr{g}', \tag{9.2, 7}$$

$$\mathscr{C}_1^2 + \mathscr{C}_2^2 = \mathscr{G}_0^2 + \mathscr{g}^2, \tag{9.2, 8}$$

$$\frac{\partial(\mathscr{G}_0, \mathscr{g})}{\partial(\boldsymbol{c}_1, \boldsymbol{c}_2)} = \frac{(\partial\mathscr{G}_0, \mathscr{g})}{\partial(\boldsymbol{G}_0, \boldsymbol{g}_{21})} \cdot \frac{\partial(\boldsymbol{G}_0, \boldsymbol{g}_{21})}{\partial(\boldsymbol{C}_1, \boldsymbol{C}_2)} = \frac{(m_1 m_2)^{\frac{3}{2}}}{(2kT)^3}. \tag{9.2, 9}$$

Also, since the angle between $\mathscr{g}$ and $\mathscr{g}'$ is the same as that between $\boldsymbol{g}_{21}$ and $\boldsymbol{g}_{21}'$,

$$\mathscr{g} \cdot \mathscr{g}' = \mathscr{g}^2 \cos\chi. \tag{9.2, 10}$$

Any function of the velocities of two molecules after encounter may be transformed into the corresponding function of the velocities before encounter by taking $\chi = 0$.

9.3. The expressions $[S(\mathscr{C}_1^2)\mathscr{C}_1, S(\mathscr{C}_2^2)\mathscr{C}_2]_{12}$ and $[S(\mathscr{C}_1^2)\overset{\circ}{\mathscr{C}_1\mathscr{C}_1}, S(\mathscr{C}_2^2)\overset{\circ}{\mathscr{C}_2\mathscr{C}_2}]_{12}$

By (4.4, 4, 9) and (3.5, 3),

$$[S_{\frac{3}{2}}^{(p)}(\mathscr{C}_1^2)\mathscr{C}_1, S_{\frac{3}{2}}^{(q)}(\mathscr{C}_2^2)\mathscr{C}_2]_{12} \tag{9.3, 1}$$

is equal to

$$\frac{1}{n_1 n_2}\iiiint f_1^{(0)} f_2^{(0)} \{S_{\frac{3}{2}}^{(p)}(\mathscr{C}_1^2)\mathscr{C}_1 - S_{\frac{3}{2}}^{(p)}(\mathscr{C}_1'^2)\mathscr{C}_1'\} . S_{\frac{3}{2}}^{(q)}(\mathscr{C}_2^2)\mathscr{C}_2 g b\,db\,d\epsilon\,d\boldsymbol{c}_1\,d\boldsymbol{c}_2,$$

which, by the definition (7.5) of $S_m^{(n)}(x)$, is the coefficient of $s^p t^q$ in the expansion of

$$\frac{(ST/st)^{\frac{5}{2}}}{n_1 n_2}\iiiint f_1^{(0)} f_2^{(0)} \{\exp(-S\mathscr{C}_1^2)\mathscr{C}_1 - \exp(-S\mathscr{C}_1'^2)\mathscr{C}_1'\} . \mathscr{C}_2 \times \exp(-T\mathscr{C}_2^2)\, g b\,db\,d\epsilon\,d\boldsymbol{c}_1\,d\boldsymbol{c}_2, \tag{9.3, 2}$$

where

$$S \equiv \frac{s}{1-s}, \quad T \equiv \frac{t}{1-t}. \tag{9.3, 3}$$

Inserting the values of $f_1^{(0)}, f_2^{(0)}$ in (9.3, 2), we get

$$\left(\frac{ST}{st}\right)^{\frac{5}{2}}\frac{(m_1 m_2)^{\frac{3}{2}}}{(2\pi kT)^3}\iiiint \exp(-\mathscr{C}_1^2-\mathscr{C}_2^2)\{\exp(-s\mathscr{C}_1^2)\mathscr{C}_1-\exp(-s\mathscr{C}_1'^2)\mathscr{C}_1'\}.\mathscr{C}_2 \times \exp(-T\mathscr{C}_2^2)\, gb\,db\,d\epsilon\,d\boldsymbol{c}_1\,d\boldsymbol{c}_2,$$

or, using (9.2, 8, 9),

$$(ST/st)^{\frac{5}{2}}\pi^{-3}\iiiint \exp(-\mathscr{G}_0^2-g^2)\{\exp(-s\mathscr{C}_1^2)\mathscr{C}_1-\exp(-s\mathscr{C}_1'^2)\mathscr{C}_1'\}.\mathscr{C}_2 \times \exp(-T\mathscr{C}_2^2)\, gb\,db\,d\epsilon\,d\mathscr{G}_0\,d\boldsymbol{g}.$$

Let $$H_{12}(\chi) \equiv \int \exp(-\mathscr{G}_0^2-g^2-s\mathscr{C}_1'^2-T\mathscr{C}_2^2)\mathscr{C}_1'.\mathscr{C}_2\,d\mathscr{G}_0. \qquad (9.3, 4)$$

Since any function of $\mathscr{C}_1', \mathscr{C}_2$ is transformed into the corresponding function of $\mathscr{C}_1, \mathscr{C}_2$ by putting $\chi = 0$,

$$H_{12}(0) = \int \exp(-\mathscr{G}_0^2-g^2-s\mathscr{C}_1^2-T\mathscr{C}_2^2)\mathscr{C}_1\mathscr{C}_2\,d\mathscr{G}_0.$$

Thus the expression (9.3, 2) is equal to

$$(ST/st)^{\frac{5}{2}}\pi^{-3}\iiint\{H_{12}(0)-H_{12}(\chi)\}gb\,db\,d\epsilon\,d\boldsymbol{g}. \qquad (9.3, 5)$$

Similarly, it may be shown that

$$[S_{\frac{5}{2}}^{(p)}(\mathscr{C}_1^2)\overset{\circ}{\mathscr{C}_1}\mathscr{C}_1,\ S_{\frac{5}{2}}^{(q)}(\mathscr{C}_2^2)\overset{\circ}{\mathscr{C}_2}\mathscr{C}_2]_{12} \qquad (9.3, 6)$$

is the coefficient of $s^p t^q$ in the expansion of

$$(ST/st)^{\frac{7}{2}}\pi^{-3}\iiint\{L_{12}(0)-L_{12}(\chi)\}gb\,db\,d\epsilon\,d\boldsymbol{g}, \qquad (9.3, 7)$$

where $$L_{12}(\chi) \equiv \int \exp(-\mathscr{G}_0^2-g^2-s\mathscr{C}_1'^2-T\mathscr{C}_2^2)\overset{\circ}{\mathscr{C}_1'}\mathscr{C}_1':\overset{\circ}{\mathscr{C}_2}\mathscr{C}_2\,d\mathscr{G}_0. \qquad (9.3, 8)$$

9.31. The integrals $H_{12}(\chi)$ and $L_{12}(\chi)$

By (9.2, 7) $$\mathscr{C}_2^2 = (M_2^{\frac{1}{2}}\mathscr{G}_0+M_1^{\frac{1}{2}}\boldsymbol{g}).(M_2^{\frac{1}{2}}\mathscr{G}_0+M_1^{\frac{1}{2}}\boldsymbol{g})$$
$$= M_2\mathscr{G}_0^2+M_1 g^2+2(M_1M_2)^{\frac{1}{2}}\mathscr{G}_0.\boldsymbol{g}, \qquad (9.31, 1)$$

and similarly $$\mathscr{C}_1'^2 = M_1\mathscr{G}_0^2+M_2 g^2-2(M_1M_2)^{\frac{1}{2}}\mathscr{G}_0.\boldsymbol{g}'. \qquad (9.31, 2)$$

Hence

$$\mathscr{G}_0^2+g^2+s\mathscr{C}_1'^2+T\mathscr{C}_2^2 = i_{12}\mathscr{G}_0^2+i_{21}g^2+2(M_1M_2)^{\frac{1}{2}}\mathscr{G}_0.(T\boldsymbol{g}-s\boldsymbol{g}'),$$

where $$i_{12} \equiv 1+M_1 s+M_2 T,\quad i_{21} \equiv 1+M_2 s+M_1 T. \qquad (9.31, 3)$$

Let $$\boldsymbol{v} \equiv \mathscr{G}_0+\frac{1}{i_{12}}(M_1M_2)^{\frac{1}{2}}(T\boldsymbol{g}-s\boldsymbol{g}'), \qquad (9.31, 4)$$

so that a change of variable from $\mathscr{G}_0$ to $\boldsymbol{v}$ is equivalent to a change of origin in the $\mathscr{G}_0$-space. Then

$$\mathscr{G}_0^2+g^2+s\mathscr{C}_1'^2+T\mathscr{C}_2^2 = i_{12}v^2+i_{21}g^2-(M_1M_2/i_{12})(T\boldsymbol{g}-s\boldsymbol{g}').(T\boldsymbol{g}-s\boldsymbol{g}')$$
$$= i_{12}v^2+j_{12}g^2, \qquad (9.31, 5)$$

where $$j_{12} \equiv i_{21} - (M_1 M_2/i_{12})(s^2 + T^2 - 2sT\cos\chi). \tag{9.31, 6}$$

Again, writing $$\left.\begin{aligned} \boldsymbol{v}_1 &\equiv (M_1/i_{12})(T\boldsymbol{g} - s\boldsymbol{g}') + \boldsymbol{g}', \\ \boldsymbol{v}_2 &\equiv (M_2/i_{12})(T\boldsymbol{g} - s\boldsymbol{g}') - \boldsymbol{g}, \end{aligned}\right\} \tag{9.31, 7}$$

we have, by (9.31, 4) and (9.2, 7),

$$\left.\begin{aligned} \mathscr{C}_1' &= M_1^{\frac{1}{2}}\boldsymbol{v} - M_2^{\frac{1}{2}}\boldsymbol{v}_1, \\ \mathscr{C}_2 &= M_2^{\frac{1}{2}}\boldsymbol{v} - M_1^{\frac{1}{2}}\boldsymbol{v}_2. \end{aligned}\right\} \tag{9.31, 8}$$

Hence

$$\mathscr{C}_1'.\mathscr{C}_2 = (M_1 M_2)^{\frac{1}{2}} v^2 - \boldsymbol{v}.(M_2\boldsymbol{v}_1 + M_1\boldsymbol{v}_2) + (M_1 M_2)^{\frac{1}{2}} \boldsymbol{v}_1.\boldsymbol{v}_2, \tag{9.31, 9}$$

$$\mathscr{C}_1'^2 = M_1 v^2 - 2(M_1 M_2)^{\frac{1}{2}} \boldsymbol{v}.\boldsymbol{v}_1 + M_2 v_1^2,$$

$$\mathscr{C}_2^2 = M_2 v^2 - 2(M_1 M_2)^{\frac{1}{2}} \boldsymbol{v}.\boldsymbol{v}_2 + M_1 v_2^2,$$

and so

$$\begin{aligned} \mathscr{C}_1'^2\mathscr{C}_2^2 = M_1 M_2 v^4 + v^2(M_1^2 v_2^2 + M_2^2 v_1^2) + 4M_1 M_2(\boldsymbol{v}.\boldsymbol{v}_1)(\boldsymbol{v}.\boldsymbol{v}_2) \\ + M_1 M_2 v_1^2 v_2^2 + \text{odd powers of } \boldsymbol{v}, \end{aligned}$$

$$\begin{aligned} (\mathscr{C}_1'.\mathscr{C}_2)^2 = M_1 M_2 v^4 + 2M_1 M_2 v^2 \boldsymbol{v}_1.\boldsymbol{v}_2 + \{\boldsymbol{v}.(M_2\boldsymbol{v}_1 + M_1\boldsymbol{v}_2)\}^2 \\ + M_1 M_2(\boldsymbol{v}_1.\boldsymbol{v}_2)^2 + \text{odd powers of } \boldsymbol{v}. \end{aligned}$$

Thus, using (1.32, 9), (1.42, 2), and the theorem of 1.421, we may write (9.3, 8) in the form

$$\begin{aligned} L_{12}(\chi) &= \int e^{-i_{12}v^2 - j_{12}g^2}\{(\mathscr{C}_1'.\mathscr{C}_2)^2 - \tfrac{1}{3}\mathscr{C}_1'^2\mathscr{C}_2^2\}\, d\boldsymbol{v} \\ &= 4\pi M_1 M_2 \int_0^\infty e^{-i_{12}v^2 - j_{12}g^2}\{\tfrac{2}{3}v^4 + \tfrac{20}{9}v^2\boldsymbol{v}_1.\boldsymbol{v}_2 + (\boldsymbol{v}_1.\boldsymbol{v}_2)^2 - \tfrac{1}{3}v_1^2 v_2^2\}\, v^2\, dv \\ &= \pi^{\frac{3}{2}} M_1 M_2 e^{-j_{12}g^2} i_{12}^{-\frac{7}{2}}\{\tfrac{5}{2} + \tfrac{10}{3}i_{12}\boldsymbol{v}_1.\boldsymbol{v}_2 + i_{12}^2(\boldsymbol{v}_1.\boldsymbol{v}_2)^2 - \tfrac{1}{3}i_{12}^2 v_1^2 v_2^2\}. \end{aligned} \tag{9.31, 10}$$

Again, by (9.31, 9),

$$\begin{aligned} H_{12}(\chi) &= (M_1 M_2)^{\frac{1}{2}} \int e^{-i_{12}v^2 - j_{12}g^2}(v^2 + \boldsymbol{v}_1.\boldsymbol{v}_2)\, d\boldsymbol{v} \\ &= 4\pi(M_1 M_2)^{\frac{1}{2}} \int_0^\infty e^{-i_{12}v^2 - j_{12}g^2}(v^2 + \boldsymbol{v}_1.\boldsymbol{v}_2)\, v^2\, dv \\ &= \pi^{\frac{3}{2}}(M_1 M_2)^{\frac{1}{2}} e^{-j_{12}g^2} i_{12}^{-\frac{5}{2}}(\tfrac{3}{2} + i_{12}\boldsymbol{v}_1.\boldsymbol{v}_2). \end{aligned} \tag{9.31, 11}$$

It remains to evaluate $\boldsymbol{v}_1.\boldsymbol{v}_2$ and $v_1^2 v_2^2$. From (9.31, 7), using (9.31, 3, 6) and (9.2, 10), we find

$$\begin{aligned} &\boldsymbol{v}_1.\boldsymbol{v}_2 \\ &= (M_1 M_2/i_{12}^2)(T\boldsymbol{g} - s\boldsymbol{g}').(T\boldsymbol{g} - s\boldsymbol{g}') + (1/i_{12})(M_2\boldsymbol{g}' - M_1\boldsymbol{g}).(T\boldsymbol{g} - s\boldsymbol{g}') - \boldsymbol{g}.\boldsymbol{g}' \\ &= (M_1 M_2/i_{12}^2)\, g^2(s^2 + T^2 - 2sT\cos\chi) + (g^2/i_{12})\{-i_{21} + 1 + (i_{12} - 1)\cos\chi\} \\ &\qquad - g^2\cos\chi \\ &= (g^2/i_{12})(1 - j_{12} - \cos\chi). \end{aligned} \tag{9.31, 12}$$

Again,
$$|\boldsymbol{v}_1 \wedge \boldsymbol{v}_2|^2 = v_1^2 v_2^2 - (\boldsymbol{v}_1 . \boldsymbol{v}_2)^2,$$
and since $\boldsymbol{g}' \wedge \boldsymbol{g} = -\boldsymbol{g} \wedge \boldsymbol{g}'$, $\boldsymbol{g} \wedge \boldsymbol{g} = 0 = \boldsymbol{g}' \wedge \boldsymbol{g}'$, we find finally, using (9.31, 3),
$$\boldsymbol{v}_1 \wedge \boldsymbol{v}_2 = (\boldsymbol{g} \wedge \boldsymbol{g}')/i_{12}.$$
Hence, since the magnitude of $\boldsymbol{g} \wedge \boldsymbol{g}'$ is $g^2 \sin \chi$,
$$v_1^2 v_2^2 - (\boldsymbol{v}_1 . \boldsymbol{v}_2)^2 = (g^4/i_{12}^2) \sin^2 \chi. \tag{9.31, 13}$$
By using (9.31, 12, 13) in (9.31, 10, 11) we find
$$H_{12}(\chi) = \pi^{\frac{3}{2}} (M_1 M_2)^{\frac{1}{2}} e^{-j_{12} g^2} i_{12}^{-\frac{5}{2}} \{\tfrac{3}{2} + (1 - j_{12} - \cos \chi) g^2\}, \tag{9.31, 14}$$
$$L_{12}(\chi) = \tfrac{2}{3} \pi^{\frac{3}{2}} M_1 M_2 e^{-j_{12} g^2} i_{12}^{-\frac{7}{2}} [\tfrac{15}{4} + 5(1 - j_{12} - \cos \chi) g^2 + \{(1 - j_{12} - \cos \chi)^2 - \tfrac{1}{2} \sin^2 \chi\} g^4]. \tag{9.31, 15}$$

9.32. $H_{12}(\chi)$ and $L_{12}(\chi)$ as functions of s and t

Write now
$$y_{12} = M_2 s + M_1 t, \quad z_{12} = 2 M_1 M_2 s t (1 - \cos \chi). \tag{9.32, 1}$$
Then, since $S^{-1} = s^{-1} - 1$, $T^{-1} = t^{-1} - 1$, we find from (9.31, 3, 6),
$$i_{12}/ST = (1 - y_{12})/st, \tag{9.32, 2}$$
$$\begin{aligned} j_{12} &= \{(1+S)(1+T) - 2 M_1 M_2 ST (1 - \cos \chi)\}/i_{12} \\ &= (1 - z_{12})/(1 - y_{12}). \end{aligned} \tag{9.32, 3}$$
If y_{12} and z_{12} are regarded as independent,
$$\frac{\partial j_{12}}{\partial y_{12}} = \frac{j_{12}}{1 - y_{12}}. \tag{9.32, 4}$$
Hence, from (9.31, 14),
$$\begin{aligned} &(ST/st)^{\frac{5}{2}} (M_1 M_2)^{-\frac{1}{2}} \pi^{-\frac{3}{2}} H_{12}(\chi) \\ &\qquad = e^{-j_{12} g^2} (1 - y_{12})^{-\frac{5}{2}} \{\tfrac{3}{2} + (1 - j_{12} - \cos \chi) g^2\} \\ &\qquad = \frac{\partial}{\partial y_{12}} \{e^{-j_{12} g^2} (1 - y_{12})^{-\frac{3}{2}}\} + e^{-j_{12} g^2} (1 - y_{12})^{-\frac{5}{2}} g^2 (1 - \cos \chi). \end{aligned}$$
But, by (9.32, 3) and the expansion defining Sonine polynomials,
$$\begin{aligned} e^{-j_{12} g^2} (1 - y_{12})^{-\frac{5}{2}} &= e^{-g^2/(1 - y_{12})} \sum_r \frac{(z_{12} g^2)^r}{r!} (1 - y_{12})^{-r - \frac{5}{2}} \\ &= e^{-g^2} \sum_r \sum_n \frac{(z_{12} g^2)^r}{r!} y_{12}^n S_{r+\frac{3}{2}}^{(n)}(g^2), \end{aligned}$$
and from a similar relation
$$\frac{\partial}{\partial y_{12}} \{e^{-j_{12} g^2} (1 - y_{12})^{-\frac{3}{2}}\} = e^{-g^2} \sum_r \sum_n \frac{(z_{12} g^2)^r}{r!} (n+1) y_{12}^n S_{r+\frac{1}{2}}^{(n+1)}(g^2).$$

Combining these results and using (9.32, 1), we find

$$(ST/st)^{\frac{3}{2}}(M_1M_2)^{-\frac{1}{2}}\pi^{-\frac{3}{2}}H_{12}(\chi)$$
$$= e^{-g^2}\sum_r\sum_n\{2M_1M_2st(1-\cos\chi)\}^r\frac{g^{2r}}{r!}(M_2s+M_1t)^n$$
$$\times\{(n+1)S^{(n+1)}_{r+\frac{1}{2}}(g^2)+g^2(1-\cos\chi)S^{(n)}_{r+\frac{3}{2}}(g^2)\}. \quad (9.32, 5)$$

Using the binomial theorem we can expand the sum on the right of (9.32, 5) in a series of powers of s and t. Since M_1, M_2 appear only in the combinations M_1t, M_2s, the coefficient of s^pt^q in this series has $M_2^pM_1^q$ as a factor, but is otherwise independent of M_1, M_2. It is in general a polynomial in g^2 and $\cos\chi$, of degree $p+q+1$ in g^2, and of degree in $\cos\chi$ equal to the lesser of $p+1$ and $q+1$. Thus, on expanding, (9.32, 5) becomes

$$\left(\frac{ST}{st}\right)^{\frac{3}{2}}(M_1M_2)^{-\frac{1}{2}}\pi^{-\frac{3}{2}}H_{12}(\chi) = e^{-g^2}\sum_{p,q,r,l}A_{pqrl}(M_2s)^p(M_1t)^qg^{2r}\cos^l\chi, \quad (9.32, 6)$$

where A_{pqrl} is a pure number, independent of M_1 and M_2.

Similarly

$$\tfrac{3}{2}(ST/st)^{\frac{7}{2}}(M_1M_2)^{-1}\pi^{-\frac{3}{2}}L_{12}(\chi)$$
$$= \frac{\partial^2}{\partial y_{12}^2}\{e^{-j_{12}g^2}(1-y_{12})^{-\frac{3}{2}}\}+2g^2(1-\cos\chi)\frac{\partial}{\partial y_{12}}\{e^{-j_{12}g^2}(1-y_{12})^{-\frac{5}{2}}\}$$
$$+g^4\{(1-\cos\chi)^2-\tfrac{1}{2}\sin^2\chi\}e^{-j_{12}g^2}(1-y_{12})^{-\frac{7}{2}}$$
$$= e^{-g^2}\sum_r\sum_n\{2M_1M_2st(1-\cos\chi)\}^r\frac{g^{2r}}{r!}(M_2s+M_1t)^n$$
$$\times[(n+1)(n+2)S^{(n+2)}_{r+\frac{1}{2}}(g^2)+2(n+1)g^2(1-\cos\chi)S^{(n+1)}_{r+\frac{3}{2}}(g^2)$$
$$+g^4\{(1-\cos\chi)^2-\tfrac{1}{2}\sin^2\chi\}S^{(n)}_{r+\frac{5}{2}}(g^2)]. \quad (9.32, 7)$$

On expanding this in powers of s, t, the coefficient of s^pt^q is found to be $e^{-g^2}M_2^pM_1^q$ multiplied by a polynomial in g^2 and $\cos\chi$, of degree $p+q+2$ in g^2, and of degree in $\cos\chi$ equal to the lesser of $p+2$ and $q+2$. Thus

$$\tfrac{3}{2}(ST/st)^{\frac{7}{2}}(M_1M_2)^{-1}\pi^{-\frac{3}{2}}L_{12}(\chi) = e^{-g^2}\sum_{p,q,r,l}B_{pqrl}(M_2s)^p(M_1t)^qg^{2r}\cos^l\chi, \quad (9.32, 8)$$

where B_{pqrl} is a pure number, independent of M_1 and M_2.

Explicit expressions for A_{pqrl}, B_{pqrl} can be obtained from (9.32, 5, 7), using the expression (7.5, 3) for $S_m^{(n)}(x)$. In view of the complication of these expressions it is, however, better in practice to calculate any desired values of A_{pqrl}, B_{pqrl} directly from (9.32, 5, 7).

9.33. The evaluation of

$$[S(\mathscr{C}_1^2)\mathscr{C}_1, S(\mathscr{C}_2^2)\mathscr{C}_2]_{12} \quad \text{and} \quad [S(\mathscr{C}_1^2)\mathscr{C}_1^{\circ}\mathscr{C}_1, S(\mathscr{C}_2^2)\mathscr{C}_2^{\circ}\mathscr{C}_2]_{12}$$

By 9.3,
$$[S_{\frac{3}{2}}^{(p)}(\mathscr{C}_1^2)\mathscr{C}_1, S_{\frac{3}{2}}^{(q)}(\mathscr{C}_2^2)\mathscr{C}_2]_{12}$$

is the coefficient of $s^p t^q$ in the expansion of (9.3, 5), that is, of

$$(ST/st)^{\frac{5}{2}}\pi^{-3}\iiint\{H_{12}(0)-H_{12}(\chi)\}g\,b\,db\,d\epsilon\,d\boldsymbol{g}.$$

Hence, using (9.32, 6)

$$[S_{\frac{3}{2}}^{(p)}(\mathscr{C}_1^2)\mathscr{C}_1, S_{\frac{3}{2}}^{(q)}(\mathscr{C}_2^2)\mathscr{C}_2]_{12}$$
$$=\pi^{-\frac{3}{2}}M_2^{p+\frac{1}{2}}M_1^{q+\frac{1}{2}}\iiint e^{-g^2}\sum_{r,l}A_{pqrl}g^{2r}(1-\cos^l\chi)g\,b\,db\,d\epsilon\,d\boldsymbol{g}.$$

The integrand here is independent of ϵ and of the direction of $\boldsymbol{g}$; thus, on integrating over all values of ϵ and all directions of $\boldsymbol{g}$, we find that

$$[S_{\frac{3}{2}}^{(p)}(\mathscr{C}_1^2)\mathscr{C}_1, S_{\frac{3}{2}}^{(q)}(\mathscr{C}_2^2)\mathscr{C}_2]_{12}$$
$$=8\pi^{\frac{1}{2}}M_2^{p+\frac{1}{2}}M_1^{q+\frac{1}{2}}\iint e^{-g^2}\sum_{r,l}A_{pqrl}g^{2r+2}(1-\cos^l\chi)g\,b\,db\,dg$$
$$=8M_2^{p+\frac{1}{2}}M_1^{q+\frac{1}{2}}\sum_{r,l}A_{pqrl}\Omega_{12}^{(l)}(r), \qquad (9.33, 1)$$

where $\Omega_{12}^{(l)}(r)$ is defined by the equation

$$\Omega_{12}^{(l)}(r) \equiv \pi^{\frac{1}{2}}\iint e^{-g^2}g^{2r+2}(1-\cos^l\chi)g\,b\,db\,dg, \qquad (9.33, 2)$$

so that
$$\Omega_{12}^{(0)}(r)=0, \quad \Omega_{12}^{(l)}(r)>0 \quad \text{if} \quad l>0.$$

Similarly, by (9.32, 8) and (9.3, 6–8),

$$[S_{\frac{3}{2}}^{(p)}(\mathscr{C}_1^2)\mathscr{C}_1^{\circ}\mathscr{C}_1, S_{\frac{3}{2}}^{(q)}(\mathscr{C}_2^2)\mathscr{C}_2^{\circ}\mathscr{C}_2]_{12}$$
$$=\tfrac{2}{3}\pi^{-\frac{3}{2}}M_2^{p+1}M_1^{q+1}\iiint e^{-g^2}\sum_{r,l}B_{pqrl}g^{2r}(1-\cos^l\chi)g\,b\,db\,d\epsilon\,d\boldsymbol{g}$$
$$=\tfrac{16}{3}\pi^{\frac{1}{2}}M_2^{p+1}M_1^{q+1}\iint e^{-g^2}\sum_{r,l}B_{pqrl}g^{2r+2}(1-\cos^l\chi)g\,b\,db\,dg$$
$$=\tfrac{16}{3}M_2^{p+1}M_1^{q+1}\sum_{r,l}B_{pqrl}\Omega_{12}^{(l)}(r). \qquad (9.33, 3)$$

In addition to $\Omega_{12}^{(l)}(r)$, it is convenient to introduce $\phi_{12}^{(l)}$, defined by*

$$\phi_{12}^{(l)}=2\pi\int(1-\cos^l\chi)\,b\,db. \qquad (9.33, 4)$$

In terms of this, by (9.2, 6),

$$\Omega_{12}^{(l)}(r)=(kT/2\pi m_0M_1M_2)^{\frac{1}{2}}\int_0^\infty e^{-g^2}g^{2r+3}\phi_{12}^{(l)}\,dg. \qquad (9.33, 5)$$

* In earlier editions of this book, $\phi_{12}^{(l)}$ was defined as $\int(1-\cos^l\chi)\,g\,b\,db$; however, it is advantageous to give $\phi_{12}^{(l)}$ the dimensions of an area, so that it can be related to a molecular cross-section.

Expressions for

$$[S_{\frac{3}{2}}^{(p)}(\mathscr{C}_1^2)\mathscr{C}_1, S_{\frac{3}{2}}^{(q)}(\mathscr{C}_2^2)\mathscr{C}_2]_{12} \quad \text{and} \quad [S_{\frac{5}{2}}^{(p)}(\mathscr{C}_1^2)\overset{\circ}{\mathscr{C}_1\mathscr{C}_1}, S_{\frac{5}{2}}^{(q)}(\mathscr{C}_2^2)\overset{\circ}{\mathscr{C}_2\mathscr{C}_2}]_{12}$$

in terms of the functions $\Omega_{12}^{(l)}(r)$ are given in 9.6 for some special values of p and q.

9.4. The evaluation of

$$[S(\mathscr{C}_1^2)\mathscr{C}_1, S(\mathscr{C}_1^2)\mathscr{C}_1]_{12} \quad \text{and} \quad [S(\mathscr{C}_1^2)\overset{\circ}{\mathscr{C}_1\mathscr{C}_1}, S(\mathscr{C}_1^2)\overset{\circ}{\mathscr{C}_1\mathscr{C}_1}]_{12}$$

The methods here required are similar to those of 9.3–9.33, so that only the main steps need be indicated.

The expression

$$[S_{\frac{3}{2}}^{(p)}(\mathscr{C}_1^2)\mathscr{C}_1, S_{\frac{3}{2}}^{(q)}(\mathscr{C}_1^2)\mathscr{C}_1]_{12} \tag{9.4, 1}$$

is equal to the coefficient of $s^p t^q$ in the expansion of

$$(ST/st)^{\frac{3}{2}}\pi^{-3}\iiint\{H_1(0)-H_1(\chi)\}gb\,db\,d\epsilon\,dg \tag{9.4, 2}$$

in powers of s and t, where

$$H_1(\chi) \equiv \int \exp(-\mathscr{G}_0^2-g^2-S\mathscr{C}_1'^2-T\mathscr{C}_1^2)\mathscr{C}_1'.\mathscr{C}_1\,d\mathscr{G}_0, \tag{9.4, 3}$$

and S and T are defined by (9.3, 3).

The expression

$$[S_{\frac{5}{2}}^{(p)}(\mathscr{C}_1^2)\overset{\circ}{\mathscr{C}_1\mathscr{C}_1}, S_{\frac{5}{2}}^{(q)}(\mathscr{C}_1^2)\overset{\circ}{\mathscr{C}_1\mathscr{C}_1}]_{12} \tag{9.4, 4}$$

is likewise equal to the coefficient of $s^p t^q$ in the expansion of

$$(ST/st)^{\frac{5}{2}}\pi^{-3}\iiint\{L_1(0)-L_1(\chi)\}gb\,db\,d\epsilon\,dg,$$

where $$L_1(\chi) \equiv \int \exp(-\mathscr{G}_0^2-g^2-S\mathscr{C}_1'^2-T\mathscr{C}_1^2)\overset{\circ}{\mathscr{C}_1'\mathscr{C}_1'}:\overset{\circ}{\mathscr{C}_1\mathscr{C}_1}\,d\mathscr{G}_0. \tag{9.4, 5}$$

Using relations analogous to (9.31, 1, 2) we find that

$$\mathscr{G}_0^2+g^2+S\mathscr{C}_1'^2+T\mathscr{C}_1^2 = i_1\mathscr{G}_0^2+i_2g^2-2(M_1M_2)^{\frac{1}{2}}\mathscr{G}_0.(Sg'+Tg),$$

where $$i_1 \equiv 1+M_1(S+T), \quad i_2 \equiv 1+M_2(S+T). \tag{9.4, 6}$$

Hence (9.4, 3, 5) can be evaluated by making the substitution

$$v \equiv \mathscr{G}_0-(M_1M_2)^{\frac{1}{2}}(Sg'+Tg)/i_1.$$

Integrating as in 9.31, we find

$$H_1(\chi) = \pi^{\frac{3}{2}}i_1^{-\frac{5}{2}}e^{-j_1g^2}[\tfrac{3}{2}M_1+\{M_1(1-j_1)+M_2\cos\chi\}g^2], \tag{9.4, 7}$$

$$L_1(\chi) = \tfrac{2}{3}\pi^{\frac{3}{2}}i_1^{-\frac{7}{2}}e^{-j_1g^2}[\tfrac{15}{4}M_1^2+5M_1\{M_1(1-j_1)+M_2\cos\chi\}g^2$$
$$+\{M_1(1-j_1)+M_2\cos\chi\}^2g^4-\tfrac{1}{2}M_2^2g^4\sin^2\chi], \tag{9.4, 8}$$

where $$j_1 \equiv i_2-(M_1M_2/i_1)(S^2+T^2+2ST\cos\chi). \tag{9.4, 9}$$

By (9.4, 6, 9) and (9.3, 3),

$$i_1 = 1 + M_1\frac{s}{1-s} + M_1\frac{t}{1-t} = \frac{1 - M_2(s+t) + (M_2 - M_1)st}{(1-s)(1-t)}, \qquad (9.4, 10)$$

$$\begin{aligned} j_1 &= 1 + M_2(S+T) - M_1M_2(S^2+T^2+2ST\cos\chi)/\{1+M_1(S+T)\} \\ &= \{(1+S)(1+T) - ST(M_1^2+M_2^2+2M_1M_2\cos\chi)\}/\{1+M_1(S+T)\} \\ &= \{1 - st(M_1^2+M_2^2+2M_1M_2\cos\chi)\}/\{1 - M_2(s+t) + (M_2-M_1)st\}, \end{aligned} \qquad (9.4, 11)$$

so that $H_1(\chi)$, $L_1(\chi)$ can be expressed as functions of s, t. The expansions corresponding to (9.32, 5, 7) are

$$\begin{aligned} (ST/st)^{\frac{1}{2}}\pi^{-\frac{1}{2}}H_1(\chi) = e^{-g^2}\sum_r\sum_n &\{st(M_1^2+M_2^2+2M_1M_2\cos\chi)\}^r(g^{2r}/r!) \\ &\times\{M_2(s+t)-(M_2-M_1)st\}^n\{M_1(n+1)S_{r+\frac{1}{2}}^{(n+1)}(g^2) \\ &+(M_1+M_2\cos\chi)g^2S_{r+\frac{1}{2}}^{(n)}(g^2)\}, \end{aligned} \qquad (9.4, 12)$$

$$\begin{aligned} \tfrac{2}{3}(ST/st)^{\frac{3}{2}}\pi^{-\frac{1}{2}}L_1(\chi) = e^{-g^2}\sum_r\sum_n &[st(M_1^2+M_2^2+2M_1M_2\cos\chi)]^r(g^{2r}/r!) \\ &\times[M_2(s+t)-(M_2-M_1)st]^n[M_1^2(n+1)(n+2)S_{r+\frac{3}{2}}^{(n+2)}(g^2) \\ &+2(n+1)g^2M_1(M_1+M_2\cos\chi)S_{r+\frac{3}{2}}^{(n+1)}(g^2) \\ &+g^4\{(M_1+M_2\cos\chi)^2-\tfrac{1}{2}M_2^2\sin^2\chi\}S_{r+\frac{3}{2}}^{(n)}(g^2)]. \end{aligned} \qquad (9.4, 13)$$

If the expansion of (9.4, 12) in powers of s and t is expressed in the form

$$e^{-g^2}\sum_{p,q,r,l}A'_{pqrl}s^pt^qg^{2r}\cos^l\chi, \qquad (9.4, 14)$$

then
$$[S_{\frac{3}{2}}^{(p)}(\mathscr{C}_1^2)\mathscr{C}_1, S_{\frac{3}{2}}^{(q)}(\mathscr{C}_1^2)\mathscr{C}_1]_{12} = 8\sum_{r,l}A'_{pqrl}\Omega_{12}^{(l)}(r), \qquad (9.4, 15)$$

which is analogous to (9.33, 1).

Likewise the value of (9.4, 4) may be derived from the coefficient of s^pt^q in the expansion of (9.4, 13). Expressions for (9.4, 1 and 4) in terms of the functions $\Omega_{12}^{(l)}(r)$ are given in 9.6 for special values of p and q.

9.5. The evaluation of

$$[S(\mathscr{C}_1^2)\mathscr{C}_1, S(\mathscr{C}_1^2)\mathscr{C}_1]_1 \quad\text{and}\quad [S(\mathscr{C}_1^2)\overset{\circ}{\mathscr{C}_1\mathscr{C}_1}, S(\mathscr{C}_1^2)\overset{\circ}{\mathscr{C}_1\mathscr{C}_1}]_1$$

It is clear from (4.4, 11) that $[S_{\frac{3}{2}}^{(p)}(\mathscr{C}_1^2)\mathscr{C}_1, S_{\frac{3}{2}}^{(q)}(\mathscr{C}_1^2)\mathscr{C}_1]_1$ can be derived from

$$[S_{\frac{3}{2}}^{(p)}(\mathscr{C}_1^2)\mathscr{C}_1, S_{\frac{3}{2}}^{(q)}(\mathscr{C}_2^2)\mathscr{C}_2]_{12} + [S_{\frac{3}{2}}^{(p)}(\mathscr{C}_1^2)\mathscr{C}_1, S_{\frac{3}{2}}^{(q)}(\mathscr{C}_1^2)\mathscr{C}_1]_{12}$$

by taking m_2 equal to m_1 and adopting the law of interaction between pairs of molecules of the first gas instead of the law for the interaction of pairs of unlike molecules. The effect of these changes is that in every result M_1, M_2

must each be replaced by $1/2$, and $\Omega_{12}^{(l)}(r)$ must be replaced by the similar integral $\Omega_1^{(l)}(r)$, relating to the encounter of pairs of molecules of the first gas.

On putting $M_1 = M_2 = \frac{1}{2}$, the expression (9.31, 14) for $H_{12}(\chi)$ becomes identical with the expression (9.4, 7) for $H_1(\chi)$, except that $\cos\chi$ has the opposite sign (cf. (9.31, 6) and (9.4, 9)). This implies that the coefficient of $\Omega_{12}^{(l)}(r)$ in the expansion (9.33, 1) is $(-1)^l$ times that in the expansion (9.4, 13). Thus

$$[S_{\frac{3}{2}}^{(p)}(\mathscr{C}_1^2)\mathscr{C}_1,\ S_{\frac{3}{2}}^{(q)}(\mathscr{C}_2^2)\mathscr{C}_2]_{12}+[S_{\frac{3}{2}}^{(p)}(\mathscr{C}_1^2)\mathscr{C}_1,\ S_{\frac{3}{2}}^{(q)}(\mathscr{C}_1^2)\mathscr{C}_1]_{12}$$

is equal to the expression derived from $[S_{\frac{3}{2}}^{(p)}(\mathscr{C}_1^2)\mathscr{C}_1,\ S_{\frac{3}{2}}^{(q)}(\mathscr{C}_2^2)\mathscr{C}_2]_{12}$ by suppressing those terms involving $\Omega_{12}^{(l)}(r)$ for odd values of l and doubling the remaining terms. Hence $[S_{\frac{3}{2}}^{(p)}(\mathscr{C}_1^2)\mathscr{C}_1,\ S_{\frac{3}{2}}^{(q)}(\mathscr{C}_1^2)\mathscr{C}_1]_1$ involves only expressions $\Omega_1^{(l)}(r)$ for even values of l.

The value of $[S_{\frac{5}{2}}^{(p)}(\mathscr{C}_1^2)\mathring{\mathscr{C}_1\mathscr{C}_1},\ S_{\frac{5}{2}}^{(q)}(\mathscr{C}_1^2)\mathring{\mathscr{C}_1\mathscr{C}_1}]_1$ is similarly derived from

$$[S_{\frac{5}{2}}^{(p)}(\mathscr{C}_1^2)\mathring{\mathscr{C}_1\mathscr{C}_1},\ S_{\frac{5}{2}}^{(q)}(\mathscr{C}_2^2)\mathring{\mathscr{C}_2\mathscr{C}_2}]_{12}.$$

Expressions for

$$[S_{\frac{3}{2}}^{(p)}(\mathscr{C}_1^2)\mathscr{C}_1,\ S_{\frac{3}{2}}^{(q)}(\mathscr{C}_1^2)\mathscr{C}_1]_1 \quad\text{and}\quad [S_{\frac{5}{2}}^{(p)}(\mathscr{C}_1^2)\mathring{\mathscr{C}_1\mathscr{C}_1},\ S_{\frac{5}{2}}^{(q)}(\mathscr{C}_1^2)\mathring{\mathscr{C}_1\mathscr{C}_1}]_1$$

in terms of the expressions $\Omega_1^{(l)}(r)$ are given in 9.6 for certain special values of p, q.

9.6. Table of formulae

The following special cases of the results of 9.33, 9.4, 9.5 are tabulated for convenience of reference:

$$[\mathscr{C}_1, \mathscr{C}_2]_{12} = -8(M_1M_2)^{\frac{1}{2}}\Omega_{12}^{(1)}(1), \tag{9.6, 1}$$

$$[\mathscr{C}_1, S_{\frac{3}{2}}^{(1)}(\mathscr{C}_2^2)\mathscr{C}_2]_{12} = 8(M_1^3M_2)^{\frac{1}{2}}\{\Omega_{12}^{(1)}(2)-\tfrac{5}{2}\Omega_{12}^{(1)}(1)\}, \tag{9.6, 2}$$

$$[S_{\frac{3}{2}}^{(1)}(\mathscr{C}_1^2)\mathscr{C}_1, S_{\frac{3}{2}}^{(1)}(\mathscr{C}_2^2)\mathscr{C}_2]_{12} = -8(M_1M_2)^{\frac{3}{2}} \times\{\tfrac{55}{4}\Omega_{12}^{(1)}(1)-5\Omega_{12}^{(1)}(2)+\Omega_{12}^{(1)}(3)-2\Omega_{12}^{(2)}(2)\}, \tag{9.6, 3}$$

$$[\mathscr{C}_1, \mathscr{C}_1]_{12} = 8M_2\Omega_{12}^{(1)}(1), \tag{9.6, 4}$$

$$[\mathscr{C}_1, S_{\frac{3}{2}}^{(1)}(\mathscr{C}_1^2)\mathscr{C}_1]_{12} = -8M_2^2\{\Omega_{12}^{(1)}(2)-\tfrac{5}{2}\Omega_{12}^{(1)}(1)\}, \tag{9.6, 5}$$

$$[S_{\frac{3}{2}}^{(1)}(\mathscr{C}_1^2)\mathscr{C}_1, S_{\frac{3}{2}}^{(1)}(\mathscr{C}_1^2)\mathscr{C}_1]_{12} = 8M_2\{\tfrac{5}{4}(6M_1^2+5M_2^2)\Omega_{12}^{(1)}(1)-5M_2^2\Omega_{12}^{(1)}(2) +M_2^2\Omega_{12}^{(1)}(3)+2M_1M_2\Omega_{12}^{(2)}(2)\}, \tag{9.6, 6}$$

$$[\mathscr{C}_1, S_{\frac{3}{2}}^{(q)}(\mathscr{C}_1^2)\mathscr{C}_1]_1 = 0. \tag{9.6, 7}$$

The last result also follows directly from (4.4, 8), by the principle of conservation of momentum. Again,

$$[S_{\frac{3}{2}}^{(1)}(\mathscr{C}_1^2)\mathscr{C}_1, S_{\frac{3}{2}}^{(1)}(\mathscr{C}_1^2)\mathscr{C}_1]_1 = 4\Omega_1^{(2)}(2), \tag{9.6, 8}$$

$$[S_{\frac{3}{2}}^{(1)}(\mathscr{C}_1^2)\mathscr{C}_1, S_{\frac{3}{2}}^{(2)}(\mathscr{C}_1^2)\mathscr{C}_1]_1 = 7\Omega_1^{(2)}(2)-2\Omega_1^{(2)}(3), \tag{9.6, 9}$$

$$[S_{\frac{3}{2}}^{(2)}(\mathscr{C}_1^2)\mathscr{C}_1, S_{\frac{3}{2}}^{(2)}(\mathscr{C}_1^2)\mathscr{C}_1]_1 = \tfrac{77}{4}\Omega_1^{(2)}(2)-7\Omega_1^{(2)}(3)+\Omega_1^{(2)}(4), \tag{9.6, 10}$$

$$[S_{\frac{3}{2}}^{(1)}(\mathscr{C}_1^2)\mathscr{C}_1, S_{\frac{3}{2}}^{(3)}(\mathscr{C}_1^2)\mathscr{C}_1]_1 = \tfrac{63}{8}\Omega_1^{(2)}(2) - \tfrac{9}{2}\Omega_1^{(2)}(3) + \tfrac{1}{2}\Omega_1^{(2)}(4), \tag{9.6, 11}$$

$$[S_{\frac{3}{2}}^{(2)}(\mathscr{C}_1^2)\mathscr{C}_1, S_{\frac{3}{2}}^{(3)}(\mathscr{C}_1^2)\mathscr{C}_1]_1 = \tfrac{945}{32}\Omega_1^{(2)}(2) - \tfrac{261}{16}\Omega_1^{(2)}(3) + \tfrac{25}{8}\Omega_1^{(2)}(4) - \tfrac{1}{4}\Omega_1^{(2)}(5), \tag{9.6, 12}$$

$$[S_{\frac{3}{2}}^{(3)}(\mathscr{C}_1^2)\mathscr{C}_1, S_{\frac{3}{2}}^{(3)}(\mathscr{C}_1^2)\mathscr{C}_1]_1 = \tfrac{14553}{256}\Omega_1^{(2)}(2) - \tfrac{1215}{32}\Omega_1^{(2)}(3) + \tfrac{313}{32}\Omega_1^{(2)}(4) - \tfrac{9}{8}\Omega_1^{(2)}(5) + \tfrac{1}{16}\Omega_1^{(2)}(6) + \tfrac{1}{6}\Omega_1^{(4)}(4), \tag{9.6, 13}$$

$$[\mathscr{C}_1\overset{\circ}{\mathscr{C}}_1, \mathscr{C}_2\overset{\circ}{\mathscr{C}}_2]_{12} = -\tfrac{16}{3}M_1M_2\{5\Omega_{12}^{(1)}(1) - \tfrac{3}{2}\Omega_{12}^{(2)}(2)\}, \tag{9.6, 14}$$

$$[\mathscr{C}_1\overset{\circ}{\mathscr{C}}_1, \mathscr{C}_1\overset{\circ}{\mathscr{C}}_1]_{12} = \tfrac{16}{3}M_2\{5M_1\Omega_{12}^{(1)}(1) + \tfrac{3}{2}M_2\Omega_{12}^{(2)}(2)\}, \tag{9.6, 15}$$

$$[\mathscr{C}_1\overset{\circ}{\mathscr{C}}_1, \mathscr{C}_1\overset{\circ}{\mathscr{C}}_1]_1 = 4\Omega_1^{(2)}(2), \tag{9.6, 16}$$

$$[S_{\frac{5}{2}}^{(1)}(\mathscr{C}_1^2)\mathscr{C}_1\overset{\circ}{\mathscr{C}}_1, \mathscr{C}_1\overset{\circ}{\mathscr{C}}_1]_1 = 7\Omega_1^{(2)}(2) - 2\Omega_1^{(2)}(3), \tag{9.6, 17}$$

$$[S_{\frac{5}{2}}^{(1)}(\mathscr{C}_1^2)\mathscr{C}_1\overset{\circ}{\mathscr{C}}_1, S_{\frac{5}{2}}^{(1)}(\mathscr{C}_1^2)\mathscr{C}_1\overset{\circ}{\mathscr{C}}_1]_1 = \tfrac{301}{12}\Omega_1^{(2)}(2) - 7\Omega_1^{(2)}(3) + \Omega_1^{(2)}(4), \tag{9.6, 18}$$

$$[S_{\frac{5}{2}}^{(2)}(\mathscr{C}_1^2)\mathscr{C}_1\overset{\circ}{\mathscr{C}}_1, \mathscr{C}_1\overset{\circ}{\mathscr{C}}_1]_1 = \tfrac{63}{8}\Omega_1^{(2)}(2) - \tfrac{9}{2}\Omega_1^{(2)}(3) + \tfrac{1}{2}\Omega_1^{(2)}(4), \tag{9.6, 19}$$

$$[S_{\frac{5}{2}}^{(2)}(\mathscr{C}_1^2)\mathscr{C}_1\overset{\circ}{\mathscr{C}}_1, S_{\frac{5}{2}}^{(1)}(\mathscr{C}_1^2)\mathscr{C}_1\overset{\circ}{\mathscr{C}}_1] = \tfrac{1365}{32}\Omega_1^{(2)}(2) - \tfrac{321}{16}\Omega_1^{(2)}(3) + \tfrac{25}{8}\Omega_1^{(2)}(4) - \tfrac{1}{4}\Omega_1^{(2)}(5), \tag{9.6, 20}$$

$$[S_{\frac{5}{2}}^{(2)}(\mathscr{C}_1^2)\mathscr{C}_1\overset{\circ}{\mathscr{C}}_1, S_{\frac{5}{2}}^{(2)}(\mathscr{C}_1^2)\mathscr{C}_1\overset{\circ}{\mathscr{C}}_1]_1 = \tfrac{25137}{256}\Omega_1^{(2)}(2) - \tfrac{1755}{32}\Omega_1^{(2)}(3) + \tfrac{381}{32}\Omega_1^{(2)}(4) - \tfrac{9}{8}\Omega_1^{(2)}(5) + \tfrac{1}{16}\Omega_1^{(2)}(6) + \tfrac{1}{2}\Omega_1^{(4)}(4). \tag{9.6, 21}$$

9.7. Viscosity and thermal conduction in a simple gas

In the case of a simple gas the elements a_{pq} of the determinants $\mathscr{A}$, $\mathscr{A}^{(m)}$ of 7.51 are defined as equal to $[\boldsymbol{a}^{(p)}, \boldsymbol{a}^{(q)}]$; that is, by (7.51, 2) to

$$[S_{\frac{3}{2}}^{(p)}(\mathscr{C}^2)\mathscr{C}, S_{\frac{3}{2}}^{(q)}(\mathscr{C}^2)\mathscr{C}].$$

Hence (9.6, 8–13) give the values of a_{pq} for values of p, q from 1 to 3. Similarly equations (9.6, 16–21) give the values of the elements b_{pq} of (7.52, 5) for the same range of values of p, q (cf. (7.52, 2)).

In (7.51, 14) and (7.52, 8) the coefficients λ and μ are expressed as infinite series. By taking one, two, ... terms of this series we obtain first, second, ... approximations to λ or μ. These will be denoted by $[\lambda]_1$, $[\lambda]_2$, ..., or $[\mu]_1$, $[\mu]_2$, ..., so that, for example,

$$[\mu]_1 = \frac{5kT}{2b_{11}}, \quad [\lambda]_1 = \frac{25c_v kT}{4a_{11}}, \tag{9.7, 1}$$

$$[\lambda]_2 = \frac{25}{4}c_v kT\left(\frac{1}{a_{11}} + \frac{(\mathscr{A}_{12}^{(2)})^2}{\mathscr{A}^{(1)}\mathscr{A}^{(2)}}\right) = \frac{25}{4}c_v kT\frac{a_{22}}{a_{11}a_{22} - a_{12}^2}. \tag{9.7, 2}$$

Similar notation will be used for successive approximations to D_{12}, D_T, λ, μ and k_T for a mixture. The first non-vanishing approximations are obtained by deleting the limit sign in such expressions as (8.51, 16–18) and (8.52, 8) and giving to m the value 0 in (8.51, 16), and 1 in the other cases. Similarly for the ratio f of (6.3, 3), so that we write $[\mathrm{f}]_1 \equiv [\lambda]_1/c_v[\mu]_1$.

This manner of indicating successive approximations differs from that used for the velocity-distribution function f, the successive approximations to which are $f^{(0)}, f^{(0)}+f^{(1)}, \ldots$; but there is no reason why a similar notation should be employed. All the quantities λ, μ, D_{12}, D_T, f, k_T depend on the second approximation to f, and approximations to them are really sub-approximations related to phenomena depending on $f^{(0)}+f^{(1)}$.

From (9.6, 8, 16) it follows that

$$[\lambda]_1 = 25c_v kT/16\Omega_1^{(2)}(2), \quad [\mu]_1 = 5kT/8\Omega_1^{(2)}(2). \tag{9.7, 3}$$

Thus, whatever the nature of the interaction between molecules,

$$[\lambda]_1 = \tfrac{5}{2}c_v[\mu]_1. \tag{9.7, 4}$$

This shows that the first approximation $[f]_1$ to the ratio $\lambda/\mu c_v$ is $\frac{5}{2}$ for all spherically symmetrical non-rotating molecules.

In considering the second approximations $[\lambda]_2$ and $[\mu]_2$ we make use of the results (9.6, 8, 16 and 9, 17, 10, 18), giving

$$b_{11} = a_{11}, \quad b_{12} = a_{12}, \quad b_{22} = a_{22} + \tfrac{35}{24}a_{11};$$

we thus find
$$[\lambda]_2/[\lambda]_1 = (1 - a_{12}^2/a_{11}a_{22})^{-1}$$
$$\geqslant (1 - b_{12}^2/b_{11}b_{22})^{-1} = [\mu]_2/[\mu]_1;$$

the sign of equality corresponds to the case when $a_{12} = b_{12} = 0$. Thus

$$[\lambda]_2 \geqslant \tfrac{5}{2}c_v[\mu]_2.$$

No similar result valid for *all* degrees of approximation has been obtained: but numerical calculations for special molecular models suggest that the ratio $\lambda/c_v\mu$ increases as successive degrees of approximation are taken into account, and that the limit is only slightly greater than $\frac{5}{2}$.

9.71. Kihara's approximations

Simplified second approximations to λ and μ were derived by Kihara,* using the fact that for most molecular models the non-diagonal elements a_{pq}, b_{pq} of the determinants $\mathscr{A}^{(m)}$, $\mathscr{B}^{(m)}$ are considerably smaller than the diagonal elements a_{pp}, b_{pp}.

From (9.33, 2), since $\mathscr{g}^2 = m_0 M_1 M_2 g^2/2kT$,

$$\Omega_{12}^{(l)}(r) = \pi^{\frac{1}{2}}\left(\frac{m_0 M_1 M_2}{2kT}\right)^{r+\frac{3}{2}} \iint e^{-m_0 M_1 M_2 g^2/2kT} g^{2r+3}(1 - \cos^l \chi)\, b\, db\, dg.$$

* T. Kihara, *Imperfect Gases* (Asakusa Bookstore, Tokyo, 1949; English translation by U.S. Office of Air Research, Wright–Patterson Air Base): see also *Rev. Mod. Phys.* **25**, 831 (1953). Kihara's approximations were suggested by results for Maxwellian molecules (cf. 10.32, below); they amount to giving certain ratios of collision-integrals $\Omega_{12}^{(l)}(r)$ the values which they take for such molecules.

Here χ is a function of g and b (cf. 3.42) but not of T. Hence

$$T\frac{d\Omega_{12}^{(l)}(r)}{dT} = -(r+\tfrac{3}{2})\Omega_{12}^{(l)}(r)+\Omega_{12}^{(l)}(r+1) \qquad (9.71, 1)$$

with a similar relation between $\Omega_1^{(l)}(r)$ and $\Omega_1^{(l)}(r+1)$. Using this relation and (9.6, 8–10, 16–18) we find

$$a_{11} = b_{11}, \quad a_{12} = b_{12} = -\tfrac{1}{2}T\frac{da_{11}}{dT},$$

$$a_{22} = \frac{1}{4}\left(T^2\frac{d^2a_{11}}{dT^2}+2T\frac{da_{11}}{dT}+\frac{21}{2}a_{11}\right), \quad b_{22} = \frac{1}{4}\left(T^2\frac{d^2a_{11}}{dT^2}+2T\frac{da_{11}}{dT}+\frac{49}{3}a_{11}\right). \qquad (9.71, 2)$$

For all physically reasonable models for electrically neutral molecules, a_{11} ($\equiv 4\Omega_1^{(2)}(2)$) varies over small temperature-ranges roughly as T^n, where n lies between about 0 and $\frac{1}{2}$ (cf. 12.31). Thus in the expressions (9.71, 2) for a_{22} and b_{22} the terms involving da_{11}/dT and d^2a_{11}/dT^2 are fairly small compared with the a_{11} terms; likewise a_{12} and b_{12}, depending on da_{11}/dT, are fairly small compared with a_{11}, a_{22} and b_{11}, b_{22} respectively.

Kihara obtained his simplified second approximations by assuming that, in the Enskog approximations

$$[\lambda]_2 = [\lambda]_1/(1-a_{12}^2/a_{11}a_{22}), \quad [\mu]_2 = [\mu]_1/(1-b_{12}^2/b_{11}b_{22}), \qquad (9.71, 3)$$

terms of higher order than second in the temperature-derivatives can be neglected. This amounts to writing $a_{22} = \frac{21}{8}a_{11}$, $b_{22} = \frac{49}{12}a_{11}$ in the second-order expressions $a_{12}^2/a_{11}a_{22}$, $b_{12}^2/b_{11}b_{22}$, so that approximately

$$[\lambda]_2 = [\lambda]_1\bigg/\left(1-\frac{8}{21}\left(\frac{a_{12}}{a_{11}}\right)^2\right) \doteqdot [\lambda]_1\left(1+\frac{2}{21}\left(\frac{\Omega_1^{(2)}(3)}{\Omega_1^{(2)}(2)}-\frac{7}{2}\right)^2\right), \qquad (9.71, 4)$$

$$[\mu]_2 = [\mu]_1\bigg/\left(1-\frac{12}{49}\left(\frac{a_{12}}{a_{11}}\right)^2\right) \doteqdot [\mu]_1\left(1+\frac{3}{49}\left(\frac{\Omega_1^{(2)}(3)}{\Omega_1^{(2)}(2)}-\frac{7}{2}\right)^2\right). \qquad (9.71, 5)$$

The formulae (9.71, 4, 5) are simpler than the exact Enskog formulae for $[\lambda]_2$ and $[\mu]_2$. They also usually give values closer to the exact values of λ and μ. The neglect of the temperature-derivatives usually decreases a_{22}, b_{22}, and so increases $[\lambda]_2$ and $[\mu]_2$; thus it is similar in effect to taking a later Enskog approximation. Mason* has indicated an extension of Kihara's method to higher approximations, based on the retention of terms of higher order in the temperature-derivatives.

* E. A. Mason, *J. Chem. Phys.* **27**, 75, 782 (1957). As regards the accuracy of the Kihara approximations, see also M. J. Offerhaus, *Physica*, **28**, 76 (1962).

9.8. The determinant elements a_{pq}, b_{pq} for a gas-mixture

The expressions for the elements a_{pq}, b_{pq} of the determinants $\mathscr{A}^{(m)}$, $\mathscr{B}^{(m)}$ discussed in 8.51 and 8.52 are simpler than those for the general complete integral $\{F, G\}$ of (4.4, 12), by reason of the conditions (8.51, 4) and (8.52, 2). Thus

$$a_{pq} \equiv \{\boldsymbol{a}^{(p)}, \boldsymbol{a}^{(q)}\} = x_1^2[\boldsymbol{a}_1^{(p)}, \boldsymbol{a}_1^{(q)}]_1 + x_1x_2[\boldsymbol{a}_1^{(p)}, \boldsymbol{a}_1^{(q)}]_{12}$$

$$\equiv x_1^2 a''_{pq} + x_1x_2 a'_{pq} \quad (p, q > 0), \tag{9.8, 1}$$

$$a_{pq} = x_1x_2[\boldsymbol{a}_2^{(p)}, \boldsymbol{a}_2^{(q)}]_{12} + x_2^2[\boldsymbol{a}_2^{(p)}, \boldsymbol{a}_2^{(q)}]_2$$

$$\equiv x_1x_2 a'_{pq} + x_2^2 a''_{pq} \quad (p, q < 0), \tag{9.8, 2}$$

$$a_{pq} = a_{qp} = x_1x_2[\boldsymbol{a}_1^{(p)}, \boldsymbol{a}_2^{(q)}]_{12}$$

$$\equiv x_1x_2 a'_{pq} \quad (p > 0 > q), \tag{9.8, 3}$$

and similarly for b_{pq}. As the quantities $\boldsymbol{a}^{(p)}$, $\boldsymbol{b}^{(p)}$ do not involve the number-densities if $p \neq 0$, these appear in the above equations only as shown explicitly.

The values adopted for $\boldsymbol{a}_1^{(0)}$, $\boldsymbol{a}_2^{(0)}$ in (8.51, 3) lead to specially simple forms of a_{0q}. By the principle of momentum it follows from (4.4, 8) that $[\mathscr{C}, \boldsymbol{F}]_1 = 0$, $[\mathscr{C}, \boldsymbol{F}]_2 = 0$, whatever the function $\boldsymbol{F}$. Thus, if $q > 0$, $a_{0q} = x_1x_2a'_{0q}$, where

$$a'_{0q} = [\boldsymbol{a}_1^{(0)} + \boldsymbol{a}_2^{(0)}, \boldsymbol{a}_1^{(q)}]_{12}.$$

Again, from (4.4, 10), $[m_1\boldsymbol{C}_1 + m_2\boldsymbol{C}_2, \boldsymbol{F}]_{12} = 0$, so that

$$M_1^{\frac{1}{2}}[\mathscr{C}_1, \boldsymbol{F}]_{12} = -M_2^{\frac{1}{2}}[\mathscr{C}_2, \boldsymbol{F}]_{12}.$$

Hence we find that
$$a'_{0q} = M_1^{\frac{1}{2}}[\mathscr{C}_1, \boldsymbol{a}_1^{(q)}]_{12}. \tag{9.8, 4}$$

Similarly, if $q < 0$,
$$a'_{0q} = -M_2^{\frac{1}{2}}[\mathscr{C}_2, \boldsymbol{a}_2^{(q)}]_{12}; \tag{9.8, 5}$$

likewise if $q = 0$, applying the same method a second time, $a_{00} = x_1x_2a'_{00}$, where

$$a'_{00} = M_1^{\frac{1}{2}}[\mathscr{C}_1, \boldsymbol{a}_1^{(0)} + \boldsymbol{a}_2^{(0)}]_{12}$$

$$= -(M_1M_2)^{\frac{1}{2}}[\mathscr{C}_1, \mathscr{C}_2]_{12}. \tag{9.8, 6}$$

Now, by (9.6, 1, 6, 8), $a'_{00} = 8M_1M_2\Omega_{12}^{(1)}(1)$,

$$a'_{11} = 8M_2[\tfrac{5}{4}(6M_1^2 + 5M_2^2)\Omega_{12}^{(1)}(1) - M_2^2\{5\Omega_{12}^{(1)}(2) - \Omega_{12}^{(1)}(3)\} + 2M_1M_2\Omega_{12}^{(2)}(2)],$$

$$a''_{11} = 4\Omega_1^{(2)}(2),$$

whence, by (9.8, 1), we know a_{11}; similarly for a'_{-1-1}, a''_{-1-1}, and a_{-1-1}, using (9.8, 2). Again by (9.6, 3, 5),

$$a'_{01} = -8M_1^{\frac{1}{2}}M_2^{\frac{3}{2}}\{\Omega_{12}^{(1)}(2) - \tfrac{5}{2}\Omega_{12}^{(1)}(1)\},$$

$$a'_{0-1} = 8M_2^{\frac{1}{2}}M_1^{\frac{3}{2}}\{\Omega_{12}^{(1)}(2) - \tfrac{5}{2}\Omega_{12}^{(1)}(1)\},$$

$$a'_{1-1} = -8(M_1M_2)^{\frac{3}{2}}[\tfrac{55}{4}\Omega_{12}^{(1)}(1) - \{5\Omega_{12}^{(1)}(2) - \Omega_{12}^{(1)}(3)\} - 2\Omega_{12}^{(2)}(2)].$$

Let four constants A, B, C, E be defined by the equations

$$\left.\begin{aligned} \text{A} \equiv \Omega_{12}^{(2)}(2)/5\Omega_{12}^{(1)}(1), \quad \text{B} \equiv \{5\Omega_{12}^{(1)}(2)-\Omega_{12}^{(1)}(3)\}/5\Omega_{12}^{(1)}(1), \\ \text{C}+1 \equiv 2\Omega_{12}^{(1)}(2)/5\Omega_{12}^{(1)}(1), \end{aligned}\right\} \tag{9.8, 7}$$

$$\text{E} \equiv kT/8M_1 M_2 \Omega_{12}^{(1)}(1), \tag{9.8, 8}$$

so that A, B, C, depending on *ratios* of the functions Ω (cf. (9.33, 2)), are pure numbers. Then

$$a'_{00} = kT/\text{E}, \tag{9.8, 9}$$

$$a'_{01} = -5\text{C}M_2 kT/2M_1^{\frac{1}{2}}\text{E}, \tag{9.8, 10}$$

$$a'_{0-1} = 5\text{C}M_1 kT/2M_2^{\frac{1}{2}}\text{E}, \tag{9.8, 11}$$

$$a'_{1-1} = -5(M_1 M_2)^{\frac{1}{2}} kT(\tfrac{11}{4}-\text{B}-2\text{A})/\text{E}, \tag{9.8, 12}$$

$$a'_{11} = 5kT\{\tfrac{1}{4}(6M_1^2+5M_2^2)-M_2^2\text{B}+2M_1 M_2 \text{A}\}/M_1 \text{E}, \tag{9.8, 13}$$

and similarly for a'_{-1-1}. Again, if $[\mu_1]_1$ denotes the first approximation to the coefficient of viscosity of the first component gas of the mixture at the temperature T, by (9.7, 3)

$$a''_{11} = 5kT/2[\mu_1]_1 \tag{9.8, 14}$$

with a similar expression for a''_{-1-1} in terms of $[\mu_2]_1$.

The values of b_{11}, b_{1-1} and b_{-1-1} may likewise be deduced from (9.6, 14, 15, 16), using the definition of b_{pq} given in 8.52. It is found that

$$\begin{aligned} b'_{1-1} &= -\tfrac{16}{3}M_1 M_2\{5\Omega_{12}^{(1)}(1)-\tfrac{3}{2}\Omega_{12}^{(2)}(2)\} \\ &= -5kT(\tfrac{2}{3}-\text{A})/\text{E}, \end{aligned} \tag{9.8, 15}$$

$$\begin{aligned} b'_{11} &= \tfrac{16}{3}M_2\{5M_1\Omega_{12}^{(1)}(1)+\tfrac{3}{2}M_2\Omega_{12}^{(2)}(2)\} \\ &= 5kT(\tfrac{2}{3}+M_2\text{A}/M_1)/\text{E}, \end{aligned} \tag{9.8, 16}$$

$$b''_{11} = 4\Omega_1^{(2)}(2) = 5kT/2[\mu_1]_1, \tag{9.8, 17}$$

with similar results for b'_{-1-1}, b''_{-1-1}.

These expressions for the quantities a_{pq}, b_{pq} give first approximations to the coefficients of viscosity, thermal conduction, and diffusion in a gas-mixture.

9.81. The coefficient of diffusion D_{12}; first and second approximations $[D_{12}]_1$, $[D_{12}]_2$

The first and second approximations to D_{12}, written $[D_{12}]_1$, $[D_{12}]_2$, are obtained (cf. 9.7) by deleting the limit sign in (8.51, 16), and putting $m = 0$ or $m = 1$. Thus

$$[D_{12}]_1 = \frac{3kT}{2nm_0 a'_{00}} = \frac{3kT}{16nm_0 M_1 M_2 \Omega_{12}^{(1)}(1)} = \frac{3\text{E}}{2nm_0}. \tag{9.81, 1}$$

The second approximation, $[D_{12}]_2$, is obtained by multiplying $[D_{12}]_1$ by the factor $a_{00}\mathscr{A}^{(1)}_{00}/\mathscr{A}^{(1)}$, i.e. by

$$a_{00}\begin{vmatrix} a_{11} & a_{1-1} \\ a_{1-1} & a_{-1-1} \end{vmatrix} \Bigg/ \begin{vmatrix} a_{11} & a_{10} & a_{1-1} \\ a_{01} & a_{00} & a_{0-1} \\ a_{-11} & a_{-10} & a_{-1-1} \end{vmatrix}, \tag{9.81, 2}$$

so that

$$[D_{12}]_2 = [D_{12}]_1/(1-\Delta), \tag{9.81, 3}$$

where $\quad a_{00}(a_{11}a_{-1-1}-a^2_{1-1})\Delta \equiv a^2_{01}a_{-1-1}+a^2_{0-1}a_{11}-2a_{-11}a_{01}a_{0-1}.$

Using the values of the quantities a_{pq} given in 9.8, Δ is found to be given by

$$\Delta = 5\text{C}^2\frac{M_1^2\text{P}_1\text{x}_1^2+M_2^2\text{P}_2\text{x}_2^2+\text{P}_{12}\text{x}_1\text{x}_2}{\text{Q}_1\text{x}_1^2+\text{Q}_2\text{x}_2^2+\text{Q}_{12}\text{x}_1\text{x}_2}, \tag{9.81, 4}$$

where

$$\text{P}_1 \equiv M_1\text{E}/[\mu_1]_1, \quad \text{P}_2 \equiv M_2\text{E}/[\mu_2]_1, \tag{9.81, 5}$$

$$\text{P}_{12} \equiv 3(M_1-M_2)^2+4M_1M_2\text{A}, \tag{9.81, 6}$$

$$\text{Q}_{12} \equiv 3(M_1-M_2)^2(5-4\text{B})+4M_1M_2\text{A}(11-4\text{B})+2\text{P}_1\text{P}_2, \tag{9.81, 7}$$

$$\text{Q}_1 \equiv \text{P}_1(6M_2^2+5M_1^2-4M_1^2\text{B}+8M_1M_2\text{A}), \tag{9.81, 8}$$

with a similar relation for Q_2.

The general expression for the third approximation to D_{12} is extremely complicated, and is not considered here.

9.82. The thermal conductivity for a gas-mixture; first approximation $[\lambda]_1$

The mth approximation to the coefficient of thermal conduction for a gas-mixture was found in (8.51, 18) to be

$$[\lambda]_m = \tfrac{75}{8}k^2T(\text{x}_1^2m_1^{-1}\mathscr{A}'^{(m)}_{11}+2\text{x}_1\text{x}_2(m_1m_2)^{-\frac{1}{2}}\mathscr{A}'^{(m)}_{1-1}+\text{x}_2^2m_2^{-1}\mathscr{A}'^{(m)}_{-1-1})/\mathscr{A}'^{(m)}.$$

The first approximation is thus

$$[\lambda]_1 = \tfrac{75}{8}k^2T(\text{x}_1^2m_1^{-1}a_{-1-1}-2\text{x}_1\text{x}_2(m_1m_2)^{-\frac{1}{2}}a_{1-1} +\text{x}_2^2m_2^{-1}a_{11})/(a_{11}a_{-1-1}-a^2_{1-1}). \tag{9.82, 1}$$

The values of a_{11}, a_{-1-1}, a_{1-1} have already been found in (9.8, 12–14) in terms of E, $[\mu_1]_1$ and $[\mu_2]_1, \ldots$, but it is convenient here to express a''_{11}, a''_{-1-1} in terms of $[\lambda_1]_1$ and $[\lambda_2]_1$, the first approximations to the coefficients of thermal conduction of the constituent gases in the mixture at the temperature T. By (9.7, 4),

$$[\lambda_1]_1 = \tfrac{5}{2}[\mu_1]_1(c_v)_1 = \frac{15k}{4m_1}[\mu_1]_1, \quad [\lambda_2]_1 = \frac{15k}{4m_2}[\mu_2]_1,$$

and the equations (9.81, 5) for P_1, P_2 can be written

$$\text{P}_1 = \frac{15k\text{E}}{4m_0[\lambda_1]_1}, \quad \text{P}_2 = \frac{15k\text{E}}{4m_0[\lambda_2]_1}. \tag{9.82, 2}$$

Thus (9.8, 14) becomes

$$a''_{11} = \frac{75k^2T}{8m_1[\lambda_1]_1} = \frac{5kT\text{P}_1}{2M_1\text{E}}, \qquad (9.82, 3)$$

with a similar equation for a''_{-1-1}. Substituting the values of a_{11}, a_{1-1}, a_{-1-1} into (9.82, 1), we obtain

$$[\lambda]_1 = \frac{\text{x}_1^2\text{Q}_1[\lambda_1]_1 + \text{x}_2^2\text{Q}_2[\lambda_2]_1 + \text{x}_1\text{x}_2\text{Q}'_{12}}{\text{x}_1^2\text{Q}_1 + \text{x}_2^2\text{Q}_2 + \text{x}_1\text{x}_2\text{Q}_{12}}, \qquad (9.82, 4)$$

where Q_1, Q_2, Q_{12} are given by (9.81, 7, 8) and

$$\text{Q}'_{12} = \frac{15k\text{E}}{2m_0}\{\text{P}_1 + \text{P}_2 + (11 - 4\text{B} - 8\text{A})M_1M_2\}. \qquad (9.82, 5)$$

If n_1 or n_2 is equal to zero, this first approximation to λ reduces, as it should, to the first approximation to the conductivity of a simple gas.

9.83. Thermal diffusion

By (8.51, 17), the first approximation $[\text{k}_T]_1$ to the thermal diffusion ratio is given by

$$[\text{k}_T]_1 = \frac{5}{2}\frac{\text{x}_1M_1^{-\frac{1}{2}}(a_{01}a_{-1-1} - a_{0-1}a_{1-1}) + \text{x}_2M_2^{-\frac{1}{2}}(a_{0-1}a_{11} - a_{01}a_{1-1})}{a_{11}a_{-1-1} - a_{1-1}^2}.$$

The first approximation to the thermal diffusion factor is $[\alpha_{12}]_1 \equiv [\text{k}_T]_1/\text{x}_1\text{x}_2$. Hence, using the values of a_{pq} given in 9.8,

$$[\alpha_{12}]_1 \equiv \frac{[\text{k}_T]_1}{\text{x}_1\text{x}_2} = 5\text{C}\frac{\text{x}_1\text{S}_1 - \text{x}_2\text{S}_2}{\text{x}_1^2\text{Q}_1 + \text{x}_2^2\text{Q}_2 + \text{x}_1\text{x}_2\text{Q}_{12}}, \qquad (9.83, 1)$$

where Q_1, Q_2, Q_{12} are as defined in (9.81, 7, 8) and

$$\text{S}_1 \equiv M_1\text{P}_1 - M_2\{3(M_2 - M_1) + 4M_1\text{A}\}, \qquad (9.83, 2)$$

$$\text{S}_2 \equiv M_2\text{P}_2 - M_1\{3(M_1 - M_2) + 4M_2\text{A}\}. \qquad (9.83, 3)$$

9.84. The coefficient of viscosity for a gas-mixture; first approximation $[\mu]_1$

By (8.52, 10), this is given by

$$[\mu]_1 = \tfrac{5}{2}kT(\text{x}_1^2b_{-1-1} + \text{x}_2^2b_{11} - 2\text{x}_1\text{x}_2b_{1-1})/(b_{11}b_{-1-1} - b_{1-1}^2). \qquad (9.84, 1)$$

With the values of b_{pq} found in 9.8, this becomes

$$[\mu]_1 = \frac{\text{x}_1^2\text{R}_1 + \text{x}_2^2\text{R}_2 + \text{x}_1\text{x}_2\text{R}'_{12}}{\text{x}_1^2\text{R}_1/[\mu_1]_1 + \text{x}_2^2\text{R}_2/[\mu_2]_1 + \text{x}_1\text{x}_2\text{R}_{12}}, \qquad (9.84, 2)$$

where $$\text{R}_1 \equiv \tfrac{2}{3} + M_1\text{A}/M_2, \quad \text{R}_2 \equiv \tfrac{2}{3} + M_2\text{A}/M_1, \qquad (9.84, 3)$$

$$\text{R}'_{12} \equiv \text{E}/2[\mu_1]_1 + \text{E}/2[\mu_2]_1 + 2(\tfrac{2}{3} - \text{A}), \quad \text{R}_{12} \equiv \text{E}/2[\mu_1]_1[\mu_2]_1 + 4\text{A}/3\text{E}M_1M_2. \qquad (9.84, 4)$$

If x_1 or x_2 is equal to zero, this expression reduces to the first approximation to μ for the corresponding simple gas.

9.85. Kihara's approximations for a gas-mixture

By (9.71, 1) and (9.8, 7)

$$\mathrm{C} = \frac{2T}{5\Omega_{12}^{(1)}(1)}\frac{d\Omega_{12}^{(1)}(1)}{dT}, \quad \mathrm{B} - \frac{3}{4} = -\frac{1}{5\Omega_{12}^{(1)}(1)}\frac{d}{dT}\left(T^2\frac{d\Omega_{12}^{(1)}(1)}{dT}\right). \tag{9.85, 1}$$

Hence, if (cf. 9.71) terms of third order in the temperature-derivatives are neglected, we can approximate in the expression (9.81, 4) for A by taking $\mathrm{B} = \frac{3}{4}$. In this case (9.81, 7, 8) take the simpler forms

$$\mathrm{Q}_{12} = 6(M_1 - M_2)^2 + 32 M_1 M_2 \mathrm{A} + 2\mathrm{P}_1\mathrm{P}_2, \tag{9.85, 2}$$

$$\mathrm{Q}_1 = 2\mathrm{P}_1(3M_2^2 + M_1^2 + 4M_1 M_2 \mathrm{A}), \tag{9.85, 3}$$

which is Kihara's approximation. Kihara took $\mathrm{B} = \frac{3}{4}$ also in the formulae for $[\lambda]_1$ and $[k_T]_1$, so that he used the values (9.85, 2, 3) for Q_{12}, Q_1, together with

$$\mathrm{Q}'_{12} = \frac{15k\mathrm{E}}{2m_0}\{\mathrm{P}_1 + \mathrm{P}_2 + 8(1-\mathrm{A})M_1 M_2\}. \tag{9.85, 4}$$

For most molecular models $\mathrm{B} < \frac{3}{4}$. It can be shown that the effect of increasing B to $\frac{3}{4}$ is to increase the magnitudes of $[D_{12}]_1$, $[\lambda]_1$ and $[k_T]_1$. The effect of taking later approximations is likewise to increase $[D_{12}]_1$ and $[\lambda]_1$, and usually $[k_T]_1$ also; hence Kihara's approximations are often closer to the true values than Enskog's, and they are easier to calculate. However, whereas in 9.81 the approximation $\mathrm{B} = \frac{3}{4}$ is made in a correcting term already known to be small, in 9.82 and 9.83 it affects the whole of $[\lambda]_1$ and $[k_T]_1$; hence in these it is a correspondingly cruder approximation. In particular, the values of $[\lambda]_1$ calculated from it are sometimes found appreciably to exceed the exact values of λ.*

* E. A. Mason and S. C. Saxena, *J. Chem. Phys.* **31**, 511 (1959).

10

VISCOSITY, THERMAL CONDUCTION, AND DIFFUSION: THEORETICAL FORMULAE FOR SPECIAL MOLECULAR MODELS

10.1. The functions $\Omega(r)$

The expressions derived in Chapter 9 apply to any type of spherically symmetrical molecule possessing only energy of translation, but they involve the functions $\Omega^{(l)}(r)$ of (9.33, 2), which can be evaluated only when the law of interaction between molecules is known. These functions will now be determined for certain special molecular models.

It is convenient first to introduce a non-dimensional variant $\mathscr{W}_{12}^{(l)}(r)$ of $\Omega_{12}^{(l)}(r)$. Let σ_{12} be a conveniently chosen length, lying within the range of b that contributes most to the integral in $\Omega_{12}^{(l)}(r)$. Then $\mathscr{W}_{12}^{(l)}(r)$ is defined by*

$$\Omega_{12}^{(l)}(r) \equiv \tfrac{1}{2}\sigma_{12}^2\left(\frac{2\pi kT}{m_0 M_1 M_2}\right)^{\frac{1}{2}}\mathscr{W}_{12}^{(l)}(r), \qquad (10.1, 1)$$

so that (cf. (9.33, 2))

$$\mathscr{W}_{12}^{(l)}(r) = \int e^{-g^2}g^{2r+2}\int(1-\cos^l\chi)\,(b/\sigma_{12})\,d(b/\sigma_{12})\,d(g^2). \qquad (10.1, 2)$$

The formula corresponding to (10.1, 1) for like molecules is

$$\Omega_1^{(l)}(r) = \sigma_1^2\left(\frac{\pi kT}{m_1}\right)^{\frac{1}{2}}\mathscr{W}_1^{(l)}(r). \qquad (10.1, 3)$$

In terms of the quantities $\mathscr{W}^{(l)}(r)$, the expressions (9.7, 1), (9.81, 1) for $[\mu]_1$, $[\lambda]_1$ and $[D_{12}]_1$ become

$$[\mu]_1 = \frac{5}{8}\frac{(kmT)^{\frac{1}{2}}}{\pi^{\frac{1}{2}}\sigma^2\mathscr{W}_1^{(2)}(2)}, \quad [\lambda]_1 = \tfrac{5}{2}c_v[\mu]_1, \qquad (10.1, 4)$$

$$[D_{12}]_1 = \frac{3}{8n\sigma_{12}^2\mathscr{W}_{12}^{(1)}(1)}\left(\frac{kT}{2\pi m_0 M_1 M_2}\right)^{\frac{1}{2}}. \qquad (10.1, 5)$$

* The quantity $\mathscr{W}_{12}^{(l)}(r)$ is twice the $W_{12}^{(l)}(r)$ used by J. O. Hirschfelder, R. B. Bird and E. L. Spotz, *J. Chem. Phys.* **16**, 968 (1948). The notation $\Omega_{12}^{*(l,r)}$ is used by J. O. Hirschfelder, C. F. Curtiss and R. B. Bird (*The Molecular Theory of Gases and Liquids*, chap. 8) to denote the ratio of $\Omega_{12}^{(l)}(r)$ to its value for rigid spherical molecules of diameter σ_{12}. The use of their notation introduces a number of additional numerical factors into the general formulae.

RIGID ELASTIC SPHERICAL MOLECULES WITHOUT FIELDS OF FORCE

10.2. An encounter between two rigid elastic spherical molecules of diameters σ_1, σ_2 occurs only if $b < \sigma_{12}$, where $\sigma_{12} = \frac{1}{2}(\sigma_1+\sigma_2)$. At an encounter, by (3.44, 2), $b = \sigma_{12}\cos\frac{1}{2}\chi$, and $b\,db = -\frac{1}{4}\sigma_{12}^2\sin\chi\,d\chi$. Hence if σ_{12} is identified with the σ_{12} of 10.1, by (10.1, 2),

$$\mathscr{W}_{12}^{(l)}(r) = \frac{1}{4}\int_0^\pi (1-\cos^l\chi)\sin\chi\,d\chi\int_0^\infty e^{-g^2}g^{2r+2}d(g^2)$$

$$= \frac{1}{4}\left[2-\frac{1}{l+1}(1+(-1)^l)\right](r+1)!. \tag{10.2, 1}$$

In particular, $\mathscr{W}_{12}^{(1)}(1) = 1$, $\mathscr{W}_{12}^{(2)}(2) = 2$. Clearly $\mathscr{W}_1^{(l)}(r) = \mathscr{W}_{12}^{(l)}(r)$.

Again, by (9.33, 4)

$$\phi_{12}^{(l)} = \tfrac{1}{2}\pi\sigma_{12}^2\int_0^\pi (1-\cos^l\chi)\sin\chi\,d\chi$$

$$= \tfrac{1}{2}\pi\sigma_{12}^2\left[2-\frac{1}{l+1}(1+(-1)^l)\right]. \tag{10.2, 2}$$

In particular, $\phi_{12}^{(1)}$ is equal to $\pi\sigma_{12}^2$, the ordinary collision cross-section.

10.21. Viscosity and conduction for a simple gas

By (10.1, 4), the first approximations $[\mu]_1$, $[\lambda]_1$ to μ and λ for a simple gas are given in terms of the molecular diameter σ by

$$[\mu]_1 = \frac{5}{16}\frac{\sqrt{(kmT)}}{\pi^{\frac{1}{2}}\sigma^2}, \quad [\lambda]_1 = \tfrac{5}{2}[\mu]_1 c_v = \frac{75}{64\sigma^2}\left(\frac{k^3T}{\pi m}\right)^{\frac{1}{2}}. \tag{10.21, 1}$$

Later approximations to λ and μ are obtained by multiplying the above values by successive partial sums of the respective series

$$1+a_{11}(\mathscr{A}_{12}^{(2)})^2/\mathscr{A}^{(1)}\mathscr{A}^{(2)}+a_{11}(\mathscr{A}_{13}^{(3)})^2/\mathscr{A}^{(2)}\mathscr{A}^{(3)}+\ldots, \tag{10.21, 2}$$

$$1+b_{11}(\mathscr{B}_{12}^{(2)})^2/\mathscr{B}^{(1)}\mathscr{B}^{(2)}+b_{11}(\mathscr{B}_{13}^{(3)})^2/\mathscr{B}^{(2)}\mathscr{B}^{(3)}+\ldots \tag{10.21, 3}$$

(cf. (7.51, 14), (7.52, 8)). In the present case each term of these series is a pure number, since the factor $1/\sigma^2\sqrt{(kT/m)}$ occurs in each element of the determinants $\mathscr{A}, \mathscr{B}$, and so cancels throughout.

From (9.6, 8–13, 16–21), using the above expression for $\Omega^{(l)}(r)$, it is easy to deduce that

$$a_{12} = -\tfrac{1}{4}a_{11},\quad a_{22} = \tfrac{45}{16}a_{11},\quad a_{13} = -\tfrac{1}{32}a_{11},\quad a_{23} = -\tfrac{103}{128}a_{11},\quad a_{33} = \tfrac{5657}{1024}a_{11},$$

$$b_{12} = -\tfrac{1}{4}b_{11},\quad b_{22} = \tfrac{205}{48}b_{11},\quad b_{13} = -\tfrac{1}{32}b_{11},\quad b_{23} = -\tfrac{163}{128}b_{11},\quad b_{33} = \tfrac{11889}{1024}b_{11}.$$

Hence the first three terms in the above series are

$$1+\frac{1}{44}+\frac{(111)^2}{88\times 66951} = 1+0\cdot02273+0\cdot00209$$

and
$$1+\frac{3}{202}+\frac{(347)^2}{808\times 145043} = 1+0\cdot01485+0\cdot00103.$$

The fourth terms of these series have also been calculated,* though the results necessary for the calculation are not given in 9.6; these terms are

$$0\cdot00031, \quad 0\cdot00012,$$

so that, to the fourth degree of approximation, the expressions for λ, μ are

$$[\lambda]_4 = 1\cdot02513[\lambda]_1, \quad [\mu]_4 = 1\cdot01600[\mu]_1, \tag{10.21, 4}$$

and
$$[\lambda]_4 = 2\cdot522[\mu]_4 c_v. \tag{10.21, 5}$$

The convergence of the above approximations is so rapid that these values may be taken as correct to within one-tenth of 1 per cent.†

10.22. Gas-mixtures; $[D_{12}]_1$, $[D_{12}]_2$, $[\lambda]_1$, $[k_T]_1$, $[\mu]_1$

The results for gas-mixtures, found in 9.81–9.84, can be completed in the present case by means of (10.2, 1), from which the following values of A, B, C and E are obtained:

$$A = 2/5, \quad B = 3/5, \quad C = 1/5, \tag{10.22, 1}$$

$$E = \left(\frac{2kTm_0}{\pi M_1 M_2}\right)^{\frac{1}{2}} \frac{1}{8\sigma_{12}^2}. \tag{10.22, 2}$$

MOLECULES THAT ARE CENTRES OF FORCE

10.3. Another important simple molecular model (first suggested by Boscovich‡ in 1758) is a point-mass surrounded by a field of force. During interaction, let two molecules m_1, m_2 exert forces $\boldsymbol{P}$, $-\boldsymbol{P}$ on each other.

Let the position-vectors of the two molecules at the time t be $\boldsymbol{r}_1$, $\boldsymbol{r}_2$; then the position-vector $\boldsymbol{r}_{21}$ of the second relative to the first is $\boldsymbol{r}_2-\boldsymbol{r}_1$. The equations of motion of the two molecules are

$$m_1\ddot{\boldsymbol{r}}_1 = -\boldsymbol{P}, \quad m_2\ddot{\boldsymbol{r}}_2 = \boldsymbol{P}.$$

Hence
$$m_1 m_2 \ddot{\boldsymbol{r}}_{21} = m_1 m_2(\ddot{\boldsymbol{r}}_2-\ddot{\boldsymbol{r}}_1) = (m_1+m_2)\boldsymbol{P} = m_0\boldsymbol{P},$$

* A numerical error in this part of Chapman's calculations was corrected by D. Burnett, *Proc. Lond. Math. Soc.* **39**, 385 (1935).

† The values of λ and μ were recalculated by C. L. Pekeris and Z. Altermann (*Proc. natn. Acad. Sci. U.S.A.* **43**, 998 (1957)) by a different method, applicable only to rigid elastic spheres; they found $\lambda = 1\cdot02513[\lambda]_1$, $\mu = 1\cdot016034[\mu]_1$.

‡ R. G. Boscovich, *Philosophiae naturalis theoria* (1758).

so that the motion of m_2 relative to m_1 is the same as the motion of a particle of unit mass about a fixed centre of force $m_0 \boldsymbol{P}/m_1 m_2$. If polar coordinates r, θ be taken in the plane of the orbit, the equations of angular momentum and of energy for such a particle have the forms

$$r^2\dot{\theta} = \text{const.} = gb, \tag{10.3, 1}$$

$$\tfrac{1}{2}(\dot{r}^2 + r^2\dot{\theta}^2) + m_0 V_{12}(r)/m_1 m_2 = \text{const.} = \tfrac{1}{2}g^2. \tag{10.3, 2}$$

Here g and b have the same meanings as in 3.41, 3.42, and $V_{12}(r)$ is the potential energy (taken to be zero at $r = \infty$) of the force $\boldsymbol{P}$.

By eliminating the time between (10.3, 1, 2), we get the differential equation of the orbit,

$$\frac{1}{2}\frac{g^2 b^2}{r^4}\left\{\left(\frac{dr}{d\theta}\right)^2 + r^2\right\} = \tfrac{1}{2}g^2 - \frac{m_0 V_{12}(r)}{m_1 m_2}.$$

This has the integral

$$\theta = \int_r^\infty \left\{\frac{r^4}{b^2}\left(1 - \frac{V_{12}(r)}{kT\mathscr{g}^2}\right) - r^2\right\}^{-\frac{1}{2}} dr \tag{10.3, 3}$$

(since $\mathscr{g}^2 = m_1 m_2 g^2/2m_0 kT$), if θ is measured from an axis parallel to the initial asymptote of the orbit. The apse $r = R$ of the orbit corresponds to $dr/d\theta = 0$, i.e.

$$\frac{V_{12}(R)}{kT\mathscr{g}^2} = 1 - \frac{b^2}{R^2}; \tag{10.3, 4}$$

if this equation has more than one root, R is the largest root.* The angle between the asymptotes of the orbit is twice the value of θ corresponding to $r = R$. Since the angle χ is the supplement of this angle,

$$\chi = \pi - 2\int_R^\infty \left\{\frac{r^4}{b^2}\left(1 - \frac{V_{12}(r)}{kT\mathscr{g}^2}\right) - r^2\right\}^{-\frac{1}{2}} dr. \tag{10.3, 5}$$

This shows that χ depends only on b, on $kT\mathscr{g}^2$ and on the force-law. Thus the integral $\mathscr{W}_{12}^{(l)}(r)$ of (10.1, 2) depends only on kT and the force-law; in particular, it does not depend on the molecular masses.

10.31. Inverse power-law force

Suppose now that $\boldsymbol{P}$ is a repulsive force given by the power laws

$$P = \kappa_{12}/r^\nu, \quad V_{12}(r) = \kappa_{12}/(\nu - 1)r^{\nu-1}. \tag{10.31, 1}$$

If pure numbers v, v_0, v_{00} are defined by

$$v \equiv b/r, \quad v_0 \equiv b(2kT\mathscr{g}^2/\kappa_{12})^{1/(\nu-1)}, \quad v_{00} \equiv b/R, \tag{10.31, 2}$$

* When (10.3, 4) has more than one root, bound states of the same energy can occur, in which the molecules remain indefinitely at a finite distance from one another. Such bound states are ignored here.

(10.3, 5) becomes, on substitution for V_{12} from (10.31, 1),

$$\chi = \pi - 2\int_0^{v_{00}} \left\{1 - v^2 - \frac{2}{\nu - 1}\left(\frac{v}{v_0}\right)^{\nu-1}\right\}^{-\frac{1}{2}} dv. \qquad (10.31, 3)$$

The quantity v_{00} is the (unique) positive root of the equation

$$1 - v^2 - \frac{2}{\nu - 1}\left(\frac{v}{v_0}\right)^{\nu-1} = 0. \qquad (10.31, 4)$$

Hence χ depends only on ν and v_0. The corresponding formulae for an attractive force (κ_{12} negative) are found by replacing κ_{12} by $|\kappa_{12}|$ in (10.31, 2), and reversing the signs of the terms involving $2/(\nu-1)$ in (10.31, 3, 4); v_{00} is the least positive root of the modified equation (10.31, 4).

The integral $\phi_{12}^{(l)}$ of (9.33, 4) can now be transformed as follows:

$$\phi_{12}^{(l)} = 2\pi\int_0^\infty (1 - \cos^l\chi)\, b\, db$$

$$= 2\pi\left(\frac{\kappa_{12}}{2kTg^2}\right)^{2/(\nu-1)} A_l(\nu), \qquad (10.31, 5)$$

where

$$A_l(\nu) \equiv \int_0^\infty (1 - \cos^l\chi)\, v_0\, dv_0; \qquad (10.31, 6)$$

$A_l(\nu)$ is a pure number depending only on l and ν. The integrals $\Omega_{12}^{(l)}(r)$ can now be expressed in terms of gamma functions, for (cf. (9.33, 5))

$$\Omega_{12}^{(l)}(r) = \left(\frac{2\pi kT}{m_0 M_1 M_2}\right)^{\frac{1}{2}}\left(\frac{\kappa_{12}}{2kT}\right)^{2/(\nu-1)} A_l(\nu)\int_0^\infty e^{-g^2} g^{2r+3-4/(\nu-1)}\, dg$$

$$= \frac{1}{2}\left(\frac{2\pi kT}{m_0 M_1 M_2}\right)^{\frac{1}{2}}\left(\frac{\kappa_{12}}{2kT}\right)^{2/(\nu-1)} A_l(\nu)\,\Gamma\left(r + 2 - \frac{2}{\nu - 1}\right). \qquad (10.31, 7)$$

The length σ_{12} introduced in 10.1 may here conveniently be defined by

$$\sigma_{12} = (\kappa_{12}/2kT)^{1/(\nu-1)} \qquad (10.31, 8)$$

so that it depends on the temperature, as well as on the law of force. Then (10.1, 1) becomes

$$\mathscr{W}_{12}^{(l)}(r) = A_l(\nu)\,\Gamma\left(r + 2 - \frac{2}{\nu - 1}\right). \qquad (10.31, 9)$$

Clearly $\mathscr{W}_1^{(l)}(r)$ has the same value as $\mathscr{W}_{12}^{(l)}(r)$ if ν has the same value in both.

From (10.31, 7) and (9.8, 7, 8) the following are found to be the special forms of A, B, C, E appropriate for molecules which interact in accordance with (10.31, 1):

$$\text{A} = \frac{(3\nu - 5)A_2(\nu)}{5(\nu - 1)A_1(\nu)}, \quad \text{B} = \frac{(3\nu - 5)(\nu + 1)}{5(\nu - 1)^2}, \quad \text{C} = \frac{\nu - 5}{5(\nu - 1)}, \qquad (10.31, 10)$$

$$\text{E} = \left(\frac{2kTm_0}{\pi M_1 M_2}\right)^{\frac{1}{2}}\left(\frac{2kT}{\kappa_{12}}\right)^{2/(\nu-1)} \Big/ 8\Gamma\left(3 - \frac{2}{\nu - 1}\right) A_1(\nu). \qquad (10.31, 11)$$

The numbers A, B and C thus depend only on ν.

The constants $A_l(\nu)$ have to be evaluated by quadrature.* The following table contains values of $A_1(\nu)$, $A_2(\nu)$ for several values of ν, and also some values of A:

Table 3

ν	$A_1(\nu)$	$A_2(\nu)$	A
5	0·422	0·436	0·517
7	0·385	0·357	0·493
9	0·382	0·332	0·478
11	0·383	0·319	0·465
15	0·393	0·309	0·450
21	—	0·307	—
25	—	0·306	—
∞	0·5	0·333	0·4

Rigid elastic spherical molecules may be regarded as a limiting case of the present model, corresponding to $\nu = \infty$; the force vanishes if $r > \sigma_{12}$ and becomes infinite at $r = \sigma_{12}$. These conditions are satisfied if

$$P \propto \lim_{\nu \to \infty} (\sigma_{12}/r)^\nu.$$

10.32. Viscosity and conduction for a simple gas

The first approximations to λ and μ for this model are given by

$$[\mu]_1 = 5(kmT/\pi)^{\frac{1}{2}} (2kT/\kappa)^{2/(\nu-1)} \Big/ 8\Gamma\left(4 - \frac{2}{\nu - 1}\right) A_2(\nu), \quad [\lambda]_1 = \tfrac{5}{2}[\mu]_1 c_v. \tag{10.32, 1}$$

Both $[\lambda]_1$ and $[\mu]_1$ are proportional to T^s, where

$$s = \frac{1}{2} + \frac{2}{\nu - 1}.$$

The limiting value of the index s as ν tends to infinity is $\frac{1}{2}$, the value already found for rigid elastic spherical molecules. For smaller values of ν the index is larger, becoming equal to unity when $\nu = 5$.

The exact values of λ and μ are obtained as in 10.21 by multiplying the above first approximations by the series (10.21, 2, 3). The sums of these series are pure numbers; by an argument similar to that used in 10.21, it may be shown that they depend neither on the temperature of the gas nor on the

* $A_1(5)$, $A_2(5)$ were calculated by J. C. Maxwell (*Collected Papers*, vol. 2, 26), and by K. Aichi and A. Tanakadate (see D. Enskog, *Arkiv. Mat. Astr. Fys.* **16**, 36 (1921)); $A_1(\nu)$ and $A_2(\nu)$ were given by S. Chapman (*Manchester Lit. and Phil. Soc. Memoirs*, **66**, 1 (1922)), for $\nu = 5, 7, 9, 11, 15$; $A_2(21)$ and $A_2(25)$ were found by J. E. (Lennard-)Jones (*Proc. R. Soc.* A, **106**, 421 (1924)), and $A_1(9)$ and $A_2(9)$ by H. R. Hassé and W. R. Cook (*Proc. R. Soc.* A, **125**, 196 (1929)). The values given in the text are those of Chapman and (Lennard-)Jones.

force-constant of the molecular fields, but only on the force-index ν. The second terms of these series can be determined, using the result

$$\frac{\Omega_1^{(l)}(r)}{\Omega_1^{(l)}(1)} = \left(r+1-\frac{2}{\nu-1}\right)_{r-1},$$

which is an immediate consequence of (10.31, 7); substituting in the equations of 9.6, we get

$$a_{12} = -\left(\frac{1}{4}-\frac{1}{\nu-1}\right)a_{11}, \quad a_{22} = \left(\frac{45}{16}-\frac{1}{\nu-1}+\frac{1}{(\nu-1)^2}\right)a_{11},$$

$$b_{12} = -\left(\frac{1}{4}-\frac{1}{\nu-1}\right)b_{11}, \quad b_{22} = \left(\frac{205}{48}-\frac{1}{\nu-1}+\frac{1}{(\nu-1)^2}\right)b_{11}.$$

This leads to the second approximations

$$[\mu]_2 = [\mu]_1\left\{1+\frac{3(\nu-5)^2}{2(\nu-1)(101\nu-113)}\right\}, \qquad (10.32, 2)$$

$$[\lambda]_2 = [\lambda]_1\left\{1+\frac{(\nu-5)^2}{4(\nu-1)(11\nu-13)}\right\}, \qquad (10.32, 3)$$

which are identical with the first when $\nu = 5$; for values of ν between 5 and infinity, the numerical multipliers of $[\mu]_1$ and $[\lambda]_1$ lie between unity and the values found above for rigid elastic spherical molecules. The ratio $\lambda/\mu c_v$ also lies between 2·5 and the value 2·522 appropriate to elastic spheres.

10.33. Maxwellian molecules

When $\nu = 5$ the formulae of 10.31, 10.32 become specially simple. The simplicity of this case was first perceived by Maxwell,* who found that it was possible to develop the theory for a gas whose molecules are centres of force repelling proportionately to r^{-5}, without determining the velocity-distribution function f. For these 'Maxwellian' molecules, by (10.31, 2),

$$gb\,db = (m_0\kappa_{12}/m_1 m_2)^{\frac{1}{2}}\, v_0\, dv_0, \qquad (10.33, 1)$$

so that the element $gb\,db$ does not depend on g, but only on v_0. The integrals of 9.3–9.5 accordingly take simpler forms: thus, for example, from (9.31, 14),

$$\iint H_{12}(\chi)\, \mathscr{g}^2 gb\,db\,d\mathscr{g}$$

$$= \pi^{\frac{3}{2}} i_{12}^{-\frac{5}{2}} \left(\frac{\kappa_{12}}{m_0}\right)^{\frac{1}{2}} \iint e^{-j_{12}\mathscr{g}^2}\{\tfrac{3}{2}+(1-j_{12}-\cos\chi)\mathscr{g}^2\}\mathscr{g}^2 v_0\,dv_0\,d\mathscr{g}$$

$$= \tfrac{3}{8}\pi^2 \left(\frac{\kappa_{12}}{m_0}\right)^{\frac{1}{2}} \int (i_{12}j_{12})^{-\frac{5}{2}}(1-\cos\chi)\, v_0\,dv_0,$$

* J. C. Maxwell, *Collected Papers*, vol. 2, 42; *Phil. Trans. R. Soc.*, **157**, 49 (1867).

and so, by (9.32, 1–3),

$$(ST/st)^{\frac{5}{2}}\pi^{-3}\iiiint\{H_{12}(0)-H_{12}(\chi)\}gb\,db\,de\,dg$$

$$=-3\pi\left(\frac{\kappa_{12}}{m_0}\right)^{\frac{1}{2}}\int\{1-2M_1M_2st(1-\cos\chi)\}^{-\frac{5}{2}}(1-\cos\chi)\,v_0\,dv_0.$$

Now the coefficient of $s^p t^q$ in the expression on the left is equal to

$$[S_{\frac{3}{2}}^{(p)}(\mathscr{C}_1^2)\mathscr{C}_1,\ S_{\frac{3}{2}}^{(q)}(\mathscr{C}_2^2)\mathscr{C}_2]_{12}:$$

this accordingly vanishes unless $p=q$. Similarly from 9.5

$$[S_{\frac{3}{2}}^{(p)}(\mathscr{C}^2)\mathscr{C},\ S_{\frac{3}{2}}^{(q)}(\mathscr{C}^2)\mathscr{C}]=0$$

unless $p=q$. An analogous argument shows that the same is true for

$$[S_{\frac{5}{2}}^{(p)}(\mathscr{C}^2)\overset{\circ}{\mathscr{C}\mathscr{C}},\ S_{\frac{5}{2}}^{(q)}(\mathscr{C}^2)\overset{\circ}{\mathscr{C}\mathscr{C}}].$$

Using these results, and the known values of α_q, β_q, the sets of equations (7.51, 8), (7.52, 4) reduce to

$$a_1a_{11}=\alpha_1,\quad a_2a_{22}=0,\quad a_3a_{33}=0,\quad \ldots,$$

$$b_1b_{11}=\beta_1,\quad b_2b_{22}=0,\quad b_3b_{33}=0,\quad \ldots,$$

so that all the coefficients a_p, b_p vanish, save a_1, b_1. The formulae for λ, μ therefore reduce to the first approximations given in 10.32, i.e.

$$\mu=\frac{1}{3\pi}\left(\frac{2m}{\kappa}\right)^{\frac{1}{2}}\frac{kT}{A_2(5)},\quad \lambda=\tfrac{5}{2}\mu c_v,\quad \mathrm{f}=\tfrac{5}{2}.$$

It may similarly be proved that all the quantities a_{pq}, b_{pq} of 8.51, 8.52 vanish unless $p=q$ or $p=-q$, and hence in the series (8.51, 1, 2), (8.52, 1) all the coefficients a_p, d_p, b_p vanish save a_1, a_{-1}, d_0, b_1 and b_{-1}. Consequently the expressions for the coefficients of diffusion, viscosity, and thermal conduction for a gas-mixture become identical with the first approximations given in 9.81, 9.82 and 9.84, while the coefficient of thermal diffusion vanishes.

These results imply that for a simple gas the expansions in series of the functions $\boldsymbol{A}$, B given in (7.51, 2), (7.52, 1) reduce to the single terms

$$\boldsymbol{A}=a_1\boldsymbol{a}^{(1)}=a_1(\tfrac{5}{2}-\mathscr{C}^2)\mathscr{C},\tag{10.33, 2}$$

$$\mathsf{B}=b_1\mathsf{b}^{(1)}=b_1\overset{\circ}{\mathscr{C}\mathscr{C}}.\tag{10.33, 3}$$

Similar results hold for a gas-mixture.

Since $\boldsymbol{A}$ and B satisfy (7.31, 2, 3) it follows that with suitable values of a_1, b_1,

$$-a_1nI\{(\mathscr{C}^2-\tfrac{5}{2})\mathscr{C}\}=f^{(0)}(\mathscr{C}^2-\tfrac{5}{2})\mathscr{C},\tag{10.33, 4}$$

$$b_1nI(\overset{\circ}{\mathscr{C}\mathscr{C}})=f^{(0)}\overset{\circ}{\mathscr{C}\mathscr{C}}.\tag{10.33, 5}$$

A direct proof of these results was given in earlier editions of this book. Similar results hold for a gas-mixture.

For a simple gas, by the results of 7.51, 7.52 and 9.7,

$$\alpha_1 = -\tfrac{15}{4}, \quad \beta_1 = \tfrac{5}{2}, \quad a_{11} = b_{11} = 5kT/2\mu.$$

Hence
$$a_1 = -\frac{3\mu}{2kT}, \quad b_1 = \frac{\mu}{kT} \tag{10.33, 6}$$

and (10.33, 2, 3) become

$$\boldsymbol{A} = \frac{3\mu}{2kT}(\mathscr{C}^2 - \tfrac{5}{2})\mathscr{C}, \quad \mathsf{B} = \frac{\mu}{kT}\overset{\circ}{\mathscr{C}\mathscr{C}}. \tag{10.33, 7}$$

Thus (cf. (7.31, 1)) for Maxwellian molecules $f^{(1)}$ is given by

$$f^{(1)} = -f^{(0)}\left\{\frac{3\mu}{2p}\left(\frac{2kT}{m}\right)^{\frac{1}{2}}(\mathscr{C}^2 - \tfrac{5}{2})\mathscr{C}.\nabla \ln T + \frac{2\mu}{p}\overset{\circ}{\mathscr{C}\mathscr{C}}:\nabla \boldsymbol{c}_0\right\}. \tag{10.33, 8}$$

10.331. Eigenvalue theory

Equations (10.33, 4, 5) are particular results of a general eigenvalue theory which has been extensively discussed, in particular by Kihara, by Wang-Chang and Uhlenbeck, and by Waldmann.* For a uniform gas under no forces, let

$$f = f^{(0)}(1 + \Phi), \tag{10.331, 1}$$

where Φ is small, be a non-normal solution of the Boltzmann equation. Then approximately

$$-n^2 I(\Phi) = f^{(0)}\frac{\partial \Phi}{\partial t}, \tag{10.331, 2}$$

as can be seen on substituting (10.331, 1) into the Boltzmann equation, and neglecting products of functions Φ. If Φ is suitably chosen, it decays exponentially like $e^{-t/\tau}$, say, so that (cf. 6.61) τ is the relaxation time for Φ. Then (10.331, 2) becomes

$$n^2 I(\Phi) = \tau^{-1} f^{(0)}\Phi. \tag{10.331, 3}$$

In (10.331, 3) τ^{-1} may be regarded as an eigenvalue of the collision operator I; Φ is the corresponding eigenfunction. Thus, for example, (10.33, 4, 5) imply that for Maxwellian molecules $(\mathscr{C}^2 - \frac{5}{2})\mathscr{C}$ and $\overset{\circ}{\mathscr{C}\mathscr{C}}$ are eigenfunctions of I; the corresponding relaxation times τ are $-a_1/n$ and b_1/n, or (cf. (10.33, 6)) $3\mu/2p$ and μ/p. These are the relaxation times effective in heat conduction and viscosity respectively (cf. (6.6, 5, 7)); the results derived by the relaxation approach of 6.6 become identical with those of the exact theory for Maxwellian molecules.

* T. Kihara, *Imperfect Gases*, section 21; C. S. Wang-Chang and G. E. Uhlenbeck, University of Michigan Project M999, 1952; L. Waldmann, *Handbuch der Physik*, vol. 12 (Springer, 1958), 367–76.

The general eigenfunction for Maxwellian molecules is found to be

$$\Phi_{l,r} = S^{(r)}_{l+\frac{1}{2}}(\mathscr{C}^2)\mathscr{C}^l Y_l(\theta,\varphi), \qquad (10.331, 4)$$

where $Y_l(\theta,\varphi)$ is a surface harmonic function, of order l, of the polar angles θ, φ that specify the direction of $\mathscr{C}$. The corresponding relaxation times are small compared with μ/p when either r or l is large.

For more general molecular models, the determination of the eigenfunctions is a matter of considerable difficulty.* However, just as (10.33, 2, 3), which are exact only for Maxwellian molecules, none the less represent first approximations yielding values of λ and μ close to the true ones for other molecular models, one may expect that the first one or two, at least, of the eigenfunctions for Maxwellian molecules give reasonable approximations to the eigenfunctions for more general models. This provides some measure of justification for the relaxation approach used in 6.6 and 6.62.

10.34. The inverse-square law of interaction

If a large proportion of the molecules of a gas are ionized, electrostatic forces play a dominant part in the molecular encounters. It is therefore of interest to consider the case when the force P between pairs of molecules satisfies the equation

$$P = e_1 e_2/r^2 \qquad (10.34, 1)$$

(where e_1, e_2 denote the electric charges of the molecules), ignoring any small part of the force which varies more rapidly with r.

If e_1, e_2 have the same sign, the variable v_0 of (10.31, 2) is given by

$$v_0 = 2kTbg^2/e_1e_2, \qquad (10.34, 2)$$

and the angle χ by

$$\chi = \pi - 2\int_0^{v_{00}} \left(1 - v^2 - \frac{2v}{v_0}\right)^{-\frac{1}{2}} dv,$$

where v_{00} is the positive root of the equation

$$1 - v_{00}^2 - \frac{2v_{00}}{v_0} = 0.$$

The expression for χ is integrable in the present case, the result being

$$\chi = 2\sin^{-1}\frac{1}{\sqrt{(1+v_0^2)}}, \qquad (10.34, 3)$$

whence it follows that

$$\cos\chi = \frac{v_0^2 - 1}{v_0^2 + 1}.$$

* In addition to discrete eigenvalues corresponding to regular eigenfunctions, there may also in special cases be a continuous spectrum of eigenvalues, possibly associated with singular eigenfunctions; see H. Grad, *3rd International Rarefied Gas Dynamics Symposium Report*, p. 26 (Academic Press, 1963). Grad also points out that for molecules repelling like $r^{-\nu}$, where $\nu < 5$, the eigenvalues may be indefinitely small, corresponding to indefinitely great relaxation times.

If e_1, e_2 have opposite signs, e_1e_2 in (10.34, 2) is replaced by $|e_1e_2|$, but the final expression for $\cos\chi$ is unaltered.

We accordingly find that, in either case,

$$\phi_{12}^{(l)} = 2\pi\left(\frac{e_1e_2}{2kTg^2}\right)^2\int\left(1-\left(\frac{v_0^2-1}{v_0^2+1}\right)^l\right)v_0\,dv_0.$$

In particular,
$$\phi_{12}^{(1)} = 2\pi\left(\frac{e_1e_2}{2kTg^2}\right)^2\ln(1+v_{01}^2), \tag{10.34, 4}$$

$$\phi_{12}^{(2)} = 4\pi\left(\frac{e_1e_2}{2kTg^2}\right)^2\left\{\ln(1+v_{01}^2)-\frac{v_{01}^2}{1+v_{01}^2}\right\}, \tag{10.34, 5}$$

where v_{01} denotes the upper limit of v_0. In the previous discussion the upper limit of v_0 was taken as infinite: but this is not a valid approximation here, as it would give infinite values for $\phi_{12}^{(1)}$, $\phi_{12}^{(2)}$, and the corresponding values of the coefficients of viscosity, diffusion, and conduction of heat would all vanish.

The difficulty arises because electrostatic forces, being proportional only to the inverse square of the distance, decrease with distance much more slowly than the ordinary forces of interaction. Hence important contributions to the integrals (10.34, 4, 5) are made by encounters in which the mutual distance of the molecules is always large. However, a molecule at a large distance from a given molecule will also be under electrostatic attraction or repulsion of many other molecules, so that these 'distant' encounters are not really binary; but throughout our analysis account is taken only of binary encounters. When 'inverse-square' forces are dominant, the molecular fields interpenetrate one another to such an extent that all encounters might be regarded as multiple. It is, however, possible, in certain circumstances, to derive approximate values for the coefficients of viscosity, conduction, and diffusion, considering only binary encounters.

The average value of g^2, or $m_0M_1M_2g^2/2kT$, for pairs of colliding molecules is 2 (cf. (5.3, 5)), so that v_0 in (10.34, 2) is in general comparable with $4kTb/|e_1e_2|$; thus, by (10.34, 3), χ is small if b is large compared with $|e_1e_2|/4kT$. The mean distance between pairs of neighbouring molecules is comparable with $n^{-\frac{1}{3}}$. Suppose now that

$$|e_1e_2|/4kT \ll n^{-\frac{1}{3}}. \tag{10.34, 6}$$

Then χ is small unless $b \ll n^{-\frac{1}{3}}$, i.e. large deflections are possible only if two molecules approach much closer than $n^{-\frac{1}{3}}$. The distant encounters producing small deflections are thus more numerous than the encounters producing large deflections; they are also more important, as the divergence of (10.34, 4, 5) for large v_{01} implies. As noted earlier, the distant encounters cannot strictly be regarded as binary. However, because the deflections due to distant encounters are small, their effects are roughly additive; that is, the change in velocity produced by one such encounter can be calculated without taking into account the forces simultaneously being exerted by other relatively distant molecules, just as if there were a succession of binary encounters.

The argument breaks down when b becomes too large; molecules at a sufficiently great distance from a given charge are shielded from its field by a polarization of the intervening gas. Debye* showed, on the basis of certain approximations, that, as a consequence of a similar polarization, the potential of a point-charge in an electrolyte varies as $(1/r)\exp(-r/d)$, where d is a distance known as the Debye shielding length. If the Debye potential is applicable in an ionized gas (it strictly applies only to the field of a charge at rest) the inverse-square force can be regarded as ceasing to be effective at about the cut-off distance d. That is, the binary-encounter formulae (10.34, 4, 5) can be used, assuming v_0 to have an effective upper limit v_{01}, whose correct order of magnitude is given by taking $b = d$ in (10.34, 2).†

In a binary mixture the Debye length is given by

$$4\pi d^2(n_1 e_1^2 + n_2 e_2^2) = kT. \qquad (10.34, 7)$$

Thus

$$d \sim \left(\frac{kT}{4\pi n|e_1 e_2|}\right)^{\frac{1}{2}} = \left(\frac{4kT}{n^{\frac{1}{3}}|e_1 e_2|}\right)^{\frac{1}{2}}\left(\frac{1}{16\pi}\right)^{\frac{1}{2}} n^{-\frac{1}{3}}. \qquad (10.34, 8)$$

Since, by (10.34, 6), $4kT/n^{\frac{1}{3}}|e_1 e_2|$ is large, $d \gg n^{-\frac{1}{3}}$; thus molecules not its immediate neighbours contribute appreciably to the deflection of a given molecule.

In actual calculations it is convenient to set

$$v_{01} = 4dkT/|e_1 e_2|, \qquad (10.34, 9)$$

equivalent to taking $b = d$ in (10.34, 2) and replacing g^2 by its mean value 2. Uncertainties in v_{01} are in any case relatively unimportant, because the functions appearing in (10.34, 4, 5) vary only slowly with v_{01}. If the cut-off distance in (10.34, 9) were taken as $n^{-\frac{1}{3}}$ instead of d, the effect would, by (10.34, 8), in no case reduce $\phi_{12}^{(1)}$ and $\phi_{12}^{(2)}$ by as much as one-third.‡

With v_{01} given by (10.34, 9), the constants $A_1(2)$, $A_2(2)$ of (10.31, 6) are found to be

$$A_1(2) = \ln(1 + v_{01}^2), \quad A_2(2) = 2\left\{\ln(1 + v_{01}^2) - \frac{v_{01}^2}{1 + v_{01}^2}\right\}. \qquad (10.34, 10)$$

* P. Debye and E. Hückel, *Phys. Z.* **24**, 185, 305 (1923).

† R. L. Liboff (*Phys. Fluids*, **2**, 40, (1959)) calculated the transport integrals for the Debye cut-off potential, obtaining results agreeing with those given here so far as the leading (logarithmic) term is concerned.

‡ The first proposal of a cut-off distance to overcome the divergence difficulty in electrostatic interaction was made by S. Chapman, *Mon. Not. R. Astr. Soc.* **82**, 294 (1922); he took the cut-off distance as $n^{-\frac{1}{3}}$. The first discussion of electrostatic shielding in ionized gases was given by E. Persico, *Mon. Not. R. Astr. Soc.* **86**, 294 (1926). The general adoption of the Debye length as the cut-off distance is largely due to Spitzer; cf. R. S. Cohen, L. Spitzer and P. McR. Routly, *Phys. Rev.* **80**, 230 (1950), L. Spitzer and R. Härm, *Phys. Rev.* **89**, 977 (1953), and papers referred to in these. At high temperatures (> 10^6 deg. K.) the cut-off distance is modified by quantum effects.

In terms of these, the first approximations to the coefficients of viscosity, conduction, and diffusion are expressible in the forms

$$[\mu]_1 = \frac{5}{8}\left(\frac{kmT}{\pi}\right)^{\frac{1}{2}}\left(\frac{2kT}{e^2}\right)^2 \Big/ A_2(2), \quad [\lambda]_1 = \tfrac{5}{2}[\mu]_1 c_v, \qquad (10.34, 11)$$

$$[D_{12}]_1 = \frac{3}{16n}\left(\frac{2kT}{\pi m_0 M_1 M_2}\right)^{\frac{1}{2}}\left(\frac{2kT}{e_1 e_2}\right)^2 \Big/ A_1(2). \qquad (10.34, 12)$$

A simple gas composed of molecules all with charges of the same sign is not met in nature. However, since $\cos\chi$ does not depend on the signs of e_1, e_2, the formulae (10.34, 11) can be taken to apply to a binary gas composed of equal numbers of ions of equal mass and charge, half positive, half negative.

From (10.32, 2, 3) $[\mu]_2 = 1{\cdot}15[\mu]_1$, $[\lambda]_2 = 1{\cdot}25[\lambda]_1$; also if $m_1 = m_2$, $e_1 = -e_2$, $n_1 = n_2$ one finds that $[D_{12}]_2 = 1{\cdot}15[D_{12}]_1$. Thus the errors of first approximations to the gas coefficients are considerably greater for the inverse-square law than for large values of the force-index ν. Further approximations have been considered by Landshoff* in the important case when m_2/m_1 is negligibly small (corresponding to ion–electron mixtures). Taking $e_1 = -Ze_2$, $n_2 = Zn_1$, corresponding to Z-fold ionized atoms in an electrically neutral mixture, he found the values shown in Table 4 for the ratio of the mth approximation to D_{12} to the first approximation. These figures show that, whereas the second approximation differs considerably from the first, the corrections added by later approximations are relatively small.

Table 4. *Values of* $[D_{12}]_m/[D_{12}]_1$ *for a gas of electrons and Z-fold ions*

m	$Z = 1$	$Z = 2$	$Z = 3$	$Z = \infty$
2	1·9320	2·3180	2·5290	3·25
3	1·9498	2·3210	2·5290	3·3906
4	1·9616	2·3256	2·5305	3·3945
5	1·9657	2·3268	2·5308	3·3950
∞	—	—	—	3·3953

MOLECULES POSSESSING BOTH ATTRACTIVE AND REPULSIVE FIELDS

10.4. The Lennard-Jones model

As noted in 3.3, in actual gases the force P between uncharged molecules is repulsive at small distances, and more weakly attractive at larger distances.

* R. Landshoff, *Phys. Rev.* **76**, 904 (1949) and **82**, 442 (1951). Landshoff's cut-off procedure differed somewhat from that described above, but the change does not greatly affect his numerical values.

This behaviour is most simply represented mathematically by the Lennard-Jones model,* for which

$$P = \kappa_{12}/r^{\nu} - \kappa'_{12}/r^{\nu'}, \tag{10.4, 1}$$

where $\nu > \nu'$; thus the interaction approximates at small distances to a repulsion κ_{12}/r^{ν}, and at large distances to an attraction $\kappa'_{12}/r^{\nu'}$. The potential energy corresponding to (10.4, 1) is

$$V(r) = \frac{\kappa_{12}}{(\nu-1)\,r^{\nu-1}} - \frac{\kappa'_{12}}{(\nu'-1)\,r^{\nu'-1}}. \tag{10.4, 2}$$

Let σ_{12}, ϵ_{12} be defined by

$$\sigma_{12}^{\nu-\nu'} \equiv \frac{\kappa_{12}(\nu'-1)}{\kappa'_{12}(\nu-1)}, \quad \epsilon_{12} \equiv \frac{\nu-\nu'}{(\nu-1)(\nu'-1)}\left(\frac{(\kappa'_{12})^{\nu-1}}{(\kappa_{12})^{\nu'-1}}\right)^{1/(\nu-\nu')}. \tag{10.4, 3}$$

Then σ_{12} denotes the 'low-velocity diameter' such that $V(\sigma_{12}) = 0$; ϵ_{12} is the depth of the potential well, i.e. $-\epsilon_{12}$ is the numerically greatest negative value attained by $V(r)$. In terms of σ_{12}, ϵ_{12}, (10.4, 2) can be written

$$V(r) = \beta\epsilon_{12}\{(\sigma_{12}/r)^{\nu-1} - (\sigma_{12}/r)^{\nu'-1}\}, \tag{10.4, 4}$$

where

$$\beta = \frac{1}{\nu-\nu'}\left\{\frac{(\nu-1)^{\nu-1}}{(\nu'-1)^{\nu'-1}}\right\}^{1/(\nu-\nu')}. \tag{10.4, 5}$$

The model is usually specified in terms of the powers appearing in the expression for $V(r)$; thus (10.4, 1) corresponds to the Lennard-Jones $\nu-1$, $\nu'-1$ model.† For a simple gas, κ_{12}, κ'_{12}, σ_{12}, ϵ_{12} in (10.4, 1–4) are replaced by κ, κ', σ, ϵ.

For fixed values of ν and ν' (10.4, 4) can be written in the form

$$V(r) = \epsilon_{12}\mathscr{V}(r/\sigma_{12}), \tag{10.4, 6}$$

where $\mathscr{V}$ is a non-dimensional function. Whenever $V(r)$ is expressible in this form, the transport coefficients conform to a law of corresponding states. The non-dimensional quantities $\mathscr{W}_{12}$ of (10.1, 1) must be functions of kT/ϵ_{12} alone, this being the one non-dimensional combination which can be formed from the parameters σ_{12}, ϵ_{12} and kT, on which alone $\mathscr{W}_{12}$ depends (cf. 10.3). There is a similar result for $\mathscr{W}_1$; thus for a simple gas, by (10.1, 4),

$$[\mu]_1 = \frac{(kmT)^{\frac{1}{2}}}{\sigma^2 F_\mu(kT/\epsilon)}, \quad [\lambda]_1 = \frac{(k^3T)^{\frac{1}{2}}}{m^{\frac{1}{2}}\sigma^2 F_\lambda(kT/\epsilon)}, \tag{10.4, 7}$$

where F_μ, F_λ are two non-dimensional functions. On dimensional grounds, later approximations can modify (10.4, 7) only by altering the functions F_μ, F_λ, so that equations similar in form to (10.4, 7) also apply to the exact

* This model, first introduced by J. E. (Lennard-)Jones (*Proc. R. Soc.* A, **106**, 441, 463 (1924)), has been widely used in discussing the equation of state of gases, and crystal properties, as well as in the theory of the transport phenomena. It was a generalization of the model (see p. 182) earlier introduced by W. Sutherland, *Phil. Mag.*, **36**, 507 (1893); **17**, 320 (1909).

† The notation differs from that of Note A (p. 392) of the 1951 edition of this book, which referred to (10.4, 1) as giving the ν, ν' model.

values of μ and λ. These equations embody the law of corresponding states for a simple gas. The quantities ϵ and σ in (10.4, 6, 7) are, of course, constants, not (like the σ_{12} of (10.31, 8)) functions of the temperature.

Similarly in a binary mixture, by (10.1, 5),

$$[D_{12}]_1 = (kT)^{\frac{1}{2}}/\{(m_0 M_1 M_2)^{\frac{1}{2}} n\sigma_{12}^2 F_D(kT/\epsilon_{12})\}, \qquad (10.4, 8)$$

where F_D is a further non-dimensional function. This gives a law of corresponding states for D_{12}, to a first approximation. However, later approximations make the function F_D depend somewhat on the parameters n_1/n_2, m_1/m_2, σ_1/σ_{12}, σ_2/σ_{12}, ϵ_1/ϵ_{12}, ϵ_2/ϵ_{12} as well as on kT/ϵ_{12}, even if the function $\mathscr{V}$ in (10.4, 6) is the same for like and unlike molecules. Thus there is no simple exact law of corresponding states for D_{12}; the same applies, but with greater force, to μ, λ and k_T for a gas-mixture.

10.41. Weak attractive fields

When in the Lennard-Jones force-law (10.4, 1) the attractive part of the force is weak, it is more convenient to replace (10.4, 7, 8) by

$$[\mu]_1 = [\mu_0]_1/\psi_\mu(\epsilon/kT), \quad [\lambda]_1 = [\lambda_0]_1/\psi_\mu(\epsilon/kT), \qquad (10.41, 1)$$

$$[D_{12}]_1 = [(D_{12})_0]_1/\psi_D(\epsilon_{12}/kT), \qquad (10.41, 2)$$

where $[\mu_0]_1$, $[\lambda_0]_1$, $[(D_{12})_0]_1$ denote the first approximations to μ, λ and D_{12} obtained when the attractive force ($\kappa'/r^{\nu'}$ or $\kappa'_{12}/r^{\nu'}$) is neglected. As the notation implies, the (non-dimensional) quantities ψ_μ and ψ_D are functions only of ϵ/kT and ϵ_{12}/kT; the same function appears in the expressions for $[\mu]_1$ and $[\lambda]_1$ because both depend on the same collision-integral $\Omega_1^{(2)}(2)$.

For small values of κ', κ'_{12}, the attractive fields can be regarded as small perturbations, whose effects on ψ_μ, ψ_D depend linearly on the strengths of those fields, i.e. on κ', κ'_{12}. In virtue of the form of the expression (10.4, 3) for ϵ_{12} we can therefore write

$$\psi_D = 1 + \frac{i_1(\nu,\nu')\,\kappa'_{12}}{(\nu'-1)\,(\kappa_{12})^{(\nu'-1)/(\nu-1)}\,(kT)^{(\nu-\nu')/(\nu-1)}} \qquad (10.41, 3)$$

and similarly $$\psi_\mu = 1 + \frac{i_2(\nu,\nu')\,\kappa'}{(\nu'-1)\,\kappa^{(\nu'-1)/(\nu-1)}\,(kT)^{(\nu-\nu')/(\nu-1)}}, \qquad (10.41, 4)$$

where $i_1(\nu,\nu')$ and $i_2(\nu,\nu')$ are pure numbers. Hence (10.41, 1, 2) reduce to the forms

$$[\mu]_1 = [\mu_0]_1/(1+ST^{-(\nu-\nu')/(\nu-1)}), \quad [\lambda]_1 = [\lambda_0]_1/(1+ST^{-(\nu-\nu')/(\nu-1)}), \qquad (10.41, 5)$$

$$[D_{12}]_1 = [(D_{12})_0]_1/(1+S_{12}T^{-(\nu-\nu')/(\nu-1)}), \qquad (10.41, 6)$$

where

$$S = \frac{i_2(\nu,\nu')\,\kappa'}{(\nu'-1)\,\kappa^{(\nu'-1)/(\nu-1)}k^{(\nu-\nu')/(\nu-1)}}, \quad S_{12} = \frac{i_1(\nu,\nu')\,\kappa'_{12}}{(\nu'-1)\,(\kappa_{12})^{(\nu'-1)/(\nu-1)}k^{(\nu-\nu')/(\nu-1)}}. \qquad (10.41, 7)$$

The evaluation of the constants i_1, i_2 is fairly simple in two special cases. The first is that of the Sutherland model, a smooth rigid elastic sphere surrounded by a field of attractive force. This may be regarded as a limiting case of the Lennard-Jones model, ν being taken as infinite, and κ_{12} so adjusted that when $r > \sigma_{12}$ (σ_{12} now being the sum of the molecular radii) the force $\kappa_{12} r^{-\nu}$ is zero, and when $r < \sigma_{12}$ it is infinite. In the limit, (10.41, 5, 6 and 7) become

$$[\mu]_1 = [\mu_0]_1/(1+S/T), \quad [\lambda]_1 = [\lambda_0]_1/(1+S/T), \qquad (10.41, 8)$$

$$[D_{12}]_1 = [(D_{12})_0]_1/(1+S_{12}/T), \qquad (10.41, 9)$$

$$S = \frac{i_2(\infty, \nu')\,\kappa'}{(\nu'-1)\,\sigma^{\nu'-1}k}, \quad S_{12} = \frac{i_1(\infty, \nu')\,\kappa'_{12}}{(\nu'-1)\,\sigma_{12}^{\nu'-1}k}. \qquad (10.41, 10)$$

Note that S is proportional to $\kappa'/(\nu'-1)\,\sigma^{\nu'-1}$, which is the potential energy of the mutual attraction of two molecules when in contact; this gives S an independent physical interest. A similar remark applies to S_{12}.

For the Sutherland model, molecules which would not have collided in the absence of the attractive field are deflected by it through an angle χ proportional to κ'_{12}; for these molecules the factor $1-\cos^l \chi$ appearing in the integrand of $\phi_{12}^{(l)}$ is of order κ'^2_{12}, and so is neglected. This means that in evaluating $\phi_{12}^{(l)}$ we need only consider encounters such that $b < \sigma_{12}$. For such encounters, the apsidal distance of the relative orbit (i.e. the least distance between the centres of the colliding molecules) is equal to σ_{12}; also in (10.4, 2) the repulsive term in $V(r)$ is to be omitted when $r > \sigma_{12}$. Hence, from (10.3, 5)

$$\chi = \pi - 2\int_0^{v_{00}} \left\{1 - v^2 + \frac{\kappa'_{12}}{(\nu'-1)kTg^2}\left(\frac{v}{b}\right)^{\nu'-1}\right\}^{-\frac{1}{2}} dv,$$

where $v = b/r$, $v_{00} = b/\sigma_{12}$. Thus to the first order in κ'_{12}

$$\chi = \chi_0 + \frac{\kappa'_{12} b^{1-\nu'}}{(\nu'-1)kTg^2}\int_0^{v_{00}} v^{\nu'-1}(1-v^2)^{-\frac{3}{2}}\, dv,$$

where χ_0 is the deflection $2\cos^{-1}(b/\sigma_{12})$ in the absence of the attractive field. Using this equation in the expression for $\Omega_{12}^{(1)}(1)$, and retaining only terms of first order in κ'_{12}, we may obtain an expression for $i_1(\infty, \nu')$; and similarly for $i_2(\infty, \nu')$.

Values of $i_1(\infty, \nu')$, $i_2(\infty, \nu')$ have been determined* for certain integral values of ν', and are given in Table 5. They are all positive; that is, the attractive force always increases the effective molecular cross-section.

The second case in which $i_1(\nu, \nu')$, $i_2(\nu, \nu')$ can be easily calculated is when $\nu' = 3$. Lennard-Jones† showed that in this case χ can be found in terms of the deflection in the absence of the attractive field. This facilitates the calculations; however, $\nu' = 3$ is not appropriate to actual gases, and so his

* D. Enskog, Uppsala Dissertation; C. G. F. James, *Proc. Camb. Phil. Soc.* **20**, 447 (1921); and R. C. Jones, *Phys. Rev.* **58**, 111 (1940).

† J. E. (Lennard-)Jones, *Proc. R. Soc.* A, **106**, 441 (1924)

Table 5. *Values of* $i_1(\infty, \nu')$ *and* $i_2(\infty, \nu')$

ν'	$i_1(\infty, \nu')$	$i_2(\infty, \nu')$
3	$\frac{1}{8}(12-\pi^2) = 0{\cdot}2663$	$\frac{1}{8}(\pi^2-8) = 0{\cdot}2337$
4	$3-4\ln 2 = 0{\cdot}2274$	$\frac{8}{3}(3\ln 2-2) = 0{\cdot}2118$
5	$\frac{1}{8}(3\pi^2-28) = 0{\cdot}2011$	$\frac{3}{2}(10-\pi^2) = 0{\cdot}1956$
7	$\frac{1}{6} = 0{\cdot}1667$	$\frac{5}{24}(9\pi^2-88) = 0{\cdot}1722$
9	$\frac{13}{90} = 0{\cdot}1444$	$\frac{7}{45} = 0{\cdot}1556$

results are not given here in detail. He found that $i_1(\nu, 3)$ and $i_2(\nu, 3)$ are positive for large values of ν (greater than about 12 for i_1, and about 16 for i_2), but negative for small values; thus the effect of the attractive force is to decrease μ and D_{12} when ν is large, but to increase them when ν is small.

10.42. Attractive forces not weak; the 12, 6 model

When the attractive force is too strong to be treated as a first-order perturbation, the calculation of the collision integrals for the Lennard-Jones model involves a succession of numerical integrations. A first integration is required to determine χ for each of a set of values of kTg^2/ϵ_{12} and b/σ_{12}; further integrations have then to be performed, with respect to b/σ_{12} and g^2, in order to determine each of the non-dimensional quantities $\mathscr{W}_{12}^{(l)}(r)$ of (10.1, 2) as a function of kT/ϵ_{12}. The early work was limited to values of ν, ν' which led to some reduction in the labour of such integrations.

Hassé and Cook* considered the Sutherland model $\nu = \infty$, $\nu' = 5$ and the Lennard-Jones 8,4 model ($\nu = 9$, $\nu' = 5$), each when the attractive force is not restricted to be small. For the first model, the deflection χ can be expressed in terms of elliptic integrals; a slight simplification of the numerical work is also possible for the 8,4 model. Clark Jones† pointed out that a similar slight simplification is possible for the 12,6 model ($\nu = 13, \nu' = 7$). In view of the power of modern electronic computers, such a simplification would now be regarded as unimportant; however, on physical grounds also the 12,6 model has some advantages.

For non-polar gases (i.e. gases whose molecules possess no permanent electric dipole moment) quantum theory‡ suggests that the principal part of the intermolecular attraction is a force κ'_{12}/r^7, due to dipole moments mutually induced in each of the two molecules; thus it is reasonable to take $\nu' = 7$. Quantum theory gives no such clear indication as to the value of ν; it actually suggests that the principal part of the repulsive force is proportional to an exponential e^{-ar} rather than to a power $r^{-\nu}$. However, the

* H. R. Hassé and W. R. Cook, *Phil. Mag.* **3**, 978 (1928), and *Proc. R. Soc.* A, **125**, 196 (1929).

† R. C. Jones, *Phys. Rev.* **59**, 1019 (1941).

‡ See, e.g., N. F. Mott and I. N. Sneddon, *Wave Mechanics and its Applications* (Oxford, 1948), pp. 164–75.

repulsive force is important only over a small range of r, within which it can be represented almost equally well by a variety of expressions, in particular by a power law $\kappa_{12}r^{-\nu}$, where $\nu > \nu'$. The choice $\nu = 13$ is suggested as reasonable, both because viscosity measurements for gases with weak attractive fields suggest a repulsive force varying at least as fast as r^{-12} (cf. 12.31, below), and because studies (by the virial method) of the deviations from Boyle's law indicate that the 12, 6 model gives a good approximation to the force-law.

For the 12, 6 model*

$$V(r) = 4\epsilon_{12}\{(\sigma_{12}/r)^{12} - (\sigma_{12}/r)^{6}\}. \tag{10.42, 1}$$

The model has been investigated independently by Kihara,† by de Boer and van Kranendonk,‡ and (more completely) by Hirschfelder, Bird and Spotz.§ Machine integrations have also been made for this model by Monchick and Mason, and by Itean, Glueck and Svehla.|| Different methods of numerical integration were used by the different workers, but in general their results agree well. Values of the quantities $\mathscr{W}_{12}^{(1)}(1)$, $\mathscr{W}_{12}^{(2)}(2)$, A, B and C based on the results of Monchick and Mason and of Itean, Glueck and Svehla¶ are listed for various values of kT/ϵ_{12} in Table 6.

Curves illustrating the variation of $\log \mathscr{W}_{12}^{(1)}(1)$, $\log \mathscr{W}_{12}^{(2)}(2)$, A, B and C with $\log(kT/\epsilon_{12})$ are given in Figs. 8, 9. In no case does the slope of the curve vary monotonically. When kT/ϵ_{12} is large, the repulsive part of the field is dominant, and the slope of the curves for $\log \mathscr{W}_{12}^{(1)}(1)$, $\log \mathscr{W}_{12}^{(2)}(2)$ is $2/(\nu-1)$, or $\frac{1}{6}$; when kT/ϵ_{12} is small, the attractive part is dominant, and the slope is $2/(\nu'-1)$, or $\frac{1}{3}$. However, the slope does not decrease steadily as kT/ϵ_{12} increases, but has a maximum greater than $\frac{1}{2}$ when kT/ϵ_{12} is about unity. Similarly C is 0·1333 when kT/ϵ_{12} is large, and 0·0667 when it is small, but takes negative values when kT/ϵ_{12} is between about 0·45 and 0·95.

Hirschfelder, Bird and Spotz proceeded as far as a second approximation to the viscosity coefficient. This differs very little from the first approximation when $kT \sim \epsilon_{12}$, but (as (10.32, 2) implies) it is about 0·168 per cent greater when kT/ϵ_{12} is small, and about 0·667 per cent greater when kT/ϵ_{12} is large.

* E. A. Mason (*J. Chem. Phys.* **22**, 169 (1954)) uses the alternative expression

$$V(r) = \epsilon_{12}\{(r_m/r)^{12} - 2(r_m/r)^{6}\},$$

where $r_m^6 = 2\sigma_{12}^6$; r_m denotes the value of r for which $-V(r)$ attains its maximum value ϵ_{12}.

† T. Kihara, *Imperfect Gases*, section 28.

‡ J. de Boer and J. van Kranendonk, *Physica*, **14**, 442 (1948).

§ J. O. Hirschfelder, R. B. Bird and E. L. Spotz, *J. Chem. Phys.* **16**, 968 (1948); *Chem. Rev.* **44**, 205 (1949).

|| L. Monchick and E. A. Mason, *J. Chem. Phys.* **35**, 1676 (1961); E. C. Itean, A. R. Glueck and R. A. Svehla, NASA Report TN D-481 (1961).

¶ A detailed tabulation of the first two expressions, based on these results, was given by P. E. Liley, *Thermophysical Properties Research Centre, Report 15* (Purdue University, 1963).

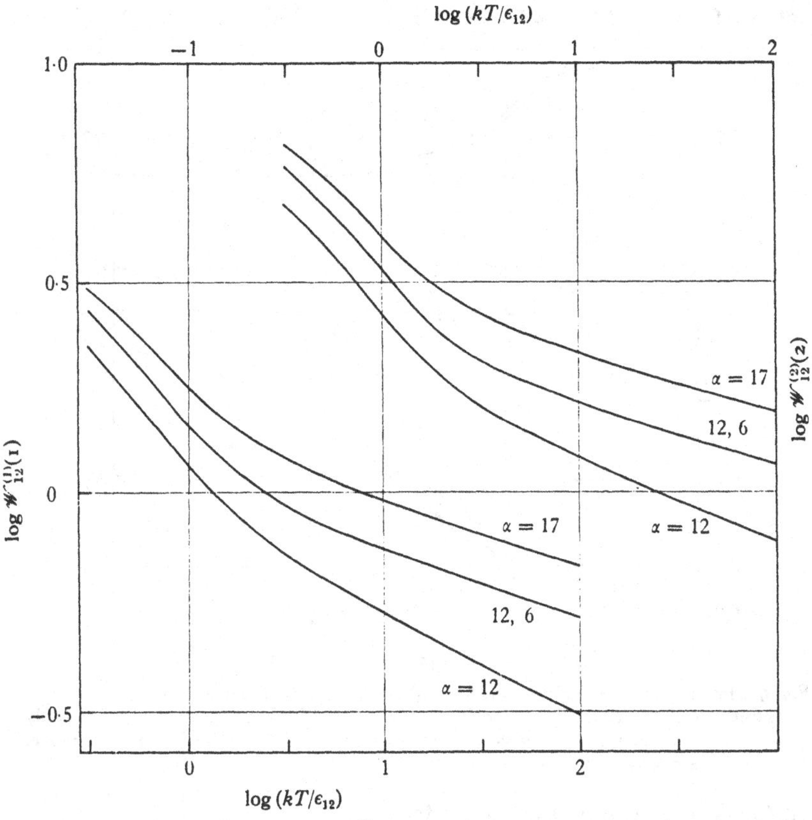

Fig. 8. Plots of log $\mathscr{W}_{12}^{(1)}(1)$ and log $\mathscr{W}_{12}^{(2)}(2)$ against log (kT/ϵ_{12}) for the 12, 6 and exp;6 models. To avoid confusion, the horizontal scale for log $\mathscr{W}_{12}^{(2)}(2)$ (top of diagram) is displaced by one unit from that for log $\mathscr{W}_{12}^{(1)}(1)$; also the curves for the exp;6 model (which would otherwise cross the 12,6 curve near log $(kT/\epsilon_{12}) = 0{\cdot}3$) are displaced vertically up (for $\alpha = 17$) or down ($\alpha = 12$) by 0·1 unit.

Table 6. *Values of* $\mathscr{W}_{12}^{(1)}(1)$, $\mathscr{W}_{12}^{(2)}(2)$, A, B, C *for the* 12,6 *model*

kT/ϵ_{12}	$\mathscr{W}_{12}^{(1)}(1)$	$\frac{1}{2}\mathscr{W}_{12}^{(2)}(2)$	A	B	C
0·3	2·648	2·841	0·4291	0·774	0·022
0·5	2·066	2·284	0·4422	0·772	−0·009
0·7	1·729	1·922	0·4446	0·747	−0·012
1·0	1·440	1·593	0·4425	0·715	0·004
1·3	1·274	1·401	0·4398	0·695	0·023
1·6	1·168	1·280	0·4383	0·682	0·042
2·0	1·075	1·176	0·4376	0·672	0·062
2·5	1·0002	1·093	0·4371	0·665	0·080
3·0	0·9500	1·0388	0·4374	0·661	0·0925
4·0	0·8845	0·9699	0·4387	0·657	0·1085
5·0	0·8428	0·9268	0·4399	0·656	0·118
7·0	0·7898	0·8728	0·4421	0·655	0·127
10·0	0·7422	0·8244	0·4443	0·655	0·133
20·0	0·6640	0·7436	0·4479	0·656	0·135
50·0	0·5761	0·6504	0·4516	0·656	0·138
100·0	0·5177	0·5869	0·4536	0·657	0·138

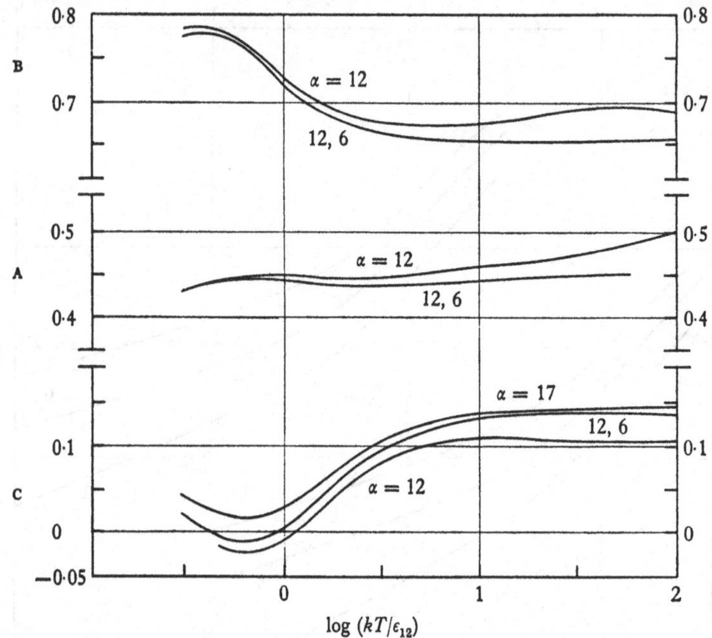

Fig. 9. Plots of A, B and C against log (kT/ϵ_{12}) for the 12,6 and exp;6 models.

The exp; 6 values of C are shown for $\alpha = 12$ and $\alpha = 17$; for A and B, curves are given only for $\alpha = 12$, since the values for $\alpha = 17$ differ only slightly from those for the 12,6 model for A, and are not available for B.

10.43. The exp; 6 and other models

Advances in computing techniques have now made it possible to discuss other and more complicated laws of force between molecules. One of the most important of these is the exp;6, or modified Buckingham law, first studied in connection with the transport phenomena by Mason.* For this law, the potential energy is

$$V(r) = \frac{\epsilon_{12}}{(1-6/\alpha)}\left[\frac{6}{\alpha}\exp\left\{\alpha\left(1-\frac{r}{r_m}\right)\right\}-\left(\frac{r_m}{r}\right)^6\right], \qquad (10.43, 1)$$

where α is a number usually in the range 12–15. As with the Lennard-Jones model, ϵ_{12} is the maximum value of $-V(r)$: r_m is the value of r for which this maximum is attained. The quantum-theory arguments in favour of (10.43, 1) have already been mentioned in 10.42. As a consequence of the finiteness of the exponential at $r = 0$, the potential energy (10.43, 1) has a positive maximum at a distance $r = r'$ $(<r_m)$, and is negative for very small r; to avoid this non-physical behaviour, Mason proposed to take $V(r) = \infty$ when $r < r'$. The effects of such an assumption are not great, since few molecular encounters are sufficiently energetic to reach $r = r'$.

* E. A. Mason, *J. Chem. Phys.* **22**, 169 (1954). The full Buckingham law includes an attractive r^{-8} term in $V(r)$. See R. A. Buckingham, *Proc. R. Soc.* A, **168**, 264 (1938); also H. Margenau, *Phys. Rev.* **38**, 747 (1931).

As with the Lennard-Jones model, the viscosity and thermal conductivity conform to a law of corresponding states (for given α). If σ_{12} again is defined as such that $V(\sigma_{12}) = 0$, σ_{12} is $0{\cdot}8761\, r_m$ for $\alpha = 12$, and $0{\cdot}8942\, r_m$ for $\alpha = 15$; for comparison, $\sigma_{12} = 0{\cdot}8909\, r_m$ for the 12, 6 model.

Detailed numerical results (for each of the values $\alpha = 12, 13, 14, 15$) were given by Mason in his original paper, and those for $\alpha = 16$ and $\alpha = 17$ were given later by Mason and Rice.* Those for $\alpha = 12$ and $\alpha = 17$ are illustrated in Figs. 8 and 9; those for $\alpha = 14$ and $\alpha = 15$ resemble fairly closely those for the 12, 6 model.

The r^{-6} variation of the attractive potential energy applies only to electrically neutral non-polar molecules. The interaction of an ion with a neutral molecule involves an r^{-4} part in $V(r)$; to discuss this type of interaction, Mason and Schamp† have calculated collision-integrals for the potential energy

$$V(r) = \tfrac{1}{2}\epsilon_{12}\left[(1+\gamma)\left(\frac{r_m}{r}\right)^{12} - 4\gamma\left(\frac{r_m}{r}\right)^{6} - 3(1-\gamma)\left(\frac{r_m}{r}\right)^{4}\right], \tag{10.43, 2}$$

which reduces to the 12, 6 model if $\gamma = 1$; γ is a parameter measuring the relative importance of the r^{-6} and r^{-4} parts of the attraction. For polar molecules, Stockmayer‡ proposed the form

$$V(r) = 4\epsilon_{12}[(\sigma_{12}/r)^{12} - (\sigma_{12}/r)^{6}] - \mathscr{M}_1\mathscr{M}_2\zeta/r^3; \tag{10.43, 3}$$

here $\mathscr{M}_1$, $\mathscr{M}_2$ are the dipole moments of the interacting molecules, and ζ is a function of their relative orientation. The energy (10.43, 3) is not spherically symmetric, and the theory of smooth molecules is strictly not applicable to it. However, Monchick and Mason § have discussed polar molecules as if they are symmetric, assuming that molecular trajectories are negligibly distorted by transfer between rotational and translational energies, and that each encounter is characterized by a constant value of ζ in (10.43, 3); to determine the collision-integrals they then average over all possible values of ζ.

Other models studied in connection with the transport properties include the pure exponential model‖ ($V(r) = \epsilon_{12}\exp(-r/\sigma_{12})$), the attractive power-law model¶ ($V(r) = -\kappa'/(\nu'-1)r^{\nu'-1}$), the exp;exp model** ($V(r) = Ae^{-r/\rho} - Be^{-r/\rho'}$) and the Lennard-Jones 9,6 and 28,7 models.†† For details of the results, the reader is referred to the original papers.

* E. A. Mason and W. E. Rice, *J. Chem. Phys.* **22**, 843 (1954).
† E. A. Mason and H. W. Schamp, *Ann. Phys.* **4**, 233 (1958).
‡ W. H. Stockmayer, *J. Chem. Phys.* **9**, 398 (1941).
§ L. Monchick and E. A. Mason, *J. Chem. Phys.* **35**, 1676 (1961).
‖ L. Monchick, *Phys. Fluids*, **2**, 695 (1959), repulsive force; R. J. Munn, E. A. Mason, and F. J. Smith, *Phys. Fluids*, **8**, 1103 (1965), attractive force.
¶ T. Kihara, M. H. Taylor and J. O. Hirschfelder, *Phys. Fluids*, **3**, 715 (1960).
** F. J. Smith and R. J. Munn, *J. Chem. Phys.* **41**, 3560 (1964).
†† F. J. Smith, E. A. Mason and R. J. Munn, *J. Chem. Phys.* **42**, 1334 (1965).

THE LORENTZ APPROXIMATION

10.5. A specially simple expression for the velocity-distribution function in a gas-mixture may be obtained if (i) the mass m_1 of the molecules of one constituent is very great compared with that of the molecules of the second constituent, and (ii) the influence of mutual encounters among the latter in altering their motions is negligible compared with that of their encounters with the heavy molecules. The latter condition is fulfilled if either (*a*) the number or (*b*) the extension of the molecular field of force is much less for the light than for the heavy molecules. The kinetic theory for a gas satisfying these special conditions (i) and (ii) was first studied by Lorentz;* hence we shall refer to such a gas as a *Lorentzian gas*.

On account of the condition (ii), the integral I_2 can be omitted from the right-hand sides of (8.31, 4–6). Also, since the mean kinetic energies of the peculiar motions of the two sets of molecules are approximately equal, the peculiar velocities of the heavy molecules will be small compared with those of the lighter molecules; hence the relative velocity g in the encounter of molecules of opposite types is roughly equal to the peculiar velocity C_2 of the lighter molecule. The problem, in fact, approximates to the determination of the distribution of motion among a set of light molecules subject to deflection by stationary obstacles.

The velocity of one of the heavier molecules is not appreciably altered by a collision with one of the lighter molecules: thus, in evaluating the integrals I_{12}, I_{21} in equations (8.31, 4–6), we can put $\boldsymbol{C}_1' = \boldsymbol{C}_1$, $\boldsymbol{A}_1' = \boldsymbol{A}_1$, $\boldsymbol{D}_1' = \boldsymbol{D}_1$, $\mathsf{B}_1' = \mathsf{B}_1$. Since, moreover, I_2 is to be neglected, and $g = C_2$, the second of equations (8.31, 4) becomes

$$f_2^{(0)}(\mathscr{C}_2^2 - \tfrac{5}{2})\,\boldsymbol{C}_2 = \iiint f_1^{(0)} f_2^{(0)}(\boldsymbol{A}_2 - \boldsymbol{A}_2')\,C_2\,b\,db\,d\epsilon\,d\boldsymbol{c}_1,$$

whence, on integration with respect to $\boldsymbol{c}_1$, we obtain

$$(\mathscr{C}_2^2 - \tfrac{5}{2})\,\boldsymbol{C}_2 = n_1 \iint (\boldsymbol{A}_2 - \boldsymbol{A}_2')\,C_2\,b\,db\,d\epsilon.$$

Since the relative velocity after encounter is equal in magnitude to that before encounter, $C_2' = C_2$. Hence this equation is equivalent to

$$(\mathscr{C}_2^2 - \tfrac{5}{2})\,\boldsymbol{C}_2 = n_1 A_2(C_2) \iint (\boldsymbol{C}_2 - \boldsymbol{C}_2')\,C_2\,b\,db\,d\epsilon. \qquad (10.5, 1)$$

Similarly the second members of the pairs of equations (8.31, 5, 6) reduce to

$$-\frac{n}{n_2}\boldsymbol{C}_2 = n_1 D_2(C_2) \iint (\boldsymbol{C}_2 - \boldsymbol{C}_2')\,C_2\,b\,db\,d\epsilon, \qquad (10.5, 2)$$

$$\frac{m_2}{2kT}\overset{\circ}{\boldsymbol{C}_2\boldsymbol{C}_2} = n_1 B_2(C_2) \iint (\overset{\circ}{\boldsymbol{C}_2\boldsymbol{C}_2} - \overset{\circ}{\boldsymbol{C}_2'\boldsymbol{C}_2'})\,C_2\,b\,db\,d\epsilon. \qquad (10.5, 3)$$

* H. A. Lorentz, *Proc. Amst. Acad.* 7, 438, 585, 684 (1905).

Since χ and ϵ now are polar angles specifying the orientation of $\mathbf{C}_2'$ about $\mathbf{C}_2$ (cf. 3.5),

$$\mathbf{C}_2' = \mathbf{C}_2 \cos\chi + C_2 \sin\chi\,(\mathbf{h}\cos\epsilon + \mathbf{i}\sin\epsilon),$$

where, $\mathbf{h}$, $\mathbf{i}$ are suitably chosen unit vectors perpendicular to $\mathbf{C}_2$ and to each other. Hence

$$\iint(\mathbf{C}_2 - \mathbf{C}_2')\,b\,db\,d\epsilon = 2\pi\mathbf{C}_2\int(1-\cos\chi)\,b\,db$$
$$= \mathbf{C}_2\,\phi_{12}^{(1)},$$

by (9.33, 4). Also, since $C_2' = C_2$,

$$\iint(\overset{\circ}{\mathbf{C}_2\mathbf{C}_2} - \overset{\circ}{\mathbf{C}_2'\mathbf{C}_2'})\,b\,db\,d\epsilon = 2\pi\int(1-\cos^2\chi)\{\mathbf{C}_2\mathbf{C}_2 - \tfrac{1}{2}C_2^2(\mathbf{hh}+\mathbf{ii})\}\,b\,db.$$

Since $\mathbf{C}_2/C_2$, $\mathbf{h}$ and $\mathbf{i}$ are mutually orthogonal unit vectors, by (1.3, 9),

$$\mathbf{C}_2\mathbf{C}_2/C_2^2 + \mathbf{hh} + \mathbf{ii} = \mathsf{U},$$

where U is the unit tensor. Hence

$$\iint(\overset{\circ}{\mathbf{C}_2\mathbf{C}_2} - \overset{\circ}{\mathbf{C}_2'\mathbf{C}_2'})\,b\,db\,d\epsilon = 3\pi\overset{\circ}{\mathbf{C}_2\mathbf{C}_2}\int(1-\cos^2\chi)\,b\,db$$
$$= \tfrac{3}{2}\overset{\circ}{\mathbf{C}_2\mathbf{C}_2}\,\phi_{12}^{(2)}.$$

Here $\phi_{12}^{(1)}$, $\phi_{12}^{(2)}$ are functions of C_2. Using these results in (10.5, 1–3) we get

$$A_2(C_2) = \frac{\mathscr{C}_2^2 - \frac{5}{2}}{n_1 C_2 \phi_{12}^{(1)}}, \tag{10.5, 4}$$

$$D_2(C_2) = -n/n_1 n_2 C_2 \phi_{12}^{(1)}, \tag{10.5, 5}$$

$$B_2(C_2) = m_2/3n_1 kTC_2\phi_{12}^{(2)}. \tag{10.5, 6}$$

In the present case equations (8.4, 1, 7) reduce to

$$\overline{\mathbf{C}}_2 = -\frac{1}{3n_2}\{\mathbf{d}_{12}\int f_2^{(0)}C_2^2 D_2(C_2)\,d\mathbf{c}_2 + \nabla\ln T\int f_2^{(0)}C_2^2 A_2(C_2)\,d\mathbf{c}_2\}$$
$$= \frac{n^2}{n_1 n_2}\{D_{12}\mathbf{d}_{12} + D_T\nabla\ln T\},$$

since here $\overline{\mathbf{C}}_1 = 0$. Thus, using (10.5, 4, 5),

$$D_{12} = \frac{1}{3n_2 n}\int f_2^{(0)}\frac{C_2}{\phi_{12}^{(1)}}\,d\mathbf{c}_2, \tag{10.5, 7}$$

$$D_T = -\frac{1}{3n^2}\int f_2^{(0)}\frac{C_2}{\phi_{12}^{(1)}}(\mathscr{C}_2^2 - \tfrac{5}{2})\,d\mathbf{c}_2. \tag{10.5, 8}$$

In the conduction of heat the lighter molecules, because of their larger velocities, are more effective, in proportion to their number, than the heavy molecules. If the conduction is assumed to be due predominantly to the light molecules, then in (8.41, 4), namely

$$\lambda = \tfrac{1}{3}kn^2[\{\mathbf{A},\mathbf{A}\} - \{\mathbf{A},\mathbf{D}\}^2/\{\mathbf{D},\mathbf{D}\}], \tag{10.5, 9}$$

we must insert the following simplified forms of (8.31, 11, 12)

$$n^2\{A, A\} = n_1^{-1}\int f_2^{(0)}(\mathscr{C}_2^2 - \tfrac{5}{2})^2 (C_2/\phi_{12}^{(1)})\, dc_2, \qquad (10.5, 10)$$

$$n\{A, D\} = -(n_1 n_2)^{-1}\int f_2^{(0)}(\mathscr{C}_2^2 - \tfrac{5}{2})(C_2/\phi_{12}^{(1)})\, dc_2, \qquad (10.5, 11)$$

$$\{D, D\} = (n_1 n_2^2)^{-1}\int f_2^{(0)}(C_2/\phi_{12}^{(1)})\, dc_2. \qquad (10.5, 12)$$

The contributions of the two sets of molecules to the hydrostatic pressure are of the same order of magnitude, if their number-densities are comparable; the viscous stresses due to the heavy molecules are, however, large compared with those due to the light. If, nevertheless, the pressure system due to the light molecules alone is considered, then (cf. 8.42) the corresponding coefficient of viscosity μ is given by

$$\mu = \tfrac{1}{5} m_2 \int f_2^{(0)} \mathring{\mathbf{C}}_2 \mathbf{C}_2 : \mathring{\mathbf{C}}_2 \mathbf{C}_2 B_2(C_2)\, dc_2$$

$$= \frac{2m_2^2}{45 n_1 kT}\int f_2^{(0)} \frac{C_2^3}{\phi_{12}^{(2)}}\, dc_2, \qquad (10.5, 13)$$

using (10.5, 6) and the theorem of 1.421.

The meaning of equations (10.5, 7–12) can perhaps be better seen if we make the substitution

$$n_1 \phi_{12}^{(1)} = 1/l(C_2).$$

If the molecules are rigid elastic spheres, $\phi_{12}^{(1)} = \pi\sigma_{12}^2$, by 10.2, and so this substitution implies that for such molecules

$$l(C_2) = 1/(\pi n_1 \sigma_{12}^2).$$

Since in the Lorentz approximation collisions between pairs of molecules m_2 can be neglected, it follows that for rigid elastic spheres $l(C_2)$ is the mean free path of molecules m_2 of peculiar speed C_2: for more general molecules we may interpret $l(C_2)$ as an equivalent mean free path for molecules m_2 of peculiar speed C_2.

For example, making the substitution, we get

$$D_{12} = \frac{n_1}{3n_2 n}\int f_2^{(0)} C_2 l(C_2)\, dc_2 = \frac{n_1}{3n}\overline{C_2 l(C_2)},$$

a relation which may be compared with (6.4, 2).

10.51. Interaction proportional to $r^{-\nu}$

These results take a specially simple form when the molecules are centres of force varying inversely as the νth power of the distance, so that, using (10.31, 5), and putting $g = \mathscr{C}_2$,

$$\phi_{12}^{(l)} = 2\pi(\kappa_{12}/2kT\mathscr{C}_2^2)^{2/(\nu-1)} A_l(\nu). \qquad (10.51, 1)$$

In this case

$$D_{12} = \left(\frac{2kT}{m_2}\right)^{\frac{1}{2}} \left(\frac{2kT}{\kappa_{12}}\right)^{2/(\nu-1)} \Gamma\left(2+\frac{2}{\nu-1}\right) \Big/ 3\pi^{\frac{1}{2}} n A_1(\nu), \quad (10.51, 2)$$

$$D_T = n_2 \left(\frac{2kT}{m_2}\right)^{\frac{1}{2}} \left(\frac{2kT}{\kappa_{12}}\right)^{2/(\nu-1)} \frac{(\nu-5)}{2(\nu-1)} \Gamma\left(2+\frac{2}{\nu-1}\right) \Big/ 3\pi^{\frac{1}{2}} n^2 A_1(\nu), \quad (10.51, 3)$$

$$\mathrm{k}_T = \frac{n_2}{n} \frac{\nu-5}{2(\nu-1)}, \quad (10.51, 4)$$

$$\lambda = kn_2 \left(\frac{2kT}{m_2}\right)^{\frac{1}{2}} \left(\frac{2kT}{\kappa_{12}}\right)^{2/(\nu-1)} \Gamma\left(3+\frac{2}{\nu-1}\right) \Big/ 3\pi^{\frac{1}{2}} n_1 A_1(\nu), \quad (10.51, 5)$$

$$\mu = 4n_2 m_2 \left(\frac{2kT}{m_2}\right)^{\frac{1}{2}} \left(\frac{2kT}{\kappa_{12}}\right)^{2/(\nu-1)} \Gamma\left(3+\frac{2}{\nu-1}\right) \Big/ 45\pi^{\frac{1}{2}} n_1 A_2(\nu), \quad (10.51, 6)$$

$$\lambda/\mu = 15kA_2(\nu)/4m_2 A_1(\nu) = (c_v)_2\, 5A_2(\nu)/2A_1(\nu), \quad (10.51, 7)$$

where $(c_v)_2$ denotes the specific heat of the lighter gas.

These equations can be made to apply also to the case of rigid elastic spherical molecules, by making ν tend to infinity, and $\kappa_{12}^{1/\nu}$ approach σ_{12}. In particular, since $A_1(\infty) = \frac{1}{2}$, $A_2(\infty) = \frac{1}{3}$, we then have

$$\mathrm{k}_T = \frac{n_2}{2n}, \quad \lambda = \tfrac{5}{3}\mu (c_v)_2. \quad (10.51, 8)$$

For the above models k_T does not depend on the temperature. This is not true for more complicated models. As can be shown by applying to (10.5, 7) the argument used in proving (9.71, 1), in general

$$\mathrm{k}_T = \frac{n_2}{n} \frac{\partial \ln (T/D_{12})}{\partial \ln T}, \quad (10.51, 9)$$

the differentiation being performed keeping n constant. Thus, for example, for both the Lennard-Jones and the exp;6 models

$$\mathrm{k}_T = \frac{n_2}{n} F(kT/\epsilon_{12}), \quad (10.51, 10)$$

where F is a function whose form depends on the particular model.

10.52. Deduction of the Lorentz results from the general formulae

The results of 10.5 can also be derived from the general solution, by making m_2/m_1 tend to zero, and neglecting the terms arising from the mutual encounters of the light molecules. The results thus derived are expressed in terms of infinite determinants, as in the general case: it may be shown* (though the proof is not given here) that these determinantal expressions are

* See S. Chapman, *J. Lond. Math. Soc.* **8**, 266 (1933).

equal to those found above, and the comparison of the two forms of the results throws interesting light on the nature of the convergence of the determinants in which the general solution is expressed.

The convergence may be illustrated by comparing the different approximations to the transport coefficients for inverse-power interaction κ_{12}/r^{ν} with the exact values given by the formulae of 10.51. In Table 7 the values of $[D_{12}]_m/[D_{12}]_1$ for $m = 2$ and 3 are compared with the limiting value $D_{12}/[D_{12}]_1$. The first line, for $\nu = \infty$, corresponds to the case of rigid elastic spheres; thus the first approximation to D_{12} for this model is in defect by about 12 per cent; the second reduces the error to less than 5 per cent, and the error of the third approximation is about 2 per cent. The errors of the various approximations for values of ν between 5 and ∞ are less than the errors for $\nu = \infty$; when $\nu = 5$ (corresponding to Maxwellian molecules) the first and all later approximations are equal to the exact value. When ν is less than 5, the first approximation falls off greatly in accuracy, though the second and third approximations are good.

Table 7. *Values of* $[D_{12}]_m/[D_{12}]_1$ *and* $D_{12}/[D_{12}]_1$

ν	$[D_{12}]_2/[D_{12}]_1$	$[D_{12}]_3/[D_{12}]_1$	$D_{12}/[D_{12}]_1$
∞	1·083	1·107	1·132
17	1·049	1·060	1·072
13	1·039	1·048	1·056
9	1·023	1·027	1·031
5	1	1	1
3	1·125	1·130	1·132
2	3·250	3·391	3·396

Table 8 gives the ratios which the approximations $[k_T]_1$, $[k_T]_2$ bear to the limiting value k_T; it shows that the first approximation to k_T is more in error than the first approximation to D_{12}. When $\nu = 5$, k_T and every approximation to it is zero, but the ratios in question tend to unity as ν tends to 5.

Table 8. *Ratios of the first and second approximations to* k_T, *to the exact value of* k_T

$\nu = \infty$	17	13	9	5	3	2
$[k_T]_1/k_T = 0·77$	0·83	0·85	0·89	1	1·11	0·77
$[k_T]_2/k_T = 0·88$	0·92	0·93	0·95	1	1·01	1·01

Next follows a table giving the ratios of $[\lambda]_1$ and $[\lambda]_2$ to the exact value λ; these are calculated on the assumption (cf. 10.5) that the conductivity due to the heavy molecules is negligible. It appears that the accuracy of the approximations to λ is about equal to that of the approximations to D_{12}.

Table 9. *Ratios of the first and second approximations to λ, to the exact value of λ*

$\nu = \infty$	17	13	9	5	3	2
(1) = 0·85	0·91	0·93	0·96	1	0·82	0·14
(2) = 0·93	0·96	0·97	0·99	1	0·99	0·92

Finally, Table 10 gives the ratios of $[\mu]_1$ and $[\mu]_2$ to μ, where μ now represents the (small) contribution to the viscosity made by the light molecules alone (cf. 10.5). These show that the approximations to μ are rather more accurate than those to D_{12}.

Table 10. *Ratios of the first and second approximations to μ, to the exact value of μ*

$\nu = \infty$	17	13	9	5	3	2
(1) = 0·92	0·95	0·96	0·98	1	0·92	0·46
(2) = 0·98	0·99	0·99	0·99	1	1·00	0·98

The inaccuracy of the first approximation when $\nu = 2$ is very marked in all the Tables 7–10. However, the Lorentz approximation is not strictly applicable to an ion–electron gas, in spite of the large mass-ratio, because the effect of electron–electron interactions is normally not small in such a mixture. These interactions reduce somewhat the inaccuracy of the first approximation in such a gas (compare Table 7 with Table 4, section 10.34).

10.53. Quasi-Lorentzian gas

Another interesting special case is that of a mixture in which m_1/m_2 is large, but n_1/n_2 is so small that encounters between pairs of molecules m_1 do not appreciably affect the velocity-distribution function f_1. This case has been studied by Kihara* and by Mason,† the latter of whom called such a gas quasi-Lorentzian.

For this case Kihara verified, by direct substitution in (8.31, 5), that the first approximation to the diffusion coefficient gives the exact value. Mason confirmed this result by examining the orders of magnitude of the determinantal elements. Unfortunately there is no corresponding simplification for the other transport coefficients.

* T. Kihara, *Imperfect Gases*, section 20.

† E. A. Mason, *J. Chem. Phys.* **27**, 782 (1957).

10.6. A mixture of mechanically similar molecules

Another specially simple case of the solution for a gas-mixture is that in which the different sets of molecules are of the same mass and obey the same law of interaction at encounter, so that they are mechanically similar. In this case the thermal conductivity and viscosity are the same as if the molecules were identical in all respects, and the coefficient of thermal diffusion vanishes; also the coefficient of diffusion is the coefficient of self-diffusion D_{11} of a simple gas.

Since all the molecules are similar, we introduce the velocity-distribution function of all the molecules. The first approximation to this is $f^{(0)}$, where $f^{(0)}/n, f_1^{(0)}/n_1, f_2^{(0)}/n_2$ are identical functions of the respective variables $\boldsymbol{C}, \boldsymbol{C}_1, \boldsymbol{C}_2$. The second approximation is $f^{(0)}+f^{(1)}$, where

$$f^{(1)}(\boldsymbol{C}) = f_1^{(1)}(\boldsymbol{C})+f_2^{(1)}(\boldsymbol{C}). \tag{10.6, 1}$$

Since f does not depend on the relative proportions of the two gases in the mixture, but only on the total number-density, $f^{(1)}$ is unaffected by the relative diffusion of the two gases, and so has no part depending on the vector $\boldsymbol{d}_{12}$ of (8.3, 8). Thus, equating the terms in (10.6, 1) depending on $\boldsymbol{d}_{12}$, we get (cf. (8.31, 9, 10))

$$n_1\boldsymbol{D}_1(\boldsymbol{C}) = -n_2\boldsymbol{D}_2(\boldsymbol{C}). \tag{10.6, 2}$$

We denote the common value of each of these expressions by $\boldsymbol{D}_0(\boldsymbol{C})$. Now, on account of the mechanical identity of the molecules of the two gases,

$$I_1\{\boldsymbol{D}_0(\boldsymbol{C}_1)\} = I_{12}\{\boldsymbol{D}_0(\boldsymbol{C}_1)+\boldsymbol{D}_0(\boldsymbol{C}_2)\},\quad I_2\{\boldsymbol{D}_0(\boldsymbol{C}_2)\} = I_{21}\{\boldsymbol{D}_0(\boldsymbol{C}_1)+\boldsymbol{D}_0(\boldsymbol{C}_2)\}.$$

Using this and (10.6, 2), the equations (8.31, 5) become

$$x_1^{-1}f_1^{(0)}\boldsymbol{C}_1 = nI_{12}\{\boldsymbol{D}_0(\boldsymbol{C}_1)\},\quad x_2^{-1}f_2^{(0)}\boldsymbol{C}_2 = nI_{21}\{\boldsymbol{D}_0(\boldsymbol{C}_2)\},$$

which are identical save for the different variable involved. Again, by (8.4, 1),

$$\overline{\boldsymbol{C}}_1-\overline{\boldsymbol{C}}_2 = -\frac{1}{3n_1n_2}\boldsymbol{d}_{12}\int f^{(0)}\boldsymbol{C}.\boldsymbol{D}_0(\boldsymbol{C})\,d\boldsymbol{c},$$

so that, by (8.4, 7),

$$D_{11} = \frac{1}{3n^2}\int f^{(0)}\boldsymbol{C}.\boldsymbol{D}_0(\boldsymbol{C})\,d\boldsymbol{c}.$$

Thus D_{11} does not depend on the proportions of the mixture, but only on its density.

Approximations to the value of D_{11} may be derived from (9.81, 1, 3) by putting $m_1 = m_2 = m$. The first approximation is $[D_{11}]_1$, where

$$[D_{11}]_1 = 3\mathrm{E}/4\rho.$$

By (9.8, 8) and (9.7, 3),

$$\mathrm{E}/[\mu]_1 = 4\Omega_1^{(2)}(2)/5\Omega_{12}^{(1)}(1),$$

or, using (9.8, 7), and remembering that in the present case $\Omega_1^{(2)}(2) = \Omega_{12}^{(2)}(2)$,

$$\mathrm{E}/[\mu]_1 = 4\mathrm{A}. \qquad (10.6, 3)$$

Hence

$$[D_{11}]_1 = 3\mathrm{A}[\mu]_1/\rho. \qquad (10.6, 4)$$

The second approximation is obtained by multiplying $[D_{11}]_1$ by $1/(1-\Delta)$, where Δ is given by (9.81, 4). Putting $M_1 = M_2$ in this equation, and using (10.6, 3), we find that

$$\Delta = 5\mathrm{C}^2/(11-4\mathrm{B}+8\mathrm{A}). \qquad (10.6, 5)$$

In the Kihara approximation ($\mathrm{B} = \frac{3}{4}$)

$$\Delta = 5\mathrm{C}^2/8(1+\mathrm{A}). \qquad (10.6, 6)$$

On substituting for A, B, C, the values of $1/(1-\Delta)$ given by (10.6, 5 and 6) for force-centres of index ν are found to be:

	$\nu = 5$	9	15	∞
(10.6, 5)	$1/(1-\Delta) = 1$	1·004	1·0085	1·017
(10.6, 6)	$1/(1-\Delta) = 1$	1·004	1·009	1·018

The true value of D_{11} will be slightly greater than the second approximation: the exact factors by which the first approximations for these models are to be multiplied may be estimated as 1, 1·005, 1·010 and 1·019 approximately. Thus, in particular, for rigid elastic spheres ($\nu = \infty$) the exact value is*

$$D_{11} = 1{\cdot}019\frac{3}{8n\sigma^2}\left(\frac{kT}{\pi m}\right)^{\frac{1}{2}} \qquad (10.6, 7)$$

$$= \frac{1{\cdot}019}{1{\cdot}016}\frac{6}{5}\frac{\mu}{\rho}$$

$$= 1{\cdot}204\mu/\rho. \qquad (10.6, 8)$$

10.61. Mixtures of isotopic molecules

In a mixture of isotopes of the same element, the law of force between all pairs of molecules is normally taken to be the same. This implies that

$$\mathscr{W}_1^{(l)}(r) = \mathscr{W}_{12}^{(l)}(r) = \mathscr{W}_2^{(l)}(r) \qquad (10.61, 1)$$

(cf. 10.1) and $\quad m_1^{\frac{1}{2}}\Omega_1^{(l)}(r) = (2m_0 M_1 M_2)^{\frac{1}{2}}\Omega_{12}^{(l)}(r) = m_2^{\frac{1}{2}}\Omega_2^{(l)}(r). \qquad (10.61, 2)$

These results can be used to simplify the general formulae for the transport coefficients in a gas mixture.

* This value of the coefficient of self-diffusion was first obtained by F. B. Pidduck, by a quite different argument (*Proc. Lond. Math. Soc.* **15**, 89 (1915)). C. L. Pekeris (*Proc. natn. Acad. Sci. U.S.A.* **41**, 661 (1955)), found the numerical factor in (10.6, 7) to be 1·01896.

Save for the lightest gases, it is sufficient to regard $(m_1 - m_2)/(m_1 + m_2)$ as a small quantity whose square can be neglected. Then, for example, the first approximation to the thermal diffusion factor $\alpha_{12} = k_T/x_1x_2$ is independent of the proportions x_1, x_2 of the isotopes in the mixture, being given* by

$$[\alpha_{12}]_1 = \frac{15C(1+A)}{A(11-4B+8A)} \frac{m_1 - m_2}{m_1 + m_2}. \qquad (10.61, 3)$$

In the Kihara approximation ($B = \frac{3}{4}$) this takes the form

$$[\alpha_{12}]_1 = \frac{15C}{8A} \cdot \frac{m_1 - m_2}{m_1 + m_2}. \qquad (10.61, 4)$$

* R. C. Jones, *Phys. Rev.* 58, 111 (1940).

11

MOLECULES WITH INTERNAL ENERGY

11.1. Communicable internal energy

Throughout the previous discussion the molecules have been regarded as possessing no internal energy, or at least none that is interchangeable with translatory energy at collisions. If interconversion of internal and translational energy is possible, extra variables specifying the internal motion have to be introduced. A general theory covering this case can be given, but the detailed discussion is very complicated for all save the simplest models. However, certain general results for all molecules with internal energy can be established. The internal energy is found to contribute, proportionately, rather less to the thermal conductivity than does the translational energy. Also the finite time required for interchange between translational and rotational energy introduces a 'volume viscosity' opposing motions of contraction or expansion of the gas. This volume viscosity can considerably increase the rate of damping of sound waves above that due to the 'shear' viscosity μ.

Detailed formulae have been derived only for one or two decidedly artificial molecular models. One is a perfectly rough, perfectly elastic and rigid spherical molecule. This model was first suggested by Bryan;* the methods developed by Chapman and Enskog for smooth spherical molecules were extended to Bryan's model by Pidduck.† The model is mathematically simpler than all other variably rotating models, in that no variables are required to specify its orientation in space.

Jeans‡ proposed as a model a smooth elastic sphere whose mass-centre does not coincide with its geometrical centre; this model has been studied by Dahler and his associates.§ Curtiss and Muckenfuss|| worked out the detailed theory of sphero-cylindrical molecules; that is, smooth cylinders with hemispherical ends. Both these models require two space variables to specify the molecular orientation, as well as variables specifying the angular velocity. The sphero-cylindrical model also has the awkward property that, especially when the molecules are long and slender, 'chattering' collisions

* G. H. Bryan, *Rep. Br. Ass. Advmt. Sci.* p. 83 (1894).

† F. B. Pidduck, *Proc. R. Soc.* A, **101**, 101 (1922).

‡ J. H. Jeans, *Phil. Trans. R. Soc.* **196**, 399 (1901); *Q. J. Math.* **25**, 224 (1904).

§ J. S. Dahler and N. F. Sather, *J. Chem. Phys.* **35**, 2029 (1961), and **38**, 2363 (1962); S. I. Sandler and J. S. Dahler, *J. Chem. Phys.* **43**, 1750 (1965); **46**, 3520 (1967), and **47**, 2621 (1967).

|| C. F. Curtiss, *J. Chem. Phys.* **24**, 225 (1956); C. F. Curtiss and C. Muckenfuss, *ibid.* **26**, 1619 (1957) and **29**, 1257 (1958).

can occur; that is, the reversal of angular velocities at one collision can lead to one or more further collisions between the same molecules immediately afterwards.

The artificial nature of such models, and the complexity of the formulae derived from them, led Mason and Monchick to attempt a simpler approach.* They based their discussion on the general theory, but determined the collision integrals approximately in terms of relaxation times.

11.2. Liouville's theorem

The general theory is based on a dynamical theorem due to Liouville, which is of fundamental importance in statistical mechanics. Let the state of a system be specified by R generalized coordinates Q^α and their conjugate momenta P^α ($\alpha = 1, 2, \ldots, R$). If H is the Hamiltonian function of the system, the Hamiltonian equations of motion are

$$\dot{Q}^\alpha = \frac{\partial \mathrm{H}}{\partial P^\alpha}, \quad \dot{P}^\alpha = -\frac{\partial \mathrm{H}}{\partial Q^\alpha}, \tag{11.2, 1}$$

so that

$$\frac{\partial \dot{Q}^\alpha}{\partial Q^\alpha} = -\frac{\partial \dot{P}^\alpha}{\partial P^\alpha}. \tag{11.2, 2}$$

The quantities Q^α, P^α may be regarded either as the components of R-dimensional vectors $\mathbf{Q}$, $\mathbf{P}$, or as the coordinates of a point in a space of $2R$ dimensions (the phase-space).

Liouville's theorem refers to the points in phase-space representative of an assembly of dynamical systems with the same H. It asserts that, if at time t these occupy an infinitesimal volume V of the phase space, at any other time they occupy an equal volume.

The proof is as follows. After a short time dt, Q^α and P^α become $Q^\alpha + \dot{Q}^\alpha dt$ and $P^\alpha + \dot{P}^\alpha dt$, and V becomes $\mathrm{V} + d\mathrm{V}$; by the rule for transforming small elements of volume

$$\frac{\mathrm{V} + d\mathrm{V}}{\mathrm{V}} = \frac{\partial(\mathbf{Q} + \dot{\mathbf{Q}}\,dt, \mathbf{P} + \dot{\mathbf{P}}\,dt)}{\partial(\mathbf{Q}, \mathbf{P})}.$$

The Jacobian on the right is a determinant whose non-diagonal elements are all proportional to dt; the diagonal elements are of form $1 + (\partial \dot{Q}^\alpha / \partial Q^\alpha)\,dt$ or $1 + (\partial \dot{P}^\alpha / \partial P^\alpha)\,dt$. Thus, neglecting squares and higher powers of dt,

$$\begin{aligned} \frac{1}{\mathrm{V}}\left(\mathrm{V} + \frac{d\mathrm{V}}{dt}\,dt\right) &= 1 + dt \sum_\alpha \left(\frac{\partial \dot{Q}^\alpha}{\partial Q^\alpha} + \frac{\partial \dot{P}^\alpha}{\partial P^\alpha}\right) \\ &= 1, \end{aligned}$$

by (11.2, 2). Hence $d\mathrm{V}/dt = 0$; that is, V does not change with time.

* E. A. Mason and L. Monchick, *J. Chem. Phys.* **36**, 1622 (1962).

11.21. The generalized Boltzmann equation

Consider a gas-mixture; a simple gas is regarded as a special case of this. The position and motion of a typical molecule m_s are specified by a set of generalized coordinates and momenta Q_s^α, P_s^α ($\alpha = 1, 2, \ldots, R$), or by the R-dimensional vectors $\boldsymbol{Q}_s$, $\boldsymbol{P}_s$ of which these are the components. The first three of the coordinates are taken to be the Cartesian coordinates x_s, y_s, z_s of the mass-centre (the components of $\boldsymbol{r}_s$): the corresponding momenta are the components $m_s u_s$, $m_s v_s$, $m_s w_s$ of the linear momentum $m_s \boldsymbol{c}_s$. The remaining coordinates and momenta, specifying the internal state, are represented collectively by $(R-3)$-dimensional vectors $\boldsymbol{Q}_s^{(i)}$, $\boldsymbol{P}_s^{(i)}$ in the phase-space specifying the internal motion; the volume-element $d\boldsymbol{Q}_s^{(i)} d\boldsymbol{P}_s^{(i)}$ in this internal phase-space is denoted by $d\boldsymbol{\Omega}_s$. The generalized velocity-distribution function $f_s(\boldsymbol{Q}_s, \boldsymbol{P}_s, t)$ is then defined as such that the probable number of molecules m_s whose coordinates and momenta $\boldsymbol{Q}_s$, $\boldsymbol{P}_s$ lie in the volume $d\boldsymbol{Q}_s d\boldsymbol{P}_s$ of the total phase-space at a time t is

$$m_s^{-3} f_s(\boldsymbol{Q}_s, \boldsymbol{P}_s, t)\, d\boldsymbol{Q}_s d\boldsymbol{P}_s \equiv f_s(\boldsymbol{Q}_s, \boldsymbol{P}_s, t)\, d\boldsymbol{r}_s d\boldsymbol{c}_s d\boldsymbol{\Omega}_s. \qquad (11.21, 1)$$

As usual, any molecule is supposed to be interacting with others during a negligible part of its total lifetime, and binary encounters alone are considered; also processes like chemical action and ionization are supposed absent. The motion of a molecule m_s between encounters is assumed to satisfy equations of the Hamiltonian form (11.2, 1). The corresponding Hamiltonian function H_s is assumed to be an even function of the momenta, not explicitly depending on the time, so that H_s is equal to the total energy of the molecule. If external forces act on the molecule, their potential energy is assumed to be a function $V_s(\boldsymbol{r}_s)$ of the position $\boldsymbol{r}_s$ only. Thus

$$\mathrm{H}_s = V_s(\boldsymbol{r}_s) + \mathrm{H}_s^{(F)}, \qquad (11.21, 2)$$

where $\mathrm{H}_s^{(F)}$ denotes the energy of free motion in the absence of the external forces.

Consider now the rate of change of f_s. Were it not for encounters, the set (11.21, 1) of molecules m_s would, after a time dt, constitute the set

$$m_s^{-3} f_s(\boldsymbol{Q}_s + \dot{\boldsymbol{Q}}_s dt, \boldsymbol{P}_s + \dot{\boldsymbol{P}}_s dt, t + dt)\, d\boldsymbol{Q}_s d\boldsymbol{P}_s$$

(the volume of phase-space which they occupy is still $d\boldsymbol{Q}_s d\boldsymbol{P}_s$, by Liouville's theorem). Let the net effect of encounters during dt be to increase the number in the set by

$$m_s^{-3} dt \frac{\partial_e f_s}{\partial t} d\boldsymbol{Q}_s d\boldsymbol{P}_s.$$

Then we find, as in 3.1, that

$$\mathscr{D}_s f_s \equiv \frac{\partial f_s}{\partial t} + \dot{\boldsymbol{Q}}_s \cdot \frac{\partial f_s}{\partial \boldsymbol{Q}_s} + \dot{\boldsymbol{P}}_s \cdot \frac{\partial f_s}{\partial \boldsymbol{P}_s} = \frac{\partial_e f_s}{\partial t}, \qquad (11.21, 3)$$

where $\partial/\partial \boldsymbol{Q}_s$, $\partial/\partial \boldsymbol{P}_s$ denote R-dimensional vector operators with components $\partial/\partial Q_s^\alpha$, $\partial/\partial P_s^\alpha$, and $\dot{\boldsymbol{Q}}_s.\partial f_s/\partial \boldsymbol{Q}_s \equiv \sum_\alpha \dot{Q}_s^\alpha \partial f_s/\partial Q_s^\alpha$, etc.

Equation (11.21, 3) is the generalized Boltzmann equation. Using the Hamiltonian equations, we may express $\mathscr{D}_s f_s$ in the form

$$\mathscr{D}_s f_s = \frac{\partial f_s}{\partial t} + \frac{\partial \mathrm{H}_s}{\partial \boldsymbol{P}_s} \cdot \frac{\partial f_s}{\partial \boldsymbol{Q}_s} - \frac{\partial \mathrm{H}_s}{\partial \boldsymbol{Q}_s} \cdot \frac{\partial f_s}{\partial \boldsymbol{P}_s}. \tag{11.21, 4}$$

11.22. The evaluation of $\partial_e f_s/\partial t$

In evaluating $\partial_e f_s/\partial t$, the encounter of a pair of molecules m_s, m_t is regarded arbitrarily as commencing at one instant t_0', and ending at another instant t_0. To avoid difficulties arising from the extremely rapid variation of the internal coordinates and momenta, the encounter is replaced by a discontinuous 'collision' as follows. The motions of the molecules prior to t_0' are supposed to be continued up to $\frac{1}{2}(t_0+t_0')$ as if no interaction were taking place, the coordinates and momenta at the latter instant being $\boldsymbol{Q}_s'$, $\boldsymbol{P}_s'$ and $\boldsymbol{Q}_t'$, $\boldsymbol{P}_t'$. Likewise the motions subsequent to t_0 are supposed to be produced backward to $\frac{1}{2}(t_0+t_0')$ as if no interaction were taking place, the coordinates and momenta at $\frac{1}{2}(t_0+t_0')$ now being $\boldsymbol{Q}_s$, $\boldsymbol{P}_s$ and $\boldsymbol{Q}_t$, $\boldsymbol{P}_t$. Then the effect of the encounter, at all times outside the (negligibly small) interval between t_0' and t_0, can be represented by replacing the actual encounter by a 'collision' occurring at the instant $\frac{1}{2}(t_0+t_0')$, at which $\boldsymbol{Q}_s'$, $\boldsymbol{P}_s'$, $\boldsymbol{Q}_t'$, $\boldsymbol{P}_t'$ change discontinuously to $\boldsymbol{Q}_s$, $\boldsymbol{P}_s$, $\boldsymbol{Q}_t$, $\boldsymbol{P}_t$.

Suppose that encounters are replaced by collisions in this fashion. The line of motion, before a collision, of the mass-centre of the molecule m_t relative to that of the molecule m_s may be specified by impact parameters b, ϵ defined as in 3.42. Consider the number of collisions occurring during dt between the $dn_s\,(\equiv m_s^{-3} f_s(\boldsymbol{Q}_s, \boldsymbol{P}_s, t)\, d\boldsymbol{Q}_s d\boldsymbol{P}_s)$ molecules m_s in the phase-element $d\boldsymbol{Q}_s d\boldsymbol{P}_s$ and molecules m_t with momenta and internal coordinates in the ranges $d\boldsymbol{P}_t$, $d\boldsymbol{Q}_t^{(i)}$, the impact parameters b, ϵ lying in the ranges db, $d\epsilon$. At the beginning of dt the molecule m_t must lie in a volume $g\,dt\,b\,db\,d\epsilon$, g being the magnitude of the relative velocity $\boldsymbol{c}_t - \boldsymbol{c}_s$ (cf. 3.5). Thus the number of such collisions is $dn_s dn_t$, where

$$dn_t = m_t^{-3} f_t(\boldsymbol{Q}_t, \boldsymbol{P}_t, t)\, g\, dt\, b\, db\, d\epsilon\, d\boldsymbol{P}_t\, d\boldsymbol{Q}_t^{(i)}.$$

Summing over all values of b, ϵ, $\boldsymbol{P}_t$, $\boldsymbol{Q}_t^{(i)}$, the total number of collisions of the dn_s molecules m_s with all molecules m_t in time dt is found to be

$$dt(m_s m_t)^{-3} f_s(\boldsymbol{Q}_s, \boldsymbol{P}_s)\, d\boldsymbol{Q}_s d\boldsymbol{P}_s \iiiint f_t(\boldsymbol{Q}_t, \boldsymbol{P}_t)\, g\, b\, db\, d\epsilon\, d\boldsymbol{Q}_t^{(i)}\, d\boldsymbol{P}_t.$$

This is the number of collisions in dt which carry molecules m_s out of the phase-element $d\boldsymbol{Q}_s d\boldsymbol{P}_s$.

The number of collisions carrying molecules m_s into the same phase-element is similarly

$$dt(m_s m_t)^{-3} \iiiint f_s(\boldsymbol{Q}_s', \boldsymbol{P}_s') f_t(\boldsymbol{Q}_t', \boldsymbol{P}_t')\, g' b'\, db'\, d\epsilon'\, d\boldsymbol{Q}_s' d\boldsymbol{P}_s' d\boldsymbol{Q}_t^{(i)\prime}\, d\boldsymbol{P}_t'.$$

Here the integration is over all collisions with initial coordinates and momenta Q'_s, P'_s, $Q_t^{(i)\prime}$, P'_t and with collision parameters b', ϵ' such that, of the final coordinates and velocities Q_s, P_s, Q_t, P_t, the pair Q_s, P_s lie in dQ_s, dP_s. By Liouville's theorem applied to the pair of molecules regarded as a single dynamical system,* if b, ϵ are impact parameters specifying the motion of the molecule m_t relative to the molecule m_s after a collision,

$$g'\,dt'\,b'\,db'\,d\epsilon'\,dQ'_s dP'_s dQ_t^{(i)\prime} dP'_t = g\,dt\,b\,db\,d\epsilon\,dQ_s dP_s dQ_t^{(i)} dP_t. \quad (11.22, 1)$$

Combining these results, we find the net gain of molecules m_s to the phase-element $dQ_s dP_s$ through collisions with molecules m_t during dt to be

$$dt(m_s m_t)^{-3} dQ_s dP_s \iiiint (f'_s f'_t - f_s f_t) g b\,db\,d\epsilon\,dQ_t^{(i)} dP_t,$$

where f'_s, f'_t denote $f_s(Q'_s, P'_s)$, $f_t(Q'_t, P'_t)$, etc. Hence, from the definition of $\partial_e f_s/\partial t$,

$$\frac{\partial_e f_s}{\partial t} = \sum_t m_t^{-3} \iiiint (f'_s f'_t - f_s f_t) g b\,db\,d\epsilon\,dQ_t^{(i)} dP_t \quad (11.22, 2)$$

$$= \sum_t \iiiint (f'_s f'_t - f_s f_t) g b\,db\,d\epsilon\,dc_t\,d\Omega_t. \quad (11.22, 3)$$

11.23. Smoothed distributions

The above derivation rests on the assumption that at any instant the proportion of the molecules m_s undergoing encounters is negligible, so that the velocity-distribution function of all molecules m_s is indistinguishable from that of molecules not undergoing encounters. The justification of the assumption is in the smallness of the duration $t_0 - t'_0$ of an encounter, which was also invoked to justify the replacement of encounters by 'collisions'.

In the Boltzmann equation (11.21, 3) the terms involving $\dot{Q}_s$, $\dot{P}_s$ give the change in f_s due to the motions of molecules between encounters. Now (cf. 11.34) the period of rotation of a molecule is in general comparable with the time taken for the molecule to travel a distance equal to the effective molecular radius, and so with the duration of an encounter. The periods of molecular vibration are in general even shorter. Thus the terms involving $\dot{Q}_s$, $\dot{P}_s$ in the Boltzmann equation introduce fluctuations in f_s on a time-scale comparable with the duration of an encounter. Since this duration is being regarded as negligibly small, it is reasonable to remove the fluctuations by averaging f_s over a time large compared with the periods of molecular rotation and vibration. So averaged, f_s ceases to depend on quantities which vary rapidly in the internal motions; it may depend on the energy of internal vibrations, and on the angular momentum and energy of rotation, but not

* Liouville's theorem has to be applied three times to the pair of molecules; first to their actual motion during an encounter, then to the (supposed undisturbed) motion from t'_0 to $\frac{1}{2}(t_0 + t'_0)$, and finally to the similar undisturbed motion from $\frac{1}{2}(t_0 + t'_0)$ to t_0. In the last two cases H is the sum of the Hamiltonians of the two molecules, ignoring interactions.

on the phase of a vibration or the changes of orientation during rotation. Corresponding to this averaging of f_s, it is necessary also to average $\partial_e f_s/\partial t$ over all states corresponding to particular values of the angular momentum and energies of rotation and vibration.

This type of averaging replaces f_s by a smoothed function which, however, suffices to give all measurable properties of the gas.*

11.24. The equations of transfer

It is here convenient to employ summational invariants measured relative to fixed axes, i.e.

$$\psi_s^{(1)} = 1, \quad \boldsymbol{\psi}_s^{(2)} = m_s \boldsymbol{c}_s, \quad \psi_s^{(3)} = \mathrm{H}_s^{(F)}, \qquad (11.24, 1)$$

where, as in 11.21, $\mathrm{H}_s^{(F)}$ is the energy of free motion in the absence of external forces. Each ψ_s is a constant of the free motion; thus

$$\dot{\boldsymbol{Q}}_s . \partial\psi_s/\partial \boldsymbol{Q}_s + \dot{\boldsymbol{P}}_s . \partial\psi_s/\partial \boldsymbol{P}_s = 0$$

in the free motion, so that

$$\frac{\partial \mathrm{H}_s^{(F)}}{\partial \boldsymbol{P}_s} \cdot \frac{\partial \psi_s}{\partial \boldsymbol{Q}_s} - \frac{\partial \mathrm{H}_s^{(F)}}{\partial \boldsymbol{Q}_s} \cdot \frac{\partial \psi_s}{\partial \boldsymbol{P}_s} = 0. \qquad (11.24, 2)$$

The equation of change of any summational invariant ψ_s is got by multiplying the Boltzmann equation (11.21, 3) by $\psi_s d\boldsymbol{P}_s d\boldsymbol{Q}_s^{(i)}$, integrating, and summing with respect to s. The summational-invariant property ensures that the $\partial_e f_s/\partial t$ terms, representing the effect of collisions, then vanish. Thus

$$\sum_s \iint \left(\frac{\partial f_s}{\partial t} + \frac{\partial \mathrm{H}_s}{\partial \boldsymbol{P}_s} \cdot \frac{\partial f_s}{\partial \boldsymbol{Q}_s} - \frac{\partial \mathrm{H}_s}{\partial \boldsymbol{Q}_s} \cdot \frac{\partial f_s}{\partial \boldsymbol{P}_s} \right) \psi_s d\boldsymbol{P}_s d\boldsymbol{Q}_s^{(i)} = 0. \qquad (11.24, 3)$$

Again, $\mathrm{H}_s = V_s(\boldsymbol{r}_s) + \mathrm{H}_s^{(F)}$ (cf. (11.21, 2)) and the force $m_s \boldsymbol{F}_s$ on a molecule m_s is given by

$$m_s \boldsymbol{F}_s = -\frac{\partial V_s}{\partial \boldsymbol{r}_s}. \qquad (11.24, 4)$$

Using (11.24, 2, 4) in (11.24, 3), we find

$$\sum_s \iint \left\{ \frac{\partial f_s}{\partial t} \psi_s + \frac{\partial}{\partial \boldsymbol{Q}_s} \cdot \left(\frac{\partial \mathrm{H}_s^{(F)}}{\partial \boldsymbol{P}_s} f_s \psi_s \right) - \frac{\partial}{\partial \boldsymbol{P}_s} \cdot \left(\frac{\partial \mathrm{H}_s^{(F)}}{\partial \boldsymbol{Q}_s} f_s \psi_s \right) \right.$$

$$\left. + \boldsymbol{F}_s \cdot \frac{\partial}{\partial \boldsymbol{c}_s} (\psi_s f_s) - \boldsymbol{F}_s \cdot \frac{\partial \psi_s}{\partial \boldsymbol{c}_s} f_s \right\} d\boldsymbol{P}_s d\boldsymbol{Q}_s^{(i)} = 0. \qquad (11.24, 5)$$

In (11.24, 5), the third term of the integrand vanishes on partial integration with respect to $\boldsymbol{P}_s$, since $f_s \to 0$ as $\boldsymbol{P}_s \to \infty$; the fourth term similarly vanishes after a partial integration with respect to $\boldsymbol{c}_s$. In the second term, the coordinates Q_s^α can be divided into three groups. First there are the Cartesian coordinates of the mass-centre; next there are Eulerian angles specifying the orientation of the molecule; finally there are coordinates giving the displace-

* The smoothing introduced here is similar in essence to that introduced in 2.21.

ments in internal vibrations. The contributions from the Eulerian coordinates to the term in question vanish on partial integration, because of periodicity; those from the vibrational coordinates similarly vanish, because $f_s \to 0$ when one of these coordinates becomes large. Since ψ_s does not explicitly depend on t, the remaining terms of (11.24, 5) give

$$\sum_s \left\{ \frac{\partial}{\partial t}(n_s \overline{\psi}_s) + \frac{\partial}{\partial \boldsymbol{r}} \cdot (n_s \overline{\boldsymbol{c}_s \psi_s}) - n_s \boldsymbol{F}_s \cdot \overline{\frac{\partial}{\partial \boldsymbol{c}_s} \psi_s} \right\} = 0. \qquad (11.24, 6)$$

Equation (11.24, 6) is identical in form with the equation obtained in the absence of internal motions. Thus, applied to the summational invariants $\psi_s^{(1)}$, $\psi_s^{(2)}$, $\psi_s^{(3)}$, it must give equations of continuity, momentum and energy identical with those of 3.21 or 8.1. In particular, the energy equation is

$$\rho \frac{D}{Dt}\left(\sum_s n_s \overline{E}_s/\rho\right) + \frac{\partial}{\partial \boldsymbol{r}} \cdot \boldsymbol{q} = \sum_s \rho_s \overline{\boldsymbol{C}}_s . \boldsymbol{F}_s - \mathsf{p} : \frac{\partial}{\partial \boldsymbol{r}} \boldsymbol{c}_0. \qquad (11.24, 7)$$

Here $\mathsf{p} = \sum_s \rho_s \overline{\boldsymbol{C}_s \boldsymbol{C}_s}$, as before, and E_s differs from $\mathrm{H}_s^{(F)}$ only by being measured relative to axes moving with velocity $\boldsymbol{c}_0$.

In addition to the summational invariants (11.24, 1), there is also a summational invariant of angular momentum, given by

$$\psi_s^{(4)} = \boldsymbol{r}_s \wedge m_s \boldsymbol{c}_s + \boldsymbol{h}_s, \qquad (11.24, 8)$$

where $\boldsymbol{h}_s$ is the angular momentum about the mass-centre. Direct substitution from (11.24, 8) into (11.24, 6) yields, after simplifying by using the momentum equation,

$$\frac{\partial}{\partial t}(\sum_s n_s \overline{\boldsymbol{h}}_s) + \frac{\partial}{\partial \boldsymbol{r}} \cdot (\sum_s n_s \overline{\boldsymbol{c}_s \boldsymbol{h}_s}) = 0. \qquad (11.24, 9)$$

However, this equation is open to the objection that $\psi_s^{(4)}$ is a summational invariant only in virtue of the difference between the values of $\boldsymbol{r}$ for the centres of colliding molecules. Thus for consistency, in considering the equation of change of $\psi_s^{(4)}$, it is necessary equally to take into account the finite distance through which the internal angular momentum $\boldsymbol{h}_s$ is displaced in being transferred from one molecule to another at collision. This collisional transfer of $\boldsymbol{h}_s$, like the corresponding transfer of heat or momentum, is very small at ordinary densities, but it modifies (11.24, 9) appreciably, because $\overline{\boldsymbol{h}}_s$ is normally very small (see 11.32).

11.3. The uniform steady state at rest

The determination of the velocity-distribution function in the uniform steady state is complicated by the non-existence, for general molecular models, of an 'inverse' collision exactly reversing the effect of a given 'direct' collision. However, a 'reverse' collision can be found, in which the course of the direct collision is exactly retraced: the final velocities of the molecules in this are *minus* the initial velocities in the direct collision. Such reverse collisions can be used (following an argument due to Lorentz*) to determine f_s in the uniform steady state, but only by invoking an additional probability assumption.

Consider the uniform steady state in a simple gas *at rest* under no external forces. In order to use reverse collisions, it is assumed that in such a state

$$f_s \equiv f_s(\boldsymbol{Q}_s, \boldsymbol{P}_s) = f_s(\boldsymbol{Q}_s, -\boldsymbol{P}_s) \equiv f_{s-}, \tag{11.3, 1}$$

i.e. that a given set of velocities is as probable as the same set reversed. This assumption is reasonable, but not self-evident.

Let the Boltzmann equation (11.21, 3) be multiplied by $\ln f_s$ and integrated over the whole range of $\boldsymbol{Q}_s^{(i)}$ and $\boldsymbol{P}_s$. In the uniform steady state, $\partial f_s/\partial t = 0$; since also f_s and H_s are both even functions of $\boldsymbol{P}_s$, the left-hand side of the resulting equation reduces to the integral of an odd function of $\boldsymbol{P}_s$ (cf. (11.21, 4)), which vanishes. Thus, on substituting for $\partial_e f_s/\partial t$ from (11.22, 2),

$$0 = \iiiiiint (f_s' f_t' - f_s f_t) \ln f_s \, g b \, db \, d\epsilon \, d\boldsymbol{Q}_t^{(i)} \, d\boldsymbol{P}_t \, d\boldsymbol{Q}_s^{(i)} \, d\boldsymbol{P}_s \tag{11.3, 2}$$

(the suffixes s, t are retained purely to distinguish between the two colliding molecules). In (11.3, 2), $\boldsymbol{Q}_s'$, $\boldsymbol{P}_s'$, $\boldsymbol{Q}_t'$, $\boldsymbol{P}_t'$ are the coordinates and momenta before collision of molecules whose coordinates and momenta after collision are $\boldsymbol{Q}_s$, $\boldsymbol{P}_s$, $\boldsymbol{Q}_t$, $\boldsymbol{P}_t$.

From (11.3, 1, 2),

$$0 = \iiiiiint (f_{s-}' f_{t-}' - f_{s-} f_{t-}) \ln f_{s-} \, g b \, db \, d\epsilon \, d\boldsymbol{Q}_t^{(i)} \, d\boldsymbol{P}_t \, d\boldsymbol{Q}_s^{(i)} \, d\boldsymbol{P}_s.$$

In this equation, $\boldsymbol{Q}_s$, $-\boldsymbol{P}_s$, $\boldsymbol{Q}_t$, $-\boldsymbol{P}_t$ can be regarded as the coordinates and momenta *before* the reverse collision which leads to the *final* coordinates and momenta $\boldsymbol{Q}_s'$, $-\boldsymbol{P}_s'$, $\boldsymbol{Q}_t'$, $-\boldsymbol{P}_t'$. Rename these quantities $\boldsymbol{Q}_s'$, $\boldsymbol{P}_s'$, $\boldsymbol{Q}_t'$, $\boldsymbol{P}_t'$ (before collision) and $\boldsymbol{Q}_s$, $\boldsymbol{P}_s$, $\boldsymbol{Q}_t$, $\boldsymbol{P}_t$ (after). Then, using (11.22, 1), one finds

$$0 = \iiiiiint (f_s f_t - f_s' f_t') \ln f_s' \, g b \, db \, d\epsilon \, d\boldsymbol{Q}_t^{(i)} \, d\boldsymbol{P}_t \, d\boldsymbol{Q}_s^{(i)} \, d\boldsymbol{P}_s. \tag{11.3, 3}$$

Add this to (11.3, 2), and to the equations got from (11.3, 2 and 3) by interchanging the role of the colliding molecules (so that $\boldsymbol{Q}_s$, $\boldsymbol{P}_s$ become $\boldsymbol{Q}_t$, $\boldsymbol{P}_t$, etc.). Then (because the space-coordinates are to be regarded as identical in $\boldsymbol{Q}_s$ and $\boldsymbol{Q}_t$)

$$0 = \iiiiiint (f_s' f_t' - f_s f_t) \ln (f_s f_t / f_s' f_t') \, g b \, db \, d\epsilon \, d\boldsymbol{Q}_t^{(i)} \, d\boldsymbol{P}_t \, d\boldsymbol{Q}_s^{(i)} \, d\boldsymbol{P}_s. \tag{11.3, 4}$$

As in 4.1, equation (11.3, 4) implies that for all possible collisions

$$\ln f_s + \ln f_t = \ln f_s' + \ln f_t'. \tag{11.3, 5}$$

When this condition is satisfied, the integrand in $\partial_e f_s/\partial t$ is identically zero, and there is detailed balancing. The interpretation to be given to detailed balancing here is, however, that collisions of a given type are balanced in detail by their *reverse* collisions.

* H. A. Lorentz, *Wien. Sitz.* 95 (2), 115 (1887).

Equation (11.3, 5) implies that $\ln f_s$ is a linear combination of summational invariants. As it is an even function of $\boldsymbol{P}_s$, it cannot depend on the linear or angular momentum; thus

$$f_s = A_s e^{-\alpha \mathrm{H}_s}, \qquad (11.3, 6)$$

where A_s and α are constants. Since H_s, the total energy, includes a part $\frac{1}{2}m_s c_s^2$, the condition $\frac{1}{2}m_s \overline{c_s^2} = \frac{3}{2}kT$ (cf. 2.41) leads as usual to $\alpha = 1/kT$; A_s can be related to the number-density. The formula (11.3, 6) specifies what is known as the Boltzmann distribution. It can readily be shown to apply also to gas-mixtures.

11.31. Boltzmann's closed chains

Lorentz's argument given in the last section determines f_s on the assumption that a steady state has been attained in which f_s is an even function of $\boldsymbol{P}_s$; it does not prove that the gas must tend to such a state. For this reason Boltzmann* gave an alternative discussion on the following lines.

He considered chains of collisions such that the coordinates and momenta of the colliding molecules, and the impact parameters, after the jth collision of a chain, lie in the same elementary ranges as do the corresponding variables (for other colliding molecules) before the $(j+1)$th collision of the chain. He argued that, if such a chain is extended indefinitely, ultimately a collision will be reached such that the coordinates and momenta after it are indistinguishable from those before the first collision, so that the chain is effectively closed. By using closed collision-chains as a generalization of inverse collisions, he was able to prove that in a uniform gas the integral

$$\iint f_s \ln f_s \, d\boldsymbol{Q}_s^{(i)} \, d\boldsymbol{P}_s$$

tends to a minimum value, and that, in the state corresponding to this minimum value, $\ln f_s$ is a summational invariant. For a gas at rest, the formula (11.3, 6) for f_s follows as before.

The weak point of Boltzmann's argument was its assumption of closed chains. Ergodic theory has gone far towards meeting objections to this assumption.† However, though there is little doubt as to its essential soundness, its detailed mathematical justification is not easy.

11.32. More general steady states

A more general steady state of the gas under no forces is obtained by assuming $\ln f_s$ to be a general summational invariant, i.e.

$$\ln f_s = \alpha_s^{(1)} + \boldsymbol{\alpha}^{(2)}.m_s \boldsymbol{c}_s - \alpha^{(3)}\mathrm{H}_s^{(F)} + \boldsymbol{\alpha}^{(4)}.(\boldsymbol{r} \wedge m_s \boldsymbol{c}_s + \boldsymbol{h}_s), \qquad (11.32, 1)$$

* L. Boltzmann, *Wien. Sitz.* **95** (2), 153 (1887). Boltzmann's argument is expounded in somewhat greater detail in R. C. Tolman's *Statistical Mechanics*, chapters 5 and 6 (Oxford, 1938). See also D. ter Haar, *Elements of Statistical Mechanics*, appendix 1 (Rinehart, New York, 1954).

† See D. ter Haar, *op. cit.*

where $\alpha_s^{(1)}$, $\boldsymbol{\alpha}^{(2)}$, $\alpha^{(3)}$, $\boldsymbol{\alpha}^{(4)}$ are independent of $\boldsymbol{Q}_s$, $\boldsymbol{P}_s$, t. Such a form makes $f_s'f_t' = f_s f_t$, so that $\partial_e f_s/\partial t = 0$;* also, since the summational invariants are constants of the free motion specified by the Hamiltonian $H_s^{(F)}$, it makes $\mathscr{D}_s f_s = 0$. As the notation implies, in a gas-mixture $\alpha_s^{(1)}$ takes different values for the separate constituent gases, but $\boldsymbol{\alpha}^{(2)}$, $\alpha^{(3)}$, $\boldsymbol{\alpha}^{(4)}$ are the same for all.

As in 11.3, since $H_s^{(F)}$ includes the part $\frac{1}{2}m_s c_s^2$, the value of $\alpha^{(3)}$ is $1/kT$. The part of (11.32, 1) depending on $\boldsymbol{c}_s$ is

$$-(m_s/2kT)\{c_s^2 - 2kT\boldsymbol{c}_s.(\boldsymbol{\alpha}^{(2)} + \boldsymbol{\alpha}^{(4)} \wedge \boldsymbol{r})\},$$

so that the mass-velocity $\boldsymbol{c}_0$ is given by

$$\boldsymbol{c}_0 = kT(\boldsymbol{\alpha}^{(2)} + \boldsymbol{\alpha}^{(4)} \wedge \boldsymbol{r}). \tag{11.32, 2}$$

Thus the whole mass of gas moves like a rigid body possessing both rotational and translatory motion, its angular velocity being $kT\boldsymbol{\alpha}^{(4)}$; the state of the gas is similar to that considered in 4.14.

To see the significance of the term $\boldsymbol{\alpha}^{(4)}.\boldsymbol{h}_s$ in (11.32, 1), consider the special case of rigid rotating spherical molecules. If I_s is the moment of inertia about a diameter, and $\boldsymbol{\omega}_s$ the angular velocity,

$$H_s^{(F)} = \tfrac{1}{2}m_s c_s^2 + \tfrac{1}{2}I_s \omega_s^2, \quad \boldsymbol{h}_s = I_s \boldsymbol{\omega}_s.$$

Thus the terms of (11.32, 1) which involve $\boldsymbol{\omega}_s$ are

$$-(I_s/2kT)(\omega_s^2 - 2kT\boldsymbol{\alpha}^{(4)}.\boldsymbol{\omega}_s).$$

Since $\boldsymbol{\omega}_s$ is a linear function of $\boldsymbol{P}_s$, it follows that the mean value of $\boldsymbol{\omega}_s$ is equal to $kT\boldsymbol{\alpha}^{(4)}$, the angular velocity of the mass as a whole. In terms of $\boldsymbol{c}_0$, by (11.32, 2),

$$2\bar{\boldsymbol{\omega}}_s = \nabla \wedge \boldsymbol{c}_0. \tag{11.32, 3}$$

This result also holds exactly for non-spherical rigid molecules, and approximately for non-rigid molecules. The velocities given to the particles composing a molecule as a result of the mean angular velocity $\bar{\boldsymbol{\omega}}_s$ are extremely small; they are comparable in order of magnitude with the change in $\boldsymbol{c}_0$ in a distance equal to the molecular radius.

11.33. Properties of the uniform steady state

For a non-rotating gas ($\boldsymbol{\alpha}^{(4)} = 0$), (11.32, 1) takes the form

$$f_s = n_s Z_s^{-1} e^{-E_s/kT}, \tag{11.33, 1}$$

where n_s is the number-density, E_s the thermal energy (differing from $H_s^{(F)}$ only by being measured relative to axes moving with the mass-velocity $\boldsymbol{c}_0$; cf. 11.24). Since

$$n_s = \iint f_s d\boldsymbol{c}_s d\boldsymbol{\Omega}_s, \tag{11.33, 2}$$

* To justify the inclusion of the $\boldsymbol{\alpha}^{(4)}$ term in (11.32, 1), it is necessary to modify the Boltzmann collision-integral (11.22, 2), to take account of the small difference between the positions of the centres of the colliding molecules (cf. 11.24).

the quantity Z_s in (11.33, 1) is the *partition function* given by

$$Z_s = \iint e^{-E_s/kT} d\mathbf{c}_s d\mathbf{\Omega}_s. \tag{11.33, 3}$$

An equation similar to (11.33, 1) also applies to a gas at rest in a conservative field of force; now $n_s \propto e^{-V_s/kT}$, where V_s is the potential energy of a molecule (cf. (4.14, 7)).

The mean value of E_s corresponding to (11.33, 1) is given by

$$\bar{E}_s = n_s^{-1}\iint f_s E_s d\mathbf{c}_s d\mathbf{\Omega}_s$$

$$= Z_s^{-1}\iint E_s e^{-E_s/kT} d\mathbf{c}_s d\mathbf{\Omega}_s \tag{11.33, 4}$$

$$= \frac{kT^2}{Z_s}\frac{dZ_s}{dT}, \tag{11.33, 5}$$

by (11.33, 3). The specific heat (per unit mass) of molecules m_s is given by

$$(c_v)_s = \frac{d}{dT}\left(\frac{\bar{E}_s}{m_s}\right) \equiv \frac{kN_s}{2m_s} \tag{11.33, 6}$$

(cf. 2.44). This specific heat can be expressed as the sum of parts $(c_v)'_s$, $(c_v)''_s$ corresponding to the translatory and internal motions, where

$$(c_v)'_s = 3k/2m_s, \quad (c_v)''_s = (N_s - 3)\,k/2m_s. \tag{11.33, 7}$$

An alternative expression for $(c_v)_s$ is obtained as follows. From (11.33, 1 and 4–6),

$$\overline{E_s^2} = Z_s^{-1}\iint E_s^2 e^{-E_s/kT} d\mathbf{c}_s d\mathbf{\Omega}_s$$

$$= kT^2 Z_s^{-1}\frac{d}{dT}(\bar{E}_s Z_s)$$

$$= kT^2\left(\frac{d\bar{E}_s}{dT} + \frac{\bar{E}_s}{Z_s}\frac{dZ_s}{dT}\right)$$

$$= kT^2 m_s (c_v)_s + (\bar{E}_s)^2. \tag{11.33, 8}$$

It is convenient to introduce the non-dimensional quantity $\mathscr{E}_s$ defined by

$$\mathscr{E}_s = E_s/kT. \tag{11.33, 9}$$

Then, combining (11.33, 6, 8),

$$\tfrac{1}{2}N_s = \overline{\mathscr{E}_s^2} - (\bar{\mathscr{E}}_s)^2. \tag{11.33, 10}$$

Similarly* if $E_s^{(i)}$ is the internal energy of a molecule, and $\mathscr{E}_s^{(i)} = E_s^{(i)}/kT$,

$$(c_v)''_s = (kT^2 m_s)^{-1}(\overline{E_s^{(i)2}} - (\overline{E_s^{(i)}})^2), \quad \tfrac{1}{2}(N_s - 3) = \overline{\mathscr{E}_s^{(i)2}} - (\overline{\mathscr{E}_s^{(i)}})^2. \tag{11.33, 11}$$

* The proof of (11.33, 11) follows the same lines as that of (11.33, 5) and (11.33, 8). It is merely necessary in the expression (11.33, 3) for Z_s to replace E_s/kT by

$$\tfrac{1}{2}m_s C_s^2/kT' + E_s^{(i)}/kT,$$

and to differentiate with respect to T only; T' is set equal to T after the differentiations.

11.34. Equipartition of energy

A simple interpretation of the number N_s defined by (11.33, 6) can often be given. Suppose that in a gas at rest E_s is expressible as a non-degenerate homogeneous quadratic function of N'_s of the coordinates and momenta. (This is the case if E_s is composed of independent parts, consisting of kinetic energies of translation and rotation, and kinetic and potential energies of small internal vibrations.) By writing $P^\alpha_s = (kT)^{\frac{1}{2}} p^\alpha_s$, $Q^\alpha_s = (kT)^{\frac{1}{2}} q^\alpha_s$ for each of these N'_s quantities, we can express the integral (11.33, 3) for Z_s as the product of $(kT)^{N'_s/2}$ and an integral independent of T. Hence $\bar{E}_s = \frac{1}{2}N'_s kT$, by (11.33, 5); and so (11.33, 6) gives $N_s = N'_s$.

The equation $\bar{E}_s = \frac{1}{2}N_s kT$ in this case expresses the general principle of equipartition of energy, which asserts that each of the coordinates and momenta appearing in the quadratic expression for E_s makes a contribution $\frac{1}{2}kT$ to the heat energy per molecule. For molecules fully free to rotate, as well as possessing translatory energy, $N_s = 6$; for diatomic molecules this is replaced by $N_s = 5$, because quantum restrictions exclude rotation about the line joining the atomic nuclei. Each full degree of vibrational freedom would increase N_s by two, corresponding to the additional P^α_s and Q^α_s involved; however, quantum restrictions normally prevent molecules from taking the full equipartition energy of their vibrational degrees of freedom, so that each such degree increases N_s by less than two. Thus the principle of equipartition of energy, as derived above for classical systems, is to be regarded as an ideal principle rather than as one applicable to real gases. However, apart from the restriction on diatomic molecules noted above, there is nearly exact equipartition of energy between the rotational and translatory degrees of freedom, save for light molecules at low temperatures (cf. 17.63).

Assuming equipartition in a gas of rigid rotating spherical molecules, if I_s is the moment of inertia about a diameter, and $\boldsymbol{\omega}_s$ is the angular velocity,

$$\tfrac{1}{2}I_s\overline{\omega_s^2} = \tfrac{1}{2}m_s\overline{C_s^2}.$$

Since I_s is comparable with $m_s \times (\text{radius})^2$, this shows that the period $2\pi/\omega_s$ of rotation of a molecule is in general comparable in order of magnitude with the time taken by the molecule to travel a distance equal to the effective molecular radius (cf. 11.23).

11.4. Non-uniform gases*

The general theory of a gas not in a steady state is here discussed for a binary gas-mixture; the results for a simple gas follow as a special case. The first

* The discussion of non-uniform gases follows (with modifications) one given by N. Taxman, *Phys. Rev.* **110**, 1235 (1956); this in turn is based on a quantum discussion by C. S. Wang-Chang and G. E. Uhlenbeck, University of Michigan Report CM-681, 1951 (cf. 17.6). The general results derived in the classical and quantum discussions are identical; differences appear only in the detailed calculation of gas coefficients.

approximation to f_s is taken to be the Boltzmann expression

$$f_s^{(0)} = n_s Z_s^{-1} e^{-E_s/kT} \tag{11.4, 1}$$

(cf. (11.33, 1, 3)). It might appear more correct to include a term corresponding to the rotational summational invariant $\psi_s^{(4)}$, as in (11.32, 1). However, this would affect only the value of $\bar{\boldsymbol{h}}_s$; a variable value of $\boldsymbol{c}_0$ is already permitted in (11.33, 1) through the term involving $\psi_s^{(2)}$. Also the mean angular velocity $\bar{\boldsymbol{\omega}}_s$ corresponding to $\bar{\boldsymbol{h}}_s$ can be expected, as in (11.32, 3), to be comparable with the space-derivatives of $\boldsymbol{c}_0$. Since these space-derivatives are normally taken into consideration only in the second approximation to f_s, it is reasonable to neglect $\bar{\boldsymbol{h}}_s$ in the first approximation, as in (11.33, 1). Actually $\bar{\boldsymbol{h}}_s$ cannot properly be taken into account in the second approximation unless account is also taken of the difference between the positions of the centres of colliding molecules, here neglected.

As in 2.41, the temperature is defined as that which would give a gas of identical composition, in a uniform steady state, the same mean energy $\bar{E}$ per molecule as the actual gas. The hydrostatic pressure is equal to knT to a first approximation only (its deviation from this value corresponds to the appearance of volume viscosity).

When $\bar{\boldsymbol{h}}_s$ is neglected, the second approximation to f_s depends only on the parameters n_1, n_2, $\boldsymbol{c}_0$, and T, and on their space-derivatives. The first approximations to Dn_s/Dt, $D\boldsymbol{c}_0/Dt$, and DT/Dt depend only on the first approximation $p_0 (\equiv nkT)$ to p; thus (cf. (8.21, 2, 4, 6))

$$\frac{D_0 n_s}{Dt} = -n_s \frac{\partial}{\partial \boldsymbol{r}} \cdot \boldsymbol{c}_0, \quad \rho \frac{D_0 \boldsymbol{c}_0}{Dt} = \sum_s \rho_s \boldsymbol{F}_s - \frac{\partial p_0}{\partial \boldsymbol{r}}, \tag{11.4, 2}$$

$$\frac{D_0 T}{Dt} = -\frac{2T}{N} \frac{\partial}{\partial \boldsymbol{r}} \cdot \boldsymbol{c}_0; \tag{11.4, 3}$$

here N is the mean value given by

$$nN = \sum_s n_s N_s. \tag{11.4, 4}$$

To the second approximation, f_s is expressible in the form

$$f_s = f_s^{(0)}(1 + \Phi_s^{(1)}). \tag{11.4, 5}$$

The equation satisfied by $\Phi_s^{(1)}$ is

$$-J_s(\Phi^{(1)}) = \frac{\partial_0 f_s^{(0)}}{\partial t} + \frac{\partial \mathrm{H}_s}{\partial \boldsymbol{P}_s} \cdot \frac{\partial f_s^{(0)}}{\partial \boldsymbol{Q}_s} - \frac{\partial \mathrm{H}_s}{\partial \boldsymbol{Q}_s} \cdot \frac{\partial f_s^{(0)}}{\partial \boldsymbol{P}_s}, \tag{11.4, 6}$$

where

$$J_s(\Phi^{(1)}) = \sum_t \iiiint f_s^{(0)} f_t^{(0)} (\Phi_s^{(1)} + \Phi_t^{(1)} - \Phi_s^{(1)\prime} - \Phi_t^{(1)\prime}) g b \, db \, d\epsilon \, d\boldsymbol{c}_t \, d\boldsymbol{\Omega}_t. \tag{11.4, 7}$$

Since the parts of E_s and H_s involving the internal coordinates and momenta are identical, by (11.4, 1) the terms involving differentiations with respect to these coordinates and momenta can be omitted on the right of (11.4, 6). Thus, if $f_s^{(0)}$ is regarded as a function of $\boldsymbol{C}_s$, $\boldsymbol{r}$, t instead of $\boldsymbol{c}_s$, $\boldsymbol{r}$, t, equation (11.4, 6) becomes

$$-J_s(\Phi^{(1)}) = \frac{D_0 f_s^{(0)}}{Dt} + \boldsymbol{C}_s \cdot \frac{\partial f_s^{(0)}}{\partial \boldsymbol{r}} + \left(\boldsymbol{F}_s - \frac{D_0 \boldsymbol{c}_0}{Dt}\right) \cdot \frac{\partial f_s^{(0)}}{\partial \boldsymbol{C}_s} - \frac{\partial f_s^{(0)}}{\partial \boldsymbol{C}_s} \boldsymbol{C}_s : \frac{\partial}{\partial \boldsymbol{r}} \boldsymbol{c}_0$$

(cf. (8.3, 3)), in which $f_s^{(0)}$ is regarded as depending on $\boldsymbol{r}$ and t only through n_s and T. Since E_s depends on $\boldsymbol{C}_s$ only through its part $\frac{1}{2}m_s C_s^2$, we get on substitution from (11.4, 1–3) and (11.33, 5),

$$\begin{aligned} -J_s(\Phi^{(1)}) &= f_s^{(0)} \left\{ \frac{1}{n_s}\left(\frac{D_0 n_s}{Dt} + \boldsymbol{C}_s \cdot \frac{\partial n_s}{\partial \boldsymbol{r}}\right) + \frac{E_s - \bar{E}_s}{kT^2}\left(\frac{D_0 T}{Dt} + \boldsymbol{C}_s \cdot \frac{\partial T}{\partial \boldsymbol{r}}\right) \right. \\ &\quad \left. - \left(\boldsymbol{F}_s - \frac{D_0 \boldsymbol{c}_0}{Dt}\right) \cdot \frac{m_s \boldsymbol{C}_s}{kT} + \frac{m_s}{kT} \boldsymbol{C}_s \boldsymbol{C}_s : \frac{\partial}{\partial \boldsymbol{r}} \boldsymbol{c}_0 \right\} \\ &= f_s^{(0)} \left\{ x_s^{-1} \boldsymbol{C}_s . \boldsymbol{d}_s + (\mathscr{E}_s - \bar{\mathscr{E}}_s - 1) \boldsymbol{C}_s . \nabla \ln T \right. \\ &\quad \left. + 2 \overset{\circ}{\mathscr{C}_s \mathscr{C}_s} : \nabla \boldsymbol{c}_0 + \left(\tfrac{2}{3}(\mathscr{C}_s^2 - \tfrac{3}{2}) - \frac{2}{N}(\mathscr{E}_s - \bar{\mathscr{E}}_s)\right) \nabla . \boldsymbol{c}_0 \right\}. \end{aligned} \quad (11.4, 8)$$

Here $\boldsymbol{d}_1$, $\boldsymbol{d}_2$ are the vectors $\boldsymbol{d}_{12}$, $\boldsymbol{d}_{21}$ ($\equiv -\boldsymbol{d}_{12}$) of (8.3, 11), with p replaced by p_0; and $\mathscr{C}_s$, $\mathscr{E}_s$, as earlier, denote the non-dimensional variables

$$\mathscr{C}_s = (m_s/2kT)^{\frac{1}{2}} \boldsymbol{C}_s, \quad \mathscr{E}_s = E_s/kT. \quad (11.4, 9)$$

The functions $\Phi_1^{(1)}$, $\Phi_2^{(1)}$ must satisfy certain extra conditions in addition to (11.4, 8), in order that, correct to the second approximation, n_s, $\boldsymbol{c}_0$ and T may represent the number-density of molecules m_s, the mass-velocity and the temperature, and that the total internal angular momentum shall be negligible, as assumed. These conditions are

$$\iint f_s^{(0)} \Phi_s^{(1)} d\boldsymbol{c}_s d\boldsymbol{\Omega}_s = 0, \quad \sum_s \iint f_s^{(0)} \Phi_s^{(1)} m_s \boldsymbol{C}_s d\boldsymbol{c}_s d\boldsymbol{\Omega}_s = 0, \quad (11.4, 10)$$

$$\sum_s \iint f_s^{(0)} \Phi_s^{(1)} E_s d\boldsymbol{c}_s d\boldsymbol{\Omega}_s = 0, \quad \sum_s \iint f_s^{(0)} \Phi_s^{(1)} \boldsymbol{h}_s d\boldsymbol{c}_s d\boldsymbol{\Omega}_s = 0. \quad (11.4, 11)$$

The conditions of solubility of (11.4, 6) can readily be found. For each of the summational invariants $\psi^{(1)}$, $\boldsymbol{\psi}^{(2)}$, $\psi^{(3)}$,

$$\sum_s \iint J_s(\Phi^{(1)}) \psi_s^{(r)} d\boldsymbol{c}_s d\boldsymbol{\Omega}_s = 0, \quad (11.4, 12)$$

since the expression on the left represents the contribution of the second approximation to the total collisional change of a summational invariant. (An explicit proof of this statement can be given along the lines of the argument of 3.53.) The conditions of solubility are obtained on substituting for $J_s(\Phi^{(1)})$ from (11.4, 8) into (11.4, 12). It is easy to show that these are in fact satisfied.

11.41. The second approximation to f_s

The solutions of (11.4, 8) must be of the form (cf. (8.31, 1, 2))

$$\Phi_1^{(1)} = -\boldsymbol{A}_1.\nabla \ln T - \boldsymbol{D}_1.\boldsymbol{d}_1 - 2\mathsf{B}_1:\nabla \boldsymbol{c}_0 - 2B_1\nabla.\boldsymbol{c}_0, \qquad (11.41, 1)$$

$$\Phi_2^{(1)} = -\boldsymbol{A}_2.\nabla \ln T - \boldsymbol{D}_2.\boldsymbol{d}_1 - 2\mathsf{B}_2:\nabla \boldsymbol{c}_0 - 2B_2\nabla.\boldsymbol{c}_0. \qquad (11.41, 2)$$

In these, $\boldsymbol{A}_s$, $\boldsymbol{D}_s$ are vectors, B_s is a non-divergent symmetric tensor, and B_s is a scalar. The quantities $\boldsymbol{A}_s$, $\boldsymbol{D}_s$ and B_s must derive their vector and tensor characteristics from vectors associated with the motion of molecules m_s. This may at first appear to allow an excessive latitude to these quantities; but if f_s is regarded as having undergone the smoothing of 11.23, it can depend only on vectors which remain constant during rotation and internal vibration. Thus $\boldsymbol{A}_s$, $\boldsymbol{D}_s$ and B_s can depend on only two vectors, $\boldsymbol{C}_s$ and the internal angular momentum $\boldsymbol{h}_s$.

Since $\boldsymbol{h}_s$ is a rotation-vector (cf. 1.11), not an ordinary vector, the only ordinary vectors that can be formed from $\boldsymbol{C}_s$ and $\boldsymbol{h}_s$ are linear combinations of $\boldsymbol{C}_s$, $\boldsymbol{h}_s \wedge \boldsymbol{C}_s$, $\boldsymbol{h}_s \wedge (\boldsymbol{h}_s \wedge \boldsymbol{C}_s)$, ... But

$$\boldsymbol{h}_s \wedge \{\boldsymbol{h}_s \wedge (\boldsymbol{h}_s \wedge \boldsymbol{C}_s)\} = -h_s^2 \boldsymbol{h}_s \wedge \boldsymbol{C}_s,$$

and so $\boldsymbol{A}_s$ can be expressed in the form

$$\boldsymbol{A}_s = A_s^{\mathrm{I}}\boldsymbol{C}_s + A_s^{\mathrm{II}}\boldsymbol{h}_s \wedge \boldsymbol{C}_s + A_s^{\mathrm{III}}\boldsymbol{h}_s \wedge (\boldsymbol{h}_s \wedge \boldsymbol{C}_s), \qquad (11.41, 3)$$

with a similar expression for $\boldsymbol{D}_s$. In (11.41, 3) A_s^{I}, A_s^{II}, A_s^{III} are scalars, which must be functions of quantities remaining constant during the internal motions, like C_s^2, h_s^2, E_s, etc. Similarly B_s is a linear function of the six symmetric non-divergent tensors that are quadratic functions of $\boldsymbol{C}_s$, $\boldsymbol{h}_s \wedge \boldsymbol{C}_s$, $\boldsymbol{h}_s \wedge (\boldsymbol{h}_s \wedge \boldsymbol{C}_s)$, together with possibly the tensor $\overset{\circ}{\boldsymbol{h}_s\boldsymbol{h}_s}$; the coefficients again are quantities not altered by the internal motions.

In virtue of the forms of $\boldsymbol{A}_s$, $\boldsymbol{D}_s$, B_s, B_s, equations (11.4, 10, 11) are found to imply that

$$\iint f_s^{(0)} B_s d\boldsymbol{c}_s d\boldsymbol{\Omega}_s = 0, \qquad (11.41, 4)$$

$$\sum_s \iint f_s^{(0)} \boldsymbol{A}_s.m_s\boldsymbol{C}_s d\boldsymbol{c}_s d\boldsymbol{\Omega}_s = 0, \qquad (11.41, 5)$$

$$\sum_s \iint f_s^{(0)} \boldsymbol{D}_s.m_s\boldsymbol{C}_s d\boldsymbol{c}_s d\boldsymbol{\Omega}_s = 0, \qquad (11.41, 6)$$

$$\sum_s \iint f_s^{(0)} B_s E_s d\boldsymbol{c}_s d\boldsymbol{\Omega}_s = 0. \qquad (11.41, 7)$$

11.5. Thermal conduction in a simple gas

The coefficients of viscosity and thermal conduction are here evaluated only for a simple gas. From (11.4, 8) and (11.41, 1) the equation satisfied by $\boldsymbol{A}$ for such a gas is

$$J(\boldsymbol{A}) = f^{(0)}(\mathscr{E} - \bar{\mathscr{E}} - 1)\boldsymbol{C}. \qquad (11.5, 1)$$

Also equations (11.41, 3, 5) become

$$A = A^{\mathrm{I}}C + A^{\mathrm{II}}h \wedge C + A^{\mathrm{III}}h \wedge (h \wedge C), \tag{11.5, 2}$$

$$\iint f^{(0)}A \,.\, mC\, dc\, d\Omega = 0. \tag{11.5, 3}$$

As a first approximation to A we adopt the form

$$nA = a_1(\mathscr{C}^2 - \tfrac{5}{2})\, C + a_2(\mathscr{E}^{(i)} - \overline{\mathscr{E}^{(i)}})\, C, \tag{11.5, 4}$$

where a_1 and a_2 are suitable constants; it may be verified that this form for A satisfies (11.5, 3). However, (11.5, 4) is not the most general expression of the form (11.5, 2) and of third degree in the velocities which satisfies (11.5, 3); it is a special case of $A^{\mathrm{I}}C$ only, and contains no terms corresponding to A^{II} and A^{III}. It is closely analogous to the first approximation to A for a gas without internal energy, but is liable to give rather less accurate results, because of the extra omitted terms. If different parts of the internal energy interchange with translational energy at altogether different rates, it may be desirable to replace the second term of (11.5, 4) by a number of separate terms depending on these different parts of the internal energy.

The expression (11.5, 4) will, of course, not satisfy (11.5, 1), since it gives only an approximation to A; it is taken to satisfy the two equations obtained by multiplying (11.5, 1) by $(\mathscr{C}^2 - \frac{5}{2})\, C\, dc\, d\Omega$ and $(\mathscr{E}^{(i)} - \overline{\mathscr{E}^{(i)}})\, C\, dc\, d\Omega$ and integrating.* This gives

$$a_1 a_{11} + a_2 a_{12} = \tfrac{15}{4}, \tag{11.5, 5}$$

$$a_1 a_{21} + a_2 a_{22} = \tfrac{3}{4}(N-3), \tag{11.5, 6}$$

where

$$a_{11} = [(\mathscr{C}^2 - \tfrac{5}{2})\, \boldsymbol{\mathscr{C}}, (\mathscr{C}^2 - \tfrac{5}{2})\, \boldsymbol{\mathscr{C}}], \qquad a_{12} = [(\mathscr{C}^2 - \tfrac{5}{2})\, \boldsymbol{\mathscr{C}}, (\mathscr{E}^{(i)} - \overline{\mathscr{E}^{(i)}})\, \boldsymbol{\mathscr{C}}], \tag{11.5, 7}$$

$$a_{21} = [(\mathscr{E}^{(i)} - \overline{\mathscr{E}^{(i)}})\, \boldsymbol{\mathscr{C}}, (\mathscr{C}^2 - \tfrac{5}{2})\, \boldsymbol{\mathscr{C}}], \quad a_{22} = [(\mathscr{E}^{(i)} - \overline{\mathscr{E}^{(i)}})\, \boldsymbol{\mathscr{C}}, (\mathscr{E}^{(i)} - \overline{\mathscr{E}^{(i)}})\, \boldsymbol{\mathscr{C}}] \tag{11.5, 8}$$

and $[\phi, \psi]$ is defined (cf. (4.4, 7) by

$$n^2[\phi, \psi] \equiv \iint \phi J(\psi)\, dc\, d\Omega. \tag{11.5, 9}$$

If ϕ and ψ are both even or both odd functions of the momenta mC, $P^{(i)}$, an argument similar to that leading to (11.3, 4), based on 'reverse' collisions, can be used to establish that in this case†

$$n^2[\phi, \psi] = \tfrac{1}{4}\iiiint\!\!\iint f^{(0)} f_1^{(0)} (\phi + \phi_1 - \phi' - \phi_1') \times (\psi + \psi_1 - \psi' - \psi_1')\, g b\, db\, d\epsilon\, dc_1\, d\Omega_1\, dc\, d\Omega \tag{11.5, 10}$$

* The use of such integrated equations is equivalent to smoothing as in 11.23.

† If one of ϕ and ψ is even in the momenta, and the other is odd, by symmetry

$$n^2[\phi, \psi] = -\iiiint\!\!\iint f^{(0)} f_1^{(0)} \phi(\psi' + \psi_1')\, g b\, db\, d\epsilon\, dc_1\, d\Omega_1\, dc\, d\Omega,$$

where, to avoid convergence difficulties, the integral is cut off where $\chi < \delta$, as in 3.6. By using reverse collisions, it follows that in this case $[\psi, \phi] = -[\phi, \psi]$.

(cf. (4.4, 8)), so that $[\phi, \psi] = [\psi, \phi]$: hence $a_{12} = a_{21}$. The factor $N-3$ in (11.5, 6) comes from (11.33, 11).

The thermal flux-vector is given by

$$\begin{aligned} \boldsymbol{q} &= -\iint f^{(0)}(\boldsymbol{A}.\nabla \ln T)E\boldsymbol{C}dc\,d\Omega \\ &= -\tfrac{1}{3}\nabla \ln T \iint f^{(0)}E\boldsymbol{A}.\boldsymbol{C}dc\,d\Omega \end{aligned}$$

by (11.5, 4) and (1.42, 4). Since $\boldsymbol{q} = -\lambda\nabla T$, the first approximation to λ is found to be

$$\begin{aligned} [\lambda]_1 &= \frac{1}{3T}\iint f^{(0)}E\boldsymbol{A}.\boldsymbol{C}dc\,d\Omega \\ &= \frac{k^2T}{2m}(5a_1+(N-3)a_2) \\ &= \frac{3k^2T}{8m}\frac{25a_{22}-10(N-3)a_{12}+(N-3)^2a_{11}}{a_{11}a_{22}-a_{12}^2} \end{aligned} \tag{11.5, 11}$$

on using (11.5, 4–6) and (11.33, 11).

In the bracket integrals a_{11}, a_{12}, a_{22} the velocities* $\boldsymbol{c}, \boldsymbol{c}_1$ are now replaced by $\mathscr{G}_0, \boldsymbol{g}$, defined as in (9.2, 6), the symmetric expression (11.5, 10) being used (since E is an even function of momenta). An integration with respect to $\mathscr{G}_0$ then follows; the resulting expressions are simplified by use of the relation

$$\mathscr{E}^{(i)}+\mathscr{E}_1^{(i)}-\mathscr{E}^{(i)\prime}-\mathscr{E}_1^{(i)\prime} = -(g^2-g'^2),$$

which expresses the conservation of the total energy at collision. The final results have the form

$$a_{11} = \tfrac{7}{2}a+a_{11}', \quad a_{12} = -\tfrac{5}{2}a, \quad a_{22} = \tfrac{3}{2}a+a_{22}'. \tag{11.5, 12}$$

Here

$$a = \frac{\pi^{\frac{1}{2}}}{Z^2}\left(\frac{kT}{m}\right)^3\iiiint\!\!\int \exp(-g^2-\mathscr{E}^{(i)}-\mathscr{E}_1^{(i)})\{g^2-g'^2\}^2 gb\,db\,de\,d\Omega\,d\Omega_1\,d\boldsymbol{g}, \tag{11.5, 13}$$

$$a_{11}' = \frac{2\pi^{\frac{1}{2}}}{Z^2}\left(\frac{kT}{m}\right)^3\iiiint\!\!\int \exp(-g^2-\mathscr{E}^{(i)}-\mathscr{E}_1^{(i)}) \times\{g^4-2g^2g'^2\cos^2\chi+g'^4\}gb\,db\,de\,d\Omega\,d\Omega_1\,d\boldsymbol{g}, \tag{11.5, 14}$$

$$a_{22}' = \frac{\pi^{\frac{1}{2}}}{Z^2}\left(\frac{kT}{m}\right)^3\iiiint\!\!\int \exp(-g^2-\mathscr{E}^{(i)}-\mathscr{E}_1^{(i)})\{g^2(\mathscr{E}^{(i)}-\mathscr{E}_1^{(i)})^2 -2gg'\cos\chi(\mathscr{E}^{(i)}-\mathscr{E}_1^{(i)})(\mathscr{E}^{(i)\prime}-\mathscr{E}_1^{(i)\prime})+g'^2(\mathscr{E}^{(i)\prime}-\mathscr{E}_1^{(i)\prime})^2\}gb\,db\,de\,d\Omega\,d\Omega_1\,d\boldsymbol{g}; \tag{11.5, 15}$$

also χ is the deflection of the relative velocity at collision, so that

$$\boldsymbol{g}.\boldsymbol{g}' = gg'\cos\chi.$$

* As in 4.1, the coordinates, velocities and momenta of the two colliding molecules are distinguished by being written as $\boldsymbol{Q}, \boldsymbol{c}, \boldsymbol{P}$ and $\boldsymbol{Q}_1, \boldsymbol{c}_1, \boldsymbol{P}_1$ respectively.

Symmetry arguments similar to those used in establishing the identity of (11.5, 9, 10) imply that the integrals are unaltered if, in the integrands, g'^4 and $g'^2(\mathscr{E}^{(i)'}-\mathscr{E}_1^{(i)'})^2$ are replaced by g^4 and $g^2(\mathscr{E}^{(i)}-\mathscr{E}_1^{(i)})^2$.

11.51. Viscosity: volume viscosity

By (11.4, 8) and (11.41, 1), the equations satisfied by B and B for a simple gas are

$$J(\mathsf{B}) = f^{(0)}\overset{\circ}{\mathscr{C}\mathscr{C}}, \tag{11.51, 1}$$

$$J(B) = f^{(0)}\left\{\tfrac{1}{3}(\mathscr{C}^2-\tfrac{3}{2})-\frac{1}{N}(\mathscr{E}-\bar{\mathscr{E}})\right\}. \tag{11.51, 2}$$

The first approximations to B and B are taken to be

$$n\mathsf{B} = b_1\overset{\circ}{\mathscr{C}\mathscr{C}}, \tag{11.51, 3}$$

$$nB = b_2\left\{\tfrac{1}{3}(\mathscr{C}^2-\tfrac{3}{2})-\frac{1}{N}(\mathscr{E}-\bar{\mathscr{E}})\right\}. \tag{11.51, 4}$$

The first of these corresponds to the first approximation of Chapter 7. The second can be shown (using (11.33, 10)) to satisfy equations (11.41, 4, 7), and it is the simplest expression that does so.

The coefficients b_1, b_2 are determined from the equations derived by multiplying (11.51, 1 and 2) by

$$\overset{\circ}{\mathscr{C}\mathscr{C}}\,d\boldsymbol{c}\,d\boldsymbol{\Omega} \quad\text{and}\quad \left\{\tfrac{1}{3}(\mathscr{C}^2-\tfrac{3}{2})-\frac{1}{N}(\mathscr{E}-\bar{\mathscr{E}})\right\}d\boldsymbol{c}\,d\boldsymbol{\Omega}$$

and integrating. This gives

$$b_1 b_{11} = \tfrac{5}{2}, \quad b_2 b_{22} = (N-3)/6N, \tag{11.51, 5}$$

where
$$b_{11} = [\overset{\circ}{\mathscr{C}\mathscr{C}}, \overset{\circ}{\mathscr{C}\mathscr{C}}],$$

$$b_{22} = \left[\tfrac{1}{3}(\mathscr{C}^2-\tfrac{3}{2})-\frac{1}{N}(\mathscr{E}-\bar{\mathscr{E}}),\ \tfrac{1}{3}(\mathscr{C}^2-\tfrac{3}{2})-\frac{1}{N}(\mathscr{E}-\bar{\mathscr{E}})\right].$$

In these bracket integrals, expressed in the symmetric form of (11.5, 10), the velocities $\boldsymbol{c}_1$, $\boldsymbol{c}_2$ are replaced by $\mathscr{G}_0$, $\boldsymbol{g}$, as in 11.5. Integration over all values of $\mathscr{G}_0$ then gives

$$b_{11} = a'_{11}-\tfrac{2}{3}a, \quad b_{22} = \tfrac{2}{9}a, \tag{11.51, 6}$$

where a and a'_{11} are the integrals of (11.5, 13, 14).

The second approximation to f makes a contribution $\mathsf{p}^{(1)}$ to the pressure tensor, where

$$\mathsf{p}^{(1)} = -2\iint f^{(0)}m\boldsymbol{CC}(\mathsf{B}:\nabla\boldsymbol{c}_0+B\nabla\cdot\boldsymbol{c}_0)\,d\boldsymbol{c}\,d\boldsymbol{\Omega}.$$

This depends on space-derivatives of $\boldsymbol{c}_0$ only in the combinations $\overset{\circ}{\overline{\overline{\nabla\boldsymbol{c}_0}}}\,(\equiv\mathsf{e})$ and $\nabla\cdot\boldsymbol{c}_0$. Since integration cannot introduce further tensors other than the unit tensor U, the integration yields a result of the form

$$\mathsf{p}^{(1)} = -2\mu\mathsf{e}-\varpi\mathsf{U}\nabla\cdot\boldsymbol{c}_0. \tag{11.51, 7}$$

Here μ denotes the usual 'shear' viscosity; ϖ is the 'volume' viscosity of 11.1, measuring the resistance of the gas to changes in density. Using (1.42, 2) and the theorem of 1.421, the first approximations to μ and ϖ corresponding to (11.51, 3, 4) are found to be

$$[\mu]_1 = \tfrac{2}{5}kT\iint f^{(0)}\mathscr{C}\mathscr{C}:\mathsf{B}\,dc\,d\Omega$$

$$= 5kT/2b_{11}, \qquad (11.51, 8)$$

$$[\varpi]_1 = \frac{4}{3}\frac{kT}{n}\iint f^{(0)}\mathscr{C}^2 b_2\left\{\tfrac{1}{3}(\mathscr{C}^2-\tfrac{3}{2})-\frac{1}{N}(\mathscr{E}-\bar{\mathscr{E}})\right\}dc\,d\Omega = \frac{1}{9}\left(\frac{N-3}{N}\right)^2\frac{kT}{b_{22}}.$$

$$(11.51, 9)$$

11.511. Volume viscosity and relaxation

The volume viscosity introduced above may be interpreted as a relaxation phenomenon. It arises because in an expansion or contraction the work done by the pressure alters the translatory energy immediately, but affects the internal energy (through inelastic collisions) only after a certain time-lag.

We can introduce temperatures T_e, T_i corresponding to the translatory and internal energies, differing slightly from T; these are such that $\tfrac{1}{2}m\overline{C^2}$ and $\overline{E^{(i)}}$ have the same values as in a gas in a uniform steady state at temperature T_e or T_i. The specific heats c_v', c_v'' corresponding to the translatory and internal energies are $3k/2m$ and $(N-3)k/2m$; hence if the temperatures T_e and T_i are to give the correct total energy for the overall temperature T,

$$3(T_e-T)+(N-3)(T_i-T) = 0, \qquad (11.511, 1)$$

$$N\frac{DT}{Dt} = 3\frac{DT_e}{Dt}+(N-3)\frac{DT_i}{Dt}. \qquad (11.511, 2)$$

Collisions may be assumed to transfer energy from translatory to internal motions at a rate proportional to T_e-T_i. Thus, ignoring the effects of heat conduction and shear viscosity, the equations of change of T_e, T_i can be written down separately in the forms

$$\tfrac{3}{2}kn\frac{DT_e}{Dt} = -p\nabla.\boldsymbol{c}_0-\frac{N-3}{2\tau}kn(T_e-T_i), \qquad (11.511, 3)$$

$$\frac{N-3}{2}kn\frac{DT_i}{Dt} = \frac{N-3}{2\tau}kn(T_e-T_i), \qquad (11.511, 4)$$

where $p = nkT_e$. The quantity τ is a time of relaxation; as (11.511, 4) implies, in a non-normal state in which T_e is maintained constant, $T-T_e$ decays exponentially, with decay time τ.

Equations (11.511, 3, 4) can be combined to give

$$\frac{D}{Dt}(T_e-T_i) = -\tfrac{2}{3}T_e\nabla.\boldsymbol{c}_0-\frac{N}{3\tau}(T_e-T_i).$$

In a normal expansion, the time-derivative on the left is small compared with the terms on the right; hence

$$T_e - T_i = -\frac{2}{N}\tau T_e \nabla . \boldsymbol{c}_0,$$

or, using (11.511, 1), and neglecting the small difference between T_e and T in the already small term on the right,

$$T_e - T = -\frac{2(N-3)}{N^2}\tau T \nabla . \boldsymbol{c}_0. \qquad (11.511, 5)$$

Thus the effect of relaxation is to alter p from $p_0 \equiv nkT$ to

$$p = nkT_e = p_0\left(1 - \frac{2(N-3)}{N^2}\tau \nabla . \boldsymbol{c}_0\right),$$

i.e. to introduce a volume viscosity ϖ such that

$$\varpi = \frac{2(N-3)}{N^2} p_0 \tau. \qquad (11.511, 6)$$

Agreement with (11.51, 9, 6) is obtained by assuming that, to a first approximation, τ is given by

$$\tau = \frac{1}{18}\frac{N-3}{nb_{22}} = \frac{N-3}{4na}. \qquad (11.511, 7)$$

Formulae like (11.511, 5) or (11.511, 7) are valid only when $T_e - T_i$ is small, i.e. when τ is small compared with the time-scale of expansion. For very rapid temperature-changes it is necessary to revert to equations like (11.511, 3) and (11.511, 4), which still remain approximately true. If τ is very large compared with the time-scale of expansion, (11.511, 3) approximates to the temperature equation for a monatomic gas.

Equation (11.511, 6) can be generalized to apply when there are several different parts of the internal energy, each with its own relaxation time. If each part of the internal energy is supposed to interact only with the translatory energy, the generalized form is

$$\varpi = \frac{2p_0}{N^2}(N_1\tau_1 + N_2\tau_2 + \ldots). \qquad (11.511, 8)$$

Here N_1, N_2, ... are the contributions to N from the separate parts of $E^{(i)}$, and τ_1, τ_2, ... are the corresponding relaxation times. As before, the interpretation of the relaxation effect simply as a volume viscosity is justified only if all the relaxation times are small compared with the time-scale of expansion.

11.52. Diffusion

The coefficient of diffusion in a binary mixture depends on the vectors $\boldsymbol{D}_s$ of (11.41, 1, 2). These satisfy the equations

$$J_1(\boldsymbol{D}) = x_1^{-1} f_1^{(0)} \boldsymbol{C}_1, \quad J_2(\boldsymbol{D}) = -x_2^{-1} f_2^{(0)} \boldsymbol{C}_2. \qquad (11.52, 1)$$

A first approximation $[D_{12}]_1$ is obtained by putting

$$x_1 \boldsymbol{D}_1 = d_0 \boldsymbol{C}_1, \quad x_2 \boldsymbol{D}_2 = -d_0 \boldsymbol{C}_2, \tag{11.52, 2}$$

where d_0 is a constant; these forms satisfy (11.41, 6). The value of d_0 is found by substituting these expressions into the equation

$$x_1^{-1}\iint \boldsymbol{C}_1 . J_1(\boldsymbol{D})\, d\boldsymbol{c}_1 d\Omega_1 - x_2^{-1}\iint \boldsymbol{C}_2 . J_2(\boldsymbol{D})\, d\boldsymbol{c}_2 d\Omega_2 = \sum_s x_s^{-2}\iint f_s^{(0)} C_s^2 d\boldsymbol{c}_s d\Omega_s, \tag{11.52, 3}$$

which replaces (11.52, 1). The right-hand side of (11.52, 3) reduces to $3kT\rho n^2/(\rho_1\rho_2)$ on integration. On the left, the conservation of momentum secures the vanishing of the contributions from the parts of $J_1(\boldsymbol{D})$, $J_2(\boldsymbol{D})$ corresponding to the collisions of like molecules. In the remaining integrals on the left, a symmetry relation like (4.4, 10) is used, and the integration with respect to $\mathscr{G}_0$ is then performed, as in 11.5 and 11.51. Then (11.52, 3) reduces to

$$d_0 = \frac{3m_0}{8\rho d_{00}}, \tag{11.52, 4}$$

where

$$d_{00} = \frac{(kT)^3 \pi^{\frac{1}{2}}}{(m_1 m_2)^{\frac{3}{2}} Z_1 Z_2}\iiiint\!\!\int \exp(-\mathscr{g}^2 - \mathscr{E}_1^{(i)} - \mathscr{E}_2^{(i)}) \times (\mathscr{g}^2 - 2\mathscr{g}\mathscr{g}'\cos\chi + \mathscr{g}'^2)\mathscr{g} b\, db\, d\epsilon\, d\Omega_1 d\Omega_2 d\boldsymbol{\mathscr{g}}. \tag{11.52, 5}$$

The velocity of diffusion $\overline{\boldsymbol{C}}_1 - \overline{\boldsymbol{C}}_2$ corresponding to (11.52, 5) is readily found to be $-n\rho d_0 kT \boldsymbol{d}_1/\rho_1\rho_2$. Hence (cf. (8.4, 7))

$$[D_{12}]_1 = \frac{\rho k T d_0}{n m_1 m_2} = \frac{3m_0 kT}{8 n m_1 m_2 d_{00}}. \tag{11.52, 6}$$

The coefficient of self-diffusion in a simple gas is obtained from (11.52, 5, 6) by setting $m_1 = m_2$, $Z_1 = Z_2$, etc.

The formulae governing thermal diffusion for molecules with internal energy are very complicated, even to a first approximation. The theory for molecules without internal energy is usually assumed (with a measure of justification) to apply approximately to molecules with internal energy also.

11.6. Rough spheres

These results are first applied to the rough elastic spherical molecules of Bryan and Pidduck, these being by definition such that at a collision the relative velocity of the spheres at their point of impact is exactly reversed. It can readily be verified that the total energy is conserved at collisions of such molecules.

The motion of a molecule m_s can now be specified completely by $\boldsymbol{c}_s$, $\boldsymbol{\omega}_s$, where $\boldsymbol{c}_s$ is the velocity of its centre and $\boldsymbol{\omega}_s$ its angular velocity. The components of $\boldsymbol{\omega}_s$ are not themselves Lagrangian velocities. Nevertheless, a theory exactly analogous to that given above can be constructed in terms of a generalized velocity-distribution function $f_s(\boldsymbol{r}, \boldsymbol{c}_s, \boldsymbol{\omega}_s)$, such that the number of molecules m_s per unit volume with velocities and angular velocities in the ranges $d\boldsymbol{c}_s$, $d\boldsymbol{\omega}_s$ is $f_s d\boldsymbol{c}_s d\boldsymbol{\omega}_s$. The Boltzmann equation satisfied by f_s in a binary gas-mixture takes the form

$$\frac{\partial f_s}{\partial t}+\boldsymbol{c}_s.\frac{\partial f_s}{\partial \boldsymbol{r}}+\boldsymbol{F}_s.\frac{\partial f_s}{\partial \boldsymbol{c}_s} = \sum_t \iiint (f_s' f_t' - f_s f_t)\, \sigma_{st}^2 \boldsymbol{g}_{ts}.\mathbf{k}\, d\mathbf{k}\, d\boldsymbol{c}_t\, d\boldsymbol{\omega}_t. \tag{11.6, 1}$$

Here $\boldsymbol{g}_{ts} = \boldsymbol{c}_t - \boldsymbol{c}_s$, $\sigma_{st} = \frac{1}{2}(\sigma_s+\sigma_t)$, where σ_s is the diameter of a molecule m_s; also $f_s' = f_s(\boldsymbol{c}_s', \boldsymbol{\omega}_s')$, $f_t' = f_t(\boldsymbol{c}_t', \boldsymbol{\omega}_t')$, where $\boldsymbol{c}_s'$, $\boldsymbol{\omega}_s'$ and $\boldsymbol{c}_t'$, $\boldsymbol{\omega}_t'$ are the velocities and angular velocities before a collision of a pair of molecules whose velocities and angular velocities after collision are $\boldsymbol{c}_s$, $\boldsymbol{\omega}_s$ and $\boldsymbol{c}_t$, $\boldsymbol{\omega}_t$; and $\mathbf{k}$ denotes a unit vector in the direction of the line from the centre of the molecule m_s to that of the molecule m_t at collision, so that $\boldsymbol{g}_{ts}.\mathbf{k} > 0$.

Let I_1, I_2 be the moments of inertia of molecules m_1, m_2 about their diameters. We write

$$\kappa_1 = 4I_1/m_1\sigma_1^2, \quad \kappa_2 = 4I_2/m_2\sigma_2^2, \quad m_0\kappa_0 = m_1\kappa_1 + m_2\kappa_2. \tag{11.6, 2}$$

The values of κ_1, κ_2 are 0·4 for uniform spheres. As the effective radius of an actual molecule is determined by its outworks, and the moment of inertia by the disposition of the atomic nuclei, κ_1, κ_2 are likely to be less than 0·4 in all practical cases. In terms of κ_0, κ_1, κ_2 it can be shown that, in a collision of molecules m_1 and m_2,

$$\boldsymbol{c}_1' = \boldsymbol{c}_1 + 2M_2\{\kappa_1\kappa_2\mathbf{V} + \kappa_0\mathbf{k}(\mathbf{k}.\mathbf{V})\}/(\kappa_1\kappa_2+\kappa_0), \tag{11.6, 3}$$

$$\boldsymbol{c}_2' = \boldsymbol{c}_2 - 2M_1\{\kappa_1\kappa_2\mathbf{V} + \kappa_0\mathbf{k}(\mathbf{k}.\mathbf{V})\}/(\kappa_1\kappa_2+\kappa_0), \tag{11.6, 4}$$

$$\boldsymbol{\omega}_1' = \boldsymbol{\omega}_1 + 4M_2\kappa_2\mathbf{k}\wedge\mathbf{V}/\sigma_1(\kappa_1\kappa_2+\kappa_0), \tag{11.6, 5}$$

$$\boldsymbol{\omega}_2' = \boldsymbol{\omega}_2 + 4M_1\kappa_1\mathbf{k}\wedge\mathbf{V}/\sigma_2(\kappa_1\kappa_2+\kappa_0). \tag{11.6, 6}$$

In these equations $\mathbf{V}$ is the relative velocity after collision of the points of the spheres which come into contact, i.e.

$$\mathbf{V} = \boldsymbol{c}_2 + \tfrac{1}{2}\sigma_2\mathbf{k}\wedge\boldsymbol{\omega}_2 - \boldsymbol{c}_1 + \tfrac{1}{2}\sigma_1\mathbf{k}\wedge\boldsymbol{\omega}_1. \tag{11.6, 7}$$

In the uniform steady state,

$$f_s = n_s\frac{(m_s I_s)^{\frac{3}{2}}}{(2\pi kT)^3}\exp\{-(m_s C_s^2 + I_s\omega_s^2)/2kT\}, \tag{11.6, 8}$$

and so (cf. (11.34)), $N_s' = N_s = 6$, and $\gamma = \frac{4}{3}$. This value is closely approached by methane and ammonia, for which γ is 1·310 and 1·318 at ordinary temperatures; methane is well represented by a spherical molecule, because of the symmetric arrangement of its hydrogen atoms.

In the limit as κ_1, κ_2 (and in consequence also I_1, I_2) tend to zero, the energies of translation and rotation cease to be interconvertible. For, according to the equipartition principle, $\frac{1}{2}m_s\overline{C_s^2} = \frac{1}{2}I_s\overline{\omega_s^2} = \frac{1}{8}m_s\kappa_s\sigma_s^2\overline{\omega_s^2}$; thus in the limit as $\kappa_s \to 0$ we must regard $\boldsymbol{c}_s$ as small compared with $\frac{1}{2}\sigma_s\boldsymbol{\omega}_s$, but large compared with $\frac{1}{2}\kappa_s\sigma_s\boldsymbol{\omega}_s$. Thus, after substitution from (11.6, 7), equations (11.6, 3–6) approximate to

$$\boldsymbol{c}_1' = \boldsymbol{c}_1 + 2M_2\mathbf{k}(\mathbf{k}.\boldsymbol{g}_{21}), \quad \boldsymbol{c}_2' = \boldsymbol{c}_2 - 2M_1\mathbf{k}(\mathbf{k}.\boldsymbol{g}_{21}),$$

$$\boldsymbol{\omega}_1' = \boldsymbol{\omega}_1 + (2M_2\kappa_2/\kappa_0\sigma_1)\mathbf{k}\wedge[\mathbf{k}\wedge(\sigma_1\boldsymbol{\omega}_1+\sigma_2\boldsymbol{\omega}_2)],$$

$$\boldsymbol{\omega}_2' = \boldsymbol{\omega}_2 + (2M_1\kappa_1/\kappa_0\sigma_2)\mathbf{k}\wedge[\mathbf{k}\wedge(\sigma_1\boldsymbol{\omega}_1+\sigma_2\boldsymbol{\omega}_2)],$$

showing that the linear and angular velocities have become uncoupled. However, the changes in angular velocity remain significant, even though they do not affect the linear motions.

11.61. Transport coefficients for rough spheres

The collision integrals a_{11}', a, a_{22}' and d_{00} of 11.5 and 11.52 can readily be calculated for rough spheres,* using (11.6, 3–6), with $gb\,db\,d\epsilon$ replaced by $\sigma^2\boldsymbol{g}_{21}.\mathbf{k}\,d\mathbf{k}$ or $\sigma_{12}^2\boldsymbol{g}_{21}.\mathbf{k}\,d\mathbf{k}$. Their values are

$$a_{11}' = \sigma^2\left(\frac{\pi kT}{m}\right)^{\frac{1}{2}}\frac{4(2+5\kappa)}{(1+\kappa)^2}, \qquad a = \sigma^2\left(\frac{\pi kT}{m}\right)^{\frac{1}{2}}\frac{4\kappa}{(1+\kappa)^2},$$

$$a_{22}' = \sigma^2\left(\frac{\pi kT}{m}\right)^{\frac{1}{2}}\frac{6(1+\kappa+2\kappa^2)}{(1+\kappa)^2}, \quad d_{00} = \sigma_{12}^2\left(\frac{2\pi kTm_0}{m_1m_2}\right)^{\frac{1}{2}}\frac{\kappa_0+2\kappa_1\kappa_2}{\kappa_0+\kappa_1\kappa_2}.$$

The corresponding first approximations to the transport coefficients are given by

$$[\lambda]_1 = \frac{9}{16\sigma^2}\left(\frac{k^3T}{\pi m}\right)^{\frac{1}{2}}\frac{(1+\kappa)^2(37+151\kappa+50\kappa^2)}{12+75\kappa+101\kappa^2+102\kappa^3}, \tag{11.61, 1}$$

$$[\mu]_1 = \frac{15}{8\sigma^2}\left(\frac{kTm}{\pi}\right)^{\frac{1}{2}}\frac{(1+\kappa)^2}{6+13\kappa}, \tag{11.61, 2}$$

$$[\varpi]_1 = \frac{1}{32\sigma^2}\left(\frac{kTm}{\pi}\right)^{\frac{1}{2}}\frac{(1+\kappa)^2}{\kappa}, \tag{11.61, 3}$$

$$[D_{12}]_1 = \frac{3}{8n\sigma_{12}^2}\left(\frac{kTm_0}{2\pi m_1m_2}\right)^{\frac{1}{2}}\frac{\kappa_0+\kappa_1\kappa_2}{\kappa_0+2\kappa_1\kappa_2}. \tag{11.61, 4}$$

The viscosity given by (11.61, 2) equals that for smooth spheres if $\kappa = 0$ (when the translatory and rotational motions become independent); it is 1·05 times as great for $\kappa = 0{\cdot}4$, but the two differ by less than one per cent if $\kappa < 0{\cdot}2$. The thermal conductivity given by (11.61, 1) is always greater than that for smooth spheres, because of the transport of internal energy; the

* In practice it is normally best to use the unsymmetric expression (11.5, 9) in calculating the bracket integrals.

ratio of the two increases steadily from 1·48 to 1·533 as κ increases from zero to 0·4. The increase comes from the increased transport of rotational energy, due to the tendency of rough spheres to knock back molecules with which they collide, and at the same time to transfer part of their rotational energy to them. In fact, the part of $[\lambda]_1$ due to the transport of translatory energy (corresponding to the a_1 term in (11.5, 4)) is found to decrease in the ratio 0·916 as κ increases from zero to 0·4, whereas the part due to the rotational energy increases in the ratio 1·28. For rough spheres $N = 6$ and so $c_v = 3k/m$ (cf. (11.33, 6)). Using this, the ratio $[f]_1 = [\lambda]_1/\{[\mu]_1 c_v\}$ is found to equal 1·85 for $\kappa = 0$, and it lies between 1·87 and 1·825 for values of κ between zero and 0·4.

As Kohler* first pointed out, the volume viscosity $[\varpi]_1$ given by (11.61, 3) is not small compared with $[\mu]_1$; it is $7[\mu]_1/15$ when κ has its greatest value 0·4, and becomes large compared with $[\mu]_1$ when κ is small. This agrees with the relaxation expression (11.511, 6), because the interchange between rotational and translatory energies becomes indefinitely slow as $\kappa \to 0$.

Finally, the diffusion coefficient given by (11.61, 4) agrees with that for smooth spheres if either κ_1 or κ_2 is zero, but becomes smaller as κ_1 and κ_2 increase; if $\kappa_1 = \kappa_2 = \kappa_0 = 0{\cdot}4$, it is only seven-ninths as large. Its decrease is due to the increased deflections of the relative velocity at collision as κ_1 and κ_2 increase.

The terms involving $\boldsymbol{h}_s \wedge \boldsymbol{C}_s$ and $\boldsymbol{h}_s \wedge (\boldsymbol{h}_s \wedge \boldsymbol{C}_s)$ in (11.41, 3) and similar relations may appreciably increase the calculated values of the transport coefficients above the first approximations (11.61, 1–4). Condiff, Lu and Dahler,† using an approximation to $\boldsymbol{A}$ involving terms up to the fifth degree in $\boldsymbol{C}$ and $\boldsymbol{\omega}$, found values of λ ranging from $1{\cdot}066[\lambda]_1$ to about $1{\cdot}1[\lambda]_1$ as κ increases from 0 to 0·4. Similarly, taking into account terms of fourth degree in B, they found values of μ ranging from $1{\cdot}015[\mu]_1$ to about $1{\cdot}06[\mu]_1$ as κ increases from 0 to 0·4. They also calculated second and third approximations to ϖ; the third approximation is equal to $[\varpi]_1$ when κ is negligibly small, but about 3 per cent greater when $\kappa = 0{\cdot}4$. They remark, however, that the effect of the $\boldsymbol{h}_s \wedge \boldsymbol{C}_s$ and $\boldsymbol{h}_s \wedge (\boldsymbol{h}_s \wedge \boldsymbol{C}_s)$ terms is likely to be greater for rough spheres than for more realistic models.

11.62. Defects of the model

The Bryan–Pidduck rough-sphere model is unsatisfactory in more than one respect. First, as with all rigid models, the deflection of the relative velocity at the collision of two molecules depends only on the ratio of their velocities, and not on the actual speeds. Secondly, because of the reversal of the relative

* M. Kohler, *Z. Phys.* **124**, 757 (1947); **125**, 715 (1949).

† D. W. Condiff, W.-K. Lu and J. S. Dahler, *J. Chem. Phys.* **42**, 3445 (1965). L. Waldmann (*Z. Naturf.* **18a**, 1033 (1963)), retaining only the term $\boldsymbol{h}_s \wedge \boldsymbol{C}_s$, found for a Lorentzian gas a decrease in D_{12} below $[D_{12}]_1$; however, this is replaced by an increase if the $\boldsymbol{h}_s \wedge (\boldsymbol{h}_s \wedge \boldsymbol{C}_s)$ term is also retained.

velocity of the points coming into contact, even a grazing collision can produce a large deflection. Finally, experiment suggests a relatively slow interchange between translatory and rotational energies; such a slow interchange is obtained by taking κ_1, κ_2 small, but (contrary to what one would expect of actual molecules) this still permits free interchange of rotational energy between the molecules.

Chapman and Hainsworth* attempted to modify the model to meet the first of these objections, but their attempt was itself open to serious objection. The third objection can be met by assuming the molecules to be rough in only a fraction of the collisions, and smooth in the rest; but this assumption is artificial. The second objection appears to be insuperable. Thus, despite its interest as giving almost the only exact results for gases whose molecules possess internal energy, the rough-sphere model cannot represent the behaviour of actual gases at all closely.

11.7. Spherocylinders

Curtiss and Muckenfuss† have given formulae from which the transport coefficients can be calculated for any smooth *rigid* convex form of molecule without an external field of force. They worked out explicit expressions for spherocylinders—cylinders of length L and diameter σ, with hemispheres, also of diameter σ, fixed to the plane ends. Their formulae involve two parameters, $\beta \equiv L/\sigma$ and $\kappa \equiv 4I/m\sigma^2$, where I is the total moment of inertia about a line through the centre of the cylinder and perpendicular to its axis. When $\beta = 0$ the molecules are spherical.

The formulae, though involving only elementary functions, are too complicated in form to be quoted here, and indeed Curtiss and Muckenfuss did not give them all explicitly. Their results can be summarized roughly as follows. The effective collision cross-sections in integrals like the a'_{11}, a'_{22} of 11.5 are larger than those for spherical molecules of the same volume, unless κ is small, by an amount which increases with increasing β (i.e. with increasing departure from the spherical form). However, if κ is small the impulsive forces between molecules at collision can produce large changes in the angular velocity without appreciably affecting the motion of the centre, and so the transport of momentum and translatory energy are both increased. Thus the transport coefficients are less than those for smooth spheres unless κ is small, but greater if κ is small (for λ the comparison is with the modified Eucken formula (11.8, 4)). The volume viscosity ϖ is always greater than the shear viscosity for smooth spheres of the same volume. The ratio of the

* S. Chapman and W. Hainsworth, *Phil. Mag.* **48**, 593 (1924). The effective diameter was supposed to vary with g', the relative velocity of approach of colliding molecules. However, since g' is not conserved at collisions, this means that direct collisions cannot be balanced against reverse collisions, as in 11.3.

† C. F. Curtiss, *J. Chem. Phys.* **24**, 225 (1956); C. F. Curtiss and C. Muckenfuss, *ibid.* **26**, 1619 (1957), and **29**, 1257 (1958).

two is very large if κ is small, and also if β is small, i.e. if the molecules are nearly spheres; in either case the interchange between rotational and translatory energies is slow.

11.71. Loaded spheres

For the loaded sphere model Dahler and his collaborators* obtained results depending on the one parameter $a = m\xi^2/2I$, where ξ is the distance between the mass-centre G and the geometrical centre O, and I is the moment of inertia about an axis through G perpendicular to OG. They found that the ratio $\mathrm{f} \equiv \lambda/\mu c_v$ has approximately the value 1·98 given by the modified Eucken formula (see 11.8, below) for all physically admissible values of a, a decrease in the part contributed by translational energy being nearly balanced by an increase in the contribution of internal energy. They also found that different eccentricities of mass-distribution can produce a non-vanishing thermal diffusion factor in a mixture of two gases whose molecules are identical in mass and radius.

11.8. Nearly smooth molecules: Eucken's formula

Nearly smooth molecules are defined as such that a large number of collisions are needed in order to produce any considerable change in the internal energies.

Eucken† suggested an approximate expression for the thermal conductivity, which strictly is applicable only to a gas of nearly smooth molecules. He divided λ into two non-interacting parts λ' and λ'', which are respectively the conductivities due to the transport of translatory and internal energy; similarly he divided c_v into parts c'_v and c''_v. For a monatomic gas, λ is very nearly equal to $\frac{5}{2}\mu c_v$ (cf. (9.7, 4)); Eucken assumed by analogy that $\lambda' = \frac{5}{2}\mu c'_v$. On the other hand, since there is little correlation between the speed of a molecule and its internal energy, the argument given at the end of 6.3 suggests that $\lambda'' = \mu c''_v$, and Eucken assumed this; the assumption is equivalent to supposing that the mean free paths effective in the transport of momentum and internal energy are equal. Thus Eucken wrote

$$\lambda = \mu(\tfrac{5}{2}c'_v + c''_v). \tag{11.8, 1}$$

Now (cf. (11.33, 6, 7)) c_v, c'_v and c''_v are given by

$$c_v = \frac{Nk}{2m}, \quad c'_v = \frac{3k}{2m}, \quad c''_v = \frac{(N-3)k}{2m}, \tag{11.8, 2}$$

* J. S. Dahler and N. F. Sather, *J. Chem. Phys.* **35**, 2029 (1961), and **38**, 2363 (1962); S. I. Sandler and J. S. Dahler, *J. Chem. Phys.* **43**, 1750 (1965), **46**, 3520 (1967), and **47**, 2621 (1967).

† A. Eucken, *Phys. Z.* **14**, 324 (1913).

and if γ is the ratio of specific heats, $N = 2/(\gamma - 1)$. Hence (11.8, 1) is equivalent to

$$\lambda = \tfrac{1}{4}(9\gamma - 5)\,\mu c_v. \tag{11.8, 3}$$

This is known as the Eucken formula.

However, for nearly smooth molecules the transport of internal energy takes place by the diffusion of molecules from one part of the gas to another, carrying with them the mean internal energy of the region in which they originate. Thus the free path effective in the transport is that effective in diffusion, and one should write $\lambda'' = \rho D_{11} c_v''$. Hence (11.8, 1) should be replaced by

$$\lambda = \tfrac{5}{2}\mu c_v' + \rho D_{11} c_v''. \tag{11.8, 4}$$

Putting $\rho D_{11} = \mathrm{u}_{11}'\mu$ (cf. (6.4, 4)) one then finds instead of (11.8, 3)

$$\lambda = \tfrac{1}{4}\{15(\gamma - 1) + 2\mathrm{u}_{11}'(5 - 3\gamma)\}\,\mu c_v. \tag{11.8, 5}$$

The numerical factor u_{11}' in general exceeds unity. For smooth rigid elastic spheres it is 1·20, for Maxwellian molecules 1·55, and for most of the usual molecular models it lies between these values. Hence (11.8, 5) indicates a somewhat greater value of $\mathrm{f} \equiv \lambda/\mu c_v$ than does (11.8, 3).

The modified Eucken formula (11.8, 4) can also be derived from the general expressions given in 11.5 and 11.52. For nearly smooth molecules we can approximate by writing $g = g'$, $\mathscr{E}^{(i)} = \mathscr{E}^{(i)\prime}$ in the collision-integrals; also we may average functions like $(\mathscr{E}^{(i)} - \mathscr{E}_1^{(i)})^2$ over the internal coordinates and momenta, because $\mathscr{E}^{(i)}$ is not correlated with the translational velocities. Hence the integral a of (11.5, 13) vanishes, and so by (11.5, 12) and (11.51, 6, 8),

$$a_{11} = b_{11} = 5kT/2[\mu]_1, \quad a_{12} = 0. \tag{11.8, 6}$$

Also $a_{22} = a_{22}'$, and by (11.5, 15),

$$a_{22}' = \overline{(\mathscr{E}^{(i)} - \mathscr{E}_1^{(i)})^2}\,\frac{\pi^{\frac{3}{2}}}{Z^2}\left(\frac{kT}{m}\right)^3 \iiiint\!\!\int \exp(-g^2 - \mathscr{E}^{(i)} - \mathscr{E}_1^{(i)})$$
$$\times (g^2 - 2gg'\cos\chi + g'^2)\,gb\,db\,d\epsilon\,dg\,d\Omega\,d\Omega_1. \tag{11.8, 7}$$

But, since $\mathscr{E}^{(i)}$ and $\mathscr{E}_1^{(i)}$ are uncorrelated,

$$\overline{(\mathscr{E}^{(i)} - \mathscr{E}_1^{(i)})^2} = 2\{\overline{(\mathscr{E}^{(i)})^2} - (\overline{\mathscr{E}^{(i)}})^2\} = N - 3,$$

by (11.33, 11); also the integral in (11.8, 7) is the same as that in the diffusion expression d_{00} of (11.52, 5, 6). Hence

$$a_{22} = a_{22}' = (N-3)(3kT/4\rho[D_{11}]_1). \tag{11.8, 8}$$

Using these values of a_{11}, a_{22} in (11.5, 11) we find

$$[\lambda]_1 = \frac{15k}{4m}[\mu]_1 + \frac{(N-3)k}{2m}\rho[D_{11}]_1$$
$$= \tfrac{5}{2}c_v'[\mu]_1 + c_v''\rho[D_{11}]_1, \tag{11.8, 9}$$

by (11.8, 2). This agrees with (11.8, 4), to the first approximation.

11.81. The Mason–Monchick theory

Mason and Monchick* have attempted to improve the modified Eucken formula, assuming that the interchange between translatory and internal energy is slow, but not completely negligible. The integral a of (11.5, 13) is then small; by (11.511, 7), it is related to the first approximation $[\tau]_1$ to the relaxation time of 11.511 by

$$a = (N-3)/4n[\tau]_1. \tag{11.81, 1}$$

According to (11.5, 12) and (11.51, 6, 8)

$$a_{12} = -\tfrac{5}{2}a = -\frac{5}{8}\frac{(N-3)}{n[\tau]_1}, \tag{11.81, 2}$$

$$a_{11} = b_{11} + \tfrac{25}{6}a = \frac{5kT}{2[\mu]_1} + \frac{25}{24}\frac{(N-3)}{n[\tau]_1}. \tag{11.81, 3}$$

In the equation $a_{22} = \frac{3}{2}a + a'_{22}$, the expression (11.8, 8) for a'_{22} is still used, but with $[D_{11}]_1$ replaced by D_{int}; D_{int} is interpreted as the first approximation to a coefficient of diffusion of internal energy. Thus

$$a_{22} = \frac{3(N-3)kT}{4\rho D_{\text{int}}} + \frac{3(N-3)}{8n[\tau]_1}. \tag{11.81, 4}$$

Using these values, and substituting for c'_v, c''_v from (11.8, 2) we find from (11.5, 11) after considerable simplification,

$$[\lambda]_1 = \tfrac{5}{2}[\mu]_1 c'_v + \rho D_{\text{int}} c''_v - \frac{\frac{1}{2}c''_v(\frac{5}{2}[\mu]_1 - \rho D_{\text{int}})^2}{p_0[\tau]_1 + \frac{1}{2}\rho D_{\text{int}} + \frac{5}{12}(N-3)[\mu]_1}. \tag{11.81, 5}$$

Apart from the replacement of $[D_{11}]_1$ by D_{int}, (11.81, 5) differs from the modified Eucken formula (11.8, 4) only through the appearance of the last term on the right. It implies that, if D_{int} is identified with $[D_{11}]_1$, then $[\lambda]_1$ and the ratio $[f]_1 \equiv [\lambda]_1/[\mu]_1 c_v$ are always smaller than is indicated by (11.8, 4).

It is questionable how far D_{int} can be identified with $[D_{11}]_1$, so that a'_{22} can still be given by (11.8, 8). As the form of (11.5, 13) indicates, the collision-integral a depends on the *square* of the energy-interchange between the translatory and internal motions at collision, and it is not clear that (11.8, 8) is correct to terms of second degree in this energy-interchange. In the integral (11.5, 15) the mean values of $(\mathscr{E}^{(i)} - \mathscr{E}_1^{(i)})^2$ and $(\mathscr{E}^{(i)\prime} - \mathscr{E}_1^{(i)\prime})^2$ are equal, because to every direct collision corresponds a reverse collision; but since $(\mathscr{E}^{(i)} - \mathscr{E}_1^{(i)} - \mathscr{E}^{(i)\prime} - \mathscr{E}_1^{(i)\prime})^2$ is essentially positive, the mean value of

$$(\mathscr{E}^{(i)} - \mathscr{E}_1^{(i)})(\mathscr{E}^{(i)\prime} - \mathscr{E}_1^{(i)\prime})$$

* E. A. Mason and L. Monchick, *J. Chem. Phys.* **36**, 1622 (1962).

must be less than that of $(\mathscr{E}^{(i)}-\mathscr{E}^{(i)}_1)^2$. Thus (11.8, 8) gives an adequate approximation to (11.5, 15) only if, for those collisions in which $\mathscr{E}^{(i)\prime}-\mathscr{E}^{(i)\prime}_1$ differs appreciably from $\mathscr{E}^{(i)}-\mathscr{E}^{(i)}_1$, the average value of $gg'\cos\chi$ is nearly zero, i.e. the mean value of χ is about $\frac{1}{2}\pi$.

For most of the realistic molecular models, encounters with $\chi < \frac{1}{2}\pi$ are more important for diffusion than those with $\chi > \frac{1}{2}\pi$. On the other hand, collisions producing appreciable changes in internal energy are likely to be weighted towards the larger values of χ. Hence, in default of detailed analysis, using $D_{\text{int}} = [D_{11}]_1$ in (11.81, 5) may in many cases be a reasonable approximation. The rough spherical model indicates a slower decrease of $[\lambda]_1$ with increasing κ than would be consistent with taking $D_{\text{int}} = [D_{11}]_1$. This is because of the peculiar properties of the model, which make the mean value of $gg'\cos\chi$ for colliding molecules negative (corresponding to a mean χ greater than $\frac{1}{2}\pi$), whereas the mean value of $(\mathscr{E}^{(i)}-\mathscr{E}^{(i)}_1)(\mathscr{E}^{(i)\prime}-\mathscr{E}^{(i)\prime}_1)gg'\cos\chi$ is positive. Such peculiarities are not likely to appear with more realistic models.

Even so, systematic deviations from $D_{\text{int}} = [D_{11}]_1$ are likely with certain molecular models. For example, as Mason and Monchick suggest, molecules with permanent dipole moments may be able to interchange internal energy without the dipole moments appreciably affecting the translatory motion at collision. This would imply a reduction in D_{int} below $[D_{11}]_1$, leading to systematically lower values of λ.

Mason and Monchick in their original paper approximated by assuming both that $D_{\text{int}} = [D_{11}]_1$, and that $[\tau]_1$ is so large that terms of order $[\tau]_1^{-2}$ can be neglected. In this case (11.81, 5) is replaced by

$$[\lambda]_1 = \tfrac{5}{2}[\mu]_1 c_v' + \rho[D_{11}]_1 c_v'' - \tfrac{1}{2}c_v''(\tfrac{5}{2}[\mu]_1 - \rho[D_{11}]_1)^2/p[\tau]_1. \quad (11.81, 6)$$

Equation (11.81, 6) can be generalized to the case when there are several different forms of internal energy, each interacting with translatory energy with its own relaxation time. The effect is simply to replace $c_v''/[\tau]_1$ in the last term on the right by

$$\frac{k}{2m}\left(\frac{N_1}{[\tau_1]_1}+\frac{N_2}{[\tau_2]_1}+\ldots\right),$$

where, as in 11.511, $N_1 k/2m$, $N_2 k/2m$, ... denote the contributions to c_v'' from the separate forms of internal energy. Energy with a large relaxation time makes only a small contribution to this sum, whereas it may make a large contribution to the volume viscosity.

12

VISCOSITY: COMPARISON OF THEORY WITH EXPERIMENT

12.1. Formulae for μ for different molecular models

The various formulae obtained for the coefficient of viscosity μ are grouped together here for convenience of reference. The formulae for $[\mu]_1$, the first approximation to μ, are as follows:

(i) Smooth rigid elastic spherical molecules of diameter σ, (10.21, 1),

$$[\mu]_1 = \frac{5}{16\sigma^2}\left(\frac{kmT}{\pi}\right)^{\frac{1}{2}}. \tag{12.1, 1}$$

(ii) Molecules repelling each other with a force κ/r^ν, (10.32, 1),

$$[\mu]_1 = \frac{5}{8}\left(\frac{kmT}{\pi}\right)^{\frac{1}{2}}\left(\frac{2kT}{\kappa}\right)^{2/(\nu-1)}\Big/ A_2(\nu)\,\Gamma\left(4-\frac{2}{\nu-1}\right). \tag{12.1, 2}$$

(iii) Attracting spheres, diameter σ (Sutherland's model, (10.41, 8)),

$$[\mu]_1 = \frac{5}{16\sigma^2}\left(\frac{kmT}{\pi}\right)^{\frac{1}{2}}\Big/\left(1+\frac{S}{T}\right). \tag{12.1, 3}$$

(iv) Lennard-Jones and exp;6 models ((10.1, 4) and (10.4, 7)),

$$[\mu]_1 = \frac{5}{8\sigma^2}\left(\frac{kmT}{\pi}\right)^{\frac{1}{2}}\Big/\mathscr{W}_1^{(2)}(2), \tag{12.1, 4}$$

where $\mathscr{W}_1^{(2)}(2)$ is a function of kT/ϵ, given numerically in Table 6 (p. 185) for the Lennard-Jones 12,6 model, and illustrated in Fig. 8 (p. 185) for this and the exp;6 model.

(v) Rough elastic spheres of diameter σ (the Bryan–Pidduck model, (11.61, 2)),

$$[\mu]_1 = \frac{15}{8\sigma^2}\left(\frac{kmT}{\pi}\right)^{\frac{1}{2}}\frac{(1+K)^2}{6+13K}, \tag{12.1, 5}$$

where K is given by (11.6, 2).

Further approximations have been determined for most of these models. For the first (cf. (10.21, 4)), the value of μ, correct to three decimal places of the numerical factor, is

$$\mu = 1{\cdot}016[\mu]_1 = 0{\cdot}1792(kmT)^{\frac{1}{2}}/\sigma^2 = 0{\cdot}499\rho\bar{C}l \tag{12.1, 6}$$

(cf. (4.11, 2), (5.21, 4) and (6.2, 1)), and for the second (cf. (10.32, 2))

$$\mu = [\mu]_1\left\{1+\frac{3(\nu-5)^2}{2(\nu-1)(101\nu-113)}+\dots\right\}.$$

Thus for rigid elastic spheres μ is 1·6 per cent in excess of $[\mu]_1$, while for Maxwellian molecules ($\nu = 5$) μ is identical with $[\mu]_1$: for values of ν between 5 and 13, μ is in excess of $[\mu]_1$ by less than 0·7 per cent. For other smooth models the correction to $[\mu]_1$ is normally a fraction of 1 per cent. We may therefore expect that no great error will be incurred for most smooth models if the first approximation $[\mu]_1$ is used in place of the true value μ.

On comparing (12.1, 1) with (12.1, 5) it appears, as in 11.61, that the possession of internal energy by the molecules does not seriously affect the rate at which they transport momentum, since the ratio of (12.1, 5) to (12.1, 1) varies only between 0·994 and 1·05 as K ranges from 0 to the uniform-sphere value 0·4. The effect of later approximations is, however, more important for rough molecules than for smooth.

12.11. The dependence of viscosity on the density

Each of the formulae of 12.1 predicts that the coefficient of viscosity of a simple gas is independent of its density: this is a general result independent of the nature of the interaction between molecules (cf. 7.52). Thus the capacity of a gas for transmitting momentum, and so for retarding the motion of a body moving in it, is not decreased when its density is diminished. This surprising law was first announced by Maxwell* on theoretical grounds, and afterwards verified experimentally by Maxwell and others.

The consequences of this law are interesting. For example, in so far as the damping of the oscillations of a pendulum is due to the viscous resistance of the gas in which it moves, the degree of damping will be independent of the density of the gas, and the oscillations will die away as rapidly in a rarefied gas as in a dense gas. This was first noted by Boyle.†

Again, according to Stokes's law, the velocity of a sphere of mass M and radius a falling under gravity (g) in a viscous fluid tends to the limiting value

$$g(M-M_0)/6\pi a\mu,$$

where M_0 is the mass of fluid displaced by the sphere. When the fluid is a gas, M_0 is negligible compared with M, so that the limiting velocity of fall should be independent of the density.

When, however, a body moves at a high velocity in a viscous fluid, the ordinary laminar motion of the fluid is unstable, and gives place to a turbulent motion. In this case the conduction of momentum away from the body is by eddies, not by the ordinary viscosity of the fluid. Thus the resistance of a gas to a body moving at a high velocity is not the ordinary viscous resistance, and may depend on the density.

Even when no turbulence is present, the viscous resistance of a gas shows some dependence on the density. The volume of the molecules, though a

* J. C. Maxwell, *Phil. Mag.* **19**, 19 (1860); **20**, 21 (1860); *Collected Works*, vol. 1, p. 391.
† See S. G. Brush, *Kinetic Theory*, vol. 1, p. 4 (Pergamon, 1965).

small fraction of the volume occupied by the gas, is not wholly negligible; this implies a correction to the preceding theory, which is investigated in Chapter 16. Also in a vapour near to condensation, the molecules form aggregations consisting of several molecules, and again the preceding theory and its results need modification. A dependence of μ upon ρ for a number of gases has, in fact, been determined experimentally* as a series

$$\mu = a+b\rho+c\rho^2+\ldots.$$

For a gas at ordinary densities the terms of this series after the first are small, and in what follows we shall ignore the dependence of μ on ρ. The phenomenon of slip at the walls at low densities, noted in 6.21, may need to be taken into account in interpreting experimental results, but it does not represent a true dependence of μ on the density.

Table 11. *Viscosities and molecular diameters of gases at* S.T.P.

Gas	Molecular weight	$10^7\times\mu$ (poise)†	$10^8\times\sigma$ (cm.)
Hydrogen	2·016	845	2·745
Deuterium	4·029	1191	2·751
Helium	4·003	1865	2·193
Methane	16·043	1024	4·187
Ammonia	17·031	923	4·477
Neon	20·183	2975	2·602
Acetylene	26·038	948	4·912
Carbon monoxide	28·011	1635	3·810
Nitrogen	28·013	1656	3·784
Ethylene	28·054	927	5·062
Air	—	1719	—
Nitric oxide	30·007	1774	3·720
Ethane	30·070	858	5·353
Oxygen	32	1919	3·636
Hydrogen sulphide	34·080	1163	4·743
Hydrogen chloride	36·461	1328	4·514
Argon	39·944	2117	3·659
Carbon dioxide	44·011	1380	4·643
Nitrous oxide	44·013	1351	4·692
Methyl chloride	50·488	968	5·737
Sulphur dioxide	64·064	1164	5·551
Chlorine	70·906	1233	5·534
Krypton	83·80	2328	4·199
Xenon	131·30	2107	4·939

* See, for example, reference (1) in the list at the end of this chapter. Later references to this list will be indicated by figures in the text, as (1).

† 1 poise = 1 g./cm. sec.

12.2. Viscosities and equivalent molecular diameters

The viscosity of a gas can be measured absolutely, or relative to that of a standard gas. The absolute determination is the more difficult, and consequently subject to greater experimental error; on the other hand, a relative determination is affected by any inaccuracy of the value adopted for the standard gas. Dry air is commonly used as a standard gas, and its viscosity has been determined with great care by a number of workers. However, even for air the probable error of a weighted mean of values of μ is about 0·1 per cent of the true value. For other gases the probable error may be several times as large. This should be borne in mind in considering the comparisons with individual runs of experimental values given in this chapter.

Table 11 gives experimental values* of μ for a number of the common gases at S.T.P., and also equivalent diameters at 0 °C. derived from them, using (12.1, 6). The table shows that the diameter increases somewhat with increasing molecular weight, and also with increasing complexity of the molecules. The viscosity itself increases much more slowly with molecular weight than would be suggested by the $m^{\frac{1}{2}}$ factor in (12.1, 6).

The molecular diameter deduced from μ in general differs from the diameter derived from other properties of a gas, such as the equation of state.

THE DEPENDENCE OF THE VISCOSITY OF THE TEMPERATURE

12.3. Rigid elastic spheres

The various formulae of 12.1 give different relations between the coefficient of viscosity and the temperature. If the molecules are rigid elastic spheres,

$$\mu \propto T^{\frac{1}{2}}, \tag{12.3, 1}$$

while all the other models imply a greater degree of variation of viscosity with temperature. Experiment shows that for all gases the actual variation of viscosity with temperature is more rapid than that given by (12.3, 1), as is to be expected. Thus if equation (12.1, 6) for the coefficient of viscosity is used, the diameter σ of a molecule must be supposed to vary with the temperature, decreasing as the temperature increases. To illustrate the order of magnitude of the variation, we give the diameters of helium molecules at different temperatures, calculated from experimental† values (5, 6) of μ, using (12.1, 6).

* The experimental values, save for those for carbon monoxide and helium, are mean values taken from *Thermophysical Properties Research Center, Data Book 2* (Wright-Patterson Air Force Base, Ohio, 1964). As regards carbon monoxide, the value quoted is derived from those used in the Data Book with the exception of those of Johnston and Grilly (2), which appear inconsistent with those of other workers. For helium, chief reliance is placed on the results of Kestin and his co-workers (1) and of Wobser and Müller (3).

† The experimental values of μ used in Tables 12 and 13 are 1 or 2 per cent higher than more modern ones, the corresponding value at 0 °C. being 1887×10^{-7} poise; the

Table 12. *Values of μ and σ for helium*

Temperature (°C.)	$10^7 \times \mu$ (poise)	$10^8 \times \sigma$ (cm.)
−258·1	294·6	2·67
−197·6	817·6	2·37
−102·6	1392	2·23
17·6	1967	2·14
183·7	2681	2·05
392	3388	2·005
815	4703	1·97

The variation of the apparent radius with temperature receives a simple explanation on the hypothesis that the molecules are centres of force, not hard spheres. The motion of the centre B of one molecule relative to the centre A of another at their encounter is then represented diagrammatically by Fig. 3*a* (p. 56). The deflection of the relative velocity g at encounter is the same as if, in that figure, the motion of B relative to A were along the lines PO, OQ instead of along the curve LMN; that is, it is the same as if the molecules were elastic spheres, and collided when B reached O: their effective diameter at the encounter is thus AO. If the elastic-sphere theory is applied to such molecules, the value of σ deduced from the experimental values of μ is the mean of the values of this effective diameter for all encounters, weighted so as to give greatest importance to encounters producing the largest deflections.

The deflection χ of the relative velocity at encounter is given in terms of the encounter-variable b and the effective diameter OA by the equation (cf. (3.44, 2))

$$b = OA \cos \tfrac{1}{2}\chi.$$

If g increases, b remaining constant, the time during which each molecule is under the influence of the other's field diminishes, and χ will also diminish: thus OA, too, must decrease. Hence an increase of temperature, which implies an increase in the average value of g at encounter, results in a decrease in the apparent value of σ.

12.31. Point-centres of force

The law of variation of viscosity with temperature for molecules repelling each other as the inverse νth power of the distance (cf. (12.1, 2)) is

$$\mu \propto T^s, \tag{12.31, 1}$$

where*

$$s = \frac{1}{2} + \frac{2}{\nu - 1}. \tag{12.31, 2}$$

temperature variations which they indicate are, however, consistent with the more modern data.

* This formula was inferred from dimensional considerations by Rayleigh (J. W. Strutt, 3rd Lord Rayleigh), *Proc. Roy. Soc.* A, **66**, 68 (1900); *Collected Papers*, vol. 4, p. 452.

Comparing this with (12.1, 6) we see that for this model the apparent radius of a molecule varies inversely as the $1/(\nu-1)$th power of the temperature; cf. (10.31, 8).

Equation (12.31, 1) can also be written in the form

$$\mu = \mu'(T/T')^s, \tag{12.31, 3}$$

where μ' is the coefficient of viscosity at an assigned temperature T'. If the value of μ is known for a second temperature, the value of s can be found. Hence the formula (12.31, 3) can always be satisfied by the experimental values of μ for two temperatures, if μ' and s are suitably chosen.

For several gases, notably hydrogen and helium, the experimental values of μ conform to an equation of the type (12.31, 3) over a large range of temperature (4, 5). The results for helium are given in Table 13; experimental values (5) of μ are given in the second column, and in the third, the corresponding values of μ calculated from (12.31, 3), taking $s = 0{\cdot}647$, and adopting the value $\mu = 1887 \times 10^{-7}$ for the viscosity at 0 °C. The fourth column gives the values calculated from Sutherland's formula (cf. 12.32).

Table 13. *Viscosity of helium*

Temperature (°C.)	$10^7 \times \mu$ (exp.)	$10^7 \times \mu$ (calc.) (force-centres)	$10^7 \times \mu$ (calc.) (Sutherland)
−258·1	294·6	288·7	92
−253·0	349·8	348·9	135
−198·4	813·2	815·5	621
−197·6	817·6	821·3	628
−183·3	918·6	918·5	745
−102·6	1392	1389	1317
−78·5	1506	1515	1460
−70·0	1564	1558	1513
−60·9	1587	1603	1563
−22·8	1788	1783	1771
17·6	1967	1965	1974
18·7	1980	1970	1979
99·8	2337	2309	2345
183·7	2681	2632	2682

For other gases, however, equation (12.31, 3) does not agree so well with the experimental results, and the value of s for which (12.31, 3) best fits observation varies with the range of temperatures considered. Even for helium a slight variation is found; for temperatures above 0 °C. the best fit with the above data is obtained by taking $s = 0{\cdot}661$ (5), and a recent study (1) makes s in the range from −240° to 800 °C. fluctuate about a mean value 0·6567. For most gases s decreases markedly as the temperature increases; thus Johnston and Grilly (2) found values of s for nitrogen, carbon monoxide, air, methane, nitric oxide, oxygen and argon, in a range from about −180°

to 20 °C., which are greater by about 0·1 than those found by Wobser and Müller (3) for the range 20–100 °C.

The decrease in s as the temperature increases implies an increase in ν, which may be interpreted as implying that the molecular repulsion varies according to a smaller power of $1/r$ when r is large than when r is small (because at low temperatures the molecules do not penetrate so far into each other's repulsive fields as at high temperatures). This is, however, not the only possible interpretation; in the derivation of (12.31, 1) no account is taken of the fact that the interaction of molecules at large distances may be an *attraction*; such an attraction would influence the molecular motions more at low temperatures, when the velocities of molecules are small, than at high.

Table 14 lists the values of s and ν for several gases, mostly for the temperature-range 20–100 °C. The values of ν range from about 13 for helium, neon and hydrogen, to values around 5 for molecules of complex structure. If ν is large, the mutual repulsion between pairs of molecules at encounter increases very rapidly as they approach, and the encounter approximates to a sharp impact: in this case the molecules are said to be hard, while for small values of ν, in the neighbourhood of 5, they are said to be soft.

Table 14. *Values of s and ν*

Gas	s	ν	Temperature range (°C.)
Hydrogen	0·668 (3)	12·9	20–100
Deuterium	0·699 (7)	11·1	−183–22
Helium	0·657 (1)	13·7	−230–800
Methane	0·836 (3)	7·0	20–100
Ammonia	1·10 (3)	4·3	20–100
Neon	0·661 (3)	13·4	20–100
Acetylene	0·998 (3)	5·0	20–100
Carbon monoxide	0·734 (3)	9·5	20–100
Nitrogen	0·738 (3)	9·4	20–100
Air	0·77 (3)	8·4	20–100
Nitric oxide	0·788 (3)	7·9	20–100
Oxygen	0·773 (3)	8·4	20–100
Hydrogen chloride	1·03 (8)	4·8	20–99
Argon	0·811 (3)	7·5	20–100
Carbon dioxide	0·933 (3)	5·6	20–100
Nitrous oxide	0·943 (3)	5·5	20–100
Chlorine	1·01 (3)	4·9	20–100
Sulphur dioxide	1·05 (3)	4·6	20–100

12.32. Sutherland's formula

The viscosity of a gas whose molecules are rigid attracting spheres is given, to a first approximation, by

$$\mu = \frac{5}{16\sigma^2}\left(\frac{kmT}{\pi}\right)^{\frac{1}{2}} \Big/ \left(1+\frac{S}{T}\right) \qquad (12.32,\ 1)$$

(cf. (12.1, 3)). Comparing this with (12.1, 1) we see that the effect of the attractive field is to increase the apparent diameter of the molecules in the ratio $\sqrt{(1+S/T)}:1$. The increase is largest if the temperature is small, and becomes negligible at very large temperatures. It results from the deflection of the molecules by the attractive fields before collision, which makes the collisions more nearly head-on than in the absence of such fields. The Sutherland constant S is a measure of the strength of the attractive fields, being proportional to the mutual potential energy of two molecules when in contact.

Equation (12.32, 1) may be written in the form

$$\mu = \mu' \left(\frac{T}{T'}\right)^{\frac{3}{2}} \frac{T'+S}{T+S}, \qquad (12.32, 2)$$

where $\mu = \mu'$ when $T = T'$. If μ is known at a second temperature T'' as well as at T', a value of S can be found such that (12.32, 2) is satisfied at both T' and T''. If T' and T'' are sufficiently close together, (12.32, 2) may be

Table 15. *Values of Sutherland's constant*

Gas	Sutherland's constant	Temp. range (°C.)
Hydrogen	66·8 (3)	20–100
Helium	72·9 (3)	20–100
Methane	169 (3)	20–100
Ammonia	503 (9)	25·1–300
Neon	64·1 (3)	20–100
Acetylene	320 (3)	20–100
Carbon monoxide	102 (3)	20–100
Ethylene	225 (9)	20–250
Nitrogen	104·7 (6)	19·8–825
Air	113 (3)	20–100
Nitric oxide	133 (3)	20–100
Ethane	252 (10)	20–250
Oxygen	125 (6)	14·8–829
Hydrogen sulphide	331 (11)	17–100
Hydrogen chloride	362 (12)	21–250
Argon	148 (3)	20–100
Carbon dioxide	253 (3)	20–100
Nitrous oxide	263 (3)	20–100
Methyl chloride	441 (13)	19·6–300·4
Sulphur dioxide	404 (3)	20–100
Chlorine	345 (3)	20–100
Krypton	188 (14)	16·3–100
Xenon	252 (14)	15·3–100·1

expected to fit the experimental values well at intermediate temperatures. Values of S for several gases at ordinary temperatures are given in Table 15.

Actually (12.32, 2) closely represents the variation of μ with temperature for several gases over fairly wide ranges of temperature, as Table 16 illustrates. The calculated values for nitrogen in this table correspond to $\mu = 1746 \times 10^{-7}$ at 0 °C., with $S = 104{\cdot}7$; those for ammonia correspond to $\mu = 1002 \times 10^{-7}$ at 25·1 °C., with $S = 503$. The Sutherland formula also represents the viscosity of other gases, such as oxygen, argon and carbon monoxide, to a similar degree of accuracy; the same is true of air, though this is a gas-mixture.

Table 16. *Viscosity of nitrogen and ammonia*

Nitrogen (6)			Ammonia (9)		
Temp. (°C.)	$10^7 \times \mu$ (exp.)	$10^7 \times \mu$ (calc.)	Temp. (°C.)	$10^7 \times \mu$ (exp.)	$10^7 \times \mu$ (calc.)
19·8	1746	1746	20	982	982
299	2797	2801	50	1092	1095
408	3141	3134	100	1279	1282
490	3374	3366	150	1463	1465
600	3664	3656	200	1645	1644
713	3930	3934	250	1813	1818
825	4192	4193	300	1986	1987

The success of (12.32, 2) for many gases does not establish the validity of Sutherland's molecular model for those gases. The formulae (12.32, 1, 2) are approximate formulae, valid only if the attractive fields are small; if this is not the case the expression $1 + S/T$ in (12.32, 1) must be replaced by a series of the form

$$1 + \frac{S}{T} + \frac{S'}{T^2} + \ldots, \qquad (12.32, 3)$$

and it is scarcely to be expected that the terms after the second will be negligible if S/T is not small compared with unity. As Table 15 shows, S/T often exceeds unity in the temperature-range in which (12.32, 2) fits the experimental values, whereas for gases like hydrogen and helium, for which S/T is fairly small, Sutherland's formula does not fit the experimental values at all well. This is illustrated by the figures for helium given in the fourth column of Table 13; S is taken as 78·2. For these gases the value of S for which (12.32, 2) best fits the observations is found to increase with the temperatures considered.

In general it is not adequate to represent the core of a molecule as a rigid sphere, or to take molecular attractions into account to a first order only. The greater rapidity of the experimental increase of μ with T, as compared with that for non-attracting rigid spheres, has to be explained as due partly to the 'softness' of the repulsive field at small distances, and partly to attractive forces which have more than a first-order effect. The chief value of Sutherland's formula seems to be as a simple interpolation formula over restricted ranges of temperature.

12.33. The Lennard-Jones 12, 6 model

The Lennard-Jones model (cf. 10.4) takes into account both the softness of the molecules and their mutual attraction at large distances. Most attention has been paid to the 12,6 model, for which the mutual potential energy of two molecules at a distance r is taken to be

$$V(r) = 4\epsilon\{(\sigma/r)^{12} - (\sigma/r)^6\} \qquad (12.33, 1)$$

$$= \epsilon\{(r_m/r)^{12} - 2(r_m/r)^6\}. \qquad (12.33, 2)$$

The viscosity of a simple gas on this model is given by an equation of the form

$$\mu = \frac{(kmT)^{\frac{1}{2}}}{\sigma^2 F(kT/\epsilon)} \qquad (12.33, 3)$$

(cf. 10.4). Equation (12.33, 3) implies that, if $\log(\mu/T^{\frac{1}{2}})$ is plotted against $\log T$, curves for different values of σ and ϵ differ only by a change in origin. Thus it is possible to determine σ and ϵ by fitting a standard $\log(\mu/T^{\frac{1}{2}})$, $\log T$ curve to a corresponding experimental curve, and determining the relative displacement in origin of the two curves.

A comparison of theory with experiment has been made for a number of gases for this model. The agreement for each gas is found to be at least as good as for the models of 12.31 and 12.32, which likewise involve two adjustable parameters.* This confirms that the 12,6 model approximates to physical reality by providing repulsive and attractive fields varying with distance roughly in the right manner, at least for non-polar gases. For polar gases the attractive part of the molecular potential energy includes a term proportional to r^{-3} (cf. (10.43, 3)), not taken into account in (12.33, 1), and the fit with experiment is less close.

Figure 10 compares the experimental and calculated viscosities for a number of gases. The values of ϵ and σ used in this figure are taken from Table 17 (below) save for carbon dioxide, for which the value $\sigma = 3{\cdot}96 \times 10^{-8}$ cm. is used. Temperatures below about 30 °K., at which quantum effects are important for helium and hydrogen (cf. 17.4 and 17.41) are not represented in the experimental values. Part of the discrepancies between experiment and the theoretical curve is due to incompatibility between the results of different experimenters.

Values of σ, r_m and ϵ/k obtained (mainly by Hirschfelder, Bird and Spotz†) from the viscosities of various gases are given in Table 17. They apply to

* Helium has so weak an attractive field that effectively the model attempts to fit it to an r^{-12} repulsive potential energy.

† J. O. Hirschfelder, R. B. Bird and E. L. Spotz, *J. Chem. Phys.* **16**, 968 (1948): *Chem. Rev.* **44**, 205 (1949). The values for D_2 are those given by J. Kestin and A. Nagashima, *Phys. Fluids*, **7**, 730 (1964). Values of ϵ/k and σ given by other authors often differ appreciably from those of Table 17.

ranges of temperatures including 0 °C. The 'low-velocity diameter' σ gives the distance of closest approach of two molecules with negligible initial relative velocity; $r_m (\equiv 2^{\frac{1}{6}}\sigma)$ is the distance at which the molecular interaction changes from a repulsion to an attraction, and $-\epsilon$ is the corresponding energy. Thus neither σ nor r_m is exactly comparable with the effective diameter of 12.2 (Table 11), which measures the distance near which most of the molecular interaction takes place. For hydrogen, helium and neon,

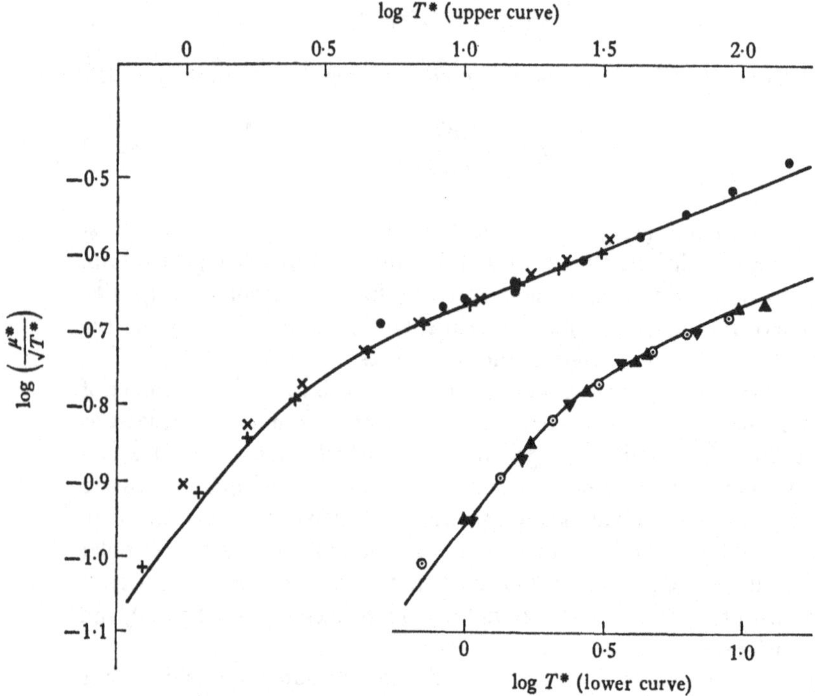

Fig. 10. Comparison for the 12, 6 model of the theoretical curve for $\log(\mu^*/\sqrt{T^*})$ with experiment, where $T^* = kT/\epsilon$, $\mu^* = \mu\sigma^2/(m\epsilon)^{\frac{1}{2}}$. Upper curve; ● = helium, + = neon, × = hydrogen. Lower curve (displaced one unit to the right); ⊙ = argon, ▲ = nitrogen, ▼ = carbon dioxide. Experimental data in the ranges 30–80, 80–290 and 290–1000 °K. are taken from the work of van Itterbeek (15), Johnston (2) and Trautz (6), (8), (16) and their respective co-workers.

whose molecules attract only weakly, the effective diameter of 12.2 is less than σ; for all other gases it is greater, and for gases with strong attractive fields (large ϵ/k) it even exceeds r_m. This indicates the importance to the viscosity of the attractive part of the field.

Different ranges of temperature may give somewhat different values of σ and ϵ. Values can also be calculated from other physical data; such values are given in the last columns of Table 17. They are mainly obtained from virial coefficients, but in the last five cases they are less accurate values derived from liquefaction data. The discrepancies arise from the inaccuracy of the physical data as well as from the limitations of the model.*

Table 17. *Values of* ϵ/k, σ *and* r_m *for the* 12, 6 *model*

Gas	Viscosity values			Other values	
	ϵ/k (°K.)	$\sigma \times 10^8$ (cm.)	$r_m \times 10^8$ (cm.)	ϵ/k (°K.)	$\sigma \times 10^8$ (cm.)
Hydrogen	33·3	2·97	3·33	37·02	2·92
Deuterium	35·2	2·95	3·31	—	—
Helium	6·03	2·70	3·03	6·03	2·63
Methane	136·5	3·82	4·29	142·7	3·81
Neon	35·7	2·80	3·14	35·7	2·74
Carbon monoxide	110·3	3·59	4·03	95·33	3·65
Nitrogen	91·46	3·68	4·13	95·9	3·72
Nitric oxide	119	3·47	3·89	131	3·17
Air	97	3·62	4·06	—	—
Oxygen	113·2	3·43	3·85	117·5	3·58
Argon	124	3·42	3·84	119·5	3·41
Carbon dioxide	190	4·00	4·49	185	4·57
Nitrous oxide	220	3·88	4·35	189	4·59
Hydrogen chloride	360	3·305	3·71	261,243	3·69
Sulphur dioxide	252	4·29	4·82	366,323	4·15
Chlorine	357	4·115	4·62	332,313	4·15
Krypton	190	3·61	4·05	169,158	3·96
Xenon	230	4·05	4·55	228,192	4·04

12.34. The exp; 6 and polar-gas models

In the exp;6 model, the mutual potential energy of two molecules is taken to be

$$V(r) = \frac{\epsilon}{(1-6/\alpha)}\left[\frac{6}{\alpha}\exp\left\{\alpha\left(1-\frac{r}{r_m}\right)\right\} - \left(\frac{r_m}{r}\right)^6\right]. \qquad (12.34, 1)$$

Here ϵ and r_m have the same meanings as for the Lennard-Jones 12, 6 model; α is a pure number normally between 12 and 16. For a given α, the model conforms to a law of corresponding states of the form (12.33, 3), and so a graphical method similar to that described for the 12, 6 model can be used to determine ϵ and r_m.

Mason and Rice† have compared experimental viscosities with results

* R. Gibert and A. Dognin (*C.R.* **246**, 2607 (1958)) point out that, over the range $0·9 < kT/\epsilon < 5$, the function $F(kT/\epsilon)$ in (12.33, 3) is well approximated by

$$C(1 + 1·077\epsilon/kT),$$

where C is a constant. This fact may explain the success of the Sutherland formula for many gases; it also suggests that an approximate value of ϵ/k for such gases is $S/1·077$.

† E. A. Mason and W. E. Rice, *J. Chem. Phys.* **22**, 522, 843 (1954); see also E. A. Mason, *J. Chem. Phys.* **32**, 1832 (1960), for a rediscussion of the krypton data.

calculated from the exp;6 potential for a number of gases. They find rather better agreement than is given by the 12,6 model. This is not surprising, since (12.34, 1) involves an extra adjustable parameter α; however, introduction of the extra parameter improves the fit surprisingly little, except for hydrogen and helium. Where marked deviations from measured viscosities are found, they are generally in the same direction for the 12,6 and exp;6 models. Values of ϵ and r_m, together with corresponding values of α and of σ, where σ is such that $V(\sigma) = 0$, are given in Table 18; except for ethane and carbon dioxide, they are due to Mason and Rice. Considering that the experimental data used are not always identical with those used in Table 17, the discrepancies between the values of ϵ and σ for the 12, 6 and exp;6 models are not large.

Table 18. *Values of ϵ/k, σ, r_m and α for the exp;6 model*

Gas	ϵ/k (°K.)	$\sigma \times 10^8$ (cm.)	$r_m \times 10^8$ (cm.)	α
Hydrogen	37·3	3·005	3·34	14
Helium	9·16	2·75	3·135	12·4
Methane	152·8	3·69	4·21	12·3
Neon	38	2·81	3·15	14·5
Carbon monoxide	119·1	3·55	3·94	17
Nitrogen	101·2	3·62	4·01	17
Ethane (17)	229·1	4·41	4·905	16
Argon	123·2	3·44	3·87	14
Carbon dioxide (1)	262	3·72	4·18	14
Krypton	158·3	3·56	4·06	12·3
Xenon	231·2	3·935	4·45	13·0

Monchick and Mason* have applied the Stockmayer potential (cf. (10.43, 3)) to discuss the viscosity of polar gases, using the method described in 10.43. For such gases the r^{-3} term in the potential leads to a more rapid variation of μ with T at low temperatures. The gases considered include ammonia, hydrogen chloride, sulphur dioxide and hydrogen sulphide; the authors state that 'the overall agreement between experiment and this model for polar gases is comparable to that of the Lennard-Jones 12,6 model for non-polar gases'.

GAS MIXTURES

12.4. The viscosity of a gas-mixture

The first approximation to the coefficient of viscosity of a gas-mixture is

$$[\mu]_1 = \frac{x_1^2 R_1 + x_2^2 R_2 + x_1 x_2 R'_{12}}{x_1^2 R_1/[\mu_1]_1 + x_2^2 R_2/[\mu_2]_1 + x_1 x_2 R_{12}} \qquad (12.4, 1)$$

* L. Monchick and E. A. Mason, *J. Chem. Phys.* **35**, 1676 (1961).

(cf. (9.84, 2)) where

$$R_1 = \tfrac{2}{3}+m_1 A/m_2, \quad R_2 = \tfrac{2}{3}+m_2 A/m_1, \tag{12.4, 2}$$

$$R'_{12} = \frac{E}{2[\mu_1]_1}+\frac{E}{2[\mu_2]_1}+2(\tfrac{2}{3}-A), \quad R_{12} = \frac{E}{2[\mu_1]_1[\mu_2]_1}+\frac{4A(m_1+m_2)^2}{3Em_1m_2}. \tag{12.4, 3}$$

Here $[\mu_1]_1$, $[\mu_2]_1$ are the first approximations to the coefficients of viscosity of the constituent gases, and A, E are quantities depending only on the interaction of molecules of different kinds. By (9.81, 1) the quantity E is equal to $\frac{2}{3}nm_0[D_{12}]_1$, where $[D_{12}]_1$ is the first approximation to the coefficient of diffusion of the two gases; A is a dimensionless quantity, values of which are listed, for molecules behaving like centres of a repulsive force $\kappa r^{-\nu}$ in Table 3 (p. 172), and for the 12,6 model in Table 6 (p. 185).

In 12.1 it has been shown that for a simple gas the first approximation $[\mu]_1$ to the coefficient of viscosity μ is in general only slightly in error. Hence the error of (12.4, 1) is certainly small when either x_1 or x_2 tends to zero. One may expect a similar degree of accuracy when neither x_1 nor x_2 is small, if A and E are correctly chosen, at least provided that the molecules of the two gases are not too dissimilar.* The error is, however, fairly considerable for inverse-square interaction (cf. 10.34).

If x_2 is very small we get the case of a gas in which a small quantity of a second gas is present as an impurity. In this case (12.4, 1) approximates to

$$[\mu]_1 = [\mu_1]_1+\frac{2x_2[\mu_1]_1^2}{R_1 E}\left\{\left(\frac{E}{2[\mu_1]_1}+\frac{2}{3}-A\right)^2-R_1R_2\right\}. \tag{12.4, 4}$$

12.41. Variation of the viscosity with composition

In comparing experimental results with the theoretical variation of μ with n_1/n_2, equation (12.4, 1) is normally employed, ignoring the difference between the first approximations and the true values of the viscosities. Values have to be assigned for the constants A, E. The constant A may be determined from the tables for the 12,6 and exp;6 models, taking the energy ϵ_{12} as $(\epsilon_1\epsilon_2)^{\frac{1}{2}}$. The 12,6 model makes A about 0·44 for most of the important range of kT/ϵ_{12}, while the exp;6 model indicates a slow increase of A from 0·44 to 0·46 as kT/ϵ_{12} increases in this range; thus the value of A is insensitive to the precise value of ϵ_{12} adopted. The constant E can be determined by a number of methods.

(I) By (9.81, 1), $[D_{12}]_1 = 3E/2nm_0$; hence E can be determined directly from experimental values of D_{12}, neglecting the difference between this and $[D_{12}]_1$. However, the error of the first approximation $[D_{12}]_1$ is sometimes

* The results for a Lorentzian gas (p. 193, Table 10) are not representative of the case when m_1/m_2 is large, since they refer only to the small part of the viscosity due to the light gas.

appreciable, and moreover the experimental determinations of coefficients of diffusion are not always very reliable.

(II) We may calculate E from viscosity data for the simple gases. For example, put

$$\Omega_1^{(2)}(2) = 2s_1^2(\pi kT/m_1)^{\frac{1}{2}}, \quad \Omega_{12}^{(2)}(2) = s_{12}^2(2\pi kT/m_0 M_1 M_2)^{\frac{1}{2}}, \qquad (12.41, 1)$$

with a similar relation for $\Omega_2^{(2)}(2)$; then (cf. (9.7, 3), (9.8, 8))

$$[\mu_1]_1 = \frac{5}{16s_1^2}\left(\frac{km_1 T}{\pi}\right)^{\frac{1}{2}}, \quad \mathrm{E} = \frac{5\mathrm{A}}{8s_{12}^2}\left(\frac{km_0 T}{2\pi M_1 M_2}\right)^{\frac{1}{2}}, \qquad (12.41, 2)$$

and s_1, s_2 and s_{12} denote effective viscosity diameters for the interaction of like or unlike molecules. Assuming that $s_{12} = \frac{1}{2}(s_1 + s_2)$, we can calculate E from experimental values of μ_1, μ_2, ignoring the difference between these and $[\mu_1]_1$, $[\mu_2]_1$. However, though s_{12} is likely to lie between s_1 and s_2, the relation $s_{12} = \frac{1}{2}(s_1 + s_2)$ applies strictly only for rigid spheres.

(III) A more sophisticated method is to calculate E from viscosity data, using the 12, 6 model, with the combination rules $\epsilon_{12} = (\epsilon_1 \epsilon_2)^{\frac{1}{2}}$ and $\sigma_{12} = \frac{1}{2}(\sigma_1 + \sigma_2)$ or $\sigma_{12} = (\sigma_1 \sigma_2)^{\frac{1}{2}}$. Similar methods can also be applied to the exp; 6 model. However, once again the combination rules, though reasonable, cannot be exact.

(IV) We may simply determine E to give the best fit of theory to experiment.

The method followed here is (IV); the values of E thus found are then checked against diffusion coefficients and the values obtained by calculation. The mixtures considered are mixtures of hydrogen and helium, and of helium and argon.

Mixtures of hydrogen and helium at 0 °*C.*

The value adopted for A is 0·458; E is then determined to be $2{\cdot}48 \times 10^{-4}$ poise, corresponding to $[D_{12}]_1 = 1{\cdot}39$ cm.^{2}/sec. at 1 atmosphere pressure. For comparison, experimental values of D_{12}, reduced to 0 °C., are 1·313 (Bunde) and 1·21 (Suetin, Shchegolev and Klestov). The values of E calculated as indicated in (II) and (III) are $2{\cdot}47 \times 10^{-4}$ and $2{\cdot}44 \times 10^{-4}$ poise.

Mixtures of helium and argon at 20 °*C.*

The value adopted for A is 0·452; E is found to be $8{\cdot}70 \times 10^{-4}$ poise, corresponding to $[D_{12}]_1 = 0{\cdot}715$ cm.^{2}/sec. at a pressure of 1 atmosphere. Recent experimental values of D_{12} are 0·697 at 14·8 °C. (Strehlow), 0·76 at 26·9 °C. (Walker and Westenberg), and 0·731 at 16 °C., 0·744 at 23 °C. (Saxena and Mason). Values of E calculated by methods II and III are $8{\cdot}125 \times 10^{-4}$ and $8{\cdot}84 \times 10^{-4}$ poise.

The calculated and experimental* values of the coefficient of viscosity are

* For helium–argon mixtures, see ref. (18); for hydrogen–helium mixtures, see (19).

Table 19. *The viscosities of two gas-mixtures*

Hydrogen–helium			Helium–argon		
H_2 (%)	$10^7\times\mu$ (exp.)	$10^7\times\mu$ (calc.)	He (%)	$10^7\times\mu$ (exp.)	$10^7\times\mu$ (calc.)
0	1892·5	[1892·5]	0	2227·5	[2227·5]
3.906	1850·0	1846	19·9	2270·7	2271
10·431	1759·6	1767	37·1	2309·5	2305
13·60	1732·7	1730	63·4	2316·1	2319
24·913	1603·2	1600	80·7	2252·8	2255
40·284	1430·6	1431	86·3	2202·7	2207
60·143	1226·7	1224	94·2	2090·2	2093
81·193	1016·5	1017	100·0	1960·4	[1960·4]
100·0	841	[841]	—	—	—

given in Table 19. As can be seen, the formula (12.4, 1) represents the variation of μ very closely.

The results for helium–argon mixtures illustrate the curious fact, first noticed by Graham,* that the addition of a moderate amount of a light and relatively inviscid gas (hydrogen, in the case mentioned by Graham) to a more viscous and heavy gas (carbon dioxide) may actually increase the viscosity of the latter. The explanation seems to be somewhat as follows. The addition of a quantity of the same gas to the heavy gas reduces the mean free path, but increases the number of carriers of momentum; these two effects just balance. When, however, a small quantity of a light gas is added to the heavy gas, the mean free path of the molecules of the latter is hardly affected, because of the large persistence of velocities after collisions with the lighter molecules (5.5): and the small additional transport of momentum by the lighter molecules may outweigh the effect of the decrease in the mean free path of the heavier.

More precisely, if $\mu_1 > \mu_2$, μ has a maximum as x_1/x_2 varies if in (12.4, 4) the coefficient of x_2 is positive. By (12.4, 2) this condition is equivalent to

$$\frac{E}{2[\mu_1]_1}+\tfrac{2}{3}-A>\left\{\left(\tfrac{2}{3}+A\frac{m_1}{m_2}\right)\left(\tfrac{2}{3}+A\frac{m_2}{m_1}\right)\right\}^{\frac{1}{2}} \qquad (12.41, 3)$$

or, in terms of the equivalent viscosity diameters s_1, s_2, s_{12} used in (12.41, 2)

$$\frac{A(m_1+m_2)^{\frac{3}{2}}}{m_1(2m_2)^{\frac{1}{2}}}\frac{s_1^2}{s_{12}^2}>\left\{\left(\tfrac{2}{3}+A\frac{m_1}{m_2}\right)\left(\tfrac{2}{3}+A\frac{m_2}{m_1}\right)\right\}^{\frac{1}{2}}-\tfrac{2}{3}+A. \qquad (12.41, 4)$$

Thus for given m_1, m_2 and A, s_1/s_{12} must exceed a certain lower limit. An effective upper limit also exists, since s_{12} can be expected not to differ too widely from $\frac{1}{2}(s_1+s_2)$ and, to ensure that $\mu_1 > \mu_2$, s_1/s_2 must be less than $(m_1/m_2)^{\frac{1}{4}}$. Take $A = \frac{4}{9}$, in rough agreement with the results for the 12, 6 and

* T. Graham, *Phil. Trans. R. Soc.* **136**, 573 (1846).

exp; 6 models; then the bounds placed on s_1/s_{12} and $s_1/\frac{1}{2}(s_1+s_2)$ are as follows, for different values of m_1/m_2:

m_1/m_2	=	$\frac{1}{16}$	$\frac{1}{9}$	$\frac{1}{4}$	1	4	9	16
s_1/s_{12}	>	0·618	0·696	0·811	1	1·15	1·205	1·235
$s_1/\frac{1}{2}(s_1+s_2)$	<	0·667	0·732	0·828	1	1·172	1·268	1·333

These figures show that a maximum of μ is likely to occur only if s_1/s_{12} lies in a relatively restricted range. It is most likely when m_1/m_2 is either large or small compared with unity, but the viscosities of the two gases are not too different (so that $s_1/\frac{1}{2}(s_1+s_2)$ is not much less than its upper bound). When m_1 and m_2 are nearly equal, a maximum can occur only when $s_{12} < s_1 < s_2$. This is most likely to be the case when s_1 and s_2 are nearly equal, and the mixture is composed of one non-polar and one polar gas; in this case the dipolar interaction increases s for the polar gas, but not s_{12}.

A maximum of μ as x_1/x_2 varies is observed in some mixtures (like He–N_2) in which $\mu_1 > \mu_2$ but $m_1 < m_2$, as well as in those displaying the original Graham phenomenon (in which $\mu_1 > \mu_2, m_1 > m_2$). A minimum of μ, though theoretically not impossible, is improbable and does not seem to have been actually observed at normal densities.

12.42. Variation of the viscosity with temperature

The complicated nature of the formula (12.4, 1) renders it difficult to predict from it the precise law of the temperature variation of the viscosity of a gas-mixture. Normally, however, the force-law between unlike molecules can be expected to be intermediate between those for like molecules of the two gases. Thus the variation of E with temperature can be expected in general to be intermediate between those of $[\mu_1]_1$ and $[\mu_2]_1$; the same is true also of $[\mu]_1$.

The viscosities of mixtures at different temperatures can be used to infer values of E at these temperatures, and from these the law of interaction between unlike molecules can in principle be found. In practice, however, because E is not determined too certainly, the process is normally inverted; the law of interaction for unlike molecules is inferred from those for like molecules, and the values of E and A calculated from it are used in (12.4, 1) to give theoretical values of the viscosity of mixtures. This method is usually employed with the 12,6 or exp;6 models, using appropriate combination rules to deduce the force constants (cf. (III) of 12.41); reasonable agreement between theory and experiment is normally obtained.

As noted in 12.41, the viscosity of a mixture of a light gas with a heavy one often has a maximum when the proportions are varied at a given temperature. It is found that, as the temperature increases, this maximum tends to disappear. Figure 11 illustrates this tendency for mixtures of hydrogen and hydrogen chloride (12).

An analytical explanation of the tendency is as follows. Suppose, as in 12.41, that the first gas is the heavier; then, as the temperature increases, μ_1 in general increases more rapidly than μ_2, and the variation of E is intermediate between those of μ_1 and μ_2, so that E/μ_1 decreases. Thus the inequality (12.41, 3) may cease to be true for sufficiently high temperatures.

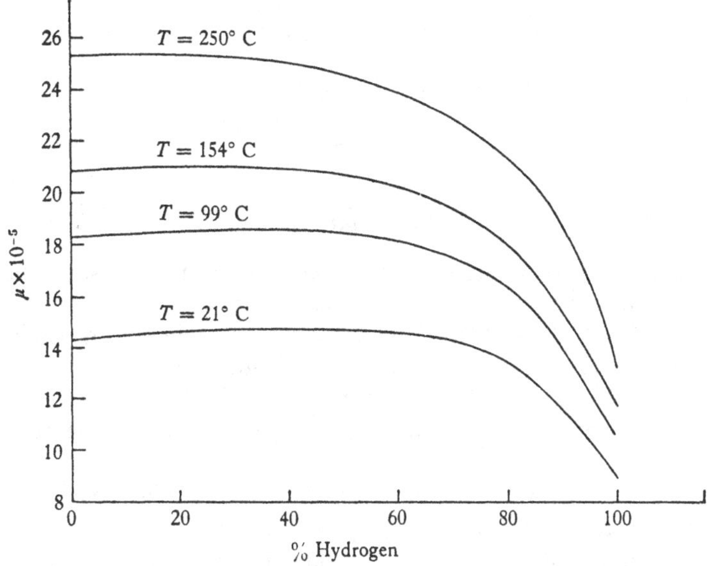

Fig. 11

12.43. Approximate formulae

Because of the relatively complicated form of (12.4, 1) a variety of simpler approximate formulae have been suggested for the viscosity of a mixture. These are in general of the Sutherland form (cf. 6.63); that is, μ is expressed as the sum of contributions $\mu_{(1)}$, $\mu_{(2)}$ arising from the two gases in the mixture, $\mu_{(1)}$ and $\mu_{(2)}$ being given by expressions of the form

$$\mu_{(1)} = \text{x}_1/\{\text{x}_1(\mu_1)^{-1}+\text{x}_2(\mu'_{12})^{-1}\}, \quad \mu_{(2)} = \text{x}_2/\{\text{x}_1(\mu'_{21})^{-1}+\text{x}_2(\mu_2)^{-1}\}. \tag{12.43, 1}$$

Here μ_1, μ_2 are, as usual, the viscosities of the two pure gases; μ'_{12}, μ'_{21} are 'mutual viscosities' whose values depend on the particular approximation introduced.

The theory of Chapters 8 and 9 can be shown to imply that the parts $\mu_{(1)}$, $\mu_{(2)}$ into which μ is divided actually satisfy, to the first approximation, the relation

$$\left[\frac{1}{[\mu_1]_1}+\frac{2\text{x}_2}{\text{E}\text{x}_1}\left(\frac{2}{3}+\frac{m_2}{m_1}\text{A}\right)\right]\mu_{(1)}-\frac{2}{\text{E}}\left(\tfrac{2}{3}-\text{A}\right)\mu_{(2)} = 1 \tag{12.43, 2}$$

and a similar equation. These equations differ from (12.43, 1) by reason of the terms involving $\frac{2}{3}-\text{A}$; these terms represent the transfer of momentum-transport from one gas to the other at collision. Since these terms are relatively small, a crude approximation is to replace $\mu_{(2)}$ in (12.43, 2) by $x_2\mu_{(1)}/x_1$, just as if the momenta transported by molecules of the two gases were proportional only to the numbers of those molecules. Then (12.43, 2) becomes

$$\mu_{(1)} = x_1 \Big/ \left\{ x_1[\mu_1]_1^{-1} + x_2 \frac{2\text{A}}{\text{E}M_1} \right\}, \tag{12.43, 3}$$

which is of the form (12.43, 1), with μ'_{12} given by

$$\mu'_{12} = \tfrac{1}{2}\text{E}M_1/\text{A}. \tag{12.43, 4}$$

In terms of D_{12},

$$\mu_{(1)} = x_1 \Big/ \left\{ \frac{x_1}{[\mu_1]_1} + \frac{3\text{A}x_2}{\rho_1[D_{12}]_1} \right\}, \tag{12.43, 5}$$

where ρ_1 is the density of the first gas, when pure, at the pressure and temperature of the actual mixture. The formula (12.43, 5), with 3A given the rather high value 1·385, and the first approximations replaced by actual values, gives the approximate formula of Buddenberg and Wilke.* The total viscosity is always somewhat overestimated by using (12.43, 5) and the corresponding equation for $\mu_{(2)}$; this is partly compensated by the high value of 3A.

A variety of other approximate formulae have been proposed, some based on further transformations of (12.43, 2) or (12.4, 1), and some more empirical.† The complication of certain of these is such that it is simpler, as well as more accurate, to use the formula (12.4, 1). However, (12.43, 1) has considerable value as a formula to represent experimental results, regarding μ'_{12} and μ'_{21} as adjustable quantities.

VOLUME VISCOSITY

12.5. As was noted in 11.51, gases whose molecules possess internal energy are subject to a volume viscosity ϖ opposing expansion or contraction; the effect of this volume viscosity is to increase the hydrostatic pressure by $-\varpi\nabla.\boldsymbol{c}_0$, or $\varpi\rho^{-1}D\rho/Dt$. The volume viscosity is a relaxation phenomenon; if the internal energy is the sum of parts independently relaxing into translatory energy with relaxation times $\tau_1, \tau_2, \ldots$

$$\varpi = 2p \sum_r N_r \tau_r / N^2 \tag{12.5, 1}$$

(cf. (11.511, 8)), where $Nk/2m$ is the total specific heat c_v of the gas, and $N_r k/2m$ is the part of this which corresponds to the relaxation time τ_r.

* J. W. Buddenberg and C. R. Wilke, *Ind. Eng. Chem.* **41**, 1345 (1949).

† R. S. Brokaw, *J. Chem. Phys.* **29**, 391 (1958), and **42**, 1140 (1965); W. E. Francis, *Trans. Faraday Soc.* **54**, 1492 (1958); T. G. Cowling, P. Gray and P. G. Wright, *Proc. R. Soc.* A, **276**, 69 (1963); D. Burnett, *J. Chem. Phys.* **42**, 2533 (1965).

The absorption coefficient (per unit length travelled) in a plane sound wave with angular frequency ω and velocity V can be shown to be*

$$\frac{\omega^2}{2\rho V^3}\left[\tfrac{4}{3}\mu+\frac{\gamma-1}{\gamma}\frac{\lambda}{c_v}+\varpi\right]. \qquad (12.5, 2)$$

Thus the dissipative effect of volume viscosity, like that of μ and λ, is especially important when the density varies very rapidly, as in ultrasonic waves (or shock waves). However, a description simply in terms of a volume viscosity is inadequate if the time-scale of density-variations is too short. If in ultrasonic waves $\omega\tau_r$ becomes large, the internal energy corresponding to τ_r does not share in the temperature-changes, nor contribute appreciably to the absorption; the effective specific heat of the gas is correspondingly less, and the sound velocity increases.

In (12.5, 2) μ and λ/c_v are comparable with $p\tau$, where τ is an ideal collision-frequency (cf. (6.6, 5, 7)). Volume viscosity is normally found to increase the absorption coefficient considerably above that due to μ and λ alone, which implies that certain of the τ_r are large compared with τ. The prime interest in ultrasonic measurements is in the determination of the τ_r. One writes

$$\tau_r = \zeta_r\tau, \qquad (12.5, 3)$$

so that ζ_r is, roughly speaking, the number of collisions required to secure energy-balance between the internal mode considered and the translational motions. To make τ precise, the formulae $\tau = l/\bar{C}$, $\mu = 0{\cdot}499\rho\bar{C}l$ appropriate to elastic spheres are used (cf. 5.21 and (12.1, 6)); thus

$$\mu = 0{\cdot}499\rho(\bar{C})^2\tau = 1{\cdot}271p\tau \qquad (12.5, 4)$$

by (4.11, 2).

Experiments at low frequencies, where the relaxation effects are adequately represented by a volume viscosity, determine the sum $\sum_r N_r\tau_r$ in (12.5, 1); they do not suffice to separate the effects of the different energy-modes. To determine the separate τ_r it is necessary to carry the experiments past the point where $\omega\tau_r = 1$; the change in effective specific heat enables one to identify the corresponding N_r, and a maximum of absorption occurs roughly at $\omega\tau_r = 1$. Experiments at such high frequencies show that the major part of the volume viscosity (12.5, 1) is normally due to vibrational relaxation. It is usually sufficient to use a single relaxation time for all the vibrational energy-modes. At ordinary temperatures ζ_r for vibrational energy takes values ranging from above 10^7, for oxygen and nitrogen, through 10^4, for other common diatomic and polyatomic gases, down to 100 or less for more complicated and polar molecules.

Hydrogen at ordinary temperatures has too little vibrational energy for this to contribute appreciably to the absorption. The rotational relaxation of

* See, e.g. K. F. Herzfeld and T. A. Litovitz, *Absorption and Dispersion of Ultrasonic Waves*, chapter 1 (Academic Press, 1959).

hydrogen and deuterium is relatively slow, ζ_r being about 300 for hydrogen and 200 for deuterium at ordinary temperatures. For other gases ζ_r for rotational energy is relatively small and difficult to measure; values in the range from 3 to 20 are normal.

For full details, and for the theory of the relaxation processes, reference should be made to specialist books on the subject.*

REFERENCES

(1) J. Kestin and W. Leidenfrost, *Physica*, **25**, 537, 1033 (1959); J. Kestin and J. H. Whitelaw, *Physica*, **29**, 335 (1963).

(2) H. L. Johnston and E. R. Grilly, *J. Phys. Chem.* **46**, 948 (1942); H. L. Johnston and K. E. McCloskey, *J. Phys. Chem.* **44**, 1038 (1940).

(3) R. Wobser and Fr. Müller, *Kolloidbeihefte*, **52**, 165 (1941).

(4) H. K. Onnes, C. Dorsman and S. Weber, *Vers. Kon. Akad. Wet. Amst.* **21**, 1375, 1913.

(5) H. K. Onnes and S. Weber, *Proc. Sect. Sci. K. ned. Akad. Wet.* **21**, 1385 (1913).

(6) M. Trautz and R. Zink, *Annln Phys.* **7**, 427 (1930).

(7) A. B. van Cleave and O. Maass, *Can. J. Res.* **13** B, 384 (1935).

(8) M. Trautz and H. E. Binkele, *Annln Phys.* **5**, 561 (1930).

(9) M. Trautz and R. Heberling, *Annln Phys.* **10**, 155 (1931).

(10) M. Trautz and K. Sorg, *Annln Phys.* **10**, 81 (1931).

(11) A. O. Rankine and C. J. Smith, *Phil. Mag.* **42**, 615 (1921).

(12) M. Trautz and A. Narath, *Annln Phys.* **79**, 637 (1926).

(13) H. Braune and R. Linke, *Z. Phys. Chem.* A, **148**, 195 (1930).

(14) A. O. Rankine, *Proc. R. Soc.* A, **84**, 188 (1910).

(15) A. van Itterbeek and O. van Paemel, *Physica*, **5**, 1009 (1938); J. M. J. Coremans, A. van Itterbeek, J. J. M. Beenakker, H. F. P. Knaap and P. Zandbergen, *Physica*, **24**, 557 (1958).

(16) M. Trautz and P. B. Baumann, *Annln Phys.* **2**, 733 (1929); M. Trautz and F. Kurz, *Annln Phys.* **9**, 981 (1931).

(17) A. G. de Rocco and J. O. Holford, *J. Chem. Phys.* **28**, 1152 (1958).

(18) H. Iwasaki and J. Kestin, *Physica*, **29**, 1345 (1963).

(19) A. Gille, *Annln Phys.* **48**, 799 (1915).

* K. F. Herzfeld and T. A. Litovitz, *op. cit.*, chapters 6 and 7; T. L. Cottrell and J. C. McCoubrey, *Molecular Energy Transfer in Gases*, chapters 5 and 6 (Butterworths, 1961).

13

THERMAL CONDUCTIVITY: COMPARISON OF THEORY WITH EXPERIMENT

13.1. Summary of the formulae

Various formulae for the coefficient of conduction of heat, λ, in a simple gas have been derived in Chapters 10 and 11. It has been found that λ is connected with the coefficient of viscosity, μ, by an equation of the form

$$\lambda = \mathrm{f}\mu c_v \tag{13.1, 1}$$

(cf. (6.3, 3)), where c_v is the specific heat at constant volume, and f is a pure number. The first approximations, $[\lambda]_1$ and $[\mu]_1$, to λ and μ are similarly connected (cf. 9.7) by a relation

$$[\lambda]_1 = [\mathrm{f}]_1 [\mu]_1 c_v,$$

where $[\mathrm{f}]_1$ is the first approximation to f.

The value of $[\mathrm{f}]_1$ has been found equal to 2·5 for all smooth spherically symmetrical molecules (cf. (9.7, 4)). Further approximations to f have been determined in a number of cases, It was shown in (10.21, 5) that for rigid elastic spheres $\mathrm{f} = 2{\cdot}522$; for molecules repelling each other with a force varying inversely as the νth power of the distance (cf. (10.32, 2, 3)), $\mathrm{f} = 2{\cdot}5$ exactly if $\nu = 5$ (Maxwellian molecules), whereas for values of ν between 5 and 13 (the important range in Table 14, p. 232), f lies between 2·5 and about 2·511. Similar results apply to the 12, 6 and exp;6 models, for which f is nearly 2·5 when kT/ϵ is comparable with unity, and rises to about 2·51 when kT/ϵ is large. We may therefore expect that

$$\lambda = \tfrac{5}{2}\mu c_v \tag{13.1, 2}$$

is very nearly true for all smooth spherically symmetric molecules.

A variety of approximate expressions for $[\mathrm{f}]_1$ have been suggested for diatomic and polyatomic gases. Assuming that the transport of momentum and translational energy is unaffected by the internal molecular motions, and that internal energy is transported at the same rate as momentum, Eucken derived the formula

$$[\mathrm{f}]_1 = \tfrac{1}{4}(9\gamma - 5) \tag{13.1, 3}$$

(cf. (11.8, 3)), where γ is the ratio of specific heats. The modified Eucken formula (11.8, 5) postulates that, when interconversion between internal and translational energy is very slow, the transport of internal energy is by a self-diffusion process. If $[D_{11}]_1$ is the self-diffusion coefficient, and $u'_{11} = \rho[D_{11}]_1/[\mu]_1$, (13.1, 3) is replaced by

$$[f]_1 = \tfrac{1}{4}\{15(\gamma-1)+2u'_{11}(5-3\gamma)\}. \qquad (13.1, 4)$$

Both (13.1, 3) and (13.1, 4) ignore interchange between the internal and translational energies at collision. The Mason–Monchick formula (11.81, 5), which takes this interchange into account, is equivalent to

$$[f]_1 = \tfrac{1}{4}\{15(\gamma-1)+2u''_{11}(5-3\gamma)\} - \tfrac{1}{4}(5-3\gamma)(\tfrac{5}{2}-u''_{11})^2 \Big/ \left\{\frac{p[\tau]_1}{[\mu]_1}+\tfrac{1}{2}u''_{11}+\frac{5(5-3\gamma)}{12(\gamma-1)}\right\}. \qquad (13.1, 5)$$

Here τ is a time of relaxation of internal energy into translational, and $u''_{11} = \rho D_{\text{int}}/[\mu]_1$; D_{int} (not now assumed identical with $[D_{11}]_1$) is the first approximation to the coefficient of diffusion of internal energy. If the interchange between internal and translational energies is slow, τ is large and D_{int}, u''_{11} approximate to $[D_{11}]_1$, u'_{11}; hence (cf. (11.81, 6)) (13.1, 5) approximates to

$$[f]_1 = \tfrac{1}{4}\{15(\gamma-1)+2u'_{11}(5-3\gamma)\} - \tfrac{1}{4}(5-3\gamma)(\tfrac{5}{2}-u'_{11})^2[\mu]_1/p[\tau]_1. \qquad (13.1, 6)$$

If different parts of the internal energy, making contributions $N_1k/2m$, $N_2k/2m$, ... to the specific heat $c_v\ (\equiv Nk/2m)$, possess different relaxation times $\tau_1, \tau_2, \ldots$, (13.1, 6) is replaced by

$$[f]_1 = \tfrac{1}{4}\{15(\gamma-1)+2u'_{11}(5-3\gamma)\} - \tfrac{1}{2}(\tfrac{5}{2}-u'_{11})^2([\mu]_1/pN)\sum_r N_r/[\tau_r]_1. \qquad (13.1, 7)$$

The results for rough spherical molecules agree with (13.1, 6) to a first approximation. However, they suggest that higher approximations to f may appreciably exceed $[f]_1$ for molecules with internal energy.

13.2. Thermal conductivities of gases at 0 °C.

Theory predicts that λ, like μ, is independent of the pressure at a given temperature. As with μ, this result is valid only to the extent that the volume of the molecules themselves is a negligible fraction of the total volume occupied by the gas; however, the variation of λ with p is relatively insignificant at ordinary pressures, in a gas not close to liquefaction. An apparent variation with p may result from the onset of convection, or from the temperature-jump at a wall (cf. 6.31), and these need to be taken into account in any experimental determination of λ. Corrections for losses of heat by radiation or by conduction in the solid parts of the apparatus also have to be made, and altogether the experimental determination of λ is a matter of considerable difficulty. The probable error (as distinct from internal inconsistencies) of a set of measured values can be estimated as not less than about one per cent.

Table 20 gives experimental values* of λ for gases at S.T.P., and values of f, or $\lambda/\mu c_v$, derived from them using the values of μ and c_v given in Tables 11 (p. 228) and 1 (p. 43). Eucken's value $\frac{1}{4}(9\gamma - 5)$ for f is also tabulated for comparison. The gases in the table are listed in three groups, first the monatomic gases, then the diatomic and polyatomic gases whose molecules possess either no permanent dipole moment or a weak one, and finally certain more strongly polar gases. The data for gases in these three groups will be discussed separately.

Table 20. *Thermal conductivities of gases at 0 °C.*

Gas	$\lambda \times 10^7$ (cal./cm. sec. deg.)	$f \equiv \lambda/\mu c_v$	$\frac{1}{4}(9\gamma - 5)$
	Monatomic gases		
Helium	3410	2·45	2·50
Neon†	1110	2·52	2·50
Argon	392	2·48	2·50
Krypton†	210	2·535	2·50
Xenon	124	2·58	2·50
	Gases with weak or no dipole moment		
Hydrogen	4130	2·02	1·92
Deuterium	3050	2·07	1·90
Methane	733	1·77	1·695
Acetylene	470	1·54	1·54
Carbon monoxide	553	1·91	1·90
Nitrogen	575	1·96	1·91
Ethylene	403	1·48	1·56
Air	577	1·96	1·90
Nitric oxide	562	1·90	1·90
Ethane	426	1·44	1·45
Oxygen	585	1·94	1·90
Carbon dioxide	348	1·64	1·68
Nitrous oxide	362	1·73	1·68
Chlorine	189	1·825	1·80
	Strongly polar gases		
Ammonia	524	1·455	1·715
Hydrogen sulphide	308	1·49	1·77
Hydrogen chloride	301	1·655	1·90
Methyl chloride	215	1·53	1·64
Sulphur dioxide	204	1·50	1·64

* In general, the values quoted are mean values derived from *Thermophysical Properties Research Center, Data Book 2* (Wright-Patterson Air Force Base, Ohio, 1964 and 1966). However, extensive reference to the original papers has also been made. In consequence, the values do not in all cases agree with those in the Data Book, though differing from them only by amounts which its authors suggest are permissible.

† The values of c_v for the monatomic gases neon and krypton are calculated from $c_v = 3k/2mJ$.

13.3. Monatomic gases

The relation $\lambda = \frac{5}{2}\mu c_v$, (13.1, 2), can be expected to apply with reasonable accuracy to monatomic gases, whose molecules have no convertible internal energy and are to a high degree spherically symmetric. The testing of this relation for such gases is of special interest, since here theory affords a definite numerical relation between three quantities all determinable directly by experiment. Moreover, the agreement of theory with experiment for the monatomic gases is crucial to the theory of other gases; the formulae (13.1, 3–5) are all based on the assumption that the transport of translational energy conforms, to a first approximation, to the formula $\lambda' = \frac{5}{2}\mu c_v'$.

Table 20 indicates that the values of $\lambda/\mu c_v$ for the inert gases are indeed in the neighbourhood of 2·5, but the deviations from this value are appreciable. For neon, argon and krypton the deviations are within the bounds of experimental error, and the difficulties of working with xenon are such that the deviation in this case cannot be regarded as serious. The deviation for helium is 2 per cent, which is more than would be expected for a gas so frequently studied; improved accuracy of measurement may remove the discrepancy.

Experiments on gas-flow through a nozzle have been used by O'Neal and Brokaw (1)* to determine directly the Prandtl number $\mu c_p/\lambda \equiv \gamma/\mathrm{f}$ for a number of gases. They found $\mathrm{f} = 2{\cdot}503$ for helium and $\mathrm{f} = 2{\cdot}506$ for argon, in satisfactory agreement with theory. However, though their method avoids the possibility of errors in the determinations of λ and μ separately, the possible sources of error in the method itself have not been fully explored.

Some runs of experiments suggest a variation of f with temperature, but different runs do not always agree even as to its sense.† It is doubtful whether this variation is real for monatomic gases. At very low temperatures (below about 20 °K.) values of λ for helium calculated from experimental values of μ using $\lambda = \frac{5}{2}\mu c_v$ still agree reasonably well with direct experimental values (3), (4), though naturally the precision is less at such low temperatures.

For mercury vapour Schleiermacher (5) gave 3·15 as the value of f. Zaitseva (6) has pointed out that if Schleiermacher's value of λ is combined with more modern values of μ, the value $\mathrm{f} = 2{\cdot}69$ is obtained, in rough agreement with her own results.

For hydrogen the value of the ratio γ of specific heats begins to increase appreciably as the temperature falls below 0 °C., and near −200 °C. it approaches the value $\frac{5}{3}$ appropriate to a monatomic gas. This indicates that at such temperatures the molecular energy is almost wholly translatory. It is therefore to be expected that f will rise towards 2·5 below −200 °C. Eucken (7) and Ubbink (8) found that this is actually so; Eucken found that

* As in Chapter 12, numbers thus written refer to a list of experimental papers given at the end of this chapter.

† An extensive review of the experimental values of λ, μ and f has been made by Keyes (2).

f = 2·25 at −192 °C., while Ubbink found that f is between 2·45 and 2·5 at temperatures below 20 °K. For deuterium Ubbink similarly found a value near 2·3 at low temperatures.

We may therefore conclude that for monatomic gases f is approximately 2·5, as theory predicts.

13.31. Non-polar gases

For diatomic and polyatomic gases, equations (13.1, 3–7) provide approximations to $[\mathrm{f}]_1$. As is shown by the values quoted in Table 20, the Eucken formula (13.1, 3) gives only a crude approximation to f.

If encounters leave the internal energies of molecules virtually unaltered, (13.1, 4) is applicable; in this case $[D_{11}]_1$ approximates to its value $3\mathrm{A}[\mu]_1/\rho$ for molecules without internal energy (cf. (10.6, 4)), and u'_{11} becomes $3\mathrm{A}$. For the 12, 6 and exp; 6 models $3\mathrm{A}$ is about 1·32 or a little larger, and here we shall use the mean value $3\mathrm{A} = \frac{4}{3}$. Then (13.1, 4) becomes

$$[\mathrm{f}]_1 = \tfrac{1}{12}(21\gamma - 5). \qquad (13.31, 1)$$

This value of $[\mathrm{f}]_1$ is always greater than the Eucken value $\frac{1}{4}(9\gamma - 5)$.

Equations (13.1, 5–7) take into account the interchange between internal and translational energies. This interchange increases the part of λ due to the transport of internal energy, but decreases that due to transport of translational energy to a greater extent, so that $[\mathrm{f}]_1$ is decreased below the value given by (13.31, 1). Equation (13.1, 6) is a simplified version of (13.1, 5), valid only if the interchange between internal and translational energies is fairly slow. Equation (13.1, 7) is preferable to (13.1, 6), since it allows for the possibility of different relaxation times for vibrational and rotational energy. The relaxation of vibrational energy is normally so slow that its effect on $[\mathrm{f}]_1$ is negligible; hence, since $\gamma = 1 + 2/N$, (13.1, 7) reduces to

$$[\mathrm{f}]_1 = \tfrac{1}{4}\{15(\gamma - 1) + 2\mathrm{u}'_{11}(5 - 3\gamma) - (\tfrac{5}{2} - \mathrm{u}'_{11})^2(\gamma - 1)\,[\mu]_1 N_r/p[\tau_r]_1\}, \qquad (13.31, 2)$$

where N_r, $[\tau_r]_1$ now refer to the rotational energy. For diatomic and other linear molecules $N_r = 2$, for non-linear polyatomic molecules $N_r = 3$; also we may write

$$[\tau_r]_1 = \zeta_r[\mu]_1/1{\cdot}271p, \qquad (13.31, 3)$$

where (cf. (12.5, 3)) ζ_r is the average number of 'collisions' occurring in the time $[\tau_r]_1$.

In (13.31, 2), the interchange of rotational and translational energy is still supposed to be fairly slow, so that $\mathrm{u}'_{11} \sim \frac{4}{3}$. If the interchange is not slow, $\mathrm{u}'_{11}\,(\equiv \rho[D_{11}]_1/[\mu]_1)$ must be replaced by $\mathrm{u}''_1\,(\equiv \rho D_{\mathrm{int}}/[\mu]_1)$; also (cf. (13.1, 5)) $p[\tau_r]_1$ is replaced by $p[\tau_r]_1 + [\mu]_1(\frac{1}{2}\mathrm{u}''_{11} + \frac{5}{12}N_r)$, so that (13.31, 2) becomes

$$[\mathrm{f}]_1 = \tfrac{1}{4}\{15(\gamma - 1) + 2\mathrm{u}''_{11}(5 - 3\gamma) - (\tfrac{5}{2} - \mathrm{u}''_{11})^2(\gamma - 1)\,N_r/(p[\tau_r]_1/[\mu]_1 + \tfrac{1}{2}\mathrm{u}''_{11} + \tfrac{5}{12}N_r)\}. \qquad (13.31, 4)$$

For hydrogen and deuterium (13.31, 4) and (13.31, 2) are virtually indistinguishable, since $\zeta_r \sim 300$ for these gases. From (13.31, 1) we find the values $[\mathrm{f}]_1 = 2{\cdot}05$ and $[\mathrm{f}]_1 = 2{\cdot}03$ for the two, agreeing satisfactorily with the experimental values $\mathrm{f} = 2{\cdot}02$ and $\mathrm{f} = 2{\cdot}07$ of Table 20.

For other non-polar gases the values of f in Table 20 often agree better with the Eucken expression $\frac{1}{4}(9\gamma-5)$ than with (13.31, 1). This corresponds to a relatively low value of $[\tau_r]_1$, and so of the collision-number ζ_r: thus (13.31, 4) must be used rather than (13.31, 2). The value of $[\mathrm{f}]_1$ given by (13.31, 4) agrees with the Eucken expression if

$$\frac{\zeta_r}{1{\cdot}271}+\tfrac{1}{2}\mathrm{u}''_{11}+\tfrac{5}{12}N_r=\frac{(\frac{5}{2}-\mathrm{u}''_{11})^2}{2(\mathrm{u}''_{11}-1)}\,\frac{N_r(\gamma-1)}{5-3\gamma}. \qquad (13.31,\,5)$$

For a diatomic gas whose internal energy is wholly rotational ($N_r = 2$, $\gamma = 1{\cdot}4$), the following table indicates the dependence of ζ_r on u''_{11} implied by (13.31, 5):

u''_{11}	...	$\frac{4}{3}$	1·3	1·25	1·2
ζ_r	...	0·688	1·165	2·12	3·55

The values of ζ_r for a polyatomic gas, or one whose molecules have vibrational energy, are even lower. Since ζ_r is likely to be well above unity, values of u''_{11} as high as 1·3 are improbable.

For methane, nitrogen and oxygen the values of f given in Table 20 are appreciably above the corresponding Eucken values; even for these, the values of ζ_r corresponding to $\mathrm{u}''_{11} = 1{\cdot}3$ are between 2 and 4. Ultrasonic experiments give distinctly higher values* (though not very accordant), indicating smaller values of u''_{11}. Thus the coefficient of diffusion of internal energy, D_{int}, is likely to be appreciably less than the self-diffusion coefficient $[D_{11}]_1$.

The values in Table 20 refer to a temperature of 0 °C. Theory† suggests that ζ_r may increase considerably with increasing temperature; correspondingly u''_{11} may also increase towards $\frac{4}{3}$. Thus at high temperatures f may approximate more closely to the value given by the modified Eucken formula (13.31, 1) or (13.1, 4). Mason and Monchick‡ give evidence that this is actually so for a number of gases.

13.32. Polar gases

As appears from Table 20, for polar gases f is well below the Eucken value. Mason and Monchick§ ascribed this to rotation changes during encounters, caused by dipole moment interactions. They calculated that for certain gases, notably hydrogen chloride and ammonia, this may reduce the diffusion

* See, for example, the values quoted by O'Neal and Brokaw (1).

† See J. G. Parker, *Phys. Fluids*, **2**, 449 (1959); also the references quoted in 12.5.

‡ E. A. Mason and L. Monchick, *J. Chem. Phys.* **36**, 1622 (1962).

§ *Loc. cit.*

coefficient D_{int} for internal energy to make u''_{11} in (13.1, 5) equal to about unity, as well as decreasing the collision-number ζ_r. For details the reader is referred to the original paper.

13.4. Monatomic gas-mixtures

The first approximation to λ for a binary mixture of gases without internal energy is (cf. 9.82)

$$[\lambda]_1 = \frac{x_1^2 Q_1 [\lambda_1]_1 + x_2^2 Q_2 [\lambda_2]_1 + x_1 x_2 Q'_{12}}{x_1^2 Q_1 + x_2^2 Q_2 + x_1 x_2 Q_{12}}, \tag{13.4, 1}$$

where
$$Q_{12} = 3(M_1 - M_2)^2 (5 - 4B) + 4M_1 M_2 A(11 - 4B) + 2P_1 P_2, \tag{13.4, 2}$$

$$Q'_{12} = 2F\{P_1 + P_2 + (11 - 4B - 8A) M_1 M_2\}, \tag{13.4, 3}$$

$$Q_1 = P_1\{6M_2^2 + (5 - 4B) M_1^2 + 8M_1 M_2 A\}, \tag{13.4, 4}$$

with a similar formula for Q_2. Here (cf. (9.82, 2)) P_1, P_2 and F are given by

$$F \equiv 15kE/4m_0, \quad P_1 = F/[\lambda_1]_1, \quad P_2 = F/[\lambda_2]_1. \tag{13.4, 5}$$

The constants A, B are given for force-centres by (10.31, 10) and Table 3; for the 12, 6 model they are given in Table 6 (10.42). For most of the significant range of ϵ_{12}/kT for the 12, 6 and exp; 6 models, A lies between about 0·44 and 0·46, and B between 0·72 and 0·66; (13.4, 1) is not very sensitive to the precise values adopted for them.

If x_2 is small, (13.4, 1) becomes the formula giving the effect of an impurity on the conductivity; it then approximates to

$$[\lambda]_1 = [\lambda_1]_1 \left\{1 + \frac{x_2}{2Q_1}\left[\{2P_1 + M_1 M_2(11 - 4B - 8A)\}^2 - \frac{Q_1 Q_2}{P_1 P_2}\right]\right\}, \tag{13.4, 6}$$

an equation which can readily be seen not to involve $[\lambda_2]_1$.

As in the case of a simple gas, (13.4, 1) is applicable only to mixtures of monatomic gases. All pairs from among the five inert gases have been studied, with reasonable agreement between theory and experiment. This is illustrated in Table 21 by results for helium–argon mixtures at 29 °C., based on the experimental data of von Ubisch (9). In applying (13.4, 1) the values $A = 0{\cdot}452$, $B = 0{\cdot}667$ are adopted, the value of A being that used in 12.41 in discussing the viscosity of the same mixture. The value of E giving the best fit with experiment is found to be $9{\cdot}103 \times 10^{-4}$ poise, as against the value $8{\cdot}70 \times 10^{-4}$ poise (at 15 °C.) used in 12.41. About half of the difference between the two may be ascribed to the different temperatures involved; the remainder is due partly to experimental error (including possible errors in the values of λ_1 and λ_2) and partly to the fact that the formulae used are not exact, but are first approximations to the exact formulae.

Table 21 gives a second set of calculated values obtained by Mason and von Ubisch* from (13.4, 1); they used the same values of A, B, but different

* E. A. Mason and H. von Ubisch, *Phys. Fluids*, 3, 355 (1960).

values of E, $[\lambda_1]_1$, $[\lambda_2]_1$, derived on the exp;6 model with the force constants for unlike molecules determined by empirical combination rules. As can be seen, the three sets of values are in reasonable agreement.

Table 21. *Thermal conductivity of mixtures of helium and argon at 29 °C.*

Helium (%)	$\lambda\times10^7$ (exp.)	$\lambda\times10^7$ (calc.)	$\lambda\times10^7$ (calc.) (Mason and von Ubisch)
0	434	434	426
29·0	843	857	854·5
45·9	1220	1205	1205
72·4	2020	2029	2030
89·4	2920	2889	2897
100	3670	3670	3702

A discussion of possible maxima and minima of $[\lambda]_1$ as x_1/x_2 varies, parallel to the corresponding discussion for $[\mu]_1$, can be based on (13.4, 1). However, such maxima and minima are not observed for mixtures of monatomic gases.

13.41. Mixtures of gases with internal energy

Hirschfelder* has generalized the modified Eucken formula (13.1, 4) to apply to mixtures of gases whose molecules possess internal energy. Translational and internal energies are regarded as conducted independently. The translational conductivity λ' is then the monatomic gas conductivity given by (13.4, 1) with $[\lambda_1]$, $[\lambda_2]_1$ replaced by $\frac{5}{2}[\mu_1]_1 c'_{v1}$, $\frac{5}{2}[\mu_2]_1 c'_{v2}$, where c'_{v1}, c'_{v2} are the translational specific heats. The contribution of internal energy to λ is assumed to arise from the diffusion of molecules through the mixture, carrying with them their internal energy. The final formula (which can readily be derived by applying to mixtures the approximations made in deriving (11.8, 9)) is

$$\lambda = \lambda' + \frac{\rho_1 c''_{v1}}{x_1 D_{11}^{-1} + x_2 D_{12}^{-1}} + \frac{\rho_2 c''_{v2}}{x_1 D_{12}^{-1} + x_2 D_{22}^{-1}}, \qquad (13.41, 1)$$

where D_{11}, D_{12}, D_{22} are the usual coefficients of diffusion and self-diffusion, at the pressure of the actual gas.

The significance of the expression $x_1 D_{11}^{-1} + x_2 D_{12}^{-1}$ in (13.41, 1) is that D_{11}^{-1}, D_{12}^{-1} denote the resistances to the diffusion of a molecule m_1 in otherwise pure gases of molecules m_1 or m_2; $x_1 D_{11}^{-1} + x_2 D_{12}^{-1}$ gives a weighted mean of these resistances appropriate to the actual mixture. The conductivity λ''_1 due

* J. O. Hirschfelder, *Sixth International Combustion Symposium*, p. 351 (Reinhold Publication Corporation, New York, 1957). See also E. A. Mason and S. C. Saxena, *Phys. Fluids*, 1, 361 (1958).

to internal energy in a pure gas of molecules m_1 at the same pressure is $c''_{v1}(\rho_1/x_1) D_{11}$, since the density of the gas m_1 in the actual mixture is x_1 times that in the pure gas. Hence (13.41, 1) can also be written

$$\lambda = \lambda' + \frac{\lambda''_1}{1 + x_2 D_{11}/x_1 D_{12}} + \frac{\lambda''_2}{1 + x_1 D_{22}/x_2 D_{12}}. \qquad (13.41, 2)$$

This equation can readily be generalized to apply to a mixture of any number of constituents.

Equation (13.41, 2) is strictly applicable only to mixtures like H_2–D_2, in which the interaction between translational and internal energies is very slow. Monchick, Pereira and Mason* have given formulae valid when this interaction is not slow, which generalize (13.1, 5) to apply to gas-mixtures. The formulae, even after a number of approximations, are highly complicated. In general they involve four relaxation times, and coefficients of diffusion for internal energy which, as in simple gases, may deviate considerably from the ordinary diffusion and self-diffusion coefficients. Monchick, Pereira and Mason had reasonable success in fitting the experimental data, both by their own formula and by the Hirschfelder formula (13.41, 2), for a number of gases. They suggest that it is usually sufficient to use (13.41, 2), with D_{11}, D_{12}, D_{22} replaced by adjustable diffusion coefficients for internal energy. In calculating λ' from (13.4, 1), however, they prefer, when the pure-gas relaxation times are known, to replace $[\lambda_1]_1$, $[\lambda_2]_1$ in that equation by translational conductivities λ'_1, λ'_2 given by

$$\lambda'_1 = \tfrac{5}{2}[\mu_1]_1 \{c'_{v1} - \tfrac{1}{2}c''_{v1}(\tfrac{5}{2}[\mu_1]_1 - \rho_1[D_{11}]_1)/p_1\tau_1\} \qquad (13.41, 3)$$

and a similar relation; (13.41, 3) gives the translational contribution to the total conductivity derived from (11.81, 6).

13.42. Approximate formulae for mixtures

As with viscosity, the complication of the formulae for λ for mixtures has led to a search for simple approximate formulae. These are normally of the Wassiljewa form (cf. 6.63), expressing λ as the sum of parts $\lambda_{(1)}$, $\lambda_{(2)}$ arising from the two gases in the mixture, where

$$\lambda_{(1)} = x_1/\{x_1(\lambda_1)^{-1} + x_2(\lambda'_{12})^{-1}\}, \quad \lambda_{(2)} = x_2/\{x_2(\lambda_2)^{-1} + x_1(\lambda'_{21})^{-1}\}, \qquad (13.42, 1)$$

and λ'_{12}, λ'_{21} are 'mutual conductivities' of the gases considered.

Such a formula can be derived for monatomic gases from (13.4, 1) by making the (unphysical) assumptions $B = \frac{3}{4}$, $A = 1$; then this equation becomes

$$[\lambda]_1 = x_1/\{x_1[\lambda_1]_1^{-1} + x_2[\lambda'_{12}]_1^{-1}\} + x_2/\{x_2[\lambda_2]_1^{-1} + x_1[\lambda'_{21}]_1^{-1}\}, \qquad (13.42, 2)$$

where $\quad [\lambda'_{12}]_1^{-1} = (3M_1 + M_2)/F, \quad [\lambda'_{21}]_1^{-1} = (3M_2 + M_1)/F, \qquad (13.42, 3)$

* L. Monchick, A. N. G. Pereira and A. E. Mason, *J. Chem. Phys.* **42**, 3241 (1965). See also L. Monchick, K. S. Yun and A. E. Mason, *J. Chem. Phys.* **39**, 654 (1963).

and $\text{F} = 15k\text{E}/4m_0 = 5p[D_{12}]_1/2T$. However, the error involved in making these substitutions is not likely to be small. A rather better approximation is obtained by applying to the conductivity the argument employed with the viscosity in deriving (12.43, 3); this gives an equation of form (13.42, 2), with

$$[\lambda'_{12}]_1^{-1} = \tfrac{1}{2}\{(M_1 - M_2)[6M_1 - (5 - 4\text{B})M_2] + 16M_1M_2\text{A}\}/\text{F} \qquad (13.42, 4)$$

and a similar equation for $[\lambda'_{21}]_1^{-1}$. This equation in general overestimates $[\lambda]_1$, but not greatly if the molecules of the two gases are not too unlike.*

Formulae of the Wassiljewa type have been used, with fair success, to represent the variation of λ with composition when the molecules possess internal energy. If no account is paid to physical interpretation, λ'_{12} and λ'_{21} can be regarded as adjustable parameters. More usually it is desired to predict conductivities for mixtures from those of the pure gases. The procedure normally followed in this case is to start with a formula of type

$$\lambda = \lambda_1 x_1/\{x_1 + \alpha_{12}x_2 D_{11}/D_{12}\} + \lambda_2 x_2/\{x_2 + \alpha_{21}x_1 D_{22}/D_{12}\}, \qquad (13.42, 5)$$

where α_{12}, α_{21} are regarded as functions of m_1/m_2, which are determined semi-empirically. In support of the assumption (13.42, 5) we may note that substitution from (13.42, 3) or (13.42, 4) into (13.42, 2) yields an expression of the form (13.42, 5), since λ_1 is a numerical multiple of pD_{11}/T; so too is the contribution to (13.41, 2) from the internal energy.

The Lindsay–Bromley formula† is one of the best known of this type. It is based on the Sutherland model, and replaces $\alpha_{12}D_{11}/D_{12}$ in (13.42, 5) by

$$\frac{1}{4}\left\{1 + \left[\frac{\mu_1}{\mu_2}\left(\frac{m_2}{m_1}\right)^{\frac{3}{4}} \frac{1 + S_1/T}{1 + S_2/T}\right]^{\frac{1}{2}}\right\}^2 \frac{1 + S_{12}/T}{1 + S_1/T}. \qquad (13.42, 6)$$

To conform exactly to the Sutherland model with $\alpha_{12} = \alpha_{21} = 1$, the power of m_2/m_1 in this should be $(m_2/m_1)^{\frac{1}{2}}$, and the whole expression should be multiplied by $\{2m_2/(m_1 + m_2)\}^{\frac{1}{2}}$; the changes made are empirical corrections. Lindsay and Bromley further suggest that the Sutherland constant S_{12} can be taken as $(S_1 S_2)^{\frac{1}{2}}$ for non-polar gases, but if one gas is strongly polar it may be only 0·733 times as large.

REFERENCES

(1) C. O'Neal and R. S. Brokaw, *Phys. Fluids*, **5**, 567 (1962), and **6**, 1675 (1963).
(2) F. G. Keyes, *Trans. Am. Soc. Mech. Eng.* **73**, 589 (1951).
(3) J. B. Ubbink, *Physica*, **13**, 629 (1947).
(4) K. Fokkens, W. Vermeer, K. W. Taconis and R. de Bruyn Ouboter, *Physica*, **30**, 2153 (1964).
(5) A. Schleiermacher, *Annln Phys.* **36**, 346 (1889).
(6) L. S. Zaitseva, *Sov. Phys. Tech. Phys.* **4**, 444 (1959).
(7) A. Eucken, *Phys. Zeit.* **14**, 324 (1913).
(8) J. B. Ubbink, *Physica*, **14**, 165 (1948).
(9) H. von Ubisch, *Ark. Fys.* **16**, 93 (1959).

* T. G. Cowling, P. Gray and P. G. Wright, *Proc. R. Soc.* A, **276**, 69 (1963).
† A. L. Lindsay and L. A. Bromley, *Ind. Eng. Chem.* **42**, 1508 (1950).

14

DIFFUSION: COMPARISON OF THEORY WITH EXPERIMENT

14.1. Causes of diffusion

The general equation of diffusion for a binary mixture, (8.4, 7), can be put in the form

$$\overline{\boldsymbol{C}}_1 - \overline{\boldsymbol{C}}_2 = -(\mathrm{x}_1\mathrm{x}_2)^{-1} D_{12}\left\{\nabla \mathrm{x}_1 + \frac{n_1 n_2(m_2-m_1)}{n\rho}\nabla \ln p - \frac{\rho_1\rho_2}{p\rho}(\boldsymbol{F}_1-\boldsymbol{F}_2) + \mathrm{k}_T \nabla \ln T\right\}, \quad (14.1, 1)$$

showing, as noted in 8.4, that the velocity of diffusion has components due to non-uniformity of the composition, the pressure and the temperature of the gas, and another component due to the different accelerative effects of the external forces on the molecules of the two constituent gases. In experimental determinations of the coefficient of diffusion, D_{12}, it is usually the diffusion due to non-uniformity of composition which is measured.

In a gas at rest (zero total *number*-flow of molecules) the equation of diffusion becomes (cf. (8.3, 13))

$$\mathrm{x}_1\overline{\boldsymbol{c}}_1 = -\mathrm{x}_2\overline{\boldsymbol{c}}_2 = -D_{12}\{p^{-1}(\nabla p_1 - \rho_1 \boldsymbol{F}_1) + \mathrm{k}_T \nabla \ln T\}. \quad (14.1, 2)$$

The first term on the right corresponds to diffusion set up by a lack of balance between the forces on the molecules of the first gas, and the gradient of its partial pressure. A similar equation exists in terms of the corresponding quantities for the second gas.

As an example of *pressure* diffusion, corresponding to the second term on the right of (14.1, 1), we may cite the process of diffusion in the atmosphere; by reason of the variation of pressure with height the various constituents tend to separate out, the heavier elements tending to sink to lower levels, and the lighter elements to rise to higher levels. This process is not immediately due to the force of gravity on the molecules, as the accelerative effect of this on all molecules is the same; it is an indirect effect, due to the pressure gradient which gravity sets up. (Actually but little variation of composition with height is observed in the atmosphere up to 100 km. height, since the mixing effect of wind currents and eddy motion counteracts the separative tendency due to diffusion.)

Pressure diffusion also occurs in a gas that is made to rotate about an axis; the heavy molecules then tend to the parts most distant from the axis, the density-distribution of each constituent in the steady state being similar to that given by (4.14, 9, 11), putting $\Psi = 0$.

The most important example of *forced* diffusion, corresponding to the third term on the right of (14.1, 1) is the diffusion of electrically charged particles in a partially ionized gas under the action of an electric field, giving rise to an electric current. This case is considered in detail in Chapter 19.

The steady state of a gas-mixture in a conservative field of force, represented by (4.3, 5) can be regarded as due to the attainment of a balance between the velocities of diffusion due to non-uniformity of composition and pressure, and to the applied forces. Thus, for example, in an isothermal atmosphere under gravity the steady state is that in which pressure diffusion exactly balances the diffusion due to non-uniformity of composition. It is possible, in fact, to derive (4.3, 5) from the condition that there is no diffusion, using (14.1, 2).

We defer consideration of thermal diffusion till 14.6.

14.2. The first approximation to D_{12}

Observations of the mutual diffusion of pairs of gases are difficult to make, and such observations as have been made are liable to a fairly large experimental error. This should be borne in mind when comparing theory with experiment.

Several formulae for the first approximation $[D_{12}]_1$ to the coefficient of diffusion have been derived in Chapters 9 and 10; they are quoted here for convenience of reference. For rigid elastic spheres of diameters σ_1, σ_2, by (10.1, 5) and (10.2, 1),

$$[D_{12}]_1 = \frac{3}{8n\sigma_{12}^2}\left\{\frac{kT(m_1+m_2)}{2\pi m_1 m_2}\right\}^{\frac{1}{2}}, \qquad (14.2, 1)$$

where $\sigma_{12} = \frac{1}{2}(\sigma_1+\sigma_2)$.

For molecules repelling each other with a force $\kappa_{12}r^{-\nu}$, by (10.1, 5) and (10.31, 8, 9),

$$[D_{12}]_1 = \frac{3}{8nA_1(\nu)\,\Gamma\left(3-\frac{2}{\nu-1}\right)}\left(\frac{kT(m_1+m_2)}{2\pi m_1 m_2}\right)^{\frac{1}{2}}\left(\frac{2kT}{\kappa_{12}}\right)^{2/(\nu-1)}. \qquad (14.2, 2)$$

For attracting spheres (Sutherland's model), by (10.41, 9),

$$[D_{12}]_1 = \frac{3}{8n\sigma_{12}^2}\left(\frac{kT(m_1+m_2)}{2\pi m_1 m_2}\right)^{\frac{1}{2}}\Big/\left(1+\frac{S_{12}}{T}\right). \qquad (14.2, 3)$$

For the 12,6 and exp;6 models, by (10.1, 5) and 10.4,

$$[D_{12}]_1 = \frac{3}{8n\sigma_{12}^2\mathscr{W}_{12}^{(1)}(1)}\left(\frac{kT(m_1+m_2)}{2\pi m_1 m_2}\right)^{\frac{1}{2}}, \qquad (14.2, 4)$$

where $\mathscr{W}_{12}^{(1)}(1)$ is a function of kT/ϵ_{12}, given in Table 6 (p. 185) for the 12,6 model, and illustrated in Fig. 8 (p. 185) for this and the exp;6 model.

For rough elastic spheres, by (11.61, 4),

$$[D_{12}]_1 = \frac{3}{8n\sigma_{12}^2}\left(\frac{kT(m_1+m_2)}{2\pi m_1 m_2}\right)^{\frac{1}{2}} \frac{K_0+K_1K_2}{K_0+2K_1K_2}, \tag{14.2, 5}$$

where K_1, K_2, K_0 are related to the radii of gyration by (11.6, 2).

This indicates that interaction between rotation and translation decreases $[D_{12}]_1$ somewhat. However, the effect of such interaction is not large; normally it is ignored in what follows, the molecules being treated as if smooth.

The first approximation to the coefficient of diffusion does not depend on the proportions of the mixture in which diffusion occurs: also it depends only on encounters between molecules of opposite types. Thus, to a first approximation, encounters between like molecules do not affect diffusion.

14.21. The second approximation to D_{12}

The second approximation obtained in (9.81, 3) is of the form

$$[D_{12}]_2 = [D_{12}]_1/(1-\Delta), \tag{14.21, 1}$$

where

$$\Delta = 5C^2 \frac{M_1^2 P_1 x_1^2 + M_2^2 P_2 x_2^2 + P_{12} x_1 x_2}{Q_1 x_1^2 + Q_2 x_2^2 + Q_{12} x_1 x_2} \tag{14.21, 2}$$

and

$$P_1 = M_1 E/[\mu_1]_1, \quad P_2 = M_2 E/[\mu_2]_1, \tag{14.21, 3}$$

$$P_{12} = 3(M_1 - M_2)^2 + 4M_1 M_2 A, \tag{14.21, 4}$$

$$Q_1 = P_1(6M_2^2 + 5M_1^2 - 4M_1^2 B + 8M_1 M_2 A), \tag{14.21, 5}$$

$$Q_{12} = 3(M_1 - M_2)^2(5-4B) + 4M_1 M_2 A(11-4B) + 2P_1 P_2, \tag{14.21, 6}$$

with a similar relation for Q_2; the quantities A, B, C, E, are defined by (9.8, 7, 8). For Maxwellian molecules (for which $\nu = 5$) $C = 0$ (cf. (10.31, 10)); hence in this case $\Delta = 0$ and the second (and every later) approximation is identical with the first, in agreement with the results of 10.33.

Further approximations to D_{12} are larger than $[D_{12}]_2$. In the special case in which the ratio m_1/m_2 of the molecular masses is very large and n_2/n_1 is very small it has been shown (cf. Table 7, 10.52) that the true value of D_{12} bears to $[D_{12}]_2$ the ratio $1{\cdot}132:1{\cdot}083 = 1{\cdot}046$ for rigid elastic spheres; for other more realistic types of interaction the ratio is usually between 1 and 1·02. In other less special cases the effect of later approximations is similar in order of magnitude, but smaller.

Kihara's approximation to $[D_{12}]_1$ is obtained by setting $B = \frac{3}{4}$ in (14.21, 5, 6). This normally increases $[D_{12}]_1$ towards the exact D_{12}; Kihara's $[D_{12}]_2$ is, in fact, often nearly as accurate as the more usual $[D_{12}]_3$, and is sufficiently accurate as an approximation to D_{12} for most purposes.*

* E. A. Mason, *J. Chem. Phys.* **27**, 75, 782 (1957).

14.3. The variation of D_{12} with the pressure and concentration-ratio

All the formulae of 14.2 and 14.21 indicate that D_{12} is inversely proportional to n, i.e. to the pressure of the gas, if the temperature and composition are constant. This proportionality was first observed by Loschmidt (1),* and has been generally confirmed at moderate pressures.

As can be seen from (14.21, 2), the quantity Δ appearing in $[D_{12}]_2$ varies with x_1/x_2, i.e. with the composition of the mixture in which diffusion is taking place. For example, the values of Δ in the limiting cases in which $x_1 \to 1$ and $x_1 \to 0$ are Δ_1, Δ_2, where

$$\Delta_1 = 5m_1^2 C^2/\{(5-4B)m_1^2+6m_2^2+8Am_1m_2\} \tag{14.3, 1}$$

with a similar relation for Δ_2. Hence $\Delta_1 > \Delta_2$ if $m_1 > m_2$; also it can be shown that Δ increases from Δ_2 to Δ_1 as x_1 increases unless m_1 and m_2 are nearly equal (in which case Δ does not vary greatly). Thus by (14.21, 1), when m_1 and m_2 are not nearly equal D_{12} can be expected to increase with increasing concentration of the heavy molecules; the increase is greatest when the molecules of the two gases have very unequal masses.

For rigid elastic spheres (14.3, 1) becomes

$$\Delta_1 = m_1^2/(13m_1^2+30m_2^2+16m_1m_2). \tag{14.3, 2}$$

Hence in this case, to a second approximation

$$\frac{[D_{12}]_{n_2=0}}{[D_{12}]_{n_1=0}} = \frac{1-\Delta_2}{1-\Delta_1} = \frac{1-m_2^2/(13m_2^2+30m_1^2+16m_1m_2)}{1-m_1^2/(13m_1^2+30m_2^2+16m_1m_2)}.$$

If $m_2/m_1 \to 0$, Δ_1, Δ_2 tend to the limits $1/13$ and zero, and the ratio of the end-values of $[D_{12}]_2$ becomes equal to $13/12$. Thus to a second approximation the maximum variation in the coefficient of diffusion for rigid elastic spheres, as the proportions vary, is $8\frac{1}{3}$ per cent. Further approximations increase this to 13·2 per cent (if $m_2/m_1 \to 0$, $[D_{12}]_{n_1=0} = [D_{12}]_1$, and $[D_{12}]_{n_2=0} = 1{\cdot}132[D_{12}]_1$ —cf. the results for the Lorentz case already quoted). The Kihara second approximation gives a variation of 1/9, or 11·1 per cent.

For more realistic molecular models, the factor C^2 in Δ is less than half as great as for elastic spheres: the other terms in Δ vary relatively little with the model. Thus for actual gases the variation of D_{12} with composition is unlikely to exceed about 6 per cent. This statement refers to uncharged molecules; for inverse-square interaction, a two- or threefold variation in D_{12} is possible (cf. Table 4, p. 179).

14.31. Comparison with experiment for different concentration-ratios

The theoretically predicted variation of D_{12} with composition is not much greater than the experimental errors, and is not easy to establish. Evidence of such a variation was first provided by a series of experiments made for this

* As in Chapters 12 and 13, figures in parentheses refer to a list of experimental papers at the end of the chapter.

purpose at Halle about the year 1908, and reported by Lonius (2). In that series, the longest runs of experiments were for H_2–CO_2 and He–A mixtures; these indicated a variation of the correct sign, similar in magnitude to that predicted by (14.21, 1, 2) assuming rigid elastic spherical molecules. However, the experimental variation was greater than the theoretical; the reverse could be expected because, even though the theoretical variation is increased by taking later approximations, this is more than offset by taking realistic values of c^2 (about 0·4 times the rigid sphere value for He–A mixtures, and about 0·25 times for H_2–CO_2). More recent experiments, using radioactive

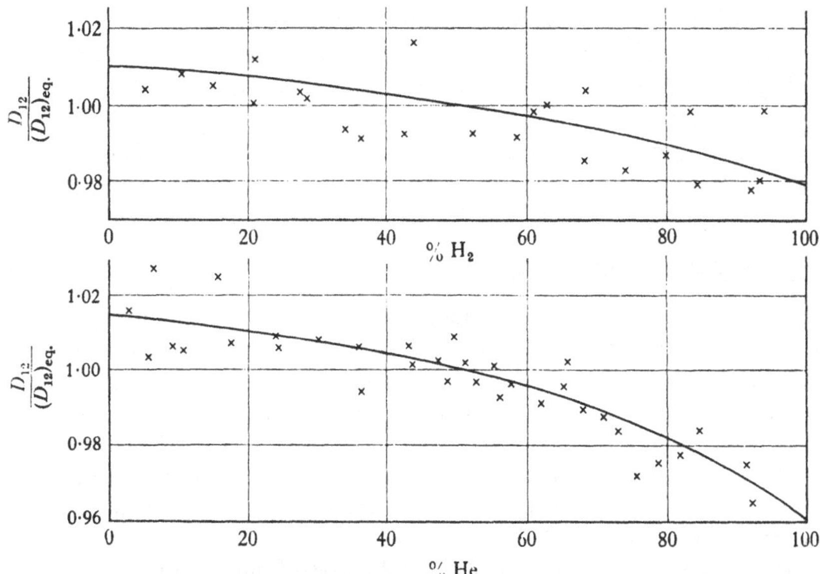

Fig. 12. Variation of $D_{12}/(D_{12})_{eq}$ with composition in H_2–N_2 mixtures at 294·8 °K. (upper curve) and He–A mixtures at 295 °K. (lower curve); $(D_{12})_{eq}$ is the value of D_{12} for equimolar mixtures. The curves give theoretical values, crosses experimental values taken from ref. (6). (We wish to acknowledge the kindness of Professor Beenakker and Dr van Heijningen, who made available their original data, used by us in constructing the figure.)

tracers (3, 4, 5), indicate a variation in D_{12} for these mixtures which is definitely smaller than that found in the Halle series.

The variation found for other mixtures, apart from discrepancies that can fairly be ascribed to experimental errors, similarly agrees with theory as regards sign and order of magnitude. The magnitude of the variation, and the experimental uncertainties, are illustrated in Fig. 12. This gives the ratio of the measured D_{12} to its value for equimolar mixtures ($x_1 = x_2 = \frac{1}{2}$) for the gas-pairs H_2–N_2, He–A, found in a careful series of experiments at Leiden (6). For comparison, theoretical curves are also given. These are based on (14.21, 1, 2), using the Kihara approximation $B = \frac{3}{4}$; A is given the value ($\sim$ 0·44) appropriate to the 12, 6 model, and C is regarded as adjustable. For N_2–H_2 mixtures the value of C agrees with that calculated on the 12, 6 model, determining the force constant ϵ_{12} either from the combination rule $\epsilon_{12} = (\epsilon_1\epsilon_2)^{\frac{1}{2}}$ or from the variation of D_{12} with the temperature (see 14.4, below). However, for He–A mixtures (and for other mixtures containing helium) the Leiden workers found that the value of C required to fit the observations was at least 20 per cent greater than that similarly calculated.

Since the error in the first approximation to the coefficient of diffusion is not very large, it is ignored in the subsequent discussion. This implies that the variation of D_{12} with the proportions of a mixture is ignored. The discussion of the experimental results is thereby the less seriously affected, since in some cases the proportions of the mixture were either not recorded, or not kept constant, by the experimenter. However, it should be remembered that the error involved in neglecting the difference between $[D_{12}]_1$ and D_{12} may be decidedly greater than the corresponding errors for μ and λ, if m_1/m_2 is very different from unity.

14.32. Molecular radii calculated from D_{12}

Table 22 gives the experimental values of D_{12} for several pairs of gases, reduced to 0 °C.* From these the values of σ_{12} given by (14.2, 1) have been calculated; these are effective values s'_{12} of σ_{12} for diffusion. The quantity s'_{12} is not identical with s_{12}, the corresponding effective value of σ_{12} in the viscosity of mixtures (cf. 12.41), but is related to it by $s'^2_{12} = (2/5\text{A})\,s^2_{12}$; taking $\text{A} = 4/9$, we see that s'_{12} is about 5 per cent smaller than s_{12}. For comparison, the table also includes values of $\frac{1}{2}(\sigma_1+\sigma_2)$, where σ_1, σ_2 are the viscosity diameters of Table 11 (p. 228).

It can be seen that the values of $\frac{1}{2}(\sigma_1+\sigma_2)$ derived from viscosity are in general larger by about 10 per cent than those of σ_{12} obtained from D_{12}. This discrepancy is largely due to the difference between s'_{12} and s_{12}, and to the fact that s'_{12} is derived by using values of D_{12} in a formula involving only the first approximation $[D_{12}]_1$, whereas σ_1, σ_2 are derived from an exact formula for μ. However, it may also arise in part from the imperfect representation of the molecules by rigid elastic spheres, and from a difference in nature between the interactions of like and unlike molecules.

The values of D_{12} quoted in Table 22 refer, where possible, to mixtures in which each constituent is present in moderate proportions. The accuracy of D_{12} measurements is not high, and the values quoted may, in unfavourable cases, be in error by several per cent.

* Here (and throughout this book) values of D_{12} are expressed in cm.²/sec.

Table 22. *Coefficients of diffusion at* S.T.P.

Gases	Ref.	D_{12}	$\sigma_{12}\times 10^8$ (cm.)	$\frac{1}{2}(\sigma_1+\sigma_2)\times 10^8$ (cm.)
H_2–D_2	(7*)	1·12	2·54	2·75
H_2–He	(8, 9, 4)	1·30	2·36	2·47
H_2–CH_4	(10)	0·625	3·17	3·47
H_2–NH_3	(8)	0·645	3·11	3·61
H_2–N_2	(4, 6, 8)	0·678	3·00	3·26
H_2–CO	—	0·651	3·07	3·28
H_2–C_2H_4	—	0·505	3·48	3·90
H_2–C_2H_6	(10)	0·460	3·64	4·05
H_2–O_2	(11*)	0·697	2·96	3·19
H_2–A	(12, 13)	0·70	2·94	3·20
H_2–CO_2	(4, 10, 13*)	0·557	3·29	3·69
H_2–N_2O	—	0·535	3·36	3·72
H_2–SO_2	—	0·480	3·54	4·15
D_2–CO_2	(3)	0·410	3·26	3·70
He–NH_3	(14)	0·665	2·65	3·335
He–Ne	(6, 15)	0·906	2·25	2·40
He–N_2	(4, 8, 11, 16)	0·607	2·72	2·99
He–O_2	(4, 16)	0·626	2·67	2·92
He–A	(3, 4, 5, 6, 12, 15, 17)	0·641	2·62	2·93
He–CO_2	(3, 9, 17)	0·540	2·85	3·42
He–Kr	(6, 15)	0·558	2·77	3·20
He–Xe	(6, 15)	0·496	2·93	3·57
CH_4–O_2	(11)	0·194	3·64	3·91
CH_4–CO_2	—	0·153	4·00	4·42
NH_3–Ne	(14)	0·296	3·05	3·54
NH_3–A	(14)	0·172	3·76	4·07
NH_3–Kr	(14)	0·140	3·99	4·34
NH_3–Xe	(14)	0·113	4·37	4·71
Ne–A	(6, 15)	0·276	2·88	3·13
Ne–CO_2	(13*)	0·232	3·12	3·62
Ne–Kr	(6, 15)	0·223	3·05	3·40
Ne–Xe	(6, 15)	0·188	3·27	3·77
N_2–CO	(18)	0·192	3·42	3·80
N_2–C_2H_4	(10)	0·142	3·97	4·42
N_2–C_2H_6	(10)	0·126	4·18	4·57
N_2–O_2	—	0·181	3·46	3·71
N_2–A	(16)	0·167	3·52	3·72
N_2–CO_2	(4, 10, 11)	0·144	3·75	4·21
N_2–Xe	(16)	0·105	4·08	4·36
CO–C_2H_4	—	0·116	4·40	4·44
CO–O_2	(11*)	0·188	3·40	3·72
CO–CO_2	—	0·137	3·85	4·23
O_2–A	(16)	0·166	3·46	3·65
O_2–CO_2	(11*)	0·138	3·76	4·14
O_2–Xe	(16)	0·099	4·09	4·28
A–CO_2	(17)	0·128	3·79	4·15
A–Kr	(6, 15)	0·119	3·68	3·93
A–Xe	(6, 15)	0·095	3·99	4·30
CO_2–N_2O	(18)	0·098	4·27	4·67
Kr–Xe	(6)	0·0641	4·28	4·57

* An asterisk attached to a reference means that other experimental values, for which no reference is given here, are quoted in the paper referred to. In addition to the references quoted above, values given in *International Critical Tables* have also been used where available.

14.4. The dependence of D_{12} on the temperature; the intermolecular force-law

The theoretical variation of D_{12} with temperature depends on the special molecular model employed. If the pressure is constant, so that nT is constant, by (9.81, 1), (9.71, 1) and (9.8, 7),

$$\frac{d\ln[D_{12}]_1}{d\ln T} = 2 - \frac{d\ln\Omega_{12}^{(1)}(1)}{d\ln T} = 2 - \tfrac{5}{2}\text{c}. \tag{14.4, 1}$$

Now $\text{c} = 0$ for Maxwellian molecules, $\text{c} = \frac{1}{5}$ for rigid elastic spheres; for physically satisfactory models of electrically neutral molecules, c in general lies between 0 and $\frac{2}{15}$. Hence in a small temperature-range $[D_{12}]_1$ varies as T^{1+s}, where s $(\equiv 1 - \frac{5}{2}\text{c})$ lies between about $\frac{2}{3}$ and 1. If unlike molecules repel each other with a force varying as $r^{-\nu_{12}}$, by (14.2, 2), $[D_{12}]_1$ varies exactly as a power T^{1+s}, where

$$s = \tfrac{1}{2} + \frac{2}{\nu_{12} - 1}; \tag{14.4, 2}$$

this is precisely similar in form to (12.31, 2). In many cases the number s can be expected to be intermediate between the two values of the s of 12.31, derived from the viscosities of the constituent gases of the mixture.

Experimental values of D_{12} over a range of temperatures can be analysed to determine force constants for unlike molecules according to various molecular models, much as viscosity data were used in Chapter 12 for like molecules. The models most commonly used are the repulsive inverse-power model, and the 12, 6 and exp; 6 models. Experimental data covering a wide range of temperatures are rare, especially below 0 °C. The constants derived from experiment for a number of gases are given in Table 23. Walker and Westenberg (11) and Saxena and Mason (3) suggest that at temperatures above 0 °C. the experimental values are as well represented by a T^{1+s} dependence as by more complicated expressions; however, the 12, 6 and exp; 6 models can be expected to give a much better representation of the actual force-law.

The values quoted for the force-constants do not always lie between those for the separate pure gases. Part of the discrepancy is likely to be experimental in origin, part may arise from indeterminacy in the molecular constants; values of D_{12} calculated on the exp; 6 or 12, 6 models, using force-constants derived from those for the pure gases by the combination rules of 12.41, in general agree fairly well with the experimental values. Where the agreement is not good, the discrepancy may be attributed to a defective molecular model, or to a difference between the interactions of like and unlike molecules.

Table 23. *Variation of D_{12} with temperature**

		Exp;6 model			12,6 model		$r^{-\nu_{12}}$ repulsion	
Gases	Ref.	ϵ_{12}/k	r_m	α	ϵ_{12}/k	σ_{12}	s	ν_{12}
CO_2–O_2	(11)	255	3·616	17	213	3·365	0·805	7·54
H_2–O_2	(11)	143	3·234	14	152	2·825	0·780	8·13
CH_4–O_2	(11)	220	3·612	17	182	3·367	0·793	7·83
CO–O_2	(11)	110	3·733	17	91	3·480	0·710	10·52
He–N_2	(11)	85	3·069	17	69	2·87	0·691	11·45
CO_2–N_2	(11)	184	3·782	17	154	3·52	0·753	8·91
He–A	(6, 11)	29·8	3·51	14	40·2	2·98	0·725	9·9
He–Ne	(6)	20·5	3·01	15	23·7	2·64	—	—
Ne–A	(6)	60·9	3·50	15	61·7	3·11	—	—
NH_3–A	(16)	—	—	—	224·65	3·286	—	—
NH_3–Kr	(16)	—	—	—	264·95	3·349	—	—
Ne–CO_2	(3)	—	—	—	82·4	3·392	0·83	7·05
H_2–CO_2	(3)	—	—	—	94	3·30	0·84	6·9
He–CO_2	(3)	—	—	—	61·5	3·19	0·84	6·9

* This list does not attempt to be exhaustive.

14.5. The coefficient of self-diffusion D_{11}

If the molecules of the two gases whose mutual diffusion is considered are identical, the coefficient of diffusion becomes the coefficient of self-diffusion of a simple gas. By (14.2, 1) the first approximation to this for rigid elastic spheres of diameter σ is given by

$$[D_{11}]_1 = \frac{3}{8n\sigma^2}\left(\frac{kT}{\pi m}\right)^{\frac{1}{2}} = \frac{6}{5}\frac{[\mu]_1}{\rho}, \tag{14.5, 1}$$

while in general, by (10.6, 4), the first approximation is given by

$$[D_{11}]_1 = 3\text{A}[\mu]_1/\rho, \tag{14.5, 2}$$

$[\mu]_1$ being the first approximation to the coefficient of viscosity, and A a numerical factor, defined in (9.8, 7). If we write

$$\rho D_{11} = 3\text{A}\mu, \tag{14.5, 3}$$

ignoring the difference between D_{11}, μ, and their first approximations, then (cf. 10.6) for rigid elastic spheres the error incurred is only $\frac{1}{3}$ per cent. The error with more realistic molecular models is at most half as great, and can be neglected.

Equation (14.5, 3), like the equation $\lambda = \frac{5}{2}\mu c_v$, provides a crucial test of the theory, and of the correctness of assumed molecular models. For Maxwellian molecules $3\text{A} = 1{\cdot}55$; for rigid elastic spheres $3\text{A} = 1{\cdot}2$; and for molecules repelling like $r^{-\nu}$, 3A varies between these values as ν increases from 5 to ∞. On the other hand, 3A is close to 1·32 for the 12,6 model, and between about 1·32 and 1·36 for the exp;6 model, if kT/ϵ lies between 0·5 and 10. Hence (14.5, 3) provides a clear test for distinguishing between models with or without attractive fields. However, the results for rough spheres suggest that unduly low values of $\rho D_{11}/\mu$ may result if interchange can occur between internal and translational energy.

Naturally it is not possible to follow the motions of individual molecules, possessing no feature distinguishing them from other molecules of the gas, so that no actual experimental determination of the self-diffusion coefficient can be made. For this reason Lord Kelvin* proceeded by calculating values of D_{11}, using the elastic-sphere model. His method was to calculate σ_{12} from D_{12}, as in Table 22 (14.32). If σ_{12} is assumed to equal the mean of the diameters of unlike molecules, from the coefficients of mutual diffusion D_{12}, D_{13}, D_{23} of the three pairs which can be chosen out of three gases the diameters σ_1, σ_2, σ_3 of molecules of the gases can be found. On substitution in (14.5, 1), the values of D_{11} are found for each of the three gases. A check on the method is afforded by the agreement or otherwise between the values of D_{11} for the same gas, derived from different triads of gas including it.

By this method, the following values of the coefficients of self-diffusion of four gases, hydrogen, oxygen, carbon monoxide and carbon dioxide, have been found, using the values of D_{12} given in Table 22:

The three gases used	D_{11} for H_2	D_{11} for O_2	D_{11} for CO	D_{11} for CO_2
H_2–O_2–CO	1·24	0·191	0·179	—
H_2–O_2–CO_2	1·30	0·182	—	0·107
H_2–CO–CO_2	1·34	—	0·170	0·105
O_2–CO–CO_2	—	0·193	0·176	0·101
Mean	1·29	0·189	0·175	0·104

The agreement between the values of D_{11} obtained from the different triads of gases is reasonable. From the means of the values of D_{11} values of $\rho D_{11}/\mu$ have been calculated; they are given in Table 24. They do not agree with the value 1·2 appropriate to rigid elastic spheres; the reason is the systematic difference between the effective radii for viscosity and diffusion, already mentioned in 14.32. Since the values were calculated on the assumption of elastic spheres, it is not legitimate to attempt to derive from them information about the actual force-law.

Table 24. *Coefficients of self-diffusion*

Gas	D_{11}	$\rho \times 10^6$	$\mu \times 10^7$	$D_{11}\rho/\mu$
H_2	1·29	89·9	850	1·37
O_2	0·189	1429	1926	1·40
CO	0·175	1250	1665	1·31
CO_2	0·104	1977	1380	1·49

* Lord Kelvin, *Baltimore Lectures*, p. 295. The formula for D_{12} used by Kelvin, however, differed from (14.2, 1) by a constant factor.

14.51. Mutual diffusion of isotopes and like molecules

The mutual diffusion of isotopes of the same gas closely resembles self-diffusion, especially when, as is the case with the heavier gases, the ratio m_1/m_2 of the molecular masses is nearly equal to unity. The diffusion is relatively easy to follow when the molecules of one isotope are radioactive. In deriving D_{11} from the mutual diffusion coefficient D_{12} of isotopes it is necessary to remember that D_{12} is proportional to the mass factor $[(m_1+m_2)/(2m_1m_2)]^{\frac{1}{2}}$, whereas D_{11} is proportional to $m_1^{-\frac{1}{2}}$.

Values of $\rho D_{11}/\mu$ derived from isotopic diffusion experiments for a number of gases are given in Table 25. The majority lie between 1·32 and 1·38, corresponding to values of A between 0·44 and 0·46. On the simple inverse-power model of 10.31, this would imply a force-index ν between about 13 and 17—irreconcilable with the viscosity results. However, values of $\rho D_{11}/\mu$ between 1·32 and 1·36 are to be expected on the 12,6 and exp;6 models. The rather higher values also found may be due partly to defects in these models, but may also be explained by a slightly smaller effective collision cross-section for unlike isotopes than for like ones. There is no evidence of a *decrease* in $\rho D_{11}/\mu$ due to interchange between internal and translational energy.

Table 25. *Self-diffusion coefficients from isotopic diffusion experiments*

Gas	Temp. (°C.)	D_{11}	$\rho D_{11}/\mu$	Ref.
Hydrogen	23	1·455	1·37	(7)
Helium	23	1·555	1·32	(7)
Methane	0	0·206	1·42	(20)
	25	0·223	1·33	(19)
Neon	0	0·452	1·35	(20)
	−195·5	0·0492	1·35	(20)
Carbon monoxide	0	0·190	1·42	(18)
Nitrogen	0	0·178	1·34	(20, 21)
	−195·5	0·0168	1·43	(20)
Oxygen	0	0·181	1·35	(20, 21)
	−195·5	0·0153	1·38	(20)
Hydrogen chloride	22	0·1246	1·33	(22)
Argon	0	0·157	1·33	(20, 23)
Carbon dioxide	0	0·0965	1·375	(18, 20, 21)
	89·5	0·1644	1·36	(18)
	−78·3	0·0505	1·31	(18, 20)
Krypton	0	0·0795	1·29	(24)
	200	0·214	1·27	(24)
Xenon	0	0·0480	1·33	(25)
	105	0·0900	1·31	(25)

The values in Table 25 mostly refer to 0 °C. or to room temperatures, but a few values at other temperatures are included for comparison. On the 12,6 and exp;6 models, $\rho D_{11}/\mu$ should increase slightly with T when $kT/\epsilon_{12} > 2$; in Table 25 any such increase is masked by random errors. Values of D_{11} calculated from $\rho D_{11} = 1{\cdot}34\mu$ in general agree with experiment as closely as any calculated by more elaborate procedures, and have as much theoretical justification.

The value of D_{11} quoted for hydrogen is derived from the mutual diffusion of hydrogen and deuterium, and of para- and ortho-hydrogen. In both cases there is reason to believe that the molecular fields of the gases involved are slightly different. Similar but smaller differences can be expected for other isotopic fields.

Values of D_{11} have also been inferred from the mutual diffusion of gases whose molecules, though chemically different, may be expected to behave similarly. For example, carbon monoxide and nitrogen have almost identical molecular weights and similar molecular structure; also their transport properties are very similar (cf. Chapter 12, Tables 11 and 17, and Chapter 13, Table 20).* Taking $D_{12} = 0{\cdot}192$, $\mu = 1{\cdot}645 \times 10^{-4}$ (the mean of the values for the two gases), we find that $\rho D_{11}/\mu = 1{\cdot}46$. In the same way, for the similar gases carbon dioxide and nitrous oxide, $\rho D_{11}/\mu = 1{\cdot}39$. However, values derived from isotopic diffusion are to be preferred to these, because different, though similar, molecules are likely to exert forces on each other different from those which each exerts on another of its own kind. In fact, experiment indicates that D_{12} for CO_2–N_2O is two per cent greater than D_{11} for CO_2, while D_{12} for N_2–CO is less than D_{11} for carbon monoxide (18).

THERMAL DIFFUSION, AND THE THERMO-DIFFUSION EFFECT

14.6. As noted in 14.1, the velocity of diffusion has a component due to non-uniformity of the temperature. This depends on the coefficients k_T, α_{12}—the thermal diffusion ratio and the thermal diffusion factor, connected by $k_T \equiv x_1 x_2 \alpha_{12}$. Thermal diffusion was first experimentally demonstrated by Chapman and Dootson in 1916.† It provides a sensitive test of intermolecular force-laws,‡ and in the separation column, developed by Clusius and Dickel § in 1938, it has supplied a powerful method of separating gases, particularly in isotopic mixtures.

* L. E. Boardman and N. E. Wild (*Proc. R. Soc.* A, **162**, 511 (1937)) first calculated $\rho D_{11}/\mu$ for the gas-pairs CO, N_2 and CO_2, N_2O. Attention was drawn to the similarity of properties of these pairs by C. J. Smith, *Proc. Phys. Soc.* **34**, 162 (1922).

† S. Chapman and F. W. Dootson, *Phil. Mag.* **33**, 248 (1917).

‡ Cf. K. E. Grew and T. L. Ibbs, *Thermal Diffusion in Gases* (Cambridge University Press, 1952).

§ K. Clusius and G. Dickel, *Z. Phys. Chem.* B, **44**, 397 (1939).

The separation column relies on the combined effects of thermal diffusion and a slow thermal convection. The theory of the column is complicated,* depending on the viscosity of the gas-mixture as well as on α_{12}, and it involves approximations which preclude the use of the column for exact determinations of α_{12}. The column gives the *sign* of α_{12}, and a rough estimate of its magnitude. For more accurate determinations, a steady-state method has been employed.

In a vessel containing a mixed gas, if different parts are maintained at different temperatures, thermal diffusion results, and tends to make the composition non-uniform. The concentration-gradients thus set up are opposed by the ordinary process of diffusion tending to equalize the composition, and in time a steady state is reached, in which the opposing influences of thermal and ordinary diffusion balance. Then, by (14.1, 1),

$$\nabla \mathrm{x}_1 = -\mathrm{x}_1\mathrm{x}_2\alpha_{12}\nabla \ln T; \tag{14.6, 1}$$

the thermal diffusion factor α_{12} is used instead of k_T because of its much smaller variation with composition. If the dependence of α_{12} on composition is ignored, integration of (14.6, 1) gives

$$\ln(\mathrm{x}_1'/\mathrm{x}_2') - \ln(\mathrm{x}_1/\mathrm{x}_2) = -\int_T^{T'} \alpha_{12}\, d\ln T \tag{14.6, 2}$$

$$= -\bar{\alpha}_{12}\ln(T'/T), \tag{14.6, 3}$$

where x_1', x_2' and x_1, x_2 refer to the composition of the gas at temperatures T', T, and $\bar{\alpha}_{12}$ is a mean value of α_{12} between the temperatures T' and T. Equation (14.6, 3) can be used to determine $\bar{\alpha}_{12}$ from the steady concentration-ratios $\mathrm{x}_1'/\mathrm{x}_2'$, $\mathrm{x}_1/\mathrm{x}_2$ ultimately reached in two vessels in free communication with each other, and maintained at different temperatures T', T. In a series of experiments the temperature T' of one vessel is often maintained constant and the temperature T of the other is varied; by taking the difference of two equations (14.6, 2), the value of $\bar{\alpha}_{12}$ between two different values of T can be found.

The degree of separation produced by thermal diffusion in an experiment of this type is not large. For non-ionized gases, values of α_{12} as large as 0·5 are rare; they are met only when a light gas has a small proportion of a much heavier gas mixed with it.† The following table relates the equilibrium concentrations x_1, x_1' in the two vessels when the temperature T of the first is half the temperature T' of the second, and $\bar{\alpha}_{12} = 0{\cdot}5$.

$\mathrm{x}_1 =$ 0·1	0·3	0·5	0·7	0·9
$\mathrm{x}_1' =$ 0·073	0·233	0·414	0·623	0·864

* L. Waldmann, *Z. Phys.* **114**, 53 (1939); W. H. Furry, R. C. Jones and L. Onsager, *Phys. Rev.* **55**, 1083 (1939); R. C. Jones and W. H. Furry, *Rev. Mod. Phys.* **18**, 151 (1946).

† See K. E. Grew and T. L. Ibbs, *op. cit.*

Because of the smallness of the separation, it was earlier thought necessary to work between widely different temperatures T, T', with a corresponding uncertainty regarding the temperature to which $\bar{\alpha}_{12}$ corresponds. This difficulty was overcome in the swing separator,* consisting of a number of separation tubes in series, in which gas in the cold section of one tube could be interchanged with gas in the hot section of the next by drawing gas to and fro through a connecting capillary. The swing separator operates effectively as a number of separation tubes in series, and requires only a small temperature-difference $T'-T$ for its operation.

As noted in 8.41, a diffusion velocity $\bar{C}_1-\bar{C}_2$ is accompanied by a heat-flow $p x_1 x_2 \alpha_{12}(\bar{C}_1-\bar{C}_2)$. This diffusion thermo-effect has been investigated both theoretically and experimentally by Waldmann.† His experiments were of two kinds. In the first (the 'stationary' effect), mixtures of the same two gases in different proportions were made to flow along two parallel tubes, connected through a slot cut along their lengths; changes in the gas temperatures on the two sides of the slot were measured by platinum wires parallel to the slot, used as resistance thermometers. In the second (the 'non-stationary' effect), two vertical cylinders filled with mixtures in different proportions were placed end to end, and diffusion was permitted between them; temperature changes in the two cylinders were measured throughout the duration of the diffusion, again by resistance thermometers. The recorded temperature changes were small (comparable with 0·5 °C.), but sufficient to be recorded with fair accuracy. As Waldmann recognized, the smallness of the temperature-range is in one respect an advantage, since the results are practically independent of any variation of α_{12} with the temperature.

Although the thermo-effect involves diffusion, α_{12} can be determined from it without prior knowledge of the diffusion coefficient D_{12}, by using the equation of conduction of heat, and the continuity equation for each of the gases. The thermal conductivity λ must, however, be known, and a heat conduction equation has to be solved, in both the 'stationary' and 'non-stationary' cases. In both cases, not only can α_{12} be determined, but values of the diffusion coefficient can be obtained as a by-product, from the ultimately exponential decrease of the temperature-deviations (with time in the 'non-stationary' case, with distance along the tubes in the 'stationary' case).

Values of α_{12} obtained from the diffusion thermo-effect are subject to several possible sources of error. They depend on approximate solutions of the heat-conduction equation, derived on the assumption that λ is constant. Again, λ for many gas-mixtures is not known accurately; sometimes no

* K. Clusius and M. Huber, *Z. Naturf.* **10a**, 230 (1955).

† L. Waldmann, *Z. Phys.* **121**, 501 (1943); **124**, 2, 30, 175 (1947–8); *Z. Naturforschung*, **1**, 59, 483 (1946); **4a**, 105 (1949); **5a**, 322, 327, 399 (1950); L. Waldmann and E. W. Becker, *Z. Naturforschung*, **3a**, 180 (1948).

experimental values are known, and λ may have to be estimated by an approximate formula from the values for the corresponding pure gases. Nevertheless, Waldmann has been able to use the thermo-effect to obtain a number of interesting results about α_{12}, particularly at low temperatures.

14.7. The thermal diffusion factor α_{12}

The first approximation to α_{12}, given in (9.83, 1), is

$$[\alpha_{12}]_1 = 5\text{C}\frac{\text{x}_1\text{s}_1 - \text{x}_2\text{s}_2}{\text{x}_1^2\text{Q}_1 + \text{x}_2^2\text{Q}_2 + \text{x}_1\text{x}_2\text{Q}_{12}}. \tag{14.7, 1}$$

Here $\text{s}_1 = M_1\text{P}_1 + 3M_2(M_1 - M_2) - 4M_1M_2\text{A}$, (14.7, 2)

$$\text{Q}_1 = \text{P}_1\{6M_2^2 + (5-4\text{B})M_1^2 + 8M_1M_2\text{A}\}, \tag{14.7, 3}$$

$$\text{Q}_{12} = 3(M_1 - M_2)^2(5-4\text{B}) + 4M_1M_2\text{A}(11-4\text{B}) + 2\text{P}_1\text{P}_2, \tag{14.7, 4}$$

with similar relations for s_2 and Q_2; also $\text{P}_1 = M_1\text{E}/[\mu_1]_1$, $\text{P}_2 = M_2\text{E}/[\mu_2]_1$, and the quantities A, B, C, E are as defined in 9.8. The Kihara first approximation $[\alpha'_{12}]_1$ is obtained by putting $\text{B} = \frac{3}{4}$ in the expressions for Q_1, Q_2 and Q_{12}. In a mixture of isotopes of nearly equal mass (cf. 10.61)

$$[\alpha_{12}]_1 = \frac{15\text{C}(1+\text{A})}{\text{A}(11-4\text{B}+8\text{A})}\frac{m_1-m_2}{m_1+m_2}, \quad [\alpha'_{12}]_1 = \frac{15\text{C}}{8\text{A}}\frac{m_1-m_2}{m_1+m_2}. \tag{14.7, 5}$$

Further approximations have been derived by Mason;* they are given by very complicated algebraic expressions. Mason calculated successive approximations for a Lorentzian gas (10.5), a quasi-Lorentzian gas (10.53) and an isotopic mixture, in each case considering power-law repulsion and the exp;6 model. On the power law model he found that $[\alpha'_{12}]_1$ is a better approximation than $[\alpha_{12}]_1$; in the Lorentz case $[\alpha'_{12}]_1$ is exact, whereas $[\alpha_{12}]_1$ is less than the true value by an amount increasing from zero to about 15 per cent as ν increases from 5 to 13; in the quasi-Lorentzian case $[\alpha'_{12}]_1$ and $[\alpha_{12}]_1$ are equal, each being in defect by an amount which increases from zero to about $3\frac{1}{2}$ per cent as ν increases from 5 to 13; in the isotopic case $[\alpha_{12}]_1$ and $[\alpha'_{12}]_1$ are in defect by amounts increasing from zero to $4\frac{1}{2}$ and $1\frac{1}{2}$ per cent respectively in the same range of ν.

For the exp;6 model $[\alpha'_{12}]_1$ has less advantage. In the Lorentzian case $[\alpha_{12}]_1$ is a good approximation when $kT/\epsilon_{12} < 2$, but is in defect by 8 per cent when $kT/\epsilon_{12} = 5$, and 13 per cent when $kT/\epsilon_{12} = 20$; $[\alpha'_{12}]_1$, on the other hand, is always in excess, by about 15 per cent when kT/ϵ_{12} is between 2 and 3, but by a much smaller amount when kT/ϵ_{12} is either large or small. In the quasi-Lorentzian case $[\alpha_{12}]_1$ and $[\alpha'_{12}]_1$ are again equal, being in defect by an amount which increases from zero to about 4 per cent as kT/ϵ_{12} increases

* E. A. Mason, *J. Chem. Phys.* **27**, 75, 782 (1957); see also S. C. Saxena and E. A. Mason, *J. Chem. Phys.* **28**, 623 (1957). Saxena and his associates have, in a number of papers, considered the case when m_2/m_1 is small, giving expansions for $[\alpha_{12}]_1$ in powers of $(m_2/m_1)^{\frac{1}{2}}$; see, e.g., S. C. Saxena and S. M. Dave, *Rev. Mod. Phys.* **33**, 148 (1961).

from 2 to 20. In the isotopic case $[\alpha_{12}]_1$ is in defect by an amount which increases from zero to about 4 per cent in the same range of kT/ϵ_{12}; $[\alpha'_{12}]_1$ is in excess by 2 per cent when kT/ϵ_{12} is between 2 and 3, in defect by about $1\frac{1}{2}$ per cent when $kT/\epsilon_{12} = 20$.

Thus for inverse-power repulsion, sufficient accuracy can usually be obtained by using $[\alpha'_{12}]_1$, but this is not true for the more important exp;6 model. The approximate nature of the formula (14.7, 1) must be borne in mind in any detailed comparison with experiment.*

14.71. The sign and composition-dependence of $[\alpha_{12}]_1$

The denominator of the fraction on the right of (14.7, 1) is essentially positive; also C is usually positive for electrically neutral molecules. Assume that $C > 0$; then the sign of $[\alpha_{12}]_1$ is that of $x_1 S_1 - x_2 S_2$. As in (12.41, 2) we write

$$[\mu_1]_1 = \frac{5}{16 s_1^2}\left(\frac{km_1 T}{\pi}\right)^{\frac{1}{2}}, \quad E = \frac{5A}{8 s_{12}^2}\left(\frac{km_0 T}{2\pi M_1 M_2}\right)^{\frac{1}{2}}, \qquad (14.71, 1)$$

so that s_1 is an equivalent viscosity diameter for molecules m_1, while s_{12} has a similar meaning for unlike molecules. Then

$$S_1 = 4M_1 M_2 A\{(2M_2)^{-\frac{3}{2}} s_1^2/s_{12}^2 - 1\} + 3M_2(M_1 - M_2). \qquad (14.71, 2)$$

The quantities s_1, s_{12} are in general not widely different. Hence if m_1 and m_2 are not too nearly equal, the sign of S_1 is that of $m_1 - m_2$ (if $m_1 > m_2$, $M_1 > \frac{1}{2} > M_2$). The sign of S_2 is similarly that of $m_2 - m_1$; thus if $m_1 > m_2$, $x_1 S_1 - x_2 S_2$ is positive, and α_{12} is positive. That is, (cf. (14.1, 1)), the heavier molecules diffuse towards the cooler regions.

If m_1 and m_2 are nearly equal, the quantities s_1^2/s_{12}^2, s_2^2/s_{12}^2 may be more important than m_1/m_2 in determining the sign of $[\alpha_{12}]_1$. If $m_1 = m_2$,

$$S_1 = A(s_1^2/s_{12}^2 - 1), \quad S_2 = A(s_2^2/s_{12}^2 - 1).$$

Thus $S_1 > 0$, $S_2 < 0$ if $s_1 > s_{12} > s_2$; in this case $[\alpha_{12}]_1 > 0$. That is, roughly, if m_1 and m_2 are nearly equal, the larger molecules tend to diffuse into the cooler regions.

In more detail,† suppose, as we may, that $m_1 \geqslant m_2$, and assume

$$A = \tfrac{4}{9} = 0{\cdot}444$$

(the argument is not sensitive to the value of A adopted). Then if $m_1/m_2 > 2{\cdot}5$, S_1 cannot be negative, and S_2 can be positive only if s_2/s_{12} has an improbably

* If in (14.7, 3), (14.7, 4) we replace B by B′, where

$$B' = 2B - \tfrac{4}{5} + C(1 + \tfrac{6}{5}C),$$

values of $[\alpha_{12}]_1$ calculated from (14.7, 1) are found to agree closely with Mason's values for the exp;6 model for isotopic and Lorentzian molecules, and are identical with those of $[\alpha'_{12}]_1$ for the power law model. This suggests that errors may in general be reduced by substituting B′ for B.

† A detailed discussion of the characteristics of thermal diffusion was given by S. Chapman, *Proc. R. Soc.* A, **177**, 38 (1940).

high value, greater than 2·4. For values of m_1/m_2 nearer to unity, the following table gives the *greatest* value of s_1/s_{12} for which s_1 can be negative, and the *least* value of s_2/s_{12} for which s_2 can be positive. The table illustrates the importance of the mass-ratio as compared with s_1/s_{12}, s_2/s_{12} in determining the signs of s_1, s_2.

M_1 ...	0·5	0·55	0·6	0·65	0·7
m_1/m_2	1	$1\frac{2}{9}$	$1\frac{1}{2}$	$1\frac{6}{7}$	$2\frac{1}{3}$
Max. s_1/s_{12}	1	0·771	0·577	0·362	0·135
Min. s_2/s_{12}	1	1·261	1·557	1·90	2·30

In special circumstances, when m_1/m_2 is not too different from unity and the heavier molecules are the smaller, s_1 and s_2 may have the same sign. For example, the table shows that this can occur when $m_1/m_2 = \frac{3}{2}$ if $s_1/s_{12} > 0·577$ and $s_2/s_{12} > 1·557$ (this gives $s_{12} < \frac{1}{2}(s_1+s_2)$, which is more probable on physical grounds than the reverse inequality). In this case $[\alpha_{12}]_1$ changes its sign as x_1 increases from 0 to 1; however, s_1, s_2 and consequently $[\alpha_{12}]_1$ are then small.

We assumed in the above that $\mathrm{C} > 0$. On the 12,6 and exp;6 models (on the latter, if $\alpha < 15$), the quantity C may just become negative at low temperatures, for values of kT/ϵ_{12} between about $\frac{1}{2}$ and 1; at such temperatures at least one of the gases in the mixture is well below its critical temperature. For inverse-square interaction—e.g. in a fully ionized gas—C is strongly negative. In either case, thermal diffusion is reversed in direction; that is, the heavier (or larger) particles diffuse towards the hotter regions. In particular, in the sun and the stars thermal diffusion reinforces pressure diffusion in tending to concentrate heavy ions in the hot central regions. This effect is not important in a star as a whole, because of the enormous times required for diffusion to produce appreciable changes of composition in so large a volume; in the outer atmosphere, however, it may be significant.

When m_1/m_2 is large, $[\alpha_{12}]_1$ may increase threefold or more as the proportion x_2 of the light molecules increases from zero to unity. In this case the thermal diffusion is due almost entirely to the motions of the light molecules m_2. In terms of M_2, regarded as a small quantity, by (14.71, 1)

$$[\mu_1]_1 = O(1), \quad [\mu_2]_1 = O(M_2^{\frac{1}{2}}), \quad \mathrm{E} = O(M_2^{-\frac{1}{2}}).$$

Hence the coefficients a_{pq} in 9.8 are such that

$$a_{11} = O(1),\ a_{-1-1} = O(M_2^{-\frac{1}{2}}),\ a_{1-1} = O(M_2),\ a_{01} = O(M_2^{\frac{3}{2}}),\ a_{0-1} = O(1).$$

Thus if quantities of order $M_2^{\frac{3}{2}}$ are neglected, the general expression for $[\mathrm{k}_T]_1$ given in 9.83 reduces to

$$[\mathrm{k}_T]_1 = \tfrac{5}{2}\mathrm{x}_2 M_2^{-\frac{1}{2}} a_{0-1}/a_{-1-1},$$

indicating the dependence on the molecules m_2. In terms of the quantities P, Q, this gives

$$[\alpha_{12}]_1 = 5\mathrm{C}/\{\mathrm{x}_1(\mathrm{Q}_1/\mathrm{P}_1) + 2\mathrm{x}_2\,\mathrm{P}_2\}, \qquad (14.71, 3)$$

so that $[\alpha_{12}]_1^{-1}$ is a linear function of the proportions x_1, x_2. This linear relationship was first found by Laranjeira* by a different approximate argument. He found that experimental results for a number of gas-pairs conform closely to such a relationship, sometimes even when m_1/m_2 is not very different from unity.

When m_1/m_2 is large, Q_1/P_1 reduces to $5-4B$, which normally is about 2·2; $2P_2$ is $2\sqrt{2}(As_2^2/s_{12}^2)$, or $1{\cdot}25s_2^2/s_{12}^2$ approximately, where s_2, s_{12} are the viscosity diameters. Since normally s_2/s_{12} is decidedly less than unity when m_1/m_2 is large, (14.71, 3) indicates a considerable increase in $[\alpha_{12}]_1$ as x_2 increases from zero to unity.

14.711. Experiment and the sign of α_{12}

Experimental values of α_{12} in general agree with the rules of sign derived from the expression (14.7, 1) for $[\alpha_{12}]_1$. That is, the heavier molecules or, when $m_1/m_2 \sim 1$, those of larger radius, tend to diffuse towards the cooler regions.

The change in sign of α_{12} at sufficiently low temperatures, due to the change in sign of C, was first observed by Watson and Woernley (26) in isotopic thermal diffusion in ammonia ($^{15}NH_3$–$^{14}NH_3$); here it occurred at about 20 °C. Waldmann (27), using the diffusion thermo-effect, found a similar change in α_{12} for N_2–A, O_2–A and N_2–CO_2 at low temperatures. The same effect has been found for other gas-pairs, e.g. H_2–D_2, H_2–He at liquid hydrogen temperatures; but Grew and Mundy (28) found a positive minimum for Kr–A, without change of sign. This behaviour is what is to be expected from the temperature-variation of C on the exp;6 model, when $\alpha > 15$; however, Grew and Mundy find that, though giving a qualitatively correct explanation, the exp;6 results require the minimum to occur at a much lower temperature than that observed.

A change in sign of α_{12} as the composition is varied was first observed by Grew (29) in Ne–NH_3 mixtures. The heavier but smaller neon molecules travel to the cold side when their concentration exceeds 75 per cent; otherwise ammonia seeks the cold side. Clusius and Huber (30) confirmed and extended Grew's results by investigating the three mixtures ^{20}Ne–NH_3, ^{22}Ne–NH_3 and ^{20}Ne–ND_3. For the ^{20}Ne–NH_3 mixtures their results agreed with those of Grew; for the ^{22}Ne–NH_3 mixtures a change in sign of α_{12} occurred when the Ne concentration was about 58 per cent. In ^{20}Ne–ND_3 mixtures, for which the molecular masses are very nearly equal, no change in sign is found—the larger ND_3 molecules always diffuse towards the cold side.

Assuming that s_1, s_2, s_{12} are identical in the two Ne–NH_3 mixtures, we can determine their ratios from the concentrations at which α_{12} vanishes, using (14.71, 2). Taking $A = 4/9$, we find that $s_{Ne}:s_{NH_3}:s_{12} = 0{\cdot}941:1{\cdot}515:1$. Thus s_{12} is intermediate between s_1 and s_2; if the same s-values apply to Ne–ND_3 mixtures, this explains why α_{12} does not reverse its sign in the latter. Also

* M. F. Laranjeira, *Physica*, **26**, 409, 417 (1960).

$s_{Ne}/s_{NH_3} = 0{\cdot}62$; the similar ratio derived from viscosity (Table 11, 12.11) is 0·59. However, s_{12} found in this way is considerably less than the value given in Table 22.

A change of sign of α_{12} was also found (31) in ^{40}A–D^{37}Cl mixtures, studied between 12° and 179 °C. Here the DCl molecules diffuse towards the cool side when their concentration exceeds 47 per cent, and in the opposite direction when their concentration is smaller.

Effects have been noted which do not appear explicable in terms of the theory of spherically symmetric molecules. When working between temperatures of 65° and 117 °C., Becker and Beyrich (32) found that the values of α_{12} for $^{16}O^{13}C^{16}O$–$^{16}O^{12}C^{16}O$ and $^{16}O^{12}C^{17}O$–$^{16}O^{12}C^{16}O$ mixtures have opposite signs; working between 98° and 180 °C., they found the same sign but significantly different values. Since the molecules $^{16}O^{13}C^{16}O$ and $^{16}O^{12}C^{17}O$ have equal masses, their differences must be due to differences in intermolecular forces or in the distributions of mass. Similar effects have been observed in mixtures of isotopes of carbon monoxide (33) and hydrogen (34). Sandler and Dahler* gave an explanation of the phenomenon, based on the loaded sphere model, in terms of the eccentric distribution of mass in certain of the isotopes.

14.72. α_{12} and the intermolecular force-law: isotopic thermal diffusion

In principle, important information about the intermolecular force-law is immediately derivable from the value of α_{12}. In (14.7, 1) $[\alpha_{12}]_1$ is expressed as the product of 5C into a factor which varies only slowly with the molecular model (through the quantities A, B, and the s_1^2/s_{12}^2, s_2^2/s_{12}^2 of (14.71, 1, 2)). A reasonably accurate value of this second factor can readily be inferred from viscosity data. Hence from an experimental value of α_{12} we can determine C; from C the value of ν can be found for inverse-power interaction, or that of ϵ_{12} for the 12,6 or exp;6 model. The advantages of using α_{12} are obvious; whereas the determination of C from D_{12} depends on the accurate measurement of its temperature-variation, here C is found from the value of α_{12} itself.

The drawback is in the large error of the first approximation $[\alpha_{12}]_1$; the use of $[\alpha_{12}]_1$ for α_{12} leads to spuriously high values of C. However, in one case, that of isotopic thermal diffusion, Mason and Saxena† have given tables from which reasonably exact values of α_{12} can be calculated for the inverse-power, 12,6 and exp;6 models. The quantity actually tabulated by them is α_0, related to α_{12} by

$$\alpha_{12} = \alpha_0(m_1 - m_2)/(m_1 + m_2) \tag{14.72, 1}$$

(cf. (14.7, 5)); this is expressed as a function of ν or kT/ϵ_{12}. Using their tables, the value of ν or ϵ_{12} can be found for any gas for which α_{12} is known.

* S. I. Sandler and J. S. Dahler, *J. Chem. Phys.* **47**, 2621 (1967).

† E. A. Mason, *J. Chem. Phys.* **27**, 782 (1957); S. C. Saxena and E. A. Mason, *J. Chem. Phys.* **28**, 623 (1958).

In practice, because of the large probable errors in determining α_{12} (which is small in isotopic mixtures), the process is inverted; a theoretical α_0 is calculated from force-constants derived from viscosity, and this is compared with the α_0 found experimentally. The results are set out in Table 26, which gives α_0 at about 25 °C. for a number of gases, together with theoretical values derived using the inverse-power, 12,6 and exp;6 models. The theoretical values show the same general trends as the experimental, but tend to be somewhat larger; the exp;6 values agree rather better with experiment than those for the other two models.

Table 26. *The isotopic thermal diffusion factor α_0 at 25 °C.*

Gas	Reference	α_0 (obs.)	α_0 (calc.) $r^{-\nu}$	12,6	exp;6
Hydrogen*	(35)	0·48	0·55	0·55	0·51
Helium*	(35, 36)	0·43	0·55	0·58	0·43
Methane	(35)	0·22	0·255	0·28	0·19
Ammonia	(35)	0	−0·1	—	—
Neon	(35, 36)	0·46	0·56	0·55	0·52
Nitrogen	(35)	0·37	0·42	0·40	0·42
Oxygen	(35)	0·35	0·36	0·34	—
Argon	(35, 36)	0·24	0·295	0·31	0·32
Krypton	(37)	0·12	—	0·17	0·23
Xenon	(37)	0·11	—	0·10	0·07

* In the case of hydrogen (H_2–D_2) and to a less extent helium (^{3}He–^{4}He) the introduction of α_0 may not be justified, since $(m_1-m_2)/(m_1+m_2)$ is not so small as to justify the neglect of its square.

The temperature-variation of α_0 is mainly due to that of c. In general terms the observed variation follows the variation of c given by the 12,6 and exp;6 models; α_0 is small for low temperatures, and reversals of sign have been observed in hydrogen (at liquid-hydrogen temperatures) and krypton (in the range 195–273 °K.) as well as in ammonia. As T increases above such temperatures, α_0 increases, becoming roughly constant at high temperatures. However, though the general behaviour accords with the 12,6 and exp;6 models, there are anomalies; it is not clear whether these are due to defects in the models, to differences in the force-law for different isotopes, or to experimental inaccuracies.

14.73. α_{12} and the intermolecular force-law; unlike molecules

For non-isotopic mixtures, only the rather inadequate first approximation $[\alpha_{12}]_1$ has in general been studied. Mason and Smith* have suggested the following way of deriving better approximations when m_1/m_2 is large. They point out that in this case α_{12} approximates to the quasi-Lorentzian value

* E. A. Mason and F. J. Smith, *J. Chem. Phys.* **44**, 3100 (1966).

when $x_1 \to 0$; they give better approximate formulae for calculating α_{12} when $x_2 \to 0$, and assume that for a general composition

$$\alpha_{12}^{-1} = x_1(\alpha_{12}^{-1})_{x_2=0} + x_2(\alpha_{12}^{-1})_{x_1=0}. \qquad (14.73, 1)$$

However, though α_{12}^{-1} does vary nearly linearly with x_1 and x_2 (cf. 14.71), it is not clear that the linear relation which gives the best fit to experiment is precisely of the form (14.73, 1).

Here we compare theory with experiment* by calculating $[\alpha_{12}]_1$ from (14.7, 1) for equimolar mixtures ($x_1 = x_2 = \frac{1}{2}$) at 0 °C. The 12, 6 model is used; the value A = 0·44 is used throughout, and B and C are found from values of kT/ϵ_{12}, using the combination rule $\epsilon_{12} = (\epsilon_1\epsilon_2)^{\frac{1}{2}}$. The expressions for the quantities P_1, P_2 appearing in (14.7, 2–4) are equivalent to

$$P_1 = \tfrac{2}{3}nm_1[D_{12}]_1/[\mu_1]_1, \quad P_2 = \tfrac{2}{3}nm_2[D_{12}]_1/[\mu_2]_1; \qquad (14.73, 2)$$

in the calculations, $[D_{12}]_1$, $[\mu_1]_1$ and $[\mu_2]_1$ in these relations are replaced by experimental values of D_{12}, μ_1 and μ_2.

The results of the calculations are given in Table 27; for comparison values given by the Kihara first approximation (B = $\frac{3}{4}$) are also given. Where the calculated values are given in brackets, either ϵ_1 (for C_2H_4–H_2) or D_{12} (for the other gas-pairs) has had to be roughly estimated, in the absence of experimental values. The experimental values of α_{12} are those quoted by Grew and Ibbs (35). Where the data are sufficient, values at 0 °C. have been estimated; otherwise a mean temperature $\overline{T} = \surd(T_1T_2)$ is given, where T_1, T_2 are the temperatures between which the apparatus is working. An increase in temperature in general increases α_{12}, a fact which should be borne in mind in a comparison with the calculated values.

Since normally $[\alpha_{12}]_1$ is less than the true α_{12}, the experimental values would be expected to exceed $[\alpha_{12}]_1$. Actually the reverse is often the case in Table 27. Part of the discrepancy may arise because the early experiments tended to underestimate α_{12}; but there appears to be an unexplained residual, due either to the values of C given by the 12, 6 model being too high, or to the theory of spherical molecules applying only approximately to thermal diffusion for gases with internal energy.†

The temperature variation of α_{12} equally suggests that this theory is only approximately applicable. The chief part of the temperature variation of $[\alpha_{12}]_1$ is that of C; on the 12, 6 and exp; 6 models, as the temperature decreases below ϵ_{12}/k, C passes through a minimum, which in certain cases may be positive. A positive minimum of α_{12} is actually observed for CO_2–A mixtures,

* In earlier editions the comparison was made by determining R_T, the ratio of α_{12} to the $[\alpha_{12}]_1$ calculated on the assumption that the molecules are rigid elastic spheres, and using the viscosity diameters. The intermolecular force-index ν in $r^{-\nu}$ interaction was then determined by comparing R_T with its Lorentzian value. However, the latter value usually overestimates the error in $[\alpha_{12}]_1$, and leads to values of ν which are spuriously large.

† The effect on α_{12} of inelastic collisions was discussed by L. Monchick, R. J. Munn and E. A. Mason, *J. Chem. Phys.* **45**, 3051 (1966), who found that the effect is small.

Table 27. *Comparison of theory and experiment for α_{12} for unlike gases*

Gases	α_{12} (exp) and $\overline{T}$ (deg K.)	$[\alpha_{12}]_1$	$[\alpha_{12}]_1$ (Kihara)
$He-H_2$	0·145	0·15	0·16
CH_4-H_2	0·25	0·31	0·34
$Ne-H_2$	0·42	(0·32)	(0·35)
$CO-H_2$	0·30	0·33	0·37
N_2-H_2	0·30	0·345	0·38
$C_2H_4-H_2$	0·28	(0·31)	(0·35)
O_2-H_2	0·19 (162°)	0·32	0·35
$A-H_2$	0·27	0·32	0·355
CO_2-H_2	0·28	0·315	0·35
N_2-D_2	0·31 (327°)	(0·31)	(0·34)
Ne-He	0·33	0·335	0·38
N_2-He	0·36 (327°)	0·38	0·42
A-He	0·38	0·38	0·425
CO_2-He	0·44 (366°)	0·375	0·42
Kr-He	0·44	0·39	0·44
Xe-He	0·43	0·40	0·45
A-Ne	0·17	0·17	0·18
Kr-Ne	0·28	0·245	0·265
Xe-Ne	0·29	0·275	0·30
O_2-N_2	0·018 (293°)	0·020	0·021
$A-N_2$	0·071 (293°)	0·056	0·058
CO_2-N_2	0·050 (339°)	0·064	0·066
N_2O-N_2	0·048 (339°)	(0·064)	(0·066)
$A-O_2$	0·050 (283°)	0·033	0·034
CO_2-O_2	0·040 (293°)	0·049	0·051
CO_2-A	0·019 (283°)	(0·019)	(0·020)
Kr-A	0·070	0·072	0·076
Xe-A	0·084	0·093	0·098
Xe-Kr	0·016 (327°)	(0·024)	(0·025)

but it occurs at a temperature well above that expected on the theoretical grounds (38); the temperature-variation of α_{12} is likewise anomalous for mixtures of carbon dioxide with other inert gases (38, 39).

REFERENCES

(1) J. Loschmidt, *Wien. Ber.* **61**, 367 (1870); **62**, 468 (1870).
(2) A. Lonius, *Annln Phys.* **29**, 664 (1909).
(3) S. C. Saxena and E. A. Mason, *Molec. Phys.* **2**, 379 (1959).
(4) J. C. Giddings and S. L. Seager, *Ind. Eng. Chem. Fund.* **1**, 277 (1962).
(5) R. E. Walker and A. A. Westenberg, *J. Chem. Phys.* **31**, 519 (1959).
(6) R. J. J. van Heijningen, A. Feberwee, A. van Oosten and J. J. M. Beenakker, *Physica*, **32**, 1649 (1966); R. J. J. van Heijningen, J. P. Harpe and J. J. M. Beenakker, *Physica*, **38**, 1 (1968).
(7) P. J. Bendt, *Phys. Rev.* **110**, 85 (1958).

(8) R. E. Bunde, *Rep. Univ. of Wisconsin Naval Res. Lab.*, CM-850 (1955).
(9) P. E. Suetin, G. T. Shchegolev and R. E. Klestov, *Sov. Phys. Tech. Phys.* **4**, 964 (1960).
(10) C. A. Boyd, N. Stein, V. Steingrimsson and W. F. Rumpel, *J. Chem. Phys.* **19**, 548 (1951).
(11) R. E. Walker and A. A. Westenberg, *J. Chem. Phys.* **29**, 1139, 1147 (1959), **32**, 436 (1960); R. E. Walker, L. Monchick, A. A. Westenberg and S. Favin, *Plan. Space Sci.* **3**, 221 (1961).
(12) R. Strehlow, *J. Chem. Phys.* **21**, 2101 (1953).
(13) S. Weissmann, S. C. Saxena and E. A. Mason, *Phys. Fluids*, **4**, 643 (1961); E. A. Mason, S. Weissmann and R. P. Wendt, *Phys. Fluids*, **7**, 174 (1964).
(14) B. N. and I. B. Srivastava, *J. Chem. Phys.* **36**, 2616 (1962); I. B. Srivastava, *Ind. J. Phys.* **36**, 193 (1962).
(15) B. N. and K. P. Srivastava, *J. Chem. Phys.* **30**, 984 (1959); K. P. Srivastava, *Physica*, **25**, 571 (1959); K. P. Srivastava and A. K. Barua, *Ind. J. Phys.* **33**, 229 (1959).
(16) R. Paul and I. B. Srivastava, *Ind. J. Phys.* **35**, 465, 523 (1961).
(17) J. N. Holsen and M. R. Strunk, *Ind. Eng. Chem. Fund.* **3**, 143 (1964).
(18) I. Amdur, J. W. Irvine, E. A. Mason and J. Ross, *J. Chem. Phys.* **20**, 436 (1952); I. Amdur and L. M. Shuler, *J. Chem. Phys.* **38**, 188 (1963).
(19) E. B. Winn and E. P. Ney, *Phys. Rev.* **72**, 77 (1947).
(20) E. B. Winn, *Phys. Rev.*, **80**, 1024 (1950).
(21) E. R. S. Winter, *Trans. Far. Soc.* **47**, 342 (1951).
(22) H. Braune and F. Zehle, *Z. Phys. Chem.* B **49**, 247 (1941).
(23) F. Hutchinson, *J. Chem. Phys.* **17**, 1081 (1949).
(24) B. N. Srivastava and R. Paul, *Physica*, **28**, 646 (1962).
(25) I. Amdur and T. F. Schatzki, *J. Chem. Phys.* **27**, 1049 (1957).
(26) W. W. Watson and D. Woernley, *Phys. Rev.* **63**, 181 (1943).
(27) L. Waldmann, *Z. Naturf.* **4a**, 105 (1949).
(28) K. E. Grew and J. N. Mundy, *Phys. Fluids*, **4**, 1325 (1961).
(29) K. E. Grew, *Phil. Mag.* **35**, 30 (1944).
(30) K. Clusius and M. Huber, *Z. Naturf.* **10a**, 556 (1955).
(31) K. Clusius and P. Flubacher, *Helv. Chim. Acta*, **41**, 2323 (1958).
(32) E. W. Becker and W. Beyrich, *J. Phys. Chem.* **56**, 911 (1952).
(33) A. E. de Vries, A. Haring and W. Slots, *Physica*, **22**, 247 (1956).
(34) J. Schirdewahn, A. Klemm and L. Waldmann, *Z. Naturf.* **16a**, 133 (1961).
(35) References quoted by K. E. Grew and T. L. Ibbs, *Thermal Diffusion in Gases*, table VA, p. 128 (Cambridge University Press, 1952).
(36) S. C. Saxena, J. G. Kelley and W. W. Watson, *Phys. Fluids*, **4**, 1216 (1961).
(37) T. I. Moran and W. W. Watson, *Phys. Rev.* **109**, 1184 (1958).
(38) J. R. Cozens and K. E. Grew, *Phys. Fluids*, **7**, 1395 (1964).
(39) S. Weissmann, S. C. Saxena and E. A. Mason, *Phys. Fluids*, **4**, 643 (1961).

15

THE THIRD APPROXIMATION TO THE VELOCITY-DISTRIBUTION FUNCTION

15.1. Successive approximations to f

In Chapter 7 it was shown how the velocity-distribution function f for a simple gas can be determined by successive approximation to any desired degree of accuracy; the first, second, third, ... approximations are $f^{(0)}$, $f^{(0)}+f^{(1)}, f^{(0)}+f^{(1)}+f^{(2)}, \ldots$. In 7.15 it was pointed out that $f^{(0)}, f^{(1)}, f^{(2)}, \ldots$ are respectively proportional to $n^1, n^0, n^{-1}, \ldots$, where n is the number-density of the molecules. Hence the later terms in the approximations become relatively more important as the density decreases.

The first approximation, $f^{(0)}$, is identical with Maxwell's function

$$f = n\left(\frac{m}{2\pi kT}\right)^{\frac{3}{2}} e^{-mC^2/2kT}$$

$$= n\left(\frac{m}{2\pi kT}\right)^{\frac{3}{2}} e^{-\mathscr{C}^2}. \tag{15.1, 1}$$

For $f^{(1)}$ the following expressions can be derived from the results of 7.3 and 7.31:

$$f^{(1)} = f^{(0)}\Phi^{(1)} = -\frac{1}{n}f^{(0)}\left\{\left(\frac{2kT}{m}\right)^{\frac{1}{2}} \boldsymbol{A}.\nabla \ln T + 2\mathsf{B}:\nabla \boldsymbol{c}_0\right\}$$

$$= -\left(\frac{m}{2\pi kT}\right)^{\frac{3}{2}} e^{-\mathscr{C}^2}\left\{\left(\frac{2kT}{m}\right)^{\frac{1}{2}} A(\mathscr{C})\boldsymbol{\mathscr{C}}.\frac{1}{T}\nabla T + 2B(\mathscr{C})\overset{\circ}{\boldsymbol{\mathscr{C}}\boldsymbol{\mathscr{C}}}:\nabla \boldsymbol{c}_0\right\}, \tag{15.1, 2}$$

and so we can write

$$f^{(1)} \equiv A'(C)\mathbf{C}.\nabla T + B'(C)\overset{\circ}{\mathbf{C}\mathbf{C}}:\nabla \boldsymbol{c}_0, \tag{15.1, 3}$$

where A', B' denote functions of C and T, defined by this equation. In virtue of (1.31, 7, 9) we can in (15.1, 2) and (15.1, 3) replace the velocity-gradient tensor $\nabla \boldsymbol{c}_0$ by the rate-of-strain tensor $\mathsf{e} \equiv \overline{\overline{\nabla \boldsymbol{c}_0}}$ or by its non-divergent part $\overset{\circ}{\mathsf{e}}$, without altering the relations.

We can readily estimate approximately the ratio $f^{(1)}/f^{(0)}$ or $\Phi^{(1)}$. In the notation of 7.51 and 7.52, first approximations to $\boldsymbol{A}$ and B are

$$\boldsymbol{A} = a_1\boldsymbol{a}^{(1)} = a_1 S_{\frac{3}{2}}^{(1)}(\mathscr{C}^2)\boldsymbol{\mathscr{C}} = a_1(\tfrac{5}{2}-\mathscr{C}^2)\boldsymbol{\mathscr{C}},$$

$$\mathsf{B} = b_1\mathsf{b}^{(1)} = b_1 S_{\frac{5}{2}}^{(0)}(\mathscr{C}^2)\overset{\circ}{\boldsymbol{\mathscr{C}}\boldsymbol{\mathscr{C}}} = b_1\overset{\circ}{\boldsymbol{\mathscr{C}}\boldsymbol{\mathscr{C}}},$$

where $a_1 = \alpha_1/a_{11} = -\frac{15}{4}/a_{11}$, $b_1 = \beta_1/b_{11} = 5/2b_{11}$.

Now (cf. 9.7) $a_{11} = b_{11}$, and the first approximation to the coefficient of viscosity is

$$[\mu]_1 = \tfrac{5}{2}kT/b_{11}.$$

Hence to a first approximation

$$a_1 = -3\mu/2kT, \quad b_1 = \mu/kT$$

and
$$\boldsymbol{A} = \frac{3\mu}{2kT}(\mathscr{C}^2 - \tfrac{5}{2})\boldsymbol{\mathscr{C}}, \quad \mathsf{B} = \frac{\mu}{kT}\overset{\circ}{\boldsymbol{\mathscr{C}}\boldsymbol{\mathscr{C}}}, \tag{15.1, 4}$$

whence it follows that (in agreement with (10.33, 8))

$$\Phi^{(1)} = -\frac{3\mu}{2p}\left(\frac{2k}{mT}\right)^{\frac{1}{2}}(\mathscr{C}^2 - \tfrac{5}{2})\boldsymbol{\mathscr{C}}.\nabla T - \frac{2\mu}{p}\overset{\circ}{\boldsymbol{\mathscr{C}}\boldsymbol{\mathscr{C}}}:\nabla \boldsymbol{c}_0.$$

We may write $\mu = 0{\cdot}499\rho\bar{C}l$, where l is an equivalent free path and $\bar{C} = (8kT/\pi m)^{\frac{1}{2}}$ (cf. (12.1, 6), (4.11, 2)). Let V be a scale mass-velocity, and L_1, L_2 be scale lengths, such that the gradients of T and $\boldsymbol{c}_0$ are comparable in magnitude with T/L_1 and V/L_2. Then if $\mathscr{C}$ is taken to be of unit order of magnitude, the two terms in $\Phi^{(1)}$ are comparable respectively with l/L_1 and $lV/L_2\bar{C}$. Now the free path l is comparable with 10^{-5} cm. in a gas at S.T.P., whereas experimental values of L_1, L_2 are normally comparable with the dimensions of the apparatus—of order 1 cm., say; also $V/\bar{C}$ is normally small or at most comparable with unity. Hence normally $\Phi^{(1)}$ is small, and $f^{(1)}$ is small compared with $f^{(0)}$. However, at densities 10^{-5} times the normal density, or in shock waves (where L_1 and L_2 are very small), $f^{(1)}$ becomes comparable with $f^{(0)}$, and it is necessary to examine whether $f^{(2)}$ is not of the same order. In so doing we confine attention for the present to a simple gas. Results for diffusion in a binary mixture are described in 15.42.

Burnett* determined the complete third approximation to f for a simple gas. Incomplete third approximations had already been given by Maxwell† and by (Lennard-)Jones,‡ who included terms involving second derivatives of T and $\boldsymbol{c}_0$, but ignored squares and products of the first derivatives (thus avoiding some of the chief mathematical difficulties of the problem); Enskog§ gave a solution which included products of the *temperature* gradients (but not those of $\boldsymbol{c}_0$).

Burnett's solution will not be given here in full; the equation to determine $f^{(2)}$ will be considered so far as to infer from it the nature of the various terms in $f^{(2)}$, but their exact expressions will not be derived. It will be shown that the new terms introduced by $f^{(2)}$ into the expressions for the thermal flux $\boldsymbol{q}$ and the pressure tensor p can be determined without solving the equation for $f^{(2)}$, and the magnitudes of the terms will be roughly estimated.

* D. Burnett, *Proc. Lond. Math. Soc.* **40**, 382 (1935).

† J. C. Maxwell, *Phil. Trans. R. Soc.* **170**, 231 (1879); *Collected Papers*, vol. 2, 681 (1890).

‡ J. E. (Lennard-)Jones, *Phil. Trans. R. Soc.* A, **223**, 1 (1923).

§ D. Enskog, Inaug. Diss. Uppsala, chapter 6 (1917).

15.2. The integral equation for $f^{(2)}$

The equation to determine $f^{(2)}$ is (cf. (7.12, 1), (7.14, 19))

$$-J(f^{(0)}f_1^{(2)})-J(f^{(2)}f_1^{(0)})=\frac{\partial_1 f^{(0)}}{\partial t}+\frac{\partial_0 f^{(1)}}{\partial t}+\mathbf{c}.\frac{\partial f^{(1)}}{\partial \mathbf{r}}+\mathbf{F}.\frac{\partial f^{(1)}}{\partial \mathbf{c}}+J(f^{(1)}f_1^{(1)}).$$

In this write
$$f^{(2)}=f^{(0)}\Phi^{(2)};\qquad (15.2,\ 1)$$
then it becomes, by (7.12, 3),

$$-n^2 I(\Phi^{(2)})=\frac{\partial_1 f^{(0)}}{\partial t}+\frac{\partial_0 f^{(1)}}{\partial t}+\mathbf{c}.\frac{\partial f^{(1)}}{\partial \mathbf{r}}+\mathbf{F}.\frac{\partial f^{(1)}}{\partial \mathbf{c}}+J(f^{(1)}f_1^{(1)}).\qquad (15.2,\ 2)$$

The various terms on the right of (15.2, 2) are developed as follows. Using (7.14, 9, 11, 13) we have*

$$\frac{\partial_1 f^{(0)}}{\partial t}=f^{(0)}\frac{\partial_1 \ln f^{(0)}}{\partial t}=f^{(0)}\frac{\partial_1}{\partial t}\left(\ln n-\tfrac{3}{2}\ln T-\frac{mC^2}{2kT}\right)$$

$$=f^{(0)}\left\{(\mathscr{C}^2-\tfrac{3}{2})\frac{1}{T}\frac{\partial_1 T}{\partial t}+\frac{m}{kT}\mathbf{C}.\frac{\partial_1 \mathbf{c}_0}{\partial t}\right\}$$

$$=-\frac{f^{(0)}}{p}\left\{(\tfrac{2}{3}\mathscr{C}^2-1)\left(\mathsf{p}^{(1)}:\frac{\partial}{\partial \mathbf{r}}\mathbf{c}_0+\frac{\partial}{\partial \mathbf{r}}.\mathbf{q}^{(1)}\right)+\mathbf{C}\frac{\partial}{\partial \mathbf{r}}:\mathsf{p}^{(1)}\right\}.\qquad (15.2,\ 3)$$

If $f^{(1)}$ is regarded as a function of $\mathbf{C}$, $\mathbf{r}$, t, the next three terms on the right of (15.2, 2) are replaced by

$$\frac{D_0 f^{(1)}}{Dt}+\mathbf{C}.\frac{\partial f^{(1)}}{\partial \mathbf{r}}+\left(\mathbf{F}-\frac{D_0\mathbf{c}_0}{Dt}\right).\frac{\partial f^{(1)}}{\partial \mathbf{C}}-\frac{\partial f^{(1)}}{\partial \mathbf{C}}\mathbf{C}:\frac{\partial}{\partial \mathbf{r}}\mathbf{c}_0$$

$$=\frac{D_0 f^{(1)}}{Dt}+\mathbf{C}.\frac{\partial f^{(1)}}{\partial \mathbf{r}}+\frac{1}{\rho}\frac{\partial p}{\partial \mathbf{r}}.\frac{\partial f^{(1)}}{\partial \mathbf{C}}-\frac{\partial f^{(1)}}{\partial \mathbf{C}}\mathbf{C}:\frac{\partial}{\partial \mathbf{r}}\mathbf{c}_0.\qquad (15.2,\ 4)$$

Now, using the expression (15.1, 3) for $f^{(1)}$,

$$\frac{\partial f^{(1)}}{\partial \mathbf{r}}=\left(\frac{\partial A'}{\partial T}\mathbf{C}.\nabla T+\frac{\partial B'}{\partial T}\overset{\circ}{\mathbf{C}\mathbf{C}}:\nabla\mathbf{c}_0+A'\mathbf{C}.\nabla\right)\nabla T+B'\nabla(\overset{\circ}{\mathbf{C}\mathbf{C}}:\nabla\mathbf{c}_0),\qquad (15.2,\ 5)$$

$$\frac{\partial f^{(1)}}{\partial \mathbf{C}}=2\left(\frac{\partial A'}{\partial C^2}\mathbf{C}.\nabla T+\frac{\partial B'}{\partial C^2}\overset{\circ}{\mathbf{C}\mathbf{C}}:\nabla\mathbf{c}_0\right)\mathbf{C}+A'\nabla T+2B'\mathbf{C}.\overset{\circ}{\overline{\overline{\nabla\mathbf{c}_0}}},\qquad (15.2,\ 6)$$

and, using (7.14, 16),

$$\frac{D_0 f^{(1)}}{Dt}=-\left(\frac{\partial A'}{\partial T}\mathbf{C}.\nabla T+\frac{\partial B'}{\partial T}\overset{\circ}{\mathbf{C}\mathbf{C}}:\nabla\mathbf{c}_0\right)\frac{2T}{3}(\nabla.\mathbf{c}_0)$$

$$+A'\mathbf{C}.\frac{D_0}{Dt}(\nabla T)+B'\overset{\circ}{\mathbf{C}\mathbf{C}}:\frac{D_0}{Dt}(\nabla\mathbf{c}_0).\qquad (15.2,\ 7)$$

* In the derivation of equations (15.2, 3–9) use is made of the tensor relations (1.2, 7, 8), (1.32, 6, 10), (1.33, 7, 8).

The time-derivatives on the right-hand side of (15.2, 7) can be determined in terms of space-derivatives; for, by (7.14, 15, 16),

$$\frac{D_0}{Dt}(\nabla T) = \left(\frac{\partial_0}{\partial t}+\boldsymbol{c}_0.\nabla\right)\nabla T$$

$$= \nabla\left(\frac{\partial_0 T}{\partial t}+\boldsymbol{c}_0.\nabla T\right)-(\nabla\boldsymbol{c}_0).\nabla T$$

$$= -\tfrac{2}{3}\nabla(T\nabla.\boldsymbol{c}_0)-(\nabla\boldsymbol{c}_0).\nabla T, \qquad (15.2, 8)$$

$$\frac{D_0}{Dt}(\nabla\boldsymbol{c}_0) = \left(\frac{\partial_0}{\partial t}+\boldsymbol{c}_0.\nabla\right)\nabla\boldsymbol{c}_0$$

$$= \nabla\left(\frac{\partial_0\boldsymbol{c}_0}{\partial t}+(\boldsymbol{c}_0.\nabla)\,\boldsymbol{c}_0\right)-(\nabla\boldsymbol{c}_0).(\nabla\boldsymbol{c}_0)$$

$$= \nabla\left(\boldsymbol{F}-\frac{1}{\rho}\nabla p\right)-(\nabla\boldsymbol{c}_0).(\nabla\boldsymbol{c}_0). \qquad (15.2, 9)$$

It is, however, more convenient for the present to retain the time-derivatives in (15.2, 7).

Again (cf. (7.11, 2) and (15.1, 3), replacing $\nabla\boldsymbol{c}_0$ by e)

$$J(f^{(1)}f_1^{(1)}) = J(A'\boldsymbol{C}.\nabla T\,A_1'\boldsymbol{C}_1.\nabla T)+J(A'\boldsymbol{C}.\nabla T\,B_1'\overset{\circ}{\boldsymbol{C}_1}\boldsymbol{C}_1:\mathsf{e})$$

$$+J(B'\overset{\circ}{\boldsymbol{C}\boldsymbol{C}}:\mathsf{e}\,A_1'\boldsymbol{C}_1.\nabla T)+J(B'\overset{\circ}{\boldsymbol{C}\boldsymbol{C}}:\mathsf{e}\,B_1'\overset{\circ}{\boldsymbol{C}_1}\boldsymbol{C}_1:\mathsf{e}). \qquad (15.2, 10)$$

The first and fourth of the expressions on the right of this equation, as may readily be seen, are even functions of $\boldsymbol{C}$; the second and third are odd functions.

Combining the above results, we obtain an expression for $-n^2I(\Phi^{(2)})$ as the sum of numerous terms which can be divided into groups as follows. First there is a part involving only the scalar C; this is

$$-\frac{1}{p}f^{(0)}(\tfrac{2}{3}\mathscr{C}^2-1)(\mathsf{p}^{(1)}:\nabla\boldsymbol{c}_0+\nabla.\boldsymbol{q}^{(1)})+\frac{1}{\rho}A'\nabla p.\nabla T. \qquad (15.2, 11)$$

Next there is a part of odd degree in $\boldsymbol{C}$, which after suitable rearrangement takes the form

$$-\frac{1}{p}f^{(0)}\boldsymbol{C}\nabla:\mathsf{p}^{(1)}-\tfrac{2}{3}\Delta\left(T\frac{\partial A'}{\partial T}+C^2\frac{\partial A'}{\partial C^2}\right)(\boldsymbol{C}.\nabla T)$$

$$+A'\boldsymbol{C}.\left(\frac{D_0}{Dt}(\nabla T)-(\nabla\boldsymbol{c}_0).\nabla T\right)+\frac{2}{\rho}\frac{\partial B'}{\partial C^2}(\boldsymbol{C}\boldsymbol{C}:\mathsf{e})(\boldsymbol{C}.\nabla p)$$

$$+\frac{2}{\rho}B'\boldsymbol{C}\nabla p:\mathsf{e}+B'\boldsymbol{C}.\nabla(\boldsymbol{C}\boldsymbol{C}:\mathsf{e})$$

$$+\left(\frac{\partial B'}{\partial T}-2\frac{\partial A'}{\partial C^2}\right)(\boldsymbol{C}\boldsymbol{C}:\mathsf{e})(\boldsymbol{C}.\nabla T)+J(A'\boldsymbol{C}.\nabla T\,B_1'\boldsymbol{C}_1\boldsymbol{C}_1:\mathsf{e})$$

$$+J(B'\boldsymbol{C}\boldsymbol{C}:\mathsf{e}\,A_1'\boldsymbol{C}_1.\nabla T), \qquad (15.2, 12)$$

where $$\Delta \equiv \nabla . \boldsymbol{c}_0 = \frac{\partial u_0}{\partial x} + \frac{\partial v_0}{\partial y} + \frac{\partial w_0}{\partial z}, \qquad (15.2, 13)$$

that is, Δ is the divergence of the mass-velocity of the gas.

Finally, there is a part of even degree in $\boldsymbol{C}$, which may be expressed as follows:

$$-\tfrac{2}{3}\Delta\left(T\frac{\partial B'}{\partial T} + C^2\frac{\partial B'}{\partial C^2}\right)(\boldsymbol{CC}:\overset{\circ}{\mathsf{e}}) + B'\boldsymbol{CC}:\left(\frac{D_0}{Dt}(\overset{\circ}{\mathsf{e}}) - 2\overset{\circ}{\mathsf{e}}.\nabla\boldsymbol{c}_0\right)$$
$$+A'\boldsymbol{CC}:\nabla(\nabla T) + \frac{2}{\rho}\frac{\partial A'}{\partial C^2}\boldsymbol{CC}:\nabla p\nabla T$$
$$+\frac{\partial A'}{\partial T}\boldsymbol{CC}:\nabla T\nabla T + J(A'\boldsymbol{C}.\nabla T\, A'_1\boldsymbol{C}_1.\nabla T)$$
$$-2\frac{\partial B'}{\partial C^2}(\boldsymbol{CC}:\overset{\circ}{\mathsf{e}})(\boldsymbol{CC}:\overset{\circ}{\mathsf{e}}) + J(B'\boldsymbol{CC}:\overset{\circ}{\mathsf{e}}\, B'_1\boldsymbol{C}_1\boldsymbol{C}_1:\overset{\circ}{\mathsf{e}}). \qquad (15.2, 14)$$

15.3. The third approximation to the thermal flux and the stress tensor

The equation (15.2, 2) for $f^{(2)}$ or $\Phi^{(2)}$ was solved by Burnett by the methods applied to obtain $f^{(1)}$ from (7.3, 7). His solution for $\Phi^{(2)}$ is in the form of a series

$$\Phi^{(2)} = \sum_{p,l} a_{pl} S^{(p)}_{l+\frac{1}{2}}(\mathscr{C}^2)\,\psi^{(l)}(\boldsymbol{C}) \equiv \sum_{p,l} a_{pl}\chi_{pl}, \qquad (15.3, 1)$$

where $\psi^{(l)}(\boldsymbol{C})$ is a spherical harmonic function of degree l in the components of $\boldsymbol{C}$; that is, a homogeneous polynomial in U, V, W, satisfying the equation

$$\frac{\partial}{\partial \boldsymbol{C}}\cdot\frac{\partial \psi^{(l)}}{\partial \boldsymbol{C}} \equiv \frac{\partial^2\psi^{(l)}}{\partial U^2} + \frac{\partial^2\psi^{(l)}}{\partial V^2} + \frac{\partial^2\psi^{(l)}}{\partial W^2} = 0. \qquad (15.3, 2)$$

In Burnett's case, only the values $l \leqslant 3$ were required. The unknown coefficients a_{pl} can be determined by methods similar to those of 7.51, 7.52.

It is, however, possible to ascertain the new terms in the thermal flux $\boldsymbol{q}$ and in the stress tensor p without obtaining $\Phi^{(2)}$; for

$$\begin{aligned}
\boldsymbol{q}^{(2)} &= \int f^{(2)}\tfrac{1}{2}mC^2\boldsymbol{C}\,d\boldsymbol{c} \\
&= \tfrac{1}{2}m\left(\frac{2kT}{m}\right)^{\frac{3}{2}}\int f^{(0)}\Phi^{(2)}\mathscr{C}^2\boldsymbol{\mathscr{C}}\,d\boldsymbol{c} \\
&= kT\left(\frac{2kT}{m}\right)^{\frac{1}{2}}\int f^{(0)}\Phi^{(2)}(\mathscr{C}^2 - \tfrac{5}{2})\boldsymbol{\mathscr{C}}\,d\boldsymbol{c} \\
&= p\left(\frac{2kT}{m}\right)^{\frac{1}{2}}\int \Phi^{(2)}I(\boldsymbol{A})\,d\boldsymbol{c} \\
&= p\left(\frac{2kT}{m}\right)^{\frac{1}{2}}[\Phi^{(2)}, \boldsymbol{A}] \\
&= p\left(\frac{2kT}{m}\right)^{\frac{1}{2}}\int \boldsymbol{A}I(\Phi^{(2)})\,d\boldsymbol{c},
\end{aligned} \qquad (15.3, 3)$$

using (7.12, 4), (7.31, 2) and (4.4, 7, 8). Similarly

$$\begin{aligned}
\mathsf{p}^{(2)} &= \int f^{(2)} m\boldsymbol{CC}\,d\boldsymbol{c} \\
&= \int f^{(0)}\Phi^{(2)} m\overset{\circ}{\boldsymbol{C}\boldsymbol{C}}\,d\boldsymbol{c} \\
&= 2kT\int f^{(0)}\Phi^{(2)}\,\overset{\circ}{\boldsymbol{\mathscr{C}}\boldsymbol{\mathscr{C}}}\,d\boldsymbol{c} \\
&= 2p\int \Phi^{(2)} I(\mathsf{B})\,d\boldsymbol{c} \\
&= 2p[\Phi^{(2)}, \mathsf{B}] \\
&= 2p\int \mathsf{B} I(\Phi^{(2)})\,d\boldsymbol{c},
\end{aligned} \tag{15.3, 4}$$

using the same relations, and (7.31, 3). Since $-n^2 I(\Phi^{(2)})$ is given as the sum of (15.2, 11, 12, 14), expressions for $\boldsymbol{q}^{(2)}$ and $\mathsf{p}^{(2)}$ can be derived.

Since

$$\boldsymbol{A} = \boldsymbol{\mathscr{C}} A(\mathscr{C}), \quad \mathsf{B} = \overset{\circ}{\boldsymbol{\mathscr{C}}\boldsymbol{\mathscr{C}}} B(\mathscr{C}) \tag{15.3, 5}$$

it follows from 1.42 that the groups of terms (15.2, 11, 14) will not contribute to $\boldsymbol{q}^{(2)}$, nor the terms (15.2, 11, 12) to $\mathsf{p}^{(2)}$. Again, the contribution to $\boldsymbol{q}^{(2)}$ from the first term of (15.2, 12) (the only one of these terms that does not depend on A' or B') is

$$\frac{p}{n^2}\left(\frac{2kT}{m}\right)^{\frac{1}{2}}\int \boldsymbol{A}\frac{1}{p} f^{(0)}\boldsymbol{C}\nabla : \mathsf{p}^{(1)} d\boldsymbol{c} = \frac{2kT}{3mn^2}\nabla . \mathsf{p}^{(1)} \int f^{(0)}\boldsymbol{A}.\boldsymbol{\mathscr{C}}\,d\boldsymbol{c},$$

which vanishes, by (7.31, 6). All the other terms of (15.2, 12) make contributions to $\boldsymbol{q}^{(2)}$; thus we can write

$$\boldsymbol{q}^{(2)} = \theta_1\frac{\mu^2}{\rho T}\Delta\nabla T + \theta_2\frac{\mu^2}{\rho T}\left\{\frac{D_0}{Dt}(\nabla T) - (\nabla \boldsymbol{c}_0).\nabla T\right\}$$
$$+\theta_3\frac{\mu^2}{\rho p}\nabla p.\mathsf{e} + \theta_4\frac{\mu^2}{\rho}\nabla.(\mathsf{e}) + \theta_5\frac{3\mu^2}{\rho T}\nabla T.\mathsf{e}, \tag{15.3, 6}$$

the vectors in this expression being the only ones which can be formed from the elements involved in the terms of (15.2, 12); the coefficients of the vectors are so chosen that θ_1, θ_2, θ_3, θ_4 and θ_5 are pure numbers. In the second term we can substitute for the time-derivative from (15.2, 8), when the expression in the bracket becomes

$$-\tfrac{2}{3}\nabla(T\Delta) - 2(\nabla \boldsymbol{c}_0).\nabla T. \tag{15.3, 7}$$

The coefficient of the last term in (15.3, 6), namely θ_5, is the only one which depends on the integrals J.

Since e and Δ both depend on the space-derivatives of $\boldsymbol{c}_0$, every term in (15.3, 6) depends on such derivatives. If the gas is at rest, or in motion uniform in space, $\boldsymbol{q}^{(2)} = 0$, and $\boldsymbol{q} = -\lambda\nabla T$ not only to the second but also to the third approximation. The third approximation, moreover, introduces no terms depending on the second or higher space-derivatives of T, nor on squares or products of derivatives of T; it does, however, supply a thermal flux even when T is uniform, if either p or Δ or $\overset{\circ}{\mathsf{e}}$ is not uniform.

In the same way, we can show that

$$\mathsf{p}^{(2)} = \varpi_1 \frac{\mu^2}{p} \Delta \overset{\circ}{\mathsf{e}} + \varpi_2 \frac{\mu^2}{p} \left(\frac{D_0}{Dt}(\overset{\circ}{\mathsf{e}}) - 2\overline{\overline{\overset{\circ}{\nabla \boldsymbol{c}_0 . \mathsf{e}}}} \right) + \varpi_3 \frac{\mu^2}{\rho T} \overline{\overline{\overset{\circ}{\nabla\nabla T}}}$$

$$+ \varpi_4 \frac{\mu^2}{\rho p T} \overline{\overline{\overset{\circ}{\nabla p \nabla T}}} + \varpi_5 \frac{\mu^2}{\rho T^2} \overline{\overline{\overset{\circ}{\nabla T \nabla T}}} + \varpi_6 \frac{\mu^2}{p} \overline{\overline{\overset{\circ}{\mathsf{e}.\mathsf{e}}}}, \qquad (15.3, 8)$$

where ϖ_1, ϖ_2, ϖ_3, ϖ_4, ϖ_5 and ϖ_6 are pure numbers; the tensors in this expression are the only symmetrical and non-divergent tensors that can be formed from the elements involved in the terms of (15.2, 14), and (cf. (15.3, 4, 5)) $\mathsf{p}^{(2)}$ is both symmetrical and non-divergent. The coefficients ϖ_5 and ϖ_6, and no others, depend on the integrals J in (15.2, 14). The expression in the bracket of the second term is, by (15.2, 9), equal to

$$\overline{\overline{\overset{\circ}{\nabla \left(\boldsymbol{F} - \frac{1}{\rho} \nabla p \right)}}} - \overline{\overline{\overset{\circ}{(\nabla \boldsymbol{c}_0).(\nabla \boldsymbol{c}_0)}}} - 2\overline{\overline{\overset{\circ}{\nabla \boldsymbol{c}_0 . \mathsf{e}}}}. \qquad (15.3, 9)$$

Thus the third approximation to the stress-system introduces terms depending on (*a*) products of first-order derivatives of the mean motion $\boldsymbol{c}_0$; (*b*) first-order derivatives of $\boldsymbol{F} - \frac{1}{\rho}\nabla p$ or $\frac{D_0 \boldsymbol{c}_0}{Dt}$; (*c*) products of first-order derivatives of the temperature with each other, and with the first-order derivatives of the pressure; and (*d*) second-order derivatives of the temperature. The terms (*a*) depend not only on the elements of the rate-of-strain tensor e, but also on the anti-symmetrical part of the velocity-gradient tensor $\nabla \boldsymbol{c}_0$, and therefore they have a part depending on the vorticity of the motion ($\frac{1}{2}$ curl $\boldsymbol{c}_0$). Similarly the contribution to $\boldsymbol{q}^{(2)}$ from the second term of (15.3, 6) includes a part proportional to $\nabla T \wedge \text{curl}\, \boldsymbol{c}_0$.

15.4. The terms in $\boldsymbol{q}^{(2)}$

Approximate values of the coefficients in the various parts of $\boldsymbol{q}^{(2)}$ and $\mathsf{p}^{(2)}$ can be found without great difficulty.

By (15.1, 2–4), to a first approximation,

$$\boldsymbol{A} = A(\mathscr{C})\boldsymbol{\mathscr{C}} = \frac{3\mu}{2kT}(\mathscr{C}^2 - \tfrac{5}{2})\boldsymbol{\mathscr{C}}, \quad \mathsf{B} = B(\mathscr{C})\overset{\circ}{\boldsymbol{\mathscr{C}}\boldsymbol{\mathscr{C}}} = \frac{\mu}{kT}\overset{\circ}{\boldsymbol{\mathscr{C}}\boldsymbol{\mathscr{C}}}, \qquad (15.4, 1)$$

$$A'(C) = -\left(\frac{m}{2\pi kT}\right)^{\frac{3}{2}} \frac{3\mu}{2kT^2} e^{-\mathscr{C}^2}(\mathscr{C}^2 - \tfrac{5}{2}), \qquad (15.4, 2)$$

$$B'(C) = -\left(\frac{m}{2\pi kT}\right)^{\frac{3}{2}} \frac{\mu m}{(kT)^2} e^{-\mathscr{C}^2}. \qquad (15.4, 3)$$

Hence, remembering that $\mathscr{C}^2 = mC^2/2kT$,

$$T\frac{\partial A'}{\partial T} = -\left(\frac{m}{2\pi kT}\right)^{\frac{3}{2}} e^{-\mathscr{C}^2}\left\{\frac{3\mu}{2kT^2}(\mathscr{C}^4 - 7\mathscr{C}^2 + \tfrac{35}{4}) + \frac{3}{2kT}\frac{d\mu}{dT}(\mathscr{C}^2 - \tfrac{5}{2})\right\},$$

$$C^2\frac{\partial A'}{\partial C^2} = \mathscr{C}^2\frac{\partial A'}{\partial \mathscr{C}^2} = \left(\frac{m}{2\pi kT}\right)^{\frac{3}{2}} e^{-\mathscr{C}^2}\frac{3\mu}{2kT^2}(\mathscr{C}^4 - \tfrac{7}{2}\mathscr{C}^2),$$

$$T\frac{\partial B'}{\partial T} = -\left(\frac{m}{2\pi kT}\right)^{\frac{3}{2}} e^{-\mathscr{C}^2}\left\{\frac{\mu m}{(kT)^2}(\mathscr{C}^2 - \tfrac{7}{2}) + \frac{m}{k^2T}\frac{d\mu}{dT}\right\},$$

$$C^2\frac{\partial B'}{\partial C^2} = \mathscr{C}^2\frac{\partial B'}{\partial \mathscr{C}^2} = \left(\frac{m}{2\pi kT}\right)^{\frac{3}{2}} e^{-\mathscr{C}^2}\frac{\mu m}{(kT)^2}\mathscr{C}^2.$$

Using these expressions, we can obtain approximations to the various coefficients in (15.3, 6, 8).

As noted above, the first term of (15.2, 12) makes no contribution to $\boldsymbol{q}^{(2)}$. By (15.3, 3), the second term contributes the expression

$$\frac{2}{3}\frac{kT}{n}\left(\frac{2kT}{m}\right)^{\frac{1}{2}}\Delta\int \boldsymbol{A}\left(T\frac{\partial A'}{\partial T} + C^2\frac{\partial A'}{\partial C^2}\right)(\boldsymbol{C}.\nabla T)\,d\boldsymbol{c}$$
$$= \frac{2}{9}\frac{kT}{n}\left(\frac{2kT}{m}\right)\Delta\nabla T\int\left(T\frac{\partial A'}{\partial T} + C^2\frac{\partial A'}{\partial C^2}\right)\boldsymbol{A}.\boldsymbol{\mathscr{C}}\,d\boldsymbol{c}$$

by (1.42, 4). Inserting the above values for $\boldsymbol{A}$ and A', we get

$$\frac{\mu^2}{\rho T}\left(\frac{7}{2} - \frac{T}{\mu}\frac{d\mu}{dT}\right)\Delta\nabla T\pi^{-\frac{1}{2}}\int e^{-\mathscr{C}^2}(\mathscr{C}^2 - \tfrac{5}{2})^2\mathscr{C}^2 d\mathscr{C} = \frac{15\mu^2}{4\rho T}\left(\frac{7}{2} - \frac{T}{\mu}\frac{d\mu}{dT}\right)\Delta\nabla T,$$

so that in (15.3, 6)
$$\theta_1 = \frac{15}{4}\left(\frac{7}{2} - \frac{T}{\mu}\frac{d\mu}{dT}\right). \qquad (15.4, 4)$$

The coefficients θ_2, θ_3 and θ_4 are similarly determined, using (1.42, 4) and 1.421: we find

$$\theta_2 = \tfrac{45}{8}, \quad \theta_3 = -3, \quad \theta_4 = 3. \qquad (15.4, 5)$$

The coefficient θ_5 depends on the integrals J in (15.2, 12). We find that the contribution of the seventh term of (15.2, 12) to $\boldsymbol{q}^{(2)}$ is, to the present approximation,

$$\frac{3\mu^2}{\rho T}\left(\frac{35}{4} + \frac{T}{\mu}\frac{d\mu}{dT}\right)\nabla T.\mathsf{e}.$$

The contribution of the terms involving the integrals J is

$$-\frac{kT}{n}\left(\frac{2kT}{m}\right)^{\frac{1}{2}}\int \boldsymbol{A}\{J(A'\boldsymbol{C}.\nabla T\,B_1'\boldsymbol{C}_1\boldsymbol{C}_1:\mathsf{e}) + J(B'\boldsymbol{C}\boldsymbol{C}:\mathsf{e}\,A_1'\boldsymbol{C}_1.\nabla T)\}\,d\boldsymbol{c}$$
$$= -\frac{kT}{n}\left(\frac{2kT}{m}\right)^{\frac{1}{2}}\iiint \boldsymbol{A}\{A'(C)\boldsymbol{C}.\nabla T\,B'(C_1)\boldsymbol{C}_1\boldsymbol{C}_1:\mathsf{e}$$
$$+ A'(C_1)\boldsymbol{C}_1.\nabla T\,B'(C)\boldsymbol{C}\boldsymbol{C}:\mathsf{e} - A'(C')\boldsymbol{C}'.\nabla T\,B'(C_1')\boldsymbol{C}_1'\boldsymbol{C}_1':\mathsf{e}$$
$$- A'(C_1')\boldsymbol{C}_1'.\nabla T\,B'(C')\boldsymbol{C}'\boldsymbol{C}':\mathsf{e}\}g\alpha_1\,d\boldsymbol{e}'\,d\boldsymbol{c}\,d\boldsymbol{c}_1,$$

using the definition of J, (7.11, 2). Transforming by methods similar to those of 3.54, this becomes

$$-\frac{kT}{n}\left(\frac{2kT}{m}\right)^{\frac{1}{2}}\iiint A'(C)\mathbf{C}.\nabla T\, B'(C_1)\mathbf{C}_1\mathbf{C}_1:\overset{\circ}{\mathsf{e}}$$
$$\times(\boldsymbol{A}+\boldsymbol{A}_1-\boldsymbol{A}'-\boldsymbol{A}_1')g\alpha_1 d\mathbf{e}'\,d\mathbf{c}\,d\mathbf{c}_1.$$

Inserting the above approximate expressions for $\boldsymbol{A}$, $A'(C)$, and $B'(C)$, and using the relation $\mathscr{C}+\mathscr{C}_1=\mathscr{C}'+\mathscr{C}_1'$, expressing the conservation of momentum, we get

$$-\frac{9\mu^3}{\pi^3\rho kT^2}\iiint e^{-\mathscr{C}^2-\mathscr{C}_1^2}(\mathscr{C}^2-\tfrac{5}{2})\mathscr{C}.\nabla T\,\mathscr{C}_1\mathscr{C}_1:\overset{\circ}{\mathsf{e}}$$
$$\times(\mathscr{C}^2\mathscr{C}+\mathscr{C}_1^2\mathscr{C}_1-\mathscr{C}'^2\mathscr{C}'-\mathscr{C}_1'^2\mathscr{C}_1')g\alpha_1 d\mathbf{e}'\,d\mathscr{C}\,d\mathscr{C}_1.$$

Let $\mathscr{G}_0$, $\mathscr{g}$ be the variables of (9.2, 6), so that for a simple gas

$$\mathscr{C}=2^{-\frac{1}{2}}(\mathscr{G}_0-\mathscr{g}),\quad \mathscr{C}_1=2^{-\frac{1}{2}}(\mathscr{G}_0+\mathscr{g})$$

and similarly for $\mathscr{C}'$, $\mathscr{C}_1'$. Then

$$\int(\mathscr{C}^2\mathscr{C}+\mathscr{C}_1^2\mathscr{C}_1-\mathscr{C}'^2\mathscr{C}'-\mathscr{C}_1'^2\mathscr{C}_1')\alpha_{12}d\mathbf{e}'$$
$$=2^{\frac{1}{2}}\iint\{(\mathscr{G}_0.\mathscr{g})\mathscr{g}-(\mathscr{G}_0.\mathscr{g}')\mathscr{g}'\}\,b\,db\,d\epsilon$$
$$=(3/\sqrt{2})\,\phi_1^{(2)}\mathscr{G}_0.\overset{\circ}{\mathscr{g}\mathscr{g}}$$

on integrating with respect to ϵ as in 10.5.

Using this result and taking $\mathscr{G}_0$ and $\mathscr{g}$ as new variables of integration we can (with some labour) complete the integration; the result is

$$-\frac{18\mu^3}{5\rho kT^2}\nabla T.\overset{\circ}{\mathsf{e}}(\Omega_1^{(2)}(3)-\tfrac{7}{2}\Omega_1^{(2)}(2)).$$

In particular, for Maxwellian molecules (for which (15.4, 1–3) are exact) the expression vanishes; thus it may be neglected to a first approximation, and we get

$$\theta_5=\left(\frac{35}{4}+\frac{T}{\mu}\frac{d\mu}{dT}\right). \qquad (15.4,\ 6)$$

15.41. The terms in $\mathsf{p}^{(2)}$

The magnitudes of the various coefficients in the expression (15.3, 8) for $\mathsf{p}^{(2)}$ can be found similarly. The four coefficients which do not involve the integrals J are, to a first approximation, given by

$$\varpi_1=\frac{4}{3}\left(\frac{7}{2}-\frac{T}{\mu}\frac{d\mu}{dT}\right),\quad \varpi_2=2,\quad \varpi_3=3,\quad \varpi_4=0. \qquad (15.41,\ 1)$$

The contributions to $\mathsf{p}^{(2)}$ from the fifth and seventh terms of (15.2, 14) are found to be

$$\frac{3\mu}{\rho T}\frac{d\mu}{dT}\overset{\circ}{\nabla T\nabla T}\quad\text{and}\quad\frac{8\mu^2}{p}\overline{\overset{\circ}{\overset{\circ}{\mathsf{e}}.\overset{\circ}{\mathsf{e}}}}$$

respectively. In determining the second of these contributions it is necessary to use the integral theorem

$$\int e^{-\mathscr{C}^2}\overset{\circ}{\mathscr{C}\mathscr{C}}(\mathscr{C}\mathscr{C}:\overset{\circ}{\mathsf{e}})(\mathscr{C}\mathscr{C}:\overset{\circ}{\mathsf{e}})\,d\mathscr{C} = \tfrac{8}{105}\,\overline{\overline{\overset{\circ}{\mathsf{e}}.\overset{\circ}{\mathsf{e}}}}\int e^{-\mathscr{C}^2}\mathscr{C}^6\,d\mathscr{C} \qquad (15.41,\ 2)$$

(cf. (1.421, 1)). The contributions to $\mathsf{p}^{(2)}$ from the sixth and eighth terms of (15.2, 14), which involve integrals J, are found in a way similar to that used in determining the contribution to $\boldsymbol{q}^{(2)}$ from the integrals J; in integrating, we use

$$\int(\mathsf{B}+\mathsf{B}_1-\mathsf{B}'-\mathsf{B}_1')\,\alpha_{12}\,de' = \iint\frac{\mu}{kT}(\overset{\circ}{\mathscr{C}\mathscr{C}}+\overset{\circ}{\mathscr{C}_1\mathscr{C}_1}-\overset{\circ}{\mathscr{C}'\mathscr{C}'}-\overset{\circ}{\mathscr{C}_1'\mathscr{C}_1'})\,b\,db\,d\epsilon$$

$$= \frac{3\mu}{kT}\phi_1^{(2)}\overset{\circ}{\boldsymbol{g}\boldsymbol{g}};$$

in the second integral we must also use (15.41, 2). The contributions are thus ultimately found to be

$$\frac{9\mu^3}{10\rho kT^3}\{\Omega_1^{(2)}(4)-9\Omega_1^{(2)}(3)+\tfrac{63}{4}\Omega_1^{(2)}(2)\}\,\overset{\circ}{\nabla T\nabla T}$$

and

$$-\frac{128\mu^3}{35pkT}\{\Omega_1^{(2)}(3)-\tfrac{7}{2}\Omega_1^{(2)}(2)\}\,\overline{\overline{\overset{\circ}{\mathsf{e}}.\overset{\circ}{\mathsf{e}}}}$$

respectively. Since both of these expressions vanish for Maxwellian molecules, they are neglected to a first approximation, and so

$$\varpi_5 = \frac{3T}{\mu}\frac{d\mu}{dT}, \quad \varpi_6 = 8. \qquad (15.41,\ 3)$$

The values of the coefficients ϖ_r, θ_r given in this and the preceding section are exact for Maxwellian molecules, and it may be expected that they will be not far from the true values for other molecular models. The coefficients ϖ_2, ϖ_3 were calculated more exactly by (Lennard-)Jones for rigid spherical molecules, using third approximations to $\boldsymbol{A}$, B (and $A'(C)$, $B'(C)$) in place of the first approximations used above; his values are respectively 1·013 and 0·800 times the values given by (15.41, 1). Burnett gave a general formula for the terms in $\mathsf{p}^{(2)}$ for the case of molecules repelling as $r^{-\nu}$ (including, as a limiting case, rigid elastic spheres), and determined the coefficients for Maxwellian molecules and for rigid elastic spheres, using fourth approximations to $\boldsymbol{A}$ and B. For Maxwellian molecules his results agree with (15.41, 1) and (15.41, 3); for rigid elastic spheres his results are equivalent to

$$\left.\begin{aligned} &\varpi_1 = 1{\cdot}014\times\frac{4}{3}\left(\frac{7}{2}-\frac{T}{\mu}\frac{d\mu}{dT}\right), \quad \varpi_2 = 1{\cdot}014\times 2, \quad \varpi_3 = 0{\cdot}806\times 3, \\ &\varpi_4 = 0{\cdot}681, \quad \varpi_5 = 0{\cdot}806\times\frac{3T}{\mu}\frac{d\mu}{dT}-0{\cdot}990, \quad \varpi_6 = 0{\cdot}928\times 8, \end{aligned}\right\} \qquad (15.41,\ 4)$$

expressions which indicate the factors by which the various first approximations are multiplied.

For molecules repelling like $r^{-\nu}$ the expressions for $\boldsymbol{q}^{(2)}$ and $\mathsf{p}^{(2)}$ are slightly simpler than the general expressions (15.3, 6, 8). In this case, because the integrals $\Omega_1^{(r)}(s)$ for different values of r and s all involve the same power of T (cf. 10.31), we can prove that $A'(C)$ and $B'(C)$ can each be expressed as the product of $\mu/T^{\frac{1}{2}}$ and a function of $\mathscr{C}$; thus

$$T\frac{\partial A'}{\partial T}+C^2\frac{\partial A'}{\partial C^2}=-\left(\frac{7}{2}-\frac{T}{\mu}\frac{d\mu}{dT}\right)A',$$

$$T\frac{\partial B'}{\partial T}+C^2\frac{\partial B'}{\partial C^2}=-\left(\frac{7}{2}-\frac{T}{\mu}\frac{d\mu}{dT}\right)B',$$

and the second and third of the terms of (15.2, 12) involve identical functions of $\boldsymbol{C}$, and so also do the first and second of the terms of (15.2, 14). It follows that

$$\theta_1=\frac{2}{3}\left(\frac{7}{2}-\frac{T}{\mu}\frac{d\mu}{dT}\right)\theta_2,\quad \varpi_1=\frac{2}{3}\left(\frac{7}{2}-\frac{T}{\mu}\frac{d\mu}{dT}\right)\varpi_2. \qquad (15.41, 5)$$

This explains the appearance of the same factor 1·014 in the expressions (15.41, 4) for ϖ_1 and ϖ_2; the reason for the appearance of 0·806 in the expressions for ϖ_3 and ϖ_5 is similar.

15.42. The velocity of diffusion

So far only simple gases have been discussed. The third approximation to the velocity distribution for a mixed gas has been considered,* in order to determine the corresponding contribution to the velocity of diffusion. Here we shall simply quote the result, that the third approximation adds to the velocity of diffusion $\overline{\boldsymbol{C}}_1-\overline{\boldsymbol{C}}_2$ nine terms, given by

$$\frac{n^2m_1m_2D_{12}}{\rho\rho x_1x_2}\left\{\epsilon_a\frac{D_T}{T}\Delta\nabla T+\epsilon_b\frac{D_T}{T}\left(\frac{D_0}{Dt}(\nabla T)-\nabla\boldsymbol{c}_0.\nabla T\right)\right.$$
$$\left.+\epsilon_c D_{12}\Delta\boldsymbol{d}_{12}+\epsilon_d D_{12}\left(\frac{D_0}{Dt}(\boldsymbol{d}_{12})-\nabla\boldsymbol{c}_0.\boldsymbol{d}_{12}\right)\right\}$$
$$+\frac{\rho D_{12}^2}{x_1x_2p}\left\{\epsilon_e\nabla x_1.\mathsf{e}+\epsilon_f\boldsymbol{d}_{12}.\mathsf{e}\right\}$$
$$+\frac{\rho D_{12}^2}{p}\left\{\epsilon_g\frac{1}{T}\nabla T.\mathsf{e}+\epsilon_h.\frac{1}{p}\nabla p.\mathsf{e}+\epsilon_i\nabla.\mathsf{e}\right\}, \qquad (15.42, 1)$$

where $\epsilon_a, \epsilon_b, \ldots, \epsilon_i$ are numerical quantities of order unity, which may, however, vary with the temperature and the properties of the mixture. Certain of the quantities ϵ have been determined correct to a first approxi-

* S. Chapman and T. G. Cowling, *Proc. R. Soc.* A, 179, 159 (1941).

mation similar to that of 15.41: their values, correct to this approximation, are

$$\epsilon_a = \frac{2}{3}\left(\frac{7}{2} - \frac{T}{D_T}\frac{\partial D_T}{\partial T}\right), \quad \epsilon_c = \frac{2}{3}\left(\frac{7}{2} - \frac{T}{D_{12}}\frac{\partial D_{12}}{\partial T}\right),$$

$$\epsilon_b = \epsilon_d = 1, \quad \epsilon_h = 0. \qquad (15.42, 2)$$

All the new terms in the diffusion velocity can be shown to depend on the velocity-gradient, and to vanish if the mass-velocity $\boldsymbol{c}_0$ is uniform. They are in general small compared with the diffusion velocities considered earlier, and are liable to be masked by turbulence when they are large. However, the last term in (15.42, 1) may be comparable with pressure diffusion in flow along a capillary tube.

15.5. The orders of magnitude of $\boldsymbol{q}^{(2)}$ and $\mathsf{p}^{(2)}$

Consider now the magnitudes of the different terms in the expressions for $\boldsymbol{q}^{(2)}$, $\mathsf{p}^{(2)}$ for a simple gas. Because $(T/\mu)(d\mu/dT)$ is a pure number whose value for ordinary gases lies between $\frac{1}{2}$ and 1 (cf. 12.31), all the coefficients θ_r and ϖ_r are less than $\frac{45}{4}$.

Let V, L_1, L_2, l have the same meanings as in 15.1, so that the first- and second-order space gradients of T, $\boldsymbol{c}_0$ are respectively comparable with T/L_1, V/L_2 and T/L_1^2, V/L_2^2. In motions involving adiabatic expansions, ∇p is comparable with p/L_2; in nearly steady motion it is normally smaller, being comparable with $\rho V^2/L_2$ or $pV^2/L_2\bar{C}^2$. Thus the first and fifth terms of (15.3, 6) are comparable with $\mu^2 V/\rho L_1 L_2$; the fourth term is comparable with $\mu^2 V/\rho L_2^2$; and the remaining terms are comparable with or smaller than these quantities. On the other hand, the thermal flux $\boldsymbol{q}^{(1)}$ given by the second approximation to f is

$$-\lambda\nabla T = -\tfrac{5}{2}\mu c_v \nabla T = -\tfrac{15}{4}\mu(k/m)\nabla T,$$

which is comparable with $\mu kT/mL_1$, or $\mu p/\rho L_1$. Since $\mu = 0{\cdot}499\rho l\bar{C}$, and p is comparable with $\rho\bar{C}^2$, the terms of $\boldsymbol{q}^{(2)}$ bear ratios to $\boldsymbol{q}^{(1)}$ comparable with $Vl/\bar{C}L_2$ or $VlL_1/\bar{C}L_2^2$. This means (cf. 15.1) that in ordinary conditions, in which l/L_2 is small, the thermal flux $\boldsymbol{q}^{(2)}$ is negligible compared with $\boldsymbol{q}^{(1)}$. The flux $\boldsymbol{q}^{(2)}$ is in any case the less important because it depends essentially on the motion of the gas, and vanishes when the gas is at rest. In normal gas-flows, the transport of heat by convection is more important than that by conduction.

Similarly in the expression (15.3, 8) for $\mathsf{p}^{(2)}$ the first, second and last terms are comparable with $\mu^2 V^2/pL_2^2$. Since $\mathsf{p}^{(1)}$ is comparable with $\mu V/L_2$, the ratio of these terms to $\mathsf{p}^{(1)}$ is similar to that of $\mathsf{p}^{(1)}$ to p; in particular, at normal pressures these terms are small compared with $\mathsf{p}^{(1)}$, and may be neglected. The thermal stresses given by the third and fifth terms of (15.3, 8), and also, in special circumstances, the stresses given by the fourth term, may be more important. They are comparable with $\mu^2/\rho L_1^2$, and bear to $\mathsf{p}^{(1)}$ ratios comparable with $\mu L_2/\rho V L_1^2$, i.e. with $\bar{C} l L_2/V L_1^2$. Thus when $\bar{C}/V$ is large and L_1 and L_2 are comparable, the temperature stresses may not be completely negligible compared with the ordinary viscous stresses, even at ordinary pressures at which l/L_1 is very small.

15.51. The range of validity of the third approximation

The regime in which $f^{(0)}+f^{(1)}$ gives a fully adequate approximation to f is known as the Navier–Stokes regime. In this regime the transport properties of a simple gas are adequately described in terms of the viscous stress $\mathsf{p}^{(1)}$ and the heat flux $\boldsymbol{q}^{(1)}$.

The cases in which $\boldsymbol{q}^{(2)}$, $\mathsf{p}^{(2)}$ are also likely to be important are those in which l is large (low pressures, of order 10^{-5} atmos.) or L_1, L_2 are very small (as in shock-wave fronts) or intermediate cases (capillary flow or ultrasonic waves at fairly low pressures). However, whenever $\boldsymbol{q}^{(2)}$ and $\mathsf{p}^{(2)}$ are important there is every reason to believe that higher-order approximations also need to be taken into account. The expansion $f = f^{(0)}+f^{(1)}+f^{(2)}+\ldots$ is, in effect, an expansion in powers of l/L_1, l/L_2; $f^{(1)}$ is linear in them, $f^{(2)}$ quadratic, and so on. Normally l/L_1, l/L_2 are small, and only one or two terms of the expansion need be considered. But if $\boldsymbol{q}^{(2)}$ and $\mathsf{p}^{(2)}$ become important, l/L_1 and l/L_2 are comparable with unity; the successive terms of the expansion decrease slowly, and the expansion may cease to converge. That is, the higher approximations can only be used when the Navier–Stokes approximation is already good, in order to give an ultra-refined fluid description.*

For example, Wang-Chang† attempted to apply the results of the third (Burnett) approximation to the propagation and attenuation of ultrasonic waves, and even gave equations corresponding to the fourth (super-Burnett) approximation. However, a comparison of these results with experiment shows that, whereas the Burnett and super-Burnett approximations may give somewhat improved agreement when l/L_2 is still fairly small (where L_2 here denotes the wave-length), they give no agreement when l/L_2 is comparable with or larger than unity.‡ This is to be expected; the first approximations to $\partial n/\partial t$ and $\partial T/\partial t$ correspond to adiabatic expansion, and when l/L_2 is large the motion is nearly isothermal.

When l/L_1 and l/L_2 are not small, a new method of approximating to f becomes necessary. Generalizing the methods given earlier, a form is assumed for f involving unknown parameters, equations to determine these

* Cf. H. Grad, *Phys. Fluids*, **6**, 147 (1963). Grad also suggests that generally the expansion $f^{(0)}+f^{(1)}+f^{(2)}+\ldots$ may be asymptotic rather than convergent; the difference is unimportant in normal conditions, when it would become apparent only at very high orders of approximation, but important when l/L_1, l/L_2 cease to be small.

† C. S. Wang-Chang, Univ. of Michigan, Dept. of Engineering Rep. UMH-3-F (APL/JHU CH-467), 1948. The first to use the Burnett equations in hydrodynamics was L. H. Thomas (*J. Chem. Phys.* **12**, 449 (1944)), in studying the profile of shock fronts.

‡ M. Greenspan, *J. Acoust. Soc. Am.* **22**, 569 (1950), and **28**, 644 (1956); E. Meyer and G. Sessler, *Z. Phys.* **149**, 15 (1957).

being then found by using an appropriate number of equations of transfer (cf. 3.11). The success of this method depends on the satisfactory choice of the form for f. Alternatively, numerical methods of solving the Boltzmann equation have been tried; these are usually applied to an approximate form of the equation. The problem has especially been studied in connection with shock-wave fronts.

15.6. The method of Grad

The method developed by Grad* is the closest to the Enskog approach, from which it differed mainly in its discussion of the time-derivatives. Because the rapidity of changes in a shock front may prevent a 'normal' solution (cf. 7.2) of the Boltzmann equation from being attained, Grad sought solutions of a more general type. He assumed an approximate expression for f, of the form

$$f = f^{(0)}\{1 + \sum_{p,\,l} a_{pl}\chi_{pl}\}, \qquad (15.6,\ 1)$$

where the functions χ_{pl} are (a finite number of) the functions of (15.3, 1), and the quantities a_{pl} are variable parameters. Arbitrary initial values of the parameters a_{pl} can be assumed; the time variation of the parameters, as well as that of n, $\boldsymbol{c}_0$ and T, is then determined from the equations of change of appropriate molecular properties (cf. (3.11, 1)), so that the values of the parameters at any subsequent time can be found. By taking a sufficiently large number of functions and parameters a_{pl}, f can be determined to any desired order of approximation.

A simple approximation, but one giving good results, is

$$f = f^{(0)}\{1 + \boldsymbol{a}.\boldsymbol{C}(\mathscr{C}^2 - \tfrac{5}{2}) + \mathsf{b}:\overset{\circ}{\boldsymbol{C}\boldsymbol{C}}\}. \qquad (15.6,\ 2)$$

Here the vector $\boldsymbol{a}$ and the tensor b are the unknown parameters; they are related to the heat flow $\boldsymbol{q}$ and the non-hydrostatic part $\mathsf{p}' = \mathsf{p} - \mathsf{U}p$ of the pressure tensor by the equations

$$\boldsymbol{a} = \frac{2}{5}\frac{\rho\boldsymbol{q}}{p^2}, \quad \mathsf{b} = \frac{\rho}{2p^2}\mathsf{p}'. \qquad (15.6,\ 3)$$

The time-derivatives of $\boldsymbol{a}$ and b are determined from the equations of change of the molecular properties $(\mathscr{C}^2 - \frac{5}{2})\boldsymbol{C}$ and $\overset{\circ}{\boldsymbol{C}\boldsymbol{C}}$. This leads to equations equivalent, in our notation, to

$$\frac{D\mathsf{p}'}{Dt} - 2\overset{\circ}{\overline{\overline{\nabla\boldsymbol{c}_0.\mathsf{p}'}}} + 4\overset{\circ}{\overline{\overline{\overset{\circ}{\mathsf{e}}.\mathsf{p}'}}} + \tfrac{7}{3}\Delta\mathsf{p}' + \tfrac{4}{5}\overset{\circ}{\overline{\overline{\nabla\boldsymbol{q}}}} + 2p\overset{\circ}{\mathsf{e}} + \frac{p}{\mu}\mathsf{p}' = 0, \qquad (15.6,\ 4)$$

$$\frac{D\boldsymbol{q}}{Dt} - \nabla\boldsymbol{c}_0.\boldsymbol{q} + \tfrac{14}{5}\overset{\circ}{\mathsf{e}}.\boldsymbol{q} + \tfrac{7}{3}\Delta\boldsymbol{q} + \frac{p}{\rho}\nabla.\mathsf{p}' + \frac{7p}{2\rho T}\nabla T.\mathsf{p}'$$

$$- \frac{\mathsf{p}'}{\rho}.(\nabla p + \nabla.\mathsf{p}') + \frac{5p^2}{2\rho T}\nabla T + \frac{2}{3}\frac{p}{\mu}\boldsymbol{q} = 0, \qquad (15.6,\ 5)$$

* H. Grad, *Communs pure appl. Maths.* **2**, 331 (1949); **5**, 257 (1952).

where μ is the first approximation to the coefficient of viscosity. These, together with the equations of continuity, motion and energy, form a complete set of differential equations giving the variation of all the physical variables. The approximation (15.6, 2) is known as the 13-moment approximation; in order to get the correct values of n, T and the 11 independent components of $\boldsymbol{c}_0$, $\boldsymbol{q}$ and p', we need the values of 13 moment-integrals $\int f\chi_{pl}d\boldsymbol{c}$.

When the parameters l/L_1, l/L_2 of 15.5 are small, (15.6, 4) and (15.6, 5) yield results equivalent to those of 15.3. As can be seen by comparing the first and last terms of each equation, p' and $\boldsymbol{q}$ may at first vary rapidly, approaching normal values (depending only on n, $\boldsymbol{c}_0$ and T and their derivatives) in a relaxation time comparable with μ/p, i.e. with the mean collision-interval. After the 'normal' state is attained, the last two terms of each equation become dominant; thus a first approximation to the solution is $\mathsf{p}' = \mathsf{p}^{(1)}$, $\boldsymbol{q} = \boldsymbol{q}^{(1)}$, where

$$\mathsf{p}^{(1)} = -2\mu\overset{\circ}{\mathsf{e}}, \quad \boldsymbol{q}^{(1)} = -\frac{15k}{4m}\mu\nabla T = -\lambda\nabla T,$$

λ being the first approximation to the thermal conductivity. These are formally equivalent to the first approximations to p' and $\boldsymbol{q}$ derived in Chapter 7; second approximations are then derived by substituting $\mathsf{p}' = \mathsf{p}^{(1)}$, $\boldsymbol{q} = \boldsymbol{q}^{(1)}$ in the earlier terms of (15.6, 4) and (15.6, 5), and $\mathsf{p}' = \mathsf{p}^{(1)}+\mathsf{p}^{(2)}$, $\boldsymbol{q} = \boldsymbol{q}^{(1)}+\boldsymbol{q}^{(2)}$ in the last term of each, and replacing D/Dt by D_0/Dt. The expressions so found for $\mathsf{p}^{(2)}$ and $\boldsymbol{q}^{(2)}$ are identical with those of 15.3, with the approximate values for the coefficients θ and ϖ given in 15.4 and 15.41.

Grad sought to apply (15.6, 4) and (15.6, 5) to the case when l/L_1, l/L_2 are not small, and D_0/Dt may not be a good approximation to D/Dt. He was able to discuss moderately strong shocks, and his theory marked a definite advance on that of Wang-Chang, based on the Burnett approximation.* However, his equations, based on the 13-moment approximation, possessed no solution when the Mach number of the inflowing gas exceeded 1·65. An approximation with further terms can extend the range of solubility, but, as Holway showed,† no solution of Boltzmann's equation of Grad's type can converge if applied to a shock-wave with Mach number greater than 1·851. This is not a wholly decisive objection to Grad's method; it is not required of an approximate solution that it should in the limit converge, but simply that it should approximate to f sufficiently well to yield meaningful results. Unfortunately, however, with increasing shock strength a polynomial expansion like (15.6, 1) or (15.6, 2) becomes progressively less able to give an adequate approximation to f.

* C. S. Wang-Chang, Univ. of Michigan, Dept. of Engineering Rep. UMH-3-F (APL/JHU CM-503), 1948.

† L. H. Holway, *Phys. Fluids*, 7, 911 (1964).

15.61. The Mott-Smith approach

Because of the extreme rapidity of the transition across a shock front, Mott-Smith* assumed that at any point within the transition zone f is the sum of two Maxwellian functions f_1, f_2 corresponding to the temperatures and mean velocities on the two sides of the shock. The number-densities n_1, n_2 of f_1, f_2 were assumed to vary across the transition zone, n_1 vanishing on one side and n_2 on the other. The variation of n_1/n_2 across the zone was determined by using the equation of transfer of u^2 or of u^3, where u is the velocity-component normal to the shock front; the equations of transfer of the summational invariants are in any case also used. Mott-Smith obtained shock-wave thicknesses agreeing moderately well with experiment for strong shocks. For weak shocks, in which the transition across the shock front is relatively slow, his results agree less well with experiment than do those obtained from the Navier–Stokes equations.

An extension of the Mott-Smith method was made by Salwen, Grosch and Ziering,† who expressed f in a shock front as the sum of parts proportional to a number of assigned functions. The proportions arising from the separate functions varied from point to point of the front, in a manner found from a number of equations of transfer. They considered in detail the case of three functions, f_1, f_2, f_3; the first two were, as before, Maxwellian functions corresponding to the T and $\mathbf{c}_0$ on the two sides of the shock, and the third was the derivative of an intermediate Maxwellian function with respect to the velocity-component u. They found that the extra function strikingly improved the agreement with experiment, especially for weak shocks.

15.62. Numerical solutions

Liepmann, Narasimha and Chahine‡ sought to avoid the convergence difficulties by using numerical methods. To simplify the problem, they used the relaxation approach of 6.6 (often called the BGK, or Bhatnagar, Gross and Krook approximation; see p. 104, first footnote), in which the Boltzmann equation takes the form

$$\frac{\partial f}{\partial t}+\mathbf{c}.\frac{\partial f}{\partial \mathbf{r}} = -\frac{f-f^{(0)}}{\tau}. \tag{15.62, 1}$$

The relaxation time τ is given by μ/p, and so is a determinate function of density and temperature for any gas.§

Take axes relative to which the shock front is at rest and parallel to $x = 0$.

* H. M. Mott-Smith, *Phys. Rev.* **82**, 885 (1951).

† H. Salwen, C. E. Grosch and S. Ziering, *Phys. Fluids*, **7**, 180 (1964).

‡ H. W. Liepmann, R. Narasimha and M. T. Chahine, *Phys. Fluids*, **5**, 1313 (1962).

§ Note, however, that the relaxation time for temperature changes is $3\mu/2p$ for a monatomic gas (cf. (6.6, 7)). The assumption of a single relaxation time is a possible source of errors.

Then $\partial f/\partial t = 0$, and (15.62, 1) reduces to

$$u\tau \frac{\partial f}{\partial x} = f^{(0)} - f. \tag{15.62, 2}$$

The solution of this, satisfying the boundary conditions that $f \to f_1^{(0)}, f_2^{(0)}$ at $x = \pm\infty$, is

$$f(\mathbf{c}, x) = \int_{-\infty}^{x} \frac{f^{(0)}(\mathbf{c}, x')}{u\tau(x')} \exp\left(-\int_{x'}^{x} \frac{dx''}{u\tau(x'')}\right) dx' \qquad (u > 0)$$

$$= \int_{\infty}^{x} \frac{f^{(0)}(\mathbf{c}, x')}{(-u)\tau(x')} \exp\left(-\int_{x'}^{x} \frac{dx''}{(-u)\tau(x'')}\right) dx' \quad (u < 0).$$

The value of f is found from this by iteration. Values of $f^{(0)}$ for all x, satisfying the boundary conditions at $x = \pm\infty$, are assumed known at the start of each iterative step, and $f(\mathbf{c}, x)$ is deduced by numerical integration; a new $f^{(0)}$ for each x is then derived corresponding to the values of n, $\mathbf{c}_0$, T given by

$$n = \int f d\mathbf{c}, \quad n\mathbf{c}_0 = \int f\mathbf{c} d\mathbf{c}, \quad \tfrac{3}{2}nkT = \int f^{(0)} \tfrac{1}{2}mC^2 d\mathbf{c}.$$

The process is then repeated; if a reasonable first approximation to $f^{(0)}$ is assumed, it is rapidly convergent.

By this method Liepmann and his colleagues were able to determine shock profiles in argon for Mach numbers up to 10. They found that the Navier–Stokes terms $\mathsf{p}^{(1)}$, $\boldsymbol{q}^{(1)}$ were dominant throughout the profile for Mach numbers less than 1·5. For higher Mach numbers, however, though the Navier–Stokes terms are still dominant on the high-density side, on the low-density side there is a considerable departure from the Navier–Stokes profile, a long 'tail' of the profile appearing upstream of the main shock. They suggest that any tendency towards a bi-modal distribution of the Mott-Smith type is mainly confined to this tail.

Their discussion rests on a simplified collision model, and to that extent is inexact. Earlier approaches were based on a more exact discussion of collisions, followed by truncation and other approximations. None can be regarded as altogether satisfactory; the discussion based on the BGK approximation has at least the advantage that it is capable of giving shock profiles valid for all Mach numbers.

The BGK approximation has also been applied with some success to solve numerically problems of the formation of shock fronts,* the efflux of gas through a small aperture at low pressures,† and the flow of rarefied gas along a cylindrical tube.‡ The last problem has also been solved more exactly by Simons§ in terms of the eigenvalues of the collision operator.

* C. K. Chu, *Phys. Fluids*, **8**, 12 and 1450 (1965).

† H. W. Liepmann, *J. Fluid Mech.* **10**, 65 (1961); R. Narasimha, *ibid.* **10**, 371 (1961).

‡ C. Cercignani and F. Sernagiotto, *Phys. Fluids*, **9**, 40 (1966).

§ S. Simons, *Proc. R. Soc.* A, **301**, 387, 401 (1967). See also C. Cercignani, *Mathematical Methods in Kinetic Theory*, chapter 7 (New York: Plenum Press, 1969).

16

DENSE GASES

16.1. Collisional transfer of molecular properties

In a gas at S.T.P., the mean free path of a molecule is large compared with the molecular dimensions. This disparity is much reduced if the gas is compressed to a density a hundred times as great, and in consequence an additional mechanism for the transfer of momentum and energy, negligible at ordinary gaseous densities, then becomes important. Hitherto the transfer of molecular properties has been regarded as due solely to the free motions of molecules *between* collisions; there is also, however, a transfer *at* collisions, over the distance separating the centres of the two colliding molecules, during the brief time of encounter. The extreme example of this is the *instantaneous* transfer of energy and momentum at the collision of two smooth elastic spherical molecules, across the distance σ_{12} between their centres. In dense gases this *collisional transfer* requires consideration; it was first studied by Enskog,* for rigid spherical molecules.

The rigid spherical model has the mathematical advantage in this connection that collisions are instantaneous, and multiple encounters can be neglected. This is not, however, the case with actual molecules; in a gas at high pressure a molecule is in the field of force of others during a large part of its motion, and multiple encounters are not rare. Enskog preferred the rigid spherical model also because he believed, on the basis of analysis by Jeans,† that for rigid elastic spheres the assumption of molecular chaos made in 3.5 remains valid at large densities. However, Jeans's argument applies strictly only to a gas in a uniform steady state. In a dense gas, even one composed of rigid spherical molecules, there may be some correlation between the velocities of neighbouring molecules, because of their recent interaction with each other or with the same neighbours. This may be important in a gas not in a uniform steady state. Thus what Enskog gained in mathematical simplicity is partially offset by inadequacy in the representation of physical reality.

An adequate discussion of dense gases for general molecular models must allow for multiple and repeated encounters, and also for the finite number of molecules which at any instant are actually undergoing encounters. Several attempts have been made to provide such a discussion. In certain respects these parallel Enskog's theory, though they also reveal inadequacies in it. They have been able to provide formal expressions for the different

* D. Enskog, *K. Svensk. Vet.-Akad. Handl.* **63**, no. 4 (1921).

† J. H. Jeans, *Dynamical Theory of Gases*, 4th ed., p. 54 (1925).

transport coefficients, but not to supply exact formulae that can be compared with experiment. We here describe Enskog's theory in detail; a brief indication of the more recent work is given in sections 16.7–16.73.

16.2. The probability of collision

Consider a simple gas composed of rigid spherical molecules of diameter σ. In discussing a collision between molecules with velocities $\boldsymbol{c}$, $\boldsymbol{c}_1$ and centres O, O_1 we specify the direction of the line of centres O_1O at collision by the unit vector $\mathbf{k}$ (cf. 3.43). Then $\sigma^2 d\mathbf{k}$ denotes a surface element on the sphere, of radius σ and centre O, on which O_1 must lie at the collision. The elementary area $b\,db\,d\epsilon$ of 3.5 is the projection of $\sigma^2 d\mathbf{k}$ on a plane normal to $\boldsymbol{g}$, where $\boldsymbol{g} = \boldsymbol{c}_1 - \boldsymbol{c}$, and if θ is the angle between $\boldsymbol{g}$ and $\mathbf{k}$,

$$g b\,db\,d\epsilon = g\sigma^2 d\mathbf{k} \cos\theta = \boldsymbol{g}.\mathbf{k}\sigma^2 d\mathbf{k}.$$

Hence (cf. (3.5, 1)), for a gas at ordinary pressure, the probability per unit time of a collision such that O lies in a volume $d\boldsymbol{r}$, the velocities of the two molecules lie in ranges $d\boldsymbol{c}$, $d\boldsymbol{c}_1$ and $\mathbf{k}$ lies in $d\mathbf{k}$ is given by

$$f(\boldsymbol{c}, \boldsymbol{r}) f(\boldsymbol{c}_1, \boldsymbol{r})\, \sigma^2 \boldsymbol{g}.\mathbf{k}\, d\mathbf{k}\, d\boldsymbol{c}\, d\boldsymbol{c}_1\, d\boldsymbol{r}.$$

This expression requires correction when the gas is dense, even though we still make the assumption of molecular chaos. First, because O is at $\boldsymbol{r}$, O_1 must be at $\boldsymbol{r} - \sigma\mathbf{k}$, so that $f(\boldsymbol{c}_1, \boldsymbol{r})$ must be replaced by $f(\boldsymbol{c}_1, \boldsymbol{r} - \sigma\mathbf{k})$. Secondly, in a dense gas the volume of the molecules is comparable with the volume occupied by the gas. The effect is to reduce the volume in which the centre of any one molecule can lie, and so to increase the probability of a collision. Thus the above expression must be multiplied by a factor χ, which is a function of position but, by the assumption of molecular chaos, not of velocity. The function χ must be evaluated at the point $\boldsymbol{r} - \frac{1}{2}\sigma\mathbf{k}$ at which the spheres actually collide: hence the corrected form of the above expression is

$$\chi(\boldsymbol{r} - \tfrac{1}{2}\sigma\mathbf{k}) f(\boldsymbol{c}, \boldsymbol{r}) f(\boldsymbol{c}_1, \boldsymbol{r} - \sigma\mathbf{k})\, \sigma^2 \boldsymbol{g}.\mathbf{k}\, d\mathbf{k}\, d\boldsymbol{c}\, d\boldsymbol{c}_1\, d\boldsymbol{r}$$
$$= \chi(\boldsymbol{r} - \tfrac{1}{2}\sigma\mathbf{k}) f(\boldsymbol{r}) f_1(\boldsymbol{r} - \sigma\mathbf{k})\, \sigma^2 \boldsymbol{g}.\mathbf{k}\, d\mathbf{k}\, d\boldsymbol{c}\, d\boldsymbol{c}_1\, d\boldsymbol{r}, \quad (16.2, 1)$$

using a notation similar to that of 3.5.

16.21. The factor χ

The quantity χ is equal to unity for a rare gas, and increases with increasing density, becoming infinite as the gas approaches the state in which the molecules are packed so closely together that motion is impossible. Approximations to its value for fairly dense gases may be obtained as follows.

When two molecules collide, the distance between their centres is σ. Let a sphere of radius σ be described about the centre of each molecule; then at a collision the centre of one molecule lies on the sphere associated with the other, and the centre of one molecule can never lie within the associated sphere of another. If the gas is fairly rare, the number of the associated spheres which intersect each other will form a small fraction of the whole, and so the volume which cannot be occupied by the centre of a molecule will be approximately the whole volume of the associated spheres, which is $\frac{4}{3}\pi n\sigma^3$ per unit volume. Hence the volume in which the centre of a molecule can lie is reduced in the ratio $1-2\mathrm{b}\rho$, where

$$\mathrm{b}\rho = \tfrac{2}{3}\pi n\sigma^3, \quad \mathrm{b} = \tfrac{2}{3}\pi\sigma^3/m. \tag{16.21, 1}$$

Correspondingly the probability of molecular collisions is increased in the ratio $1/(1-2\mathrm{b}\rho)$.

A second factor, of comparable effect, operates to reduce the probability of collisions: this is the shielding of one molecule by another. If an area S of the sphere associated with a given molecule lies within the sphere associated with a second, no other molecule can collide with the first in such a way that at collision its centre lies on S.

If the gas is fairly rare, the number of cases in which more than two associated spheres have a common volume can be neglected. Let two spheres of radii x and $x+dx$, where x lies between σ and 2σ, be drawn concentric with a given molecule. Then the probability that the centre of another molecule should lie within the shell of volume $4\pi x^2dx$ between these two spheres is $4\pi nx^2dx$. The sphere associated with such a molecule would cut off from that associated with the given molecule a cap of height $\sigma-\frac{1}{2}x$, and area $2\pi\sigma(\sigma-\frac{1}{2}x)$. Hence the total probable area cut off from the sphere associated with the given molecule by those associated with other molecules is

$$\int_\sigma^{2\sigma} 2\pi\sigma(\sigma-\tfrac{1}{2}x)\,4\pi nx^2\,dx = \tfrac{11}{3}\pi^2 n\sigma^5.$$

Since the whole area of the sphere is $4\pi\sigma^2$, the part on which the centres of molecules can lie at collision is a fraction $1-\frac{11}{12}n\pi\sigma^3$ or $1-\frac{11}{8}\mathrm{b}\rho$ of the whole. Hence the effect of shielding by other molecules is to reduce the probability of a collision in this ratio.

On combining these results, the value of χ for a fairly rare uniform gas is found to be

$$\begin{aligned}\chi &= (1-\tfrac{11}{8}\mathrm{b}\rho)/(1-2\mathrm{b}\rho)\\ &= 1+\tfrac{5}{8}\mathrm{b}\rho,\end{aligned} \tag{16.21, 2}$$

correct to the first order in $\mathrm{b}\rho$. Boltzmann and Clausius* carried the calculation as far as the second order in $\mathrm{b}\rho$, by taking into account the volume common to more than two associated spheres; further terms have been obtained by numerical methods.† The result is

$$\chi = 1+\tfrac{5}{8}\mathrm{b}\rho+0\cdot2869(\mathrm{b}\rho)^2+0\cdot1103(\mathrm{b}\rho)^3+0\cdot0386(\mathrm{b}\rho)^4+\dots \tag{16.21, 3}$$

* R. Clausius, *Mech. Wärmetheorie*, vol. 3 (2nd ed.), 57 (1879). L. Boltzmann, *Proc. Amsterdam*, p. 403 (1899); *Wiss. Abhandl.* **3**, 663.

† M. N. Rosenbluth and A. W. Rosenbluth, *J. Chem. Phys.* **22**, 881 (1954); B. J. Alder and T. E. Wainwright, *J. Chem. Phys.* **33**, 1439 (1960); F. H. Ree and W. G. Hoover, *J. Chem. Phys.* **46**, 4181 (1967).

In a non-uniform gas the value of χ may be expected to involve the space-derivatives of the density. Since, however, no invariant function of these derivatives can be constructed which does not involve either products of first derivatives or derivatives of higher order than the first, in a study of the first and second approximations the above value for χ will be used.*

16.3. Boltzmann's equation; $\partial_e f/\partial t$

Boltzmann's equation for a dense gas, as for a normal gas, can be written in the form

$$\frac{\partial f}{\partial t}+\mathbf{c}.\frac{\partial f}{\partial \boldsymbol{r}}+\boldsymbol{F}.\frac{\partial f}{\partial \mathbf{c}}=\frac{\partial_e f}{\partial t}. \tag{16.3, 1}$$

The expression for $\partial_e f/\partial t$, however, is not quite the same as before. Consider the probability per unit time of an inverse collision, such that the centre of the first molecule lies in a volume $d\boldsymbol{r}$, and the velocities of the two molecules after collision lie in ranges $d\mathbf{c}$, $d\mathbf{c}_1$, while the direction of the line of centres is $-\mathbf{k}$, where $\mathbf{k}$ lies in the range $d\mathbf{k}$. In such a collision the centre of the second molecule is at $\boldsymbol{r}+\sigma\mathbf{k}$, while the molecules actually impinge at $\boldsymbol{r}+\frac{1}{2}\sigma\mathbf{k}$. Hence, by analogy with (3.52, 7) and (16.2, 1), the probability in question will be

$$\chi(\boldsymbol{r}+\tfrac{1}{2}\sigma\mathbf{k})f(\mathbf{c}',\boldsymbol{r})f(\mathbf{c}_1',\boldsymbol{r}+\sigma\mathbf{k})\,\sigma^2\boldsymbol{g}.\mathbf{k}\,d\mathbf{k}\,d\mathbf{c}\,d\mathbf{c}_1\,d\boldsymbol{r}$$
$$=\chi(\boldsymbol{r}+\tfrac{1}{2}\sigma\mathbf{k})f'(\boldsymbol{r})f_1'(\boldsymbol{r}+\sigma\mathbf{k})\,\sigma^2\boldsymbol{g}.\mathbf{k}\,d\mathbf{k}\,d\mathbf{c}\,d\mathbf{c}_1\,d\boldsymbol{r}, \tag{16.3, 2}$$

where $\mathbf{c}'$, $\mathbf{c}_1'$ denote the velocities before collision of the two molecules, so that, as in (3.43, 2)

$$\mathbf{c}'=\mathbf{c}+\mathbf{k}(\boldsymbol{g}.\mathbf{k}),\quad \mathbf{c}_1'=\mathbf{c}_1-\mathbf{k}(\boldsymbol{g}.\mathbf{k}). \tag{16.3, 3}$$

The evaluation of $\partial_e f/\partial t$ now proceeds as in 3.52, using (16.3, 2) and (16.2, 1) in place of the corresponding expressions of that section. The result is

$$\frac{\partial_e f}{\partial t}=\iint\{\chi(\boldsymbol{r}+\tfrac{1}{2}\sigma\mathbf{k})f'(\boldsymbol{r})f_1'(\boldsymbol{r}+\sigma\mathbf{k})$$
$$-\chi(\boldsymbol{r}-\tfrac{1}{2}\sigma\mathbf{k})f(\boldsymbol{r})f_1(\boldsymbol{r}-\sigma\mathbf{k})\}\,\sigma^2\boldsymbol{g}.\mathbf{k}\,d\mathbf{k}\,d\mathbf{c}_1. \tag{16.3, 4}$$

Combining this equation with (16.3, 1), we derive the equation for f.

Since in this relation the expressions χ, f_1', f_1 are evaluated at points other than $\boldsymbol{r}$, it is not possible to show by a transformation similar to that of 3.53 that the expression

$$n\Delta\bar{\phi}\equiv\int\phi\frac{\partial_e f}{\partial t}d\mathbf{c}$$

vanishes when ϕ is a summational invariant. This, in fact, is not true in general. For though the total amount of ϕ possessed by molecules is conserved at collision, the effect of a collision is to transfer part of this total

* The possibility that χ should also include scalar combinations of $\mathbf{k}$ and of the gradients of ρ, T and $\mathbf{c}_0$ is not considered.

through a distance σ from one molecule to another. Hence, since $n\Delta\bar{\phi}$ denotes the total change by collisions in the total amount of ϕ possessed by molecules per unit volume *at a given point*, in general it will not vanish. The derivation of the equations of momentum and energy is therefore deferred till later.

16.31. The equation for $f^{(1)}$

If the gas is uniform the expressions χ and f do not depend on $\boldsymbol{r}$; hence in this case

$$\frac{\partial_e f}{\partial t} = \chi \iint (f'f'_1 - ff_1)\, \sigma^2 \mathbf{g}.\mathbf{k}\, d\mathbf{k}\, d\boldsymbol{c}_1$$

This expression differs from that obtained earlier only by the presence of the factor χ. The arguments of 4.1 can accordingly be used to show that in the uniform steady state f again takes the Maxwellian form

$$f = f^{(0)} \equiv n\left(\frac{m}{2\pi kT}\right)^{\frac{3}{2}} e^{-mC^2/2kT}. \qquad (16.31, 1)$$

When the gas is not in a uniform steady state, a first approximation to the value of f is given by (16.31, 1). A second approximation is

$$f^{(0)}(1+\Phi^{(1)}), \qquad (16.31, 2)$$

where $\Phi^{(1)}$ is a linear function of the first derivatives of n, T and the mass-velocity $\boldsymbol{c}_0$; the equation satisfied by $\Phi^{(1)}$ is obtained from that satisfied by f by neglecting all terms involving products of derivatives of these quantities, or derivatives of higher order than the first. Thus in substituting from (16.31, 2) into the left-hand side of (16.3, 1), the terms involving $\Phi^{(1)}$ are to be neglected. This equation may therefore be written in the form

$$\frac{\partial_e f}{\partial t} = f^{(0)}\left(\frac{D}{Dt} + \boldsymbol{C}.\frac{\partial}{\partial \boldsymbol{r}} + \boldsymbol{F}.\frac{\partial}{\partial \boldsymbol{C}}\right) \ln f^{(0)},$$

where, as in (3.12, 1), $$\frac{D}{Dt} \equiv \frac{\partial}{\partial t} + \boldsymbol{c}_0 . \frac{\partial}{\partial \boldsymbol{r}}.$$

In this we substitute the value of $f^{(0)}$ from (16.31, 1); then, since $\boldsymbol{C} = \boldsymbol{c} - \boldsymbol{c}_0$ and $\boldsymbol{c}_0$ is a function of $\boldsymbol{r}$, t, it becomes

$$\frac{\partial_e f}{\partial t} = f^{(0)}\left[\frac{1}{n}\frac{Dn}{Dt} + (\mathscr{C}^2 - \tfrac{3}{2})\frac{1}{T}\frac{DT}{Dt} + 2\mathscr{C}\mathscr{C} : \nabla \boldsymbol{c}_0 \right.$$
$$\left. + \boldsymbol{C}.\left\{\nabla \ln (knT) + (\mathscr{C}^2 - \tfrac{5}{2}) \nabla \ln T + \frac{m}{kT}\left(\frac{D\boldsymbol{c}_0}{Dt} - \boldsymbol{F}\right)\right\}\right], \qquad (16.31, 3)$$

where, as before, $\mathscr{C} = (m/2kT)^{\frac{1}{2}} \boldsymbol{C}$.

16.32. The second approximation to $\partial_e f/\partial t$

The approximate form of equation (16.3, 4) is obtained by expanding χ, f_1, f_1' in powers of $\sigma\mathbf{k}$ by Taylor's theorem, and retaining only the first derivatives; it is

$$\frac{\partial_e f}{\partial t} = \chi\iint (f'f_1' - ff_1)\,\sigma^2 \boldsymbol{g}.\mathbf{k}\,d\mathbf{k}\,d\boldsymbol{c}_1$$
$$+\chi\iint \mathbf{k}.(f'\nabla f_1' + f\nabla f_1)\,\sigma^3 \boldsymbol{g}.\mathbf{k}\,d\mathbf{k}\,d\boldsymbol{c}_1$$
$$+\frac{1}{2}\iint \mathbf{k}.\nabla\chi\,(f'f_1' + ff_1)\,\sigma^3 \boldsymbol{g}.\mathbf{k}\,d\mathbf{k}\,d\boldsymbol{c}_1, \qquad (16.32, 1)$$

where in this relation all the quantities are evaluated at the point $\boldsymbol{r}$.

In substituting from (16.31, 2) into the first term on the right in this equation we neglect the products $\Phi^{(1)}\Phi_1^{(1)}$, $\Phi^{(1)\prime}\Phi_1^{(1)\prime}$, as before. Hence, since

$$f^{(0)\prime}f_1^{(0)\prime} = f^{(0)}f_1^{(0)}, \qquad (16.32, 2)$$

this term becomes

$$\chi\iint f^{(0)}f_1^{(0)}(\Phi^{(1)\prime} + \Phi_1^{(1)\prime} - \Phi^{(1)} - \Phi_1^{(1)})\,\sigma^2 \boldsymbol{g}.\mathbf{k}\,d\mathbf{k}\,d\boldsymbol{c}_1. \qquad (16.32, 3)$$

The second and third terms on the right already involve space-derivatives; thus in substituting from (16.31, 2) into these we may neglect all the terms involving $\Phi^{(1)}$, and so write $f^{(0)}$ in place of f. Using (16.32, 2), these two terms then become

$$\chi\iint f^{(0)}f_1^{(0)}\mathbf{k}.\nabla\ln(f_1^{(0)\prime}f_1^{(0)}\chi)\,\sigma^3 \boldsymbol{g}.\mathbf{k}\,d\mathbf{k}\,d\boldsymbol{c}_1$$

or, after substituting the values of $f_1^{(0)}, f_1^{(0)\prime}$

$$\chi\sigma^3\iint f^{(0)}f_1^{(0)}\mathbf{k}.\left\{\nabla\ln\left(\frac{n^2\chi}{T^3}\right) + \frac{m}{2kT^2}(C_1^2 + C_1'^2)\nabla T\right.$$
$$\left. + \nabla\boldsymbol{c}_0.\frac{m}{kT}(\boldsymbol{C}_1 + \boldsymbol{C}_1')\right\}\boldsymbol{g}.\mathbf{k}\,d\mathbf{k}\,d\boldsymbol{c}_1.$$

The integration with respect to $\mathbf{k}$, using results proved later (16.8, 2, 5, 6), gives

$$\tfrac{2}{3}\pi\chi\sigma^3\int f^{(0)}f_1^{(0)}\left[\boldsymbol{g}.\nabla\ln\left(\frac{n^2\chi}{T^3}\right)\right.$$
$$+\frac{m}{2kT^2}\nabla T.\{2C_1^2\boldsymbol{g} - \tfrac{2}{5}(2\boldsymbol{g}(\boldsymbol{g}.\boldsymbol{C}_1) + \boldsymbol{C}_1 g^2) + \tfrac{3}{5}g^2\boldsymbol{g}\}$$
$$\left.+\frac{m}{kT}\{\nabla\boldsymbol{c}_0 : (2\boldsymbol{C}_1\boldsymbol{g} - \tfrac{2}{5}\boldsymbol{g}\boldsymbol{g}) - \tfrac{1}{5}g^2\nabla.\boldsymbol{c}_0\}\right]d\boldsymbol{c}_1.$$

Substituting $\boldsymbol{g} = \boldsymbol{C}_1 - \boldsymbol{C}$ and integrating with respect to $\boldsymbol{C}_1$, we get

$$-\mathrm{b}\rho\chi f^{(0)}\{\boldsymbol{C}.(\nabla\ln(n^2\chi T) + \tfrac{3}{5}(\mathscr{C}^2 - \tfrac{5}{2})\nabla\ln T)$$
$$+ \tfrac{2}{5}(2\mathscr{C}\mathscr{C} : \nabla\boldsymbol{c}_0 + (\mathscr{C}^2 - \tfrac{5}{2})\nabla.\boldsymbol{c}_0)\}, \qquad (16.32, 4)$$

since (cf. (16.21, 1)) $\mathrm{b}\rho = \tfrac{2}{3}\pi n\sigma^3$. By (16.32, 1), $\partial_e f/\partial t$ is the sum of (16.32, 3 and 4).

16.33. The value of $f^{(1)}$

The equation satisfied by the function $\Phi^{(1)}$ is found by substituting the value of $\partial_0 f/\partial t$ in (16.31, 3). It is

$$\chi\iint f^{(0)}f_1^{(0)}(\Phi^{(1)}+\Phi_1^{(1)}-\Phi^{(1)'}-\Phi_1^{(1)'})\,\sigma^2\boldsymbol{g}.\mathbf{k}\,d\mathbf{k}\,d\boldsymbol{c}_1$$

$$= -f^{(0)}\left[\frac{1}{n}\frac{Dn}{Dt}+(\mathscr{C}^2-\tfrac{3}{2})\frac{1}{T}\frac{DT}{Dt}+\boldsymbol{C}.\left\{\frac{m}{kT}\left(\frac{D\boldsymbol{c}_0}{Dt}-\boldsymbol{F}\right)\right.\right.$$

$$\left.+(nkT)^{-1}\nabla p_0+(1+\tfrac{2}{5}\mathrm{b}\rho\chi)(\mathscr{C}^2-\tfrac{5}{2})\nabla\ln T\right\}$$

$$\left.+\tfrac{2}{3}\mathrm{b}\rho\chi(\mathscr{C}^2-\tfrac{5}{2})\nabla.\boldsymbol{c}_0+2(1+\tfrac{4}{5}\mathrm{b}\rho\chi)\mathscr{C}\mathscr{C}:\nabla\boldsymbol{c}_0\right], \qquad (16.33, 1)$$

where
$$p_0 = knT(1+\mathrm{b}\rho\chi). \qquad (16.33, 2)$$

If (16.33, 1) is multiplied by $\psi\,d\boldsymbol{c}$, where ψ is a summational invariant, and integrated over all values of $\boldsymbol{c}$, the left-hand side of the resulting equation vanishes, by virtue of the transformation used in deriving (4.4, 8).* Thus when $\psi = 1$ we get the result

$$\frac{Dn}{Dt}+n\nabla.\boldsymbol{c}_0 = 0, \qquad (16.33, 3)$$

and when $\psi = m\boldsymbol{C}$,†
$$\frac{D\boldsymbol{c}_0}{Dt}-\boldsymbol{F}+\frac{1}{\rho}\nabla p_0 = 0. \qquad (16.33, 4)$$

Finally, $\psi = \frac{1}{2}mC^2$ gives

$$\frac{3kT}{2n}\frac{Dn}{Dt}+\frac{3k}{2}\frac{DT}{Dt}+\tfrac{5}{2}kT(1+\tfrac{2}{5}\mathrm{b}\rho\chi)\nabla.\boldsymbol{c}_0 = 0$$

or, using (16.33, 2, 3),
$$\frac{DT}{Dt} = -\frac{2p_0}{3nk}\nabla.\boldsymbol{c}_0. \qquad (16.33, 5)$$

Equations (16.3, 3–5) are first approximations to the equations of continuity, momentum and energy; they replace (7.14, 14–16), which would reduce to them if the hydrostatic pressure was taken to be p_0, given by (16.33, 2), instead of knT. Using the values of Dn/Dt, DT/Dt and $D\boldsymbol{c}_0/Dt$

* The proof of this transformation differs slightly from that given in 4.4, since account is taken in this chapter of the difference in position of the centres of colliding molecules. This difficulty is overcome by considering an ideal uniform gas, whose velocity-distribution function at all points is equal to $f(\boldsymbol{c}, \boldsymbol{r})$ at the point considered. The proof is then readily derived as in 4.4, provided that the summational invariant ψ is not itself a function of position.

† The function $m\boldsymbol{C}$ is not a summational invariant if $\boldsymbol{C}$ is defined as $\boldsymbol{c}-\boldsymbol{c}_0$, and if $\boldsymbol{c}_0 = \boldsymbol{c}_0(\boldsymbol{r})$ for one of the colliding molecules, and $\boldsymbol{c}_0 = \boldsymbol{c}_0(\boldsymbol{r}+\sigma\mathbf{k})$ for the other. To overcome this difficulty we put $\boldsymbol{C} = \boldsymbol{c}-\boldsymbol{c}_0'$ for each molecule, where $\boldsymbol{c}_0'$ is taken to be a vector independent of position, which coincides with the mass-velocity at the *special* point $\boldsymbol{r}$ under consideration. This ensures that $m\boldsymbol{C}$ is not a function of position. Similar remarks apply to the summational invariant $\frac{1}{2}mC^2$.

given by (16.33, 3–5), we find that (16.33, 1) reduces to

$$\iint f^{(0)}f_1^{(0)}(\Phi^{(1)}+\Phi_1^{(1)}-\Phi^{(1)\prime}-\Phi_1^{(1)\prime})\,\sigma^2\boldsymbol{g}.\mathbf{k}\,d\mathbf{k}\,d\boldsymbol{c}_1$$
$$= -\chi^{-1}f^{(0)}\{(1+\tfrac{3}{5}\mathrm{b}\rho\chi)(\mathscr{C}^2-\tfrac{5}{2})\nabla\ln T+2(1+\tfrac{2}{5}\mathrm{b}\rho\chi)\overset{\circ}{\mathscr{C}\mathscr{C}}:\nabla\boldsymbol{c}_0\}. \quad (16.33, 6)$$

This differs from the corresponding equation for a gas at ordinary densities only in that the term involving $\nabla \boldsymbol{c}_0$ is multiplied by $(1+\frac{2}{5}\mathrm{b}\rho\chi)/\chi$, and that involving $\nabla \ln T$ by $(1+\frac{3}{5}\mathrm{b}\rho\chi)/\chi$. Hence its solution can be given at once in terms of the functions $\boldsymbol{A}$, B of 7.31; it is

$$\Phi^{(1)} = -\frac{1}{n\chi}\left\{(1+\tfrac{3}{5}\mathrm{b}\rho\chi)\left(\frac{2kT}{m}\right)^{\frac{1}{2}}\boldsymbol{A}.\nabla\ln T+2(1+\tfrac{2}{5}\mathrm{b}\rho\chi)\,\mathsf{B}:\nabla\boldsymbol{c}_0\right\}. \quad (16.33, 7)$$

16.34. The mean values of ρCC and $\frac{1}{2}\rho C^2C$

Approximations to the mean values of certain functions of the velocity can now be obtained. For example, the first approximations to the values of $\rho\overline{\boldsymbol{CC}}$ and $\frac{1}{2}\rho\overline{C^2\boldsymbol{C}}$, as for a normal gas, are

$$\rho(\overline{\boldsymbol{CC}})^{(0)} = knT\mathsf{U}, \quad \tfrac{1}{2}\rho(\overline{C^2\boldsymbol{C}})^{(0)} = 0, \quad (16.34, 1)$$

where U is the unit tensor. The second approximations are obtained by adding to these the quantities $\rho(\overline{\boldsymbol{CC}})^{(1)}$, $\frac{1}{2}\rho(\overline{C^2\boldsymbol{C}})^{(1)}$, which, by (16.33, 7), are respectively $(1+\frac{2}{5}\mathrm{b}\rho\chi)/\chi$ and $(1+\frac{3}{5}\mathrm{b}\rho\chi)/\chi$ times those for a normal gas; that is,

$$\rho(\overline{\boldsymbol{CC}})^{(1)} = -\chi^{-1}(1+\tfrac{2}{5}\mathrm{b}\rho\chi).2\mu\overset{\circ}{\overline{\overline{\nabla\boldsymbol{c}_0}}}, \quad (16.34, 2)$$

$$\tfrac{1}{2}\rho(\overline{C^2\boldsymbol{C}})^{(1)} = -\chi^{-1}(1+\tfrac{3}{5}\mathrm{b}\rho\chi).\lambda\nabla T, \quad (16.34, 3)$$

where λ and μ are the conductivity and viscosity of the gas at normal densities.

The quantities $\rho\overline{\boldsymbol{CC}}$ and $\frac{1}{2}\rho\overline{C^2\boldsymbol{C}}$ give the parts of the pressure tensor p and the thermal flux $\boldsymbol{q}$ that arise from the transport of momentum and energy by the motion of molecules from point to point. To these must be added the contributions arising from the transport of momentum and energy by molecular collisions. These will now be evaluated, correct to the second approximation.

16.4. The collisional transfer of molecular properties

Consider the collisional transfer of a molecular property ψ across an element of area dS at the point $\boldsymbol{r}$. It will be supposed that ψ is a summational invariant, and does not depend on $\boldsymbol{r}$; in this case a collision does not alter the total amount of ψ possessed by the molecules, but transfers part from one molecule to the other, so that a flow of ψ results. If a collision is to produce a transfer of ψ across dS, the centres of the colliding molecules must lie on opposite sides of dS, and their join must cut it.

Let the normal to dS, drawn from its negative side to its positive, have a direction specified by the unit vector $\mathbf{n}$. In a collision between molecules with velocities $\boldsymbol{c}$ and $\boldsymbol{c}_1$, of which the first lies on the positive side of dS and the second on the negative, since $\mathbf{k}$ is the direction of the line drawn from the centre of the second to that of the first, $\mathbf{k}.\mathbf{n}$ is positive. If the line of centres at collision, of length σ, is to cut dS, the centre of the first molecule must lie within a cylinder on dS as base, with generators parallel to $\mathbf{k}$ and of length σ; the volume of this cylinder is $\sigma(\mathbf{k}.\mathbf{n})\,dS$. Again, the mean positions of the centres of the two molecules are the points $\boldsymbol{r}\pm\frac{1}{2}\sigma\mathbf{k}$, while the mean position of the point of impact is the point $\boldsymbol{r}$. Hence the probable number of such collisions per unit time, in which $\boldsymbol{c}$, $\boldsymbol{c}_1$, $\mathbf{k}$ lie respectively in ranges $d\boldsymbol{c}$, $d\boldsymbol{c}_1$, $d\mathbf{k}$, is, by analogy with (16.2, 1),

$$\chi(\boldsymbol{r})f(\boldsymbol{r}+\tfrac{1}{2}\sigma\mathbf{k})f_1(\boldsymbol{r}-\tfrac{1}{2}\sigma\mathbf{k})\,\sigma^2\boldsymbol{g}.\mathbf{k}\,d\mathbf{k}\,d\boldsymbol{c}\,d\boldsymbol{c}_1(\mathbf{k}.\mathbf{n})\,dS.$$

Each such collision causes a molecule on the positive side of dS to gain a quantity $\psi'-\psi$ of the property ψ at the expense of a molecule on the negative side. Thus the total transfer across dS by collisions of this type is

$$(\psi'-\psi)\,\chi(\boldsymbol{r})f(\boldsymbol{r}+\tfrac{1}{2}\sigma\mathbf{k})f_1(\boldsymbol{r}-\tfrac{1}{2}\sigma\mathbf{k})\,\sigma^3(\boldsymbol{g}.\mathbf{k})(\mathbf{k}.\mathbf{n})\,d\mathbf{k}\,d\boldsymbol{c}\,d\boldsymbol{c}_1\,dS,$$

and the total rate of transfer of ψ across dS by collisions, per unit time and area, is

$$\sigma^3\chi(\boldsymbol{r})\iiint(\psi'-\psi)f(\boldsymbol{r}+\tfrac{1}{2}\sigma\mathbf{k})f_1(\boldsymbol{r}-\tfrac{1}{2}\sigma\mathbf{k})(\boldsymbol{g}.\mathbf{k})(\mathbf{k}.\mathbf{n})\,d\mathbf{k}\,d\boldsymbol{c}\,d\boldsymbol{c}_1,$$

the integration being over all values of the variables such that $\boldsymbol{g}.\mathbf{k}$ and $\mathbf{k}.\mathbf{n}$ are positive. In this expression let the variables $\boldsymbol{c}$ and $\boldsymbol{c}_1$ be interchanged; this is equivalent to interchanging the roles of the two colliding molecules. Thus $\mathbf{k}$ is to be replaced by $-\mathbf{k}$, $\boldsymbol{g}$ by $-\boldsymbol{g}$, and $\psi'-\psi$ by $\psi_1'-\psi_1$, which is equal to $-(\psi'-\psi)$. On performing the interchange, we obtain an expression identical in form with the original one, but with the integration now taken over all values of the variables such that $\boldsymbol{g}.\mathbf{k}$ is positive and $\mathbf{k}.\mathbf{n}$ negative. The rate of transfer in question is therefore equal to

$$\tfrac{1}{2}\sigma^3\chi(\boldsymbol{r})\iiint(\psi'-\psi)f(\boldsymbol{r}+\tfrac{1}{2}\sigma\mathbf{k})f_1(\boldsymbol{r}-\tfrac{1}{2}\sigma\mathbf{k})(\boldsymbol{g}.\mathbf{k})(\mathbf{k}.\mathbf{n})\,d\mathbf{k}\,d\boldsymbol{c}\,d\boldsymbol{c}_1,$$

taken over all values of the variables such that $\boldsymbol{g}.\mathbf{k}$ is positive. This expression is the scalar product of $\mathbf{n}$ and another vector which, by analogy with 2.3, represents the contribution of collisions to the vector of flow of ψ.

As in 16.32, f and f_1 are expanded by Taylor's theorem, and derivatives of higher order than the first are neglected; the collisional part of the vector of flow of ψ is then found to be

$$\tfrac{1}{2}\chi\sigma^3\iiint(\psi'-\psi)ff_1\mathbf{k}(\boldsymbol{g}.\mathbf{k})\,d\mathbf{k}\,d\boldsymbol{c}\,d\boldsymbol{c}_1$$
$$+\tfrac{1}{4}\chi\sigma^4\iiint(\psi'-\psi)ff_1\mathbf{k}.\nabla\ln(f/f_1)\,\mathbf{k}(\boldsymbol{g}.\mathbf{k})\,d\mathbf{k}\,d\boldsymbol{c}\,d\boldsymbol{c}_1,\quad(16.4,1)$$

where all the quantities are evaluated at the point $\boldsymbol{r}$. Since the second term of this already involves space-derivatives, f and f_1 can here be replaced by $f^{(0)}$ and $f_1^{(0)}$.

16.41. The viscosity of a dense gas

Consider first the transfer of momentum. Let

$$\psi = mC = m(c - c_0),$$

where c_0 is evaluated at the special point r under consideration, and does not vary with the position of the molecule. Then (16.4, 1) becomes

$$\tfrac{1}{2}\chi\sigma^3\iiint m(C' - C) f f_1 \mathbf{k}(g.\mathbf{k})\, d\mathbf{k}\, dc\, dc_1$$
$$+ \tfrac{1}{4}\chi\sigma^4\iiint m(C' - C) f^{(0)} f_1^{(0)} \mathbf{k}.\nabla \ln(f^{(0)}/f_1^{(0)})\, \mathbf{k}(g.\mathbf{k})\, d\mathbf{k}\, dc\, dc_1.$$

After integration with respect to $\mathbf{k}$, using (16.8, 7, 8), this expression becomes

$$\frac{\pi}{15}\chi\sigma^3 m \iint f f_1 (2gg + \mathsf{U}g^2)\, dc\, dc_1$$
$$+ \frac{\pi}{48}\chi\sigma^4 m \iint f^{(0)} f_1^{(0)} [\{g.\nabla \ln(f^{(0)}/f_1^{(0)})\}(gg + g^2\mathsf{U})/g$$
$$+ g\{g\nabla \ln(f^{(0)}/f_1^{(0)}) + \nabla \ln(f^{(0)}/f_1^{(0)})\, g\}]\, dc\, dc_1. \quad (16.41, 1)$$

In the first integral we use the relation $g = C_1 - C$; then since, for any function ϕ,

$$\int f\phi\, dc = \int f_1 \phi_1\, dc_1 = n\bar{\phi},$$

and $\overline{C_1} = 0$, $\overline{C} = 0$, the value of this integral is

$$\tfrac{4}{5}\mathrm{b}\rho^2\chi(2\overline{CC} + \overline{C^2}\mathsf{U}). \quad (16.41, 2)$$

In the second integral, since

$$\nabla \ln(f^{(0)}/f_1^{(0)}) = \frac{m}{2kT^2}(C^2 - C_1^2)\nabla T + \frac{m}{kT}(\nabla c_0).(C - C_1),$$
$$(16.41, 3)$$

the terms involving ∇T yield odd functions of C, C_1, which vanish on integration; hence the integral is equal to

$$-\frac{\pi m^2\sigma^4\chi}{48kT}\iint f^{(0)} f_1^{(0)} [(\nabla c_0 : gg)(gg + g^2\mathsf{U})/g$$
$$+ g\{(\nabla c_0 . g) g + g(\nabla c_0 . g)\}]\, dc\, dc_1.$$

This can readily be evaluated by changing the variables of integration from c and c_1 to G_0 and g, where $G_0 = \tfrac{1}{2}(C + C_1)$. Its value is found to be

$$-\varpi\{\tfrac{6}{5}\overset{\circ}{\overline{\overline{\nabla c_0}}} + \mathsf{U}\nabla . c_0\} = -\varpi\{\tfrac{6}{5}\overset{\circ}{\mathsf{e}} + \mathsf{U}\nabla . c_0\}, \quad (16.41, 4)$$

where $\varpi \equiv \tfrac{4}{9}\pi^{\frac{1}{2}}\chi n^2\sigma^4(mkT)^{\frac{1}{2}} = \chi \mathrm{b}^2\rho^2(mkT)^{\frac{1}{2}}/\pi^{\frac{3}{2}}\sigma^2. \quad (16.41, 5)$

Alternatively, in terms of the coefficient of viscosity at ordinary pressures, namely

$$\mu = 1{\cdot}016 \times 5(mkT)^{\frac{1}{2}}/16\pi^{\frac{1}{2}}\sigma^2,$$

ϖ is given by

$$\varpi = 1{\cdot}002\mu\chi \mathrm{b}^2\rho^2. \quad (16.41, 6)$$

Adding (16.41, 2) and (16.41, 4), we obtain for the value of (16.41, 1)

$$\tfrac{1}{5}\mathrm{b}\rho^2\chi(2\overline{\boldsymbol{CC}}+\overline{C^2}\mathsf{U})-\varpi(\tfrac{6}{5}\overset{\circ}{\mathsf{e}}+\mathsf{U}\nabla\cdot\boldsymbol{c}_0).$$

This gives the part of the pressure tensor which arises from the collisional transport of momentum. To this we must add $\rho\overline{\boldsymbol{CC}}$, which gives the rate of transport by molecular motions; then the whole pressure tensor is found to be

$$\mathsf{p}=\rho(1+\tfrac{2}{5}\mathrm{b}\rho\chi)\,\overline{\boldsymbol{CC}}+\tfrac{1}{5}\mathrm{b}\rho^2\chi\overline{C^2}\mathsf{U}-\varpi(\tfrac{6}{5}\overset{\circ}{\mathsf{e}}+\mathsf{U}\nabla\cdot\boldsymbol{c}_0).$$

On substituting in this the value of $\rho\overline{\boldsymbol{CC}}$ found in 16.34, it becomes

$$\mathsf{p}=(p_0-\varpi\nabla\cdot\boldsymbol{c}_0)\,\mathsf{U}-\{2\mu\chi^{-1}(1+\tfrac{2}{5}\mathrm{b}\rho\chi)^2+\tfrac{6}{5}\varpi\}\,\overset{\circ}{\mathsf{e}}, \qquad (16.41, 7)$$

where $p_0 = knT(1+\mathrm{b}\rho\chi)$, as in (16.33, 2).

For a gas in a uniform steady state, p reduces to a hydrostatic pressure p_0—a result already implicit in the forms of equations (16.33, 4, 5). For a gas not in a uniform steady state, the hydrostatic pressure is $p_0-\varpi\nabla\cdot\boldsymbol{c}_0$. The extra term $-\varpi\nabla\cdot\boldsymbol{c}_0$ differs from zero whenever the density of the gas is varying (cf. (16.33, 3)); it represents a volume viscosity similar to that of 11.51, opposing expansion or contraction of the gas. The deviation of the pressure system from the hydrostatic is given by the last term of (16.41, 7), which represents the usual shear viscosity. Equating this term to $-2\mu'\overset{\circ}{\mathsf{e}}$, we see that in a dense gas the coefficient of viscosity μ' is given in terms of μ, the viscosity at ordinary densities (at the same temperature), by the relation

$$\begin{aligned}\mu' &= \mu\chi^{-1}(1+\tfrac{2}{5}\mathrm{b}\rho\chi)^2+\tfrac{3}{5}\varpi\\ &= \mu\mathrm{b}\rho\{(\mathrm{b}\rho\chi)^{-1}+0{\cdot}8+0{\cdot}7614\mathrm{b}\rho\chi\}\end{aligned} \qquad (16.41, 8)$$

using (16.41, 6).

16.42. The thermal conductivity of a dense gas

To evaluate the thermal flow, we write $\psi = \tfrac{1}{2}mC^2$ in (16.4, 1), where again $\boldsymbol{C}$ is taken as the velocity of a molecule relative to the mass-velocity of the gas at the special point $\boldsymbol{r}$. Then the expression becomes

$$\tfrac{1}{4}\chi m\sigma^3\iiint(C'^2-C^2)ff_1\mathbf{k}(\boldsymbol{g}\cdot\mathbf{k})\,d\mathbf{k}\,d\boldsymbol{c}\,d\boldsymbol{c}_1$$
$$+\tfrac{1}{8}\chi m\sigma^4\iiint(C'^2-C^2)f^{(0)}f_1^{(0)}\mathbf{k}\cdot\nabla\ln(f^{(0)}/f_1^{(0)})\,\mathbf{k}(\boldsymbol{g}\cdot\mathbf{k})\,d\mathbf{k}\,d\boldsymbol{c}\,d\boldsymbol{c}_1.$$

The two integrals in this can be evaluated in the same way as those in 16.41, using (16.8, 9, 10). On carrying out the integration, the collisional transport of energy is found to be

$$\tfrac{3}{10}\mathrm{b}\rho^2\chi\overline{C^2\boldsymbol{C}}-c_v\varpi\nabla T, \qquad (16.42, 1)$$

where ϖ is given by (16.41, 5), and $c_v = 3k/2m$. On adding to (16.42, 1) the transport $\tfrac{1}{2}\rho\overline{C^2\boldsymbol{C}}$ by molecular motions, the total flux-vector of energy is found to be given by

$$\boldsymbol{q} = \tfrac{1}{2}\rho\overline{C^2\boldsymbol{C}}(1+\tfrac{3}{5}\mathrm{b}\rho\chi)-c_v\varpi\,\nabla T$$

or, using the value of $\tfrac{1}{2}\rho\overline{C^2\boldsymbol{C}}$ found in 16.34, by

$$\boldsymbol{q} = -\{\chi^{-1}(1+\tfrac{3}{5}\mathrm{b}\rho\chi)^2\,\lambda+c_v\varpi\}\,\nabla T. \qquad (16.42, 2)$$

Thus at any temperature the true conductivity, λ', is given in terms of that at ordinary densities, λ, by the equation

$$\lambda' = \chi^{-1}(1+\tfrac{3}{5}b\rho\chi)^2\lambda + c_v\varpi$$
$$= b\rho\lambda\{(b\rho\chi)^{-1}+1\cdot 2+0\cdot 7574 b\rho\chi\}. \qquad (16.42, 3)$$

16.5. Comparison with experiment

Equations (16.41, 8), (16.42, 3), which give the coefficients of viscosity and heat conduction μ', λ' for a dense gas in terms of the corresponding coefficients μ, λ for a normal gas, may be written in the forms

$$\mu'/b\rho = \mu\{(b\rho\chi)^{-1}+0\cdot 8+0\cdot 7614 b\rho\chi\}, \qquad (16.5, 1)$$

$$\lambda'/b\rho = \lambda\{(b\rho\chi)^{-1}+1\cdot 2+0\cdot 7574 b\rho\chi\}. \qquad (16.5, 2)$$

Thus the theory predicts that, as ρ varies at a given temperature, μ'/ρ and λ'/ρ should have minima respectively corresponding to

$$b\rho\chi = (0\cdot 7614)^{-\frac{1}{2}} = 1\cdot 146, \quad b\rho\chi = (0\cdot 7574)^{-\frac{1}{2}} = 1\cdot 149, \qquad (16.5, 3)$$

i.e. at practically the same density. These minima are actually observed; qualitatively, at least, Enskog's formula fits the experimental variations with density at any temperature well above the critical temperature.

For quantitative comparisons with experiment we require values of $b\rho$ and χ. If the molecules were strictly rigid spheres, these could be calculated from viscosity diameters, using (16.21, 1, 3). Sengers* used viscosity diameters at ordinary pressures at the temperature concerned in a comparison of (16.5, 1, 2) with experiment. He found that the calculated values of μ' and λ' give moderate agreement with experiment for values of $b\rho$ less than 0·4, but are too high for greater values.

Enskog preferred a different procedure, based on the close relation between $b\rho\chi$ and the compressibility (cf. (16.33, 2)). He observed that, if the (spherical) molecules were surrounded by weak attractive fields of force, the equation of state

$$p_0 = knT(1+b\rho\chi) \qquad (16.5, 4)$$

would be modified to the form

$$p_0 + a\rho^2 = knT(1+b\rho\chi), \qquad (16.5, 5)$$

where a is independent of the temperature (the term $a\rho^2$ is the well-known van der Waals correction). He therefore suggested that $b\rho\chi$ should be determined from compressibility experiments, using the relation

$$T\frac{\partial p_0}{\partial T} = knT(1+b\rho\chi) \qquad (16.5, 6)$$

* J. V. Sengers, *Int. J. Heat Mass Transfer*, **8**, 1103 (1965).

instead of (16.5, 4). This means that p_0 in (16.5, 4) is replaced by the 'thermal pressure'* $T\,\partial p_0/\partial T$. For spherical molecules without attractive fields of force, (16.5, 4) and (16.5, 6) are equivalent: for actual molecules this is no longer the case, because the values calculated for $\mathrm{b}\rho\chi$ depend on T.

When $\mathrm{b}\rho\chi$ is determined from (16.5, 6), b can be found from the limit of $\mathrm{b}\chi$ as ρ tends to zero. That is, if the experimental p_0, ρ relation is expressed in the form

$$p_0 = knT(1+\rho\mathrm{B}(T)+\rho^2\mathrm{C}(T)+\ldots), \tag{16.5, 7}$$

b is given by

$$\mathrm{b} = d(T\mathrm{B})/dT. \tag{16.5, 8}$$

Alternatively, a direct determination of b can be avoided by using the minima of μ'/ρ and λ'/ρ. If the minimum values are denoted by $(\mu'/\rho)_{\min}$ and $(\lambda'/\rho)_{\min}$, (16.5, 1, 2) are equivalent to

$$2{\cdot}545\mu'/\rho = (\mu'/\rho)_{\min}((\mathrm{b}\rho\chi)^{-1}+0{\cdot}8+0{\cdot}7614\mathrm{b}\rho\chi), \tag{16.5, 9}$$

$$2{\cdot}941\lambda'/\rho = (\lambda'/\rho)_{\min}((\mathrm{b}\rho\chi)^{-1}+1{\cdot}2+0{\cdot}7574\mathrm{b}\rho\chi). \tag{16.5, 10}$$

Values of μ' and λ' calculated from these equations can be expected to fit the experimental results best near the minima of μ'/ρ and λ'/ρ, i.e. for large pressures; values calculated from (16.5, 1, 2), using (16.5, 8), can be expected to give a better fit at lower pressures.

The following table gives the experimental values of the coefficient of viscosity of nitrogen at 50 °C. obtained by Michels and Gibson,† together with values calculated using (16.5, 1 and 9). The values of $\mathrm{b}\rho\chi$ were derived

Table 28. *Viscosity of nitrogen at high pressures*

Pressure (atmos.)	$\mathrm{b}\rho\chi$	$\mu'\times 10^6$ (calc. (16.5, 1))	$\mu'\times 10^6$ (calc. (16.5, 9))	$\mu'\times 10^6$ (exp.)	ρ (g./cm.^{3})	μ'/ρ (exp.)
15·37	0·031	191	181	191·3	0·01623	0·01179
57·60	0·119	201	190	198·1	0·06049	0·003274
104·5	0·215	217	205	208·8	0·1083	0·001928
212·4	0·491	237	224	237·3	0·2067	0·001148
320·4	0·717	281	266	273·7	0·2875	0·000952
430·2	0·920	326	308	312·9	0·3528	0·000887
541·7	1·111	368	348	350·9	0·4053	0·000866
630·4	1·247	401	380	378·6	0·4409	0·000859
742·1	1·413	442	418	416·3	0·4786	0·000870
854·1	1·576	481	455	455·0	0·5117	0·000889
965·8	1·732	520	492	491·3	0·5404	0·000909

* The first approximation to the equation for DT/Dt for a general dense gas is

$$\rho c_v \frac{DT}{Dt} = -T\frac{\partial p_0}{\partial T}\nabla . \mathbf{c}_0,$$

so that the thermal pressure replaces the total pressure p_0 in this equation. On the other hand, it is p_0 which appears in the equation of motion.

† A. Michels and R. O. Gibson, *Proc. R. Soc.* A, **134**, 288 (1931).

from the experimental results of Deming and Shupe.* The values calculated from (16.5, 1), which are adjusted to agree with experiment at moderate pressures, are nearly 6 per cent too high at large pressures; those calculated from (16.5, 9) agree with experiment very well at high pressures, but are too low at normal pressures. The minimum value of μ'/ρ (last column) corresponds roughly to $b\rho\chi = 1\cdot 2$.

A similarly rough agreement with experiment was quoted by Sengers† for a number of other gases; for neon and hydrogen the theoretical μ' increases faster with density than the experimental; for helium, argon and xenon the situation is reversed. Sengers quoted similar results for λ'; for both neon and argon λ' calculated from (16.5, 2) is less than the experimental λ' at high pressures, by small amounts at pressures below 250 atmospheres, but by 6 per cent for neon and 15 per cent for argon at pressures of order 2000 atmospheres.

Michels and Botzen‡ attempted to apply (16.5, 2) to the thermal conductivity of nitrogen, obtaining calculated values 15 per cent higher than the experimental at 2000 atmospheres. This is not surprising if the transport of internal energy is by a self-diffusion process, as in 11.8, since at high densities the self-diffusion coefficient is cut down roughly in the same ratio as $1/\rho\chi$ (cf. 16.6). However, Sengers, taking the transport of translatory energy to be given by (16.5, 2) and that of internal energy to be by self-diffusion, obtained calculated values of λ' for nitrogen well below the experimental ones, suggesting that the self-diffusion approximation is only roughly valid.

To summarize, Enskog's theory approximately represents the observed variation of μ' and λ' with ρ at a given temperature. It is unable to represent their variation with temperature, because of the rigid sphere model; it is necessary to determine $b\rho\chi$ at any temperature (and density) by a semi-empirical rule. The dependence of μ' and λ' on temperature has its own features of interest; for a given density, λ'/λ and μ'/μ appear to decrease with increasing temperature, and for some gases it has been suggested that $\mu' - \mu$ is nearly a function of ρ, independent of the temperature.§ Enskog's theory cannot explain such behaviour.

Madigosky‖ in ultrasonic experiments with compressed argon, found evidence of a volume viscosity ϖ similar to that of (16.41, 6, 7). When χ and b were determined from compressibility, as in (16.5, 6), a moderate agreement with (16.41, 6) was obtained.

* W. E. Deming and L. E. Shupe, *Phys. Rev.* **37**, 638 (1931).

† *Loc. cit.*

‡ A. Michels and A. Botzen, *Physica*, **19**, 585 (1953).

§ See, e.g., G. P. Flynn, R. V. Hanks, N. A. Lemaire and J. Ross, *J. Chem. Phys.* **38**, 154 (1963). However, the result quoted is only very approximate.

‖ W. M. Madigosky, *J. Chem. Phys.* **46**, 4441 (1967).

16.6. Extension to mixed dense gases

Enskog's methods were generalized by the late H. H. Thorne (of the University of Sydney, Australia), to apply to a binary mixture. We give here only his results as they affect the first approximation to the coefficient of diffusion.*

Let m_1 and m_2 be the masses of molecules of the two gases, σ_1 and σ_2 their radii. Then b_1, b_2, b_1' and b_2' are defined by

$$\mathrm{b}_1\rho_1 = \tfrac{2}{3}\pi n_1\sigma_1^3, \quad \mathrm{b}_2\rho_2 = \tfrac{2}{3}\pi n_2\sigma_2^3, \quad \mathrm{b}_1'\rho_1 = \tfrac{2}{3}\pi n_1\sigma_{12}^3, \quad \mathrm{b}_2'\rho_2 = \tfrac{2}{3}\pi n_2\sigma_{12}^3. \tag{16.6, 1}$$

Corresponding to the factor χ of 16.2, three factors χ_1, χ_2 and χ_{12} are introduced, related respectively to collisions between pairs of molecules m_1, pairs of molecules m_2, and pairs of dissimilar molecules. It is found that

$$\left.\begin{aligned} \chi_1 &= 1+\tfrac{5}{12}\pi n_1\sigma_1^3+\frac{\pi}{12}n_2(\sigma_1^3+16\sigma_{12}^3-12\sigma_{12}^2\sigma_1)+\ldots, \\ \chi_{12} &= 1+\frac{\pi}{12}n_1\sigma_1^3(8-3\sigma_1/\sigma_{12})+\frac{\pi}{12}n_2\sigma_2^3(8-3\sigma_2/\sigma_{12})+\ldots, \end{aligned}\right\} \tag{16.6, 2}$$

with a corresponding equation for χ_2. In terms of these, the first approximation to the hydrostatic pressure is found to be $p_0 = p_{10}+p_{20}$, where

$$p_{10} = kn_1T(1+\mathrm{b}_1\rho_1\chi_1+\mathrm{b}_2'\rho_2\chi_{12}), \quad p_{20} = kn_2T(1+\mathrm{b}_2\rho_2\chi_2+\mathrm{b}_1'\rho_1\chi_{12}). \tag{16.6, 3}$$

The Boltzmann equations for the two gases are of obvious forms. First approximations to their solutions are the Maxwellian functions $f_1^{(0)}$ and $f_2^{(0)}$; second approximations are $f_1^{(0)}(1+\Phi_1^{(1)})$, $f_2^{(0)}(1+\Phi_2^{(1)})$, where $\Phi_1^{(1)}$ and $\Phi_2^{(1)}$ are found to satisfy the equations

$$\begin{aligned} &\chi_1\iint f_1^{(0)}f^{(0)}(\Phi_1^{(1)}+\Phi^{(1)}-\Phi_1^{(1)\prime}-\Phi^{(1)\prime})\,\sigma_1^2\boldsymbol{g}_1\cdot\mathbf{k}\,d\mathbf{k}\,dc \\ &\quad+\chi_{12}\iint f_1^{(0)}f_2^{(0)}(\Phi_1^{(1)}+\Phi_2^{(1)}-\Phi_1^{(1)\prime}-\Phi_2^{(1)\prime})\,\sigma_{12}^2\boldsymbol{g}_{21}\cdot\mathbf{k}\,d\mathbf{k}\,d\boldsymbol{c}_2 \\ &= -f_1^{(0)}[(1+\tfrac{3}{5}\mathrm{b}_1\rho_1\chi_1+\tfrac{12}{5}M_1M_2\mathrm{b}_2'\rho_2\chi_{12})(\mathscr{C}_1^2-\tfrac{5}{2})\boldsymbol{C}_1\cdot\nabla\ln T \\ &\quad+\mathrm{x}_1^{-1}\boldsymbol{d}_{12}\cdot\boldsymbol{C}_1+2(1+\tfrac{2}{5}\mathrm{b}_1\rho_1\chi_1+\tfrac{4}{5}M_2\mathrm{b}_2'\rho_2\chi_{12})\overset{\circ}{\mathscr{C}_1\mathscr{C}_1}:\nabla\boldsymbol{c}_0 \\ &\quad+\tfrac{2}{3}\mathrm{x}_2\{\mathrm{b}_1\rho_1\chi_1-b_2\rho_2\chi_2+2\chi_{12}(M_2\rho_2\mathrm{b}_2'-M_1\rho_1\mathrm{b}_1')\}(\mathscr{C}_1^2-\tfrac{3}{2})\nabla\cdot\boldsymbol{c}_0] \end{aligned} \tag{16.6, 4}$$

and a similar equation; $\boldsymbol{d}_{12}$ and $\boldsymbol{d}_{21}$ are given by

$$\begin{aligned} \boldsymbol{d}_{12} = -\boldsymbol{d}_{21} &= \frac{\rho_1\rho_2}{\rho knT}\left(\boldsymbol{F}_2-\boldsymbol{F}_1+\frac{1}{\rho_1}\nabla p_{10}-\frac{1}{\rho_2}\nabla p_{20}\right) \\ &\quad+\mathrm{x}_2\mathrm{b}_1'\rho_1\chi_{12}(\nabla\ln(\mathrm{x}_2/\mathrm{x}_1)+(M_1-M_2)\nabla\ln T). \end{aligned} \tag{16.6, 5}$$

For a gas at rest at uniform temperature under no forces, $\boldsymbol{d}_{12} = \nabla\mathrm{x}_1$ for

* A fuller account was given in earlier editions of this book.

self-diffusion, as in 8.4; however, $\boldsymbol{d}_{12}$ may differ considerably from ∇x_1 for a general mixed gas.

The velocity of diffusion is found to be

$$\overline{\boldsymbol{C}}_1-\overline{\boldsymbol{C}}_2 = -(x_1x_2)^{-1}D'_{12}(\boldsymbol{d}_{12}+k_T\nabla\ln T),$$

where D'_{12} is, to a first approximation, given by

$$\rho D'_{12} = \rho_0[D_{12}]_1/\chi_{12}, \qquad (16.6, 6)$$

$[D_{12}]_1$ being the first approximation to the coefficient of diffusion for the same gas at a moderate density ρ_0. Thus the effect of high pressure is simply to decrease D_{12} in the ratio $\rho_0/\rho\chi_{12}$, as a consequence of the increased number of collisions between unlike molecules. For self-diffusion similarly

$$\rho D'_{11} = \rho_0[D_{11}]_1/\chi. \qquad (16.6, 7)$$

The experimental results give, at best, only moderate agreement with (16.6, 6, 7). They suggest that dense-gas corrections are considerably overestimated by these formulae, though much depends on the method of calculation of the correction factors χ_{12}, χ.*

THE BBGKY EQUATIONS

16.7. A brief sketch will now be given of a more general theory of dense gases, based on the concept of multiple velocity-distribution functions. We consider only a simple gas whose molecules do not possess internal energy.

A generalized velocity-distribution function† $f^{(s)}$ is defined as such that the probability of finding s molecules whose centres are respectively in small volume elements $d\boldsymbol{r}_1, d\boldsymbol{r}_2, \ldots, d\boldsymbol{r}_s$, and whose respective velocities lie in the ranges $d\boldsymbol{c}_1, d\boldsymbol{c}_2, \ldots, d\boldsymbol{c}_s$, is

$$f^{(s)}(\boldsymbol{c}_1, \boldsymbol{c}_2, \ldots, \boldsymbol{c}_s, \boldsymbol{r}_1, \boldsymbol{r}_2, \ldots, \boldsymbol{r}_s, t)\, d\boldsymbol{r}_1\ldots d\boldsymbol{r}_s d\boldsymbol{c}_1\ldots d\boldsymbol{c}_s.$$

A generalized number-density $n^{(s)}$ is similarly defined; it satisfies the equation

$$n^{(s)} = \int\ldots\int f^{(s)}\, d\boldsymbol{c}_1\ldots d\boldsymbol{c}_s. \qquad (16.7, 1)$$

The quantities $n^{(1)}$, $f^{(1)}$ are identical with the usual number-density n and velocity-distribution function f. In a gas at ordinary densities one could write, to a sufficient approximation,

$$f^{(2)} = f(\boldsymbol{c}_1, \boldsymbol{r}_1)f(\boldsymbol{c}_2, \boldsymbol{r}_2), \quad f^{(3)} = f(\boldsymbol{c}_1, \boldsymbol{r}_1)f(\boldsymbol{c}_2, \boldsymbol{r}_2)f(\boldsymbol{c}_3, \boldsymbol{r}_3),$$

and so on; but in a dense gas this is no longer adequate, since any one molecule perturbs the distribution of others in its vicinity.‡

* K. D. Timmerhaus and H. G. Drickamer, *J. Chem. Phys.* **20**, 981 (1952); T. R. Mifflin and C. O. Bennett, *J. Chem. Phys.* **29**, 975 (1958); M. Lipsicas, *J. Chem. Phys.* **36**, 1235 (1962); L. Durbin and R. Kobayashi, *J. Chem. Phys.* **37**, 1643 (1962).

† The change of notation should be noted; $f^{(s)}$ does not here refer to successive approximations to f.

‡ Note that the volume elements $d\boldsymbol{r}_1, d\boldsymbol{r}_2, \ldots, d\boldsymbol{r}_s$ used in defining $f^{(s)}$ must be assumed small compared with the dimensions of the molecular fields of force.

The functions $f^{(s)}$ can be used to construct mean values in accordance with the formula

$$n^{(s)}\overline{\phi}^{(s)} = \int \ldots \int f^{(s)} \phi \, d\boldsymbol{c}_1 \ldots d\boldsymbol{c}_s. \qquad (16.7, 2)$$

A function $\phi(\boldsymbol{c}_1)$ may have several different mean values $\overline{\phi}^{(1)}, \overline{\phi}^{(2)}, \ldots$. For example, $\overline{\boldsymbol{c}}_1^{(1)}$ (or $\overline{\boldsymbol{c}}_1$) is the mean velocity of all molecules near $\boldsymbol{r}_1$, $\overline{\boldsymbol{c}}_1^{(2)}$ is the mean velocity of a molecule in $d\boldsymbol{r}_1$ when it is known that a second molecule is in $d\boldsymbol{r}_2$; when $\boldsymbol{r}_1 - \boldsymbol{r}_2$ is small, $\overline{\boldsymbol{c}}_1^{(1)}$ and $\overline{\boldsymbol{c}}_1^{(2)}$ may differ appreciably. The mean value $\overline{\phi(\boldsymbol{c})}$ is understood to be $\overline{\phi(\boldsymbol{c})}^{(1)}$.

Each function $n^{(s)}$ satisfies an equation of continuity

$$\frac{\partial n^{(s)}}{\partial t} + \sum_{i=1}^{s} \frac{\partial}{\partial \boldsymbol{r}_i} \cdot (n^{(s)} \overline{\boldsymbol{c}}_i^{(s)}) = 0. \qquad (16.7, 3)$$

Similarly $f^{(s)}$ satisfies a Boltzmann equation

$$\frac{\partial f^{(s)}}{\partial t} + \sum_{i=1}^{s} \left\{ \boldsymbol{c}_i \cdot \frac{\partial f^{(s)}}{\partial \boldsymbol{r}_i} + \frac{\partial}{\partial \boldsymbol{c}_i} \cdot (f^{(s)} \boldsymbol{\eta}_i^{(s)}) \right\} = 0, \qquad (16.7, 4)$$

where $\boldsymbol{\eta}_i^{(s)}$ is the average acceleration of the ith molecule of the group of s, the average being taken over all possible positions and velocities of molecules not belonging to the group. Let $\boldsymbol{F}_i$, $\boldsymbol{X}_{ij}$ be the accelerations of the molecule i due respectively to external forces and to the force exerted by the molecule j; $\boldsymbol{X}_{ij}$ is taken to be in the direction of the vector $\boldsymbol{r}_{ij} \equiv \boldsymbol{r}_i - \boldsymbol{r}_j$, and is given by

$$m\boldsymbol{X}_{ij} = -\frac{\partial V_{ij}}{\partial \boldsymbol{r}_i}, \qquad (16.7, 5)$$

where $V_{ij} \,(\equiv V(r_{ij}))$ is the mutual potential energy of the molecules i, j. Then*

$$\boldsymbol{\eta}_i^{(s)} = \boldsymbol{F}_i + \sum_{j=1}^{s}{}' \boldsymbol{X}_{ij} + (f^{(s)})^{-1} \iint \boldsymbol{X}_{i,s+1} f^{(s+1)} d\boldsymbol{r}_{s+1} d\boldsymbol{c}_{s+1};$$

the last term on the right represents the average acceleration due to molecules outside the group of s. Hence (16.7, 4) becomes

$$\frac{\partial f^{(s)}}{\partial t} + \sum_{i=1}^{s} \left\{ \boldsymbol{c}_i \cdot \frac{\partial f^{(s)}}{\partial \boldsymbol{r}_i} + \left(\boldsymbol{F}_i + \sum_{j=1}^{s}{}' \boldsymbol{X}_{ij} \right) \cdot \frac{\partial f^{(s)}}{\partial \boldsymbol{c}_i} \right\}$$
$$= -\iint \sum_{i=1}^{s} \boldsymbol{X}_{i,s+1} \cdot \frac{\partial f^{(s+1)}}{\partial \boldsymbol{c}_i} d\boldsymbol{r}_{s+1} d\boldsymbol{c}_{s+1}. \qquad (16.7, 6)$$

The equations (16.7, 6) were given in 1946 independently by Bogoliubov,† by Born and Green,‡ and by Kirkwood;§ they had already been introduced in 1935 by Yvon,‖ but attracted little attention at the time. In recognition of the multiple authorship, they are frequently styled the BBGKY equations.

* Here and later, the symbol Σ' is used to indicate that meaningless terms representing the interaction of a molecule with itself are omitted from the sum.

† N. N. Bogoliubov, *Problems of a Dynamical Theory in Statistical Physics* (Moscow: State Technical Press, 1946). English translation by E. K. Gora in *Studies in Statistical Mechanics*, vol. 1 (ed. J. de Boer and G. E. Uhlenbeck, 1962).

‡ M. Born and H. S. Green, *Proc. R. Soc.* A, **188**, 10 (1946); **189**, 103 (1947); **190**, 455 (1947).

§ J. G. Kirkwood, *J. Chem. Phys.* **14**, 180 (1946); **15**, 72 (1947).

‖ J. Yvon, *La Théorie des fluides et de l'équation d'état* (Paris: Hermann et Cie, 1935).

16.71. The equations of transfer

In a dense gas, the pressure tensor p can be expressed in the form

$$\mathsf{p} = \rho\overline{\boldsymbol{CC}} + \mathsf{p}'. \tag{16.71, 1}$$

Here $\rho\overline{\boldsymbol{CC}}$ denotes, as usual, the transport of momentum relative to the mass-velocity $\boldsymbol{c}_0$, and p' gives the effect of the intermolecular forces $m\boldsymbol{X}_{ij}$. By an appropriate generalization of the argument of 16.4, the value of p' at the point $\boldsymbol{r}$ can be shown to be

$$\mathsf{p}' = \tfrac{1}{2}m\int\left\{\int_0^1 n^{(2)}(\boldsymbol{r}_1, \boldsymbol{r}_2)\,dx\right\}\boldsymbol{r}_{12}\boldsymbol{X}_{12}\,d\boldsymbol{r}_{12}, \tag{16.71, 2}$$

where

$$\boldsymbol{r}_1 = \boldsymbol{r} + x\boldsymbol{r}_{12}, \quad \boldsymbol{r}_2 = \boldsymbol{r} - (1-x)\boldsymbol{r}_{12}. \tag{16.71, 3}$$

The contribution of p' to the hydrostatic pressure is

$$p' \equiv \tfrac{1}{6}m\int\left\{\int_0^1 n^{(2)}(\boldsymbol{r}_1, \boldsymbol{r}_2)\,dx\right\}\boldsymbol{r}_{12}\cdot\boldsymbol{X}_{12}\,d\boldsymbol{r}_{12}, \tag{16.71, 4}$$

and the force per unit volume at $\boldsymbol{r}$ due to p' is*

$$-\frac{\partial}{\partial\boldsymbol{r}}\cdot\mathsf{p}' = m\int n^{(2)}(\boldsymbol{r}, \boldsymbol{r}_2)\boldsymbol{X}_{12}\,d\boldsymbol{r}_{12}. \tag{16.71, 5}$$

The heat energy of the gas includes a part corresponding to intermolecular potential energy. Half of the energy V_{ij} is arbitrarily supposed to reside in each of the molecules i, j; thus the average heat energy of a molecule at $\boldsymbol{r}$ is $\bar{E}$, where

$$n\bar{E} = \tfrac{1}{2}\rho\overline{C^2} + \tfrac{1}{2}\int n^{(2)}(\boldsymbol{r}, \boldsymbol{r}_2)\,V(r_{12})\,d\boldsymbol{r}_{12}. \tag{16.71, 6}$$

The thermal flux-vector $\boldsymbol{q}$ is given by

$$\boldsymbol{q} = \tfrac{1}{2}\rho\overline{C^2\boldsymbol{C}} + \tfrac{1}{2}\int n^{(2)}(\boldsymbol{r}, \boldsymbol{r}_2)\,V(r_{12})\,\overline{\boldsymbol{C}}_1^{(2)}\,d\boldsymbol{r}_{12} + \boldsymbol{q}'. \tag{16.71, 7}$$

The first two terms on the right of (16.71, 7) give the transport of kinetic and potential energies arising from the molecular motions, and $\boldsymbol{q}'$ represents work done by intermolecular forces. In calculating $\boldsymbol{q}'$ it has to be remembered that work done in altering $\boldsymbol{c}_0$ is taken into account as work done by p'. It is found that

$$\boldsymbol{q}' = \frac{1}{2}\int\boldsymbol{r}_{12}\left\{\int_0^1 n^{(2)}(\boldsymbol{r}_1, \boldsymbol{r}_2)\,m\boldsymbol{X}_{12}\cdot\overline{\boldsymbol{c}}_1^{(2)}\,dx\right\}d\boldsymbol{r}_{12} - \mathsf{p}'\cdot\boldsymbol{c}_0(\boldsymbol{r}), \tag{16.71, 8}$$

where $\boldsymbol{r}_1$, $\boldsymbol{r}_2$ are given by (16.71, 3). The rate of heating per unit volume at $\boldsymbol{r}$ arising from $\boldsymbol{q}'$ is

$$-\frac{\partial}{\partial\boldsymbol{r}}\cdot\boldsymbol{q}' = \frac{1}{2}\int n^{(2)}(\boldsymbol{r}, \boldsymbol{r}_2)\,m\boldsymbol{X}_{12}\cdot(\overline{\boldsymbol{c}}_1^{(2)} + \overline{\boldsymbol{c}}_2^{(2)})\,d\boldsymbol{r}_{12} + \frac{\partial}{\partial\boldsymbol{r}}\cdot(\mathsf{p}'\cdot\boldsymbol{c}_0). \tag{16.71, 9}$$

* In (16.71, 5) (and (16.71, 6, 7, 9)) the molecule 1 is taken to be at the point $\boldsymbol{r}$, and to possess velocity $\boldsymbol{c}$.

The equation of transfer of momentum is obtained by multiplying by $m\mathbf{c}_1$ the equation for $f^{(1)}$ ((16.7, 6), with $s = 1$) and integrating with respect to $\mathbf{c}_1$. The energy equation is found by multiplying the equation for $f^{(1)}$ by $\frac{1}{2}mc_1^2$ and integrating with respect to $\mathbf{c}_1$, and adding the equation got by multiplying the equation of continuity (16.7, 3) for $n^{(2)}$ by $\frac{1}{2}V_{12}$ and integrating with respect to $\mathbf{r}_2$. Using (16.71, 1–8), the momentum and energy equations can be reduced to the usual forms

$$\rho\frac{D\mathbf{c}_0}{Dt} = \rho\mathbf{F} - \frac{\partial}{\partial\mathbf{r}}\cdot\mathsf{p}, \quad n\frac{d\bar{E}}{dt} = -\frac{\partial}{\partial\mathbf{r}}\cdot\mathbf{q} - \mathsf{p}:\frac{\partial}{\partial\mathbf{r}}\mathbf{c}_0. \qquad (16.71, 10)$$

The expressions p', $\mathbf{q}'$ correspond to the collisional transport of momentum and energy in Enskog's theory. The second term on the right of (16.71, 6), however, has no analogue in Enskog's work. For a gas in a uniform steady state, (16.71, 4) reduces to the expression given by the virial theorem.

16.72. The uniform steady state

In the uniform steady state under no forces, the BBGKY equations (16.7, 6) are satisfied by*

$$f^{(s)} = n^{(s)}\left(\frac{m}{2\pi kT}\right)^{3s/2}\exp\left\{-\frac{m}{2kT}\sum_{i=1}^{s}C_s^2\right\}, \qquad (16.72, 1)$$

provided that, for all $i \leqslant s$,

$$\frac{\partial n^{(s)}}{\partial\mathbf{r}_i} - \frac{m}{kT}n^{(s)}\sum_{j=1}^{s}{}'\,\mathbf{X}_{ij} = \frac{m}{kT}\int\mathbf{X}_{i,s+1}n^{(s+1)}\,d\mathbf{r}_{s+1}. \qquad (16.72, 2)$$

Since $m\mathbf{X}_{ij} = -\partial V_{ij}/\partial\mathbf{r}_i \equiv -\nabla_i V_{ij}$, we write

$$\chi_s(q) \equiv \exp\left\{-\sum_{i=1}^{s}V_{qi}/kT\right\} \quad (q > s), \qquad (16.72, 3)$$

$$n^{(s)} \equiv n^s\chi_1(2)\,\chi_2(3)\ldots\chi_{s-1}(s)\,N^{(s)}. \qquad (16.72, 4)$$

Then (16.72, 2) becomes

$$\nabla_i N^{(s)} = n\int N^{(s+1)}\nabla_i\chi_s(s+1)\,d\mathbf{r}_{s+1}. \qquad (16.72, 5)$$

This equation has to be solved subject to the boundary condition that, when the molecules $1, 2, \ldots, s$ are all sufficiently far apart, $n^{(s)} = n^s$ and so $N^{(s)} = 1$ (since then $V_{ij} = 0$, $i, j = 1, 2, \ldots, s$).

If the gas density is not too large, (16.72, 5) can be solved by successive approximations. The first approximation is got by neglecting the right-hand side; this gives $N^{(s)} = \text{const.} = 1$, using the boundary condition. The second approximation is then got by substituting $N^{(s+1)} = 1$ on the right of (16.72, 5) and solving for $N^{(s)}$: the third approximation by substituting the second

* No complete proof of the uniqueness of this solution has been given, though a number of partial proofs exist.

approximation for $N^{(s+1)}$, and again solving: and so on. Thus, for example, writing

$$z_{ij} \equiv \exp(-V_{ij}/kT) - 1 \tag{16.72, 6}$$

it can be shown that

$$\begin{aligned} n^{(2)} = n^2(1+z_{12})\{1 + n\int z_{13}z_{32}\,d\boldsymbol{r}_3 \\ + \tfrac{1}{2}n^2\iint [z_{13}z_{34}z_{42}(2+2z_{14}+2z_{23}+z_{14}z_{23}) \\ + z_{13}z_{23}z_{14}z_{24}]\,d\boldsymbol{r}_3\,d\boldsymbol{r}_4 + \ldots\}. \end{aligned} \tag{16.72, 7}$$

Using (16.71, 4, 6), the contributions p_0', $\bar{E}'$ from the intermolecular forces to the hydrostatic pressure and the mean heat energy per molecule in the uniform steady state can be shown to be

$$p_0' = -kn^2T\frac{\partial\Psi}{\partial n}, \quad \bar{E}' = kT^2\frac{\partial\Psi}{\partial T}, \tag{16.72, 8}$$

where*

$$\begin{aligned} \Psi \equiv \tfrac{1}{2}n\int z_{12}\,d\boldsymbol{r}_2 + \tfrac{1}{6}n^2\iint z_{12}z_{23}z_{31}\,d\boldsymbol{r}_2\,d\boldsymbol{r}_3 \\ + \tfrac{1}{24}n^3\iiint z_{12}z_{23}z_{34}z_{41}(3+6z_{13}+z_{13}z_{24})\,d\boldsymbol{r}_2\,d\boldsymbol{r}_3\,d\boldsymbol{r}_4 + \ldots. \end{aligned} \tag{16.72, 9}$$

The expression $-n\,\partial\Psi/\partial n$ in the equation for p_0' is a series of powers of the density, equivalent to the expression $\mathrm{b}\rho\chi$ in (16.33, 2). If we write

$$p_0' = knT(\rho \mathrm{B}(T) + \rho^2\mathrm{C}(T) + \ldots) \tag{16.72, 10}$$

(cf. (16.5, 7)), the coefficients B, C, ... give the second, third, ... virial coefficients. The coefficients arise respectively from the interactions of molecules in groups of two, three, ...

16.73. The transport phenomena

For a gas not in a uniform steady state, the local temperature T is defined as in 2.41, i.e. as the temperature at which the same gas, when in a uniform steady state at the same density, would have the same mean energy $\bar{E}$ per molecule.

For simplicity we consider a gas subject to no external forces ($\boldsymbol{F} = 0$). In order to determine the transport coefficients, we need to solve the BBGKY equations for $f^{(s)}$, correct to terms of first order in the space gradients of T and $\boldsymbol{c}_0$. Bogoliubov proposed the following generalization of the method of successive approximations used in 16.72 to determine $n^{(s)}$ in the uniform steady state.

A first approximation to $f^{(s)}$, where $s > 1$, is obtained from (16.7, 6) by neglecting the integral on the right. The equation then asserts that $f^{(s)}$ is

* The results (16.72, 8, 9) are usually derived by a rather different method, depending on ideas of statistical mechanics. See, e.g., J. E. Mayer and M. G. Mayer, *Statistical Mechanics*, chapter 13 (Wiley, 1940).

constant along a trajectory of the s molecules as they interact; since their velocities were uncorrelated before they began to interact, we may write to this approximation

$$f^{(s)} = f^{(1)}(\boldsymbol{c}_1', \boldsymbol{r}_1') f^{(1)}(\boldsymbol{c}_2', \boldsymbol{r}_2') \dots f^{(1)}(\boldsymbol{c}_s', \boldsymbol{r}_s'). \qquad (16.73, 1)$$

However, we cannot simply assert that $\boldsymbol{c}_1'$, $\boldsymbol{r}_1'$, etc., are the velocities and positions 'before encounter', since the precise position is important, and 'before encounter' has no precise meaning. Bogoliubov assumed that $\boldsymbol{c}_1'$, $\boldsymbol{r}_1'$ could be obtained by first following the motion backward for a time τ, until an instant is reached before the molecules began to interact, and then following the motion forward with constant velocity for the same time τ. Then the first approximation to $f^{(s)}$ is known in terms of $f^{(1)}$.

When the first approximation has been found, a second approximation to $f^{(s)}$ is got by substituting the first approximation to $f^{(s+1)}$ on the right of (16.7, 6) and solving for $f^{(s)}$. The equation now indicates how fast $f^{(s)}$ varies along a trajectory as a consequence of $(s+1)$-molecule interactions; once more it is integrated using the boundary condition (16.73, 1), with $\boldsymbol{c}_1'$, $\boldsymbol{r}_1'$, etc., obtained by the same process of backward and forward projection of the s-molecule motion. Further approximations are similarly derived. Thus each $f^{(s)}$ $(s > 1)$ is in principle determinate to any order of approximation in terms of $f^{(1)}$, the successive approximations introducing terms depending respectively on the encounters of s, $s+1$, $s+2$, ... molecules. When $f^{(2)}$ is known in terms of $f^{(1)}$ to any order of approximation, $f^{(1)}$ can in principle be determined to a similar order from the first of equations (16.7, 6).

Essentially, Bogoliubov's approach replaces the first of equations (16.7, 6) by an equation of the form

$$\frac{\partial f^{(1)}}{\partial t} + \boldsymbol{c} \cdot \frac{\partial f^{(1)}}{\partial \boldsymbol{r}} = J_2(f^{(1)} f^{(1)}) + J_3(f^{(1)} f^{(1)} f^{(1)}) + \dots, \qquad (16.73, 2)$$

where J_s is a functional depending on products of s functions $f^{(1)}$ (with different arguments). In a first approximation we retain only J_2, which now reduces to $J_2^{(0)}$, the Boltzmann collision-integral for $\partial_e f/\partial t$. In the second approximation we retain J_3, which represents the effect of ternary encounters, to a first approximation $J_3^{(0)}$; also J_2 becomes $J_2^{(0)} + J_2^{(1)}$, where $J_2^{(1)}$ represents the effect of differences of position of molecules undergoing binary encounters (cf. 16.3, 2). We may write

$$f^{(1)} = f^{(1,0)} + f^{(1,1)}, \qquad (16.73, 3)$$

where $f^{(1,0)}$ is the usual Maxwellian function, and $f^{(1,1)}$ is linear in the gradients of T and $\boldsymbol{c}_0$. Correct to terms of first order in these gradients, we can replace $f^{(1)}$ on the left of (16.73, 2) and in $J_2^{(1)}$ by $f^{(1,0)}$, and neglect terms involving products of functions $f^{(1,1)}$. Hence (16.73, 2) becomes, to the second approximation,

$$\frac{\partial_0 f^{(1,0)}}{\partial t}+\boldsymbol{c}\cdot\frac{\partial f^{(1,0)}}{\partial \boldsymbol{r}} = J_2^{(0)}(f^{(1,1)}f^{(1,0)})+J_2^{(0)}(f^{(1,0)}f^{(1,1)})$$
$$+J_2^{(1)}(f^{(1,0)}f^{(1,0)})+J_3^{(0)}(f^{(1,1)}f^{(1,0)}f^{(1,0)})$$
$$+J_3^{(0)}(f^{(1,0)}f^{(1,1)}f^{(1,0)})+J_3^{(0)}(f^{(1,0)}f^{(1,0)}f^{(1,1)}), \quad (16.73, 4)$$

since $J_2^{(0)}$, $J_3^{(0)}$ vanish if $f^{(1)} = f^{(1,0)}$. This equation may be compared with (16.31, 3), (16.32, 3, 4); the terms in $J_2^{(1)}$ in (16.73, 4) correspond to those appearing in (16.32, 4), taking $\chi = 1$.

Further approximations to (16.73, 2) can be similarly obtained. However, most workers have not proceeded beyond (16.73, 4). This is not only because of difficulties in discussing many-body encounters; there is also a doubt whether the later approximations are correct. Indeed, (16.73, 4) itself may not be exact.

The expression $J_3^{(0)}$ in (16.73, 4) represents all types of ternary effects, including (cf. 16.21) the effect of one molecule being shielded from colliding with a second by the intervention of a third. Also, in view of uncertainty where exactly an encounter begins and ends, and of the degree of correlation of the velocities of two molecules after encounter, it is necessary to regard as a ternary encounter three successive binary encounters of three molecules 1, 2, 3, e.g. the successions (1, 2)–(2, 3)–(1, 2) or (1, 2)–(2, 3)–(1, 3). In such successions, the later encounters may occur at points well separated from the first. In practice, such distant successions are ruled out because at least one of the molecules would collide with a fourth molecule before the succession is complete; but this correction cannot be made until one introduces J_4. The correction is not large, since three successive binary encounters involving three molecules are very unlikely when the distance between them is large. But when one considers successions of four binary encounters the case is altered; distant encounters become so numerous that the integral giving J_4 is found to diverge. Clearly this divergence does not correspond to any physical reality, but its removal poses a question of considerable difficulty.*

The difficulties of a discussion of three body encounters have so far prevented the derivation of exact formulae permitting a detailed comparison with experiment. For rigid elastic spheres, results equivalent to those of Enskog can be obtained from (16.73, 4), but only by making additional assumptions.† Sengers‡ has found that, if no such assumptions are made,

* The first discussion of equation (16.73, 4) was given by S. T. Choh and G. E. Uhlenbeck, *The Kinetic Theory of Dense Gases* (Univ. of Michigan Report, 1958). See also M. S. Green, *J. Chem. Phys.* **25**, 836 (1956); E. G. D. Cohen, *Physica*, **28**, 1025, 1045, 1060 (1962); L. S. García-Colín, M. S. Green and F. Chaos, *Physica*, **32**, 450 (1966); J. R. Dorfman and E. G. D. Cohen, *J. Math. Phys.* **8**, 282 (1967).

† J. T. O'Toole and J. S. Dahler, *J. Chem. Phys.* **32**, 1097 (1960); R. F. Snider and C. F. Curtiss, *Phys. Fluids*, **3**, 903 (1960).

‡ J. V. Sengers, *Phys. Fluids*, **9**, 1333 (1966).

the results derived from (16.73, 4) may differ appreciably from those given by Enskog's theory, retaining in that theory only the terms linear in $b\rho$.

Curtiss and his co-workers* have discussed more general molecular models by a method based on (16.73, 4), but making an extra assumption similar to that made in deducing Enskog's theory. Even after this assumption, the results obtained are extremely complicated, and they are not given here. The authors state that they give satisfactory agreement with experiment.

16.8. The evaluation of certain integrals

Certain results of integration with respect to **k**, assumed earlier in this chapter, will be proved here.

Let **h**, **i**, **j** denote three mutually perpendicular unit vectors, **h** being in the direction of $\boldsymbol{g}$. Let θ, φ be the polar angles of **k** with respect to **h** as axis and the plane of **h** and **i** as initial plane; then

$$\mathbf{k} = \mathbf{h}\cos\theta + \mathbf{i}\sin\theta\cos\varphi + \mathbf{j}\sin\theta\sin\varphi, \qquad (16.8,\ 1)$$

so that $\boldsymbol{g}.\mathbf{k} = g\cos\theta$, and the element of solid angle $d\mathbf{k}$ can be expressed in the form

$$d\mathbf{k} = \sin\theta\, d\theta\, d\varphi.$$

In all the integrals to be evaluated the integration is over all values of **k** for which $\boldsymbol{g}.\mathbf{k}$ is positive. Thus the limits of integration are, for θ, 0 and $\pi/2$, and for φ, 0 and 2π, and so all terms of the integrand containing odd powers of $\sin\varphi$ or $\cos\varphi$ may be neglected. Hence in particular

$$\int \mathbf{k}(\boldsymbol{g}.\mathbf{k})\,d\mathbf{k} = \iint \mathbf{h}\cos\theta . g\cos\theta . \sin\theta\, d\theta\, d\varphi$$

$$= \frac{2\pi}{3} g\mathbf{h} = \frac{2\pi}{3}\boldsymbol{g}. \qquad (16.8,\ 2)$$

Again,

$$\int \mathbf{kk}(\boldsymbol{g}.\mathbf{k})^2 d\mathbf{k} = \iint (\mathbf{hh}\cos^2\theta + \mathbf{ii}\sin^2\theta\cos^2\varphi + \mathbf{jj}\sin^2\theta\sin^2\varphi) g^2\cos^2\theta\sin\theta\, d\theta\, d\varphi$$

$$= \frac{2\pi}{15} g^2(3\mathbf{hh} + \mathbf{ii} + \mathbf{jj})$$

$$= \frac{2\pi}{15} g^2(2\mathbf{hh} + \mathsf{U}) = \frac{2\pi}{15}(2\boldsymbol{gg} + g^2\mathsf{U}), \qquad (16.8,\ 3)$$

using (1.3, 9).

* R. F. Snider and C. F. Curtiss, *Phys. Fluids*, **1**, 122 (1958), and **3**, 903 (1960); R. F. Snider and F. R. McCourt, *Phys. Fluids*, **6**, 1020 (1963); D. K. Hoffmann and C. F. Curtiss, *Phys. Fluids*, **7**, 1887 (1964), and **8**, 667 (1965).

Also, if $\mathbf{v}$ is any vector independent of θ, φ,

$$\int \mathbf{kk}(\mathbf{v}.\mathbf{k})(\boldsymbol{g}.\mathbf{k})^2 d\mathbf{k} = \int \mathbf{kk}(\mathbf{v}.\mathbf{h}\cos\theta + \mathbf{v}.\mathbf{i}\sin\theta\cos\varphi + \mathbf{v}.\mathbf{j}\sin\theta\sin\varphi)(\boldsymbol{g}.\mathbf{k})^2 d\mathbf{k}$$

$$= \iint \{\mathbf{v}.\mathbf{h}\cos\theta(\mathbf{hh}\cos^2\theta + \mathbf{ii}\sin^2\theta\cos^2\varphi + \mathbf{jj}\sin^2\theta\sin^2\varphi) + \mathbf{v}.\mathbf{i}(\mathbf{hi}+\mathbf{ih})\cos\theta\sin^2\theta\cos^2\varphi + \mathbf{v}.\mathbf{j}(\mathbf{hj}+\mathbf{jh})\cos\theta\sin^2\theta\sin^2\varphi\} g^2\cos^2\theta\sin\theta\, d\theta\, d\varphi$$

$$= \frac{\pi}{12} g^2\{\mathbf{v}.\mathbf{h}(4\mathbf{hh}+\mathbf{ii}+\mathbf{jj}) + \mathbf{v}.\mathbf{i}(\mathbf{hi}+\mathbf{ih}) + \mathbf{v}.\mathbf{j}(\mathbf{hj}+\mathbf{jh})\}$$

$$= \frac{\pi}{12} g^2\{\mathbf{v}.\mathbf{h}(\mathbf{hh}+\mathsf{U}) + \mathbf{v}.(\mathbf{hh}+\mathbf{ii}+\mathbf{jj})\mathbf{h} + \mathbf{h}(\mathbf{hh}+\mathbf{ii}+\mathbf{jj}).\mathbf{v}\}$$

$$= \frac{\pi}{12} g^2\{\mathbf{v}.\mathbf{h}(\mathbf{hh}+\mathsf{U}) + \mathbf{vh} + \mathbf{hv}\}$$

$$= \frac{\pi}{12}\{\mathbf{v}.\boldsymbol{g}(\boldsymbol{gg}+g^2\mathsf{U})/g + g(\mathbf{v}\boldsymbol{g}+\boldsymbol{g}\mathbf{v})\}. \tag{16.8, 4}$$

Now, by (16.3, 3),

$$\boldsymbol{C}' = \boldsymbol{C} + \mathbf{k}(\mathbf{k}.\boldsymbol{g}), \quad \boldsymbol{C}_1' = \boldsymbol{C}_1 - \mathbf{k}(\mathbf{k}.\boldsymbol{g}).$$

Hence
$$C'^2 = \{\boldsymbol{C}+\mathbf{k}(\mathbf{k}.\boldsymbol{g})\}.\{\boldsymbol{C}+\mathbf{k}(\mathbf{k}.\boldsymbol{g})\}$$
$$= C^2 + 2(\mathbf{k}.\boldsymbol{C})(\mathbf{k}.\boldsymbol{g}) + (\mathbf{k}.\boldsymbol{g})^2,$$

and similarly
$$C_1'^2 = C_1^2 - 2(\mathbf{k}.\boldsymbol{C}_1)(\mathbf{k}.\boldsymbol{g}) + (\mathbf{k}.\boldsymbol{g})^2.$$

Thus, using (16.8, 2–4),

$$\int \mathbf{k}(C_1^2 + C_1'^2)\boldsymbol{g}.\mathbf{k}\, d\mathbf{k} = \int \mathbf{k}\{2C_1^2 - 2(\mathbf{k}.\boldsymbol{C}_1)(\mathbf{k}.\boldsymbol{g}) + (\mathbf{k}.\boldsymbol{g})^2\}(\boldsymbol{g}.\mathbf{k})\, d\mathbf{k}$$

$$= \frac{2\pi}{3}[2C_1^2\boldsymbol{g} - \tfrac{2}{5}\{2\boldsymbol{g}(\boldsymbol{g}.\boldsymbol{C}_1) + g^2\boldsymbol{C}_1\} + \tfrac{3}{5}g^2\boldsymbol{g}], \tag{16.8, 5}$$

$$\int \mathbf{k}(\boldsymbol{C}_1 + \boldsymbol{C}_1')\boldsymbol{g}.\mathbf{k}\, d\mathbf{k} = \int \mathbf{k}\{2\boldsymbol{C}_1 - \mathbf{k}(\mathbf{k}.\boldsymbol{g})\}(\boldsymbol{g}.\mathbf{k})\, d\mathbf{k}$$

$$= \frac{2\pi}{3}\{2\boldsymbol{C}_1\boldsymbol{g} - \tfrac{1}{5}(2\boldsymbol{gg} + g^2\mathsf{U})\}, \tag{16.8, 6}$$

$$\int (\boldsymbol{C}' - \boldsymbol{C}) \mathbf{k} (\boldsymbol{g} . \mathbf{k}) \, d\mathbf{k} = \int \mathbf{k}\mathbf{k} (\boldsymbol{g} . \mathbf{k})^2 \, d\mathbf{k}$$

$$= \frac{2\pi}{15} (2\boldsymbol{g}\boldsymbol{g} + g^2 \mathsf{U}), \tag{16.8, 7}$$

$$\int (\boldsymbol{C}' - \boldsymbol{C}) \mathbf{k} (\mathbf{k} . \boldsymbol{\nu}) (\boldsymbol{g} . \mathbf{k}) \, d\mathbf{k} = \int \mathbf{k}\mathbf{k} (\boldsymbol{\nu} . \mathbf{k}) (\boldsymbol{g} . \mathbf{k})^2 \, d\mathbf{k}$$

$$= \frac{\pi}{12} \{\boldsymbol{\nu} . \boldsymbol{g} (\boldsymbol{g}\boldsymbol{g} + g^2 \mathsf{U})/g + g(\boldsymbol{\nu}\boldsymbol{g} + \boldsymbol{g}\boldsymbol{\nu})\}, \tag{16.8, 8}$$

$$\int (C'^2 - C^2) \mathbf{k} (\boldsymbol{g} . \mathbf{k}) \, d\mathbf{k} = \int \mathbf{k} . (\boldsymbol{g} + 2\boldsymbol{C}) \mathbf{k} (\boldsymbol{g} . \mathbf{k})^2 \, d\mathbf{k}$$

$$= 2 \int (\mathbf{k} . \boldsymbol{G}_0) \, \mathbf{k} (\boldsymbol{g} . \mathbf{k})^2 \, d\mathbf{k}$$

$$= \frac{4\pi}{15} \{2\boldsymbol{g}(\boldsymbol{g} . \boldsymbol{G}_0) + g^2 \boldsymbol{G}_0\}, \tag{16.8, 9}$$

$$\int (C'^2 - C^2) \mathbf{k}\mathbf{k} (\boldsymbol{g} . \mathbf{k}) \, d\mathbf{k} = 2 \int (\mathbf{k} . \boldsymbol{G}_0) \, \mathbf{k}\mathbf{k} (\boldsymbol{g} . \mathbf{k})^2 \, d\mathbf{k}$$

$$= \frac{\pi}{6} \{(\boldsymbol{G}_0 . \boldsymbol{g}) (\boldsymbol{g}\boldsymbol{g} + g^2 \mathsf{U})/g + g(\boldsymbol{g}\boldsymbol{G}_0 + \boldsymbol{G}_0 \boldsymbol{g})\}. \tag{16.8, 10}$$

17

QUANTUM THEORY AND THE TRANSPORT PHENOMENA

THE QUANTUM THEORY OF MOLECULAR COLLISIONS

17.1. The wave fields of molecules

When quantum methods are applied to molecular encounters some divergence from the classical results is found. The deflection of the relative velocity $\boldsymbol{g}$, resulting from the encounter of a pair of molecules of masses m_1 and m_2, is much the same as if, on the classical theory, each molecule were supposed surrounded by a 'wave' field whose linear extension is of the order of the 'wave-length'

$$h(m_1+m_2)/2\pi m_1 m_2 g,$$

h being Planck's constant, equal to $6{\cdot}626\times 10^{-27}$ g. cm.^{2}/sec. Thus, for example, rigid spherical molecules behave as if such wave fields produce a deflection even when the spheres do not actually collide; in general, the effective diameters of rigid spherical molecules are larger according to quantum theory than according to classical theory. For more realistic molecular models, on the other hand, the wave fields may decrease the effective diameters. The quantum effects are appreciable if the extension of the wave fields is not small compared with the molecular collision-distance σ_{12}; that is, using a mean value for g, if

$$h(m_1+m_2)^{\frac{1}{2}}/2\pi\sigma_{12}(m_1 m_2 kT)^{\frac{1}{2}} \qquad (17.1, 1)$$

is not small compared with unity. This quantity is largest for light molecules and low temperatures; for helium at 0 °C. it is 0·13, and for hydrogen it is somewhat larger; hence for these two gases the quantum correction is appreciable even at ordinary temperatures, and becomes very important at low temperatures. Molecules of other gases have much larger masses, and their radii are in most cases larger; moreover, it is not necessary to consider such low temperatures for the other gases, because they liquefy more readily. In consequence, the modifications introduced by the use of quantum methods are relatively unimportant except for hydrogen and helium.

A second modification of the classical theory may be required at very low temperatures, when the wave fields associated with the molecules may increase the effective size of the molecules, so that there may be a state of congestion similar to that considered in the last chapter, even when the total volume of the molecules is small compared with that of the containing vessel.

A gas in such a state of congestion is said to be degenerate. In ordinary gases the congestion is very slight even at the lowest temperatures considered, but in an electron gas, such as is encountered in a metal or in a dense star, the congestion is extreme.

Naturally it is not possible in this book to give a full account of the relevant sections of quantum theory; we shall simply indicate its applications to our subject.

17.2. Interaction of two molecular streams

It is necessary first to put certain of the earlier results into forms suitable for use in connection with the quantum theory of collisions.* The uncertainty principle states that the precise position and velocity of a molecule cannot be determined simultaneously. It is therefore not possible to determine exactly the deflection χ of the relative velocity at the encounter of two molecules m_1, m_2 of velocities $\boldsymbol{c}_1$, $\boldsymbol{c}_2$. It is, however, possible, when the set of molecules m_1 with velocities in the range $\boldsymbol{c}_1$, $d\boldsymbol{c}_1$ streams through the set of molecules m_2 with velocities in the range $\boldsymbol{c}_2$, $d\boldsymbol{c}_2$, to determine the probable number of encounters per unit volume and time, such that the unit vector $\mathbf{e}'$ in the direction of the relative velocity $\boldsymbol{g}'_{21}$ after collision lies in the solid angle $d\mathbf{e}'$. This number is proportional to the number-densities $f_1 d\boldsymbol{c}_1$, $f_2 d\boldsymbol{c}_2$ of the two sets, and depends on χ and on the magnitude g of the relative velocity. If it is written as

$$f_1 f_2 g\, \alpha_{12}(g, \chi)\, d\mathbf{e}'\, d\boldsymbol{c}_1\, d\boldsymbol{c}_2 \tag{17.2, 1}$$

the quantity α_{12} is identical with that introduced in 3.5, save that the variables on which it depends are now g and χ instead of g and b. Since in 3.5

$$b\,db\,d\epsilon = \alpha_{12}\,d\mathbf{e}' = \alpha_{12} \sin\chi\, d\chi\, d\epsilon$$

the earlier theory has simply to be modified by replacing $b\,db$ by

$$\alpha_{12}(g, \chi) \sin\chi\, d\chi$$

in all integrals involving collision variables.

17.3. The distribution of molecular deflections

We proceed to obtain an expression for $\alpha_{12}(g, \chi)$.† Let the mutual potential energy of a pair of molecules m_1, m_2 at a distance r be V. Then the wave equation governing the relative motion when a molecule m_1 of velocity $\boldsymbol{c}_1$ encounters a molecule m_2 of velocity $\boldsymbol{c}_2$ is the same as that governing the motion of a molecule of mass $m_1 m_2/(m_1 + m_2)$ and initial velocity $\boldsymbol{g}_{21}$ under

* This modification was first made in gas-theory by E. A. Uehling and G. E. Uhlenbeck, *Phys. Rev.* **43**, 552 (1933), and H. S. W. Massey and C. B. O. Mohr, *Proc. R. Soc.* A, **141**, 434 (1933). Similar changes are, however, implicit in earlier discussions of electron-theory.

† The results of this section are due to H. Faxén and J. Holtsmark, *Z. Phys.* **45**, 307 (1927).

the action of a stationary centre of force whose potential energy is V at a distance r. Suppose that the force-centre is at the origin, and that a beam of molecules of velocity g and of unit number-density is incident on it in the direction of Oz. Then the number of scattered molecules crossing a large sphere whose centre is the origin, with velocities whose directions lie in a small solid angle $\mathbf{e}'$, de', is $g\alpha_{12}(g, \chi)\, de'$ per unit time, where χ is the angle between $\mathbf{e}'$ and Oz.

The wave equation governing the motion is

$$\frac{h^2 m_0}{8\pi^2 m_1 m_2}\nabla^2\psi + \left(\frac{m_1 m_2 g^2}{2m_0} - V\right)\psi = 0, \qquad (17.3, 1)$$

where h is Planck's constant and $m_0 = m_1 + m_2$. In this we write

$$j \equiv \frac{2\pi m_1 m_2 g}{m_0 h}, \qquad (17.3, 2)$$

when it reduces to

$$\nabla^2\psi + \left(j^2 - \frac{8\pi^2 m_1 m_2 V}{h^2 m_0}\right)\psi = 0. \qquad (17.3, 3)$$

The solution we require, representing the beam of particles scattered by the force-centre, is symmetrical about Oz and finite at the origin. If $z = r\cos\theta$, one solution possessing these properties is

$$\psi = u_n(r)P_n(\cos\theta)/r, \qquad (17.3, 4)$$

where $P_n(\cos\theta)$ is the Legendre polynomial of order n, and $u_n(r)$ is a real function of r, satisfying the equation

$$\frac{d^2u_n}{dr^2} + \left(j^2 - \frac{8\pi^2 m_1 m_2 V}{h^2 m_0} - \frac{n(n+1)}{r^2}\right)u_n = 0, \qquad (17.3, 5)$$

and such that $u_n(0) = 0$. These conditions determine u_n save for an arbitrary factor; if this is suitably adjusted, the asymptotic expression for u_n corresponding to large values of r is

$$u_n \sim \sin(jr - \tfrac{1}{2}n\pi + \delta_n), \qquad (17.3, 6)$$

where δ_n is a constant depending on the form of the function V, and on the values of j and n.

The solution of (17.3, 3) appropriate to the present problem is a combination of solutions of the type (17.3, 4), say

$$\psi = \frac{1}{r}\sum_{n=0}^{\infty}\eta_n u_n(r)P_n(\cos\theta), \qquad (17.3, 7)$$

where the quantities η_n are constants, real or complex. This may be divided into two parts, ψ' and ψ''; the first represents the incident beam, of unit number-density, advancing parallel to Oz; for this we have

$$\psi' = e^{\iota jz} = e^{\iota jr\cos\theta} \quad (\iota = \sqrt{-1}), \qquad (17.3, 8)$$

which, as can readily be seen, is asymptotic to a solution of (17.3, 3) for large r. Using known results in the theory of Bessel functions,* this becomes

$$\psi' = \sum_{n=0}^{\infty} (2n+1)\iota^n \sqrt{\left(\frac{\pi}{2jr}\right)} J_{n+\frac{1}{2}}(jr) P_n(\cos\theta)$$
$$= \sum_{n=0}^{\infty} (2n+1)(-2jr\iota)^n \left(\frac{d}{d(j^2r^2)}\right)^n \left(\frac{\sin jr}{jr}\right) P_n(\cos\theta),$$

and so, for large values of r,

$$\psi' \sim \frac{1}{jr}\sum_{n=0}^{\infty} (2n+1)\iota^n \sin(jr - \tfrac{1}{2}n\pi) P_n(\cos\theta). \qquad (17.3, 9)$$

Since ψ is a solution of (17.3, 1), and ψ' is asymptotic to a solution, the remaining part ψ'' of ψ must also be asymptotic to a solution. It represents the scattered molecules, which at a large distance r from the origin are moving radially outward; hence it must be such that $\partial\psi''/\partial r \sim \iota j\psi''$ when r is large. These conditions are satisfied if

$$\psi'' \sim \frac{1}{r} e^{\iota jr} \sum_{n=0}^{\infty} \eta''_n P_n(\cos\theta), \qquad (17.3, 10)$$

where the quantities η''_n are constants.

Combining the results (17.3, 6, 7, 9, 10) we have

$$\sum_{n=0}^{\infty} \{(2n+1)\iota^n \sin(jr - \tfrac{1}{2}n\pi)/jr + \eta''_n e^{\iota jr}/r\} P_n(\cos\theta)$$
$$= \frac{1}{r}\sum_{n=0}^{\infty} \eta_n \sin(jr - \tfrac{1}{2}n\pi + \delta_n) P_n(\cos\theta),$$

whence, equating the coefficients of $P_n(\cos\theta)$, we obtain

$$(2n+1)\iota^n \sin(jr - \tfrac{1}{2}n\pi)/j + \eta''_n e^{\iota jr} = \eta_n \sin(jr - \tfrac{1}{2}n\pi + \delta_n).$$

This equation is an identity in r. Suppose that

$$jr - \tfrac{1}{2}n\pi + \delta_n = 0.$$

Then it is found that

$$\eta''_n = \frac{1}{2j\iota}(2n+1)(e^{2\iota\delta_n} - 1),$$

and so (17.3, 10) becomes

$$\psi'' \sim \frac{e^{\iota jr}}{2\iota jr}\sum_{n=0}^{\infty} (2n+1)(e^{2\iota\delta_n} - 1) P_n(\cos\theta).$$

The number-density of scattered molecules is $|\psi''|^2$. At a large distance from the origin these are moving radially outward with velocity g. Hence the number per unit time which cross a sphere of large radius r with velocities

* G. N. Watson, *Bessel Functions*, pp. 56, 128, 368 (Cambridge, 1944).

whose directions lie in a solid angle $\mathbf{e}'$, de' is $|\psi''|^2 gr^2 de'$. Since this is the same as $g\alpha_{12} de'$,

$$\alpha_{12}(g,\chi) = \frac{1}{4j^2}\left|\sum_{n=0}^{\infty}(2n+1)(e^{2\iota\delta_n}-1)P_n(\cos\chi)\right|^2. \qquad (17.3, 11)$$

This expression gives the distribution of scattered molecules when encounters of unlike molecules are considered. If the molecules are identical, their interchangeability requires the replacement of (17.3, 8) by

$$\psi' = 2^{-\frac{1}{2}}(e^{\iota jz} \pm e^{-\iota jz}). \qquad (17.3, 12)$$

This corresponds to superposed beams moving with velocities g, $-g$ parallel to Oz; the numerical factor is chosen so as to make the average number-density unity. The upper and lower signs in (17.3, 12) correspond respectively to Bose–Einstein and Fermi–Dirac molecules, i.e. molecules composed respectively of even and odd numbers of elementary particles (electrons, protons and neutrons). From (17.3, 12),

$$\psi' = 2^{\frac{1}{2}}\Sigma'(2n+1)(-2jr\iota)^n\left(\frac{d}{d(j^2r^2)}\right)^n\frac{\sin jr}{jr}P_n(\cos\theta),$$

where the sum Σ' is one taken over even values of n (including zero) or odd values, according as Bose–Einstein or Fermi–Dirac molecules are considered. We must therefore replace (17.3, 11) by

$$\alpha_1(g,\chi) = \frac{1}{2j^2}\left|\Sigma'(2n+1)(e^{2\iota\delta_n}-1)P_n(\cos\chi)\right|^2. \qquad (17.3, 13)$$

According to quantum theory,* molecules of the same substance cannot be regarded as identical unless they possess the same nuclear spins and the same rotational quantum numbers. In consequence, even in helium the formula for unlike molecules must be used for the collisions of two molecules of oppositely directed spins; and in the majority of collisions occurring in even a simple diatomic or polyatomic gas, the scattering probability is that for unlike molecules.

17.31. The collision-probability and mean free path

The total probability per unit time that a molecule of velocity $\mathbf{c}_1$, moving in a stream of unlike molecules of velocity $\mathbf{c}_2$ whose number-density is unity, should undergo a collision, is

$$\begin{aligned} g\int\alpha_{12}(g,\chi)\,de' &= 2\pi g\int_0^\pi \alpha_{12}(g,\chi)\sin\chi\,d\chi \\ &= \frac{\pi g}{2j^2}\int_{-1}^{1}\left|\sum_{n=0}^{\infty}(2n+1)(e^{2\iota\delta_n}-1)P_n(x)\right|^2 dx \\ &= \frac{4\pi g}{j^2}\sum_{n=0}^{\infty}(2n+1)\sin^2\delta_n, \end{aligned} \qquad (17.31, 1)$$

* O. Halpern and E. Gwathmey, *Phys. Rev.* 52, 944 (1937).

using results in the theory of Legendre polynomials.* It is found that this series converges (so that there is a finite probability of collision), when V tends to zero, as $r\to\infty$, like $r^{-\nu}$, where $\nu > 3$. Thus when the field satisfies this condition there is a natural definition of a free path in quantum theory, whereas on the classical theory we can define a free path, when the molecules are not rigid, only by making arbitrary assumptions as to the limits of the molecular fields. The molecules undergo as many collisions as if they had a definite joint collision-distance σ_{12} and collision cross-section Q given by

$$Q = \pi\sigma_{12}^2 = \frac{4\pi}{j^2}\sum_{n=0}^{\infty}(2n+1)\sin^2\delta_n. \qquad (17.31, 2)$$

The similar relation for identical molecules involves a sum Σ' over either even or odd values of n, multiplied by the factor $8\pi/j^2$.

The free path corresponding to the cross-section Q is, however, very different from that appropriate to viscosity and the other transport phenomena. Even on the quantum theory the majority of the 'collisions' are between relatively distant molecules, and result in comparatively small deflections of the relative velocity. The small attractive forces at large distances are found to affect the free path more than the larger repulsions at close encounters. Hence the persistence of velocities at collision is nearly unity; in fact the free path corresponding to (17.31, 2) is important only in experiments in which one can study the motion of individual molecules, and distinguish whether they undergo a deflection or not in a given distance. Such experiments have not been carried out with uncharged molecules, but measurements of free paths of electrons and positive ions in gases have been made. These give values for the collision-radius of molecules which are often many times as large as the values appropriate for viscosity, but which vary widely with the speed of the ions.

The first approximations to the coefficients of viscosity, thermal conduction and diffusion depend on the integrals $\phi_{12}^{(1)}$, $\phi_{12}^{(2)}$, now given by

$$\phi_{12}^{(l)} = 2\pi\int_0^\pi (1-\cos^l\chi)\,\alpha_{12}(g,\chi)\sin\chi\,d\chi$$

(cf. (9.33, 4)). Thus for unlike particles†

$$\phi_{12}^{(1)} = \frac{\pi}{2j^2}\int_{-1}^{1}(1-x)\left|\sum_{n=0}^{\infty}(2n+1)(e^{2i\delta_n}-1)P_n(x)\right|^2 dx$$

$$= \frac{4\pi}{j^2}\sum_{n=0}^{\infty}(n+1)\sin^2(\delta_{n+1}-\delta_n), \qquad (17.31, 3)$$

* See Massey and Mohr, *loc. cit.*

† The forms (17.31, 3, 4) (simpler than those originally given by Massey and Mohr) are due to H. A. Kramers; see J. de Boer and E. G. D. Cohen, *Physica*, 17, 993 (1951).

using the properties of Legendre polynomials. Similarly

$$\phi_{12}^{(2)} = \frac{4\pi}{j^2}\sum_{n=0}^{\infty}\frac{(n+1)(n+2)}{2n+3}\sin^2(\delta_{n+2}-\delta_n). \qquad (17.31, 4)$$

For identical molecules,*

$$\phi_1^{(2)} = \frac{8\pi}{j^2}\Sigma'\frac{(n+1)(n+2)}{2n+3}\sin^2(\delta_{n+2}-\delta_n) \qquad (17.31, 5)$$

(the sum Σ' being over even or odd values of n), with a corresponding expression for $\phi_1^{(1)}$.†

17.32. The phase-angles δ_n

It is in general not possible to give an exact expression for the phases δ_n. Approximate expressions have, however, been given by Mott‡ and Jeffreys§ for the cases in which δ_n is small or large compared with unity. It is found that δ_n is small if $8\pi^2 m_1 m_2 V/h^2(m_1+m_2)$ is small compared with $n(n+1)/r^2$ for all values of r such that jr is not small compared with n. In this case Mott has shown by a perturbation method that

$$\delta_n = \frac{4\pi^3 m_1 m_2}{h^2(m_1+m_2)}\int_0^\infty V(r)\{J_{n+\frac{1}{2}}(jr)\}^2 r\,dr \qquad (17.32, 1)$$

approximately. When δ_n is large, we can use Jeffreys's approximation for u_n, which is

$$u_n = \sin\left\{\frac{\pi}{4}+\int_{r_0}^{r}\{f(x)\}^{\frac{1}{2}}dx\right\},$$

where $$f(r) = j^2 - \frac{8\pi^2 m_1 m_2}{h^2(m_1+m_2)}V(r) - \frac{n(n+1)}{r^2},$$

and r_0 is the largest zero of $f(r)$; this approximation is valid if $f'/f^{\frac{3}{2}}$ is small in the range of integration. The corresponding value of δ_n is

$$\delta_n = \frac{n\pi}{2}+\frac{\pi}{4}-jr_0+\int_{r_0}^{\infty}\{[f(r)]^{\frac{1}{2}}-j\}\,dr. \qquad (17.32, 2)$$

* Massey and Mohr give an expression for the cross-section of molecules appropriate to self-diffusion; they find that, even at high temperatures, this cross-section is twice the classical one. This fact is, however, not of direct importance, since, on account of the interchangeability of the molecules, the two molecules lose their identity at an encounter; it is therefore not possible to trace the diffusion of a selected set of the molecules through the rest, i.e. it is impossible to measure self-diffusion. The closest we can get to self-diffusion is by considering the mutual diffusion of two sets of similar, but distinguishable, molecules.

† For an elegant proof that, in the limit when j is large, (17.31, 3–5) reduce to the corresponding classical expressions, see L. Waldmann, *Handbuch der Physik*, vol. 12, 460 (1958).

‡ N. F. Mott, *Proc. Camb. Phil. Soc.* **25**, 304 (1929).

§ H. Jeffreys, *Proc. Lond. Math. Soc.* **23**, 428 (1924). In books on quantum theory Jeffreys' expression is often known as the WKB (Wentzel–Kramers–Brillouin) approximation.

When δ_n is neither small nor large compared with unity, neither of the approximations (17.32, 1, 2) is valid. Normally only a few values of δ_n are near unity, and these can be found with tolerable accuracy by interpolation. More exact values can be obtained by using electronic computers.

For one model it is possible to obtain an explicit expression for δ_n. Suppose that the molecules are rigid elastic spheres of diameter σ. Then in the wave equation (17.3, 3) we have $V = 0$ if $r > \sigma$ and $V = \infty$ if $r < \sigma$. This is equivalent to taking u_n as the solution of the equation

$$\frac{d^2u_n}{dr^2} + \left(j^2 - \frac{n(n+1)}{r^2}\right)u_n = 0$$

such that $u_n = 0$ if $r = \sigma$. The solution is of the form

$$u_n \propto r^{\frac{1}{2}}\left\{\frac{J_{n+\frac{1}{2}}(jr)}{J_{n+\frac{1}{2}}(j\sigma)} - \frac{J_{-n-\frac{1}{2}}(jr)}{J_{-n-\frac{1}{2}}(j\sigma)}\right\}.$$

Since for large values of jr

$$J_{n+\frac{1}{2}}(jr) \sim \left(\frac{2}{\pi jr}\right)^{\frac{1}{2}} \sin\left(jr - \frac{n\pi}{2}\right), \quad J_{-n-\frac{1}{2}}(jr) \sim \left(\frac{2}{\pi jr}\right)^{\frac{1}{2}} \cos\left(jr + \frac{n\pi}{2}\right),$$

a comparison with (17.3, 6) shows that

$$\delta_n = \tan^{-1}\{(-1)^{n+1} J_{n+\frac{1}{2}}(j\sigma)/J_{-n-\frac{1}{2}}(j\sigma)\}. \qquad (17.32, 3)$$

In particular, $\delta_0 = -j\sigma$.

For more realistic models, numerical methods have to be used throughout. The values of δ_n, and hence of the integrals $\phi^{(l)}$, have first to be found for a variety of values of j, i.e. of g (cf. (17.3, 2)); the integrals $\Omega^{(l)}(r)$ are then determined by numerical integration.

17.4. Comparison with experiment for helium

As noted in 17.1, quantum corrections to the classical formulae for the transport coefficients are unimportant for gases other than hydrogen and helium. Detailed calculations by de Boer and Bird* show that for other gases the corrections do not in general exceed the experimental errors. Even for hydrogen and helium, Mason and Rice† estimate that the corrections to the viscosity do not exceed 0·6 per cent for helium above 200° K. and for hydrogen above 250° K.

Massey and Mohr‡ compared theory with experiment for helium and hydrogen, supposing the molecules to be rigid elastic spheres. As the general considerations of 17.1 suggest, the effect of quantum interaction is to increase the apparent radius of the molecules; the increase is fairly small at ordinary temperatures, but becomes large as the temperature decreases; as the

* J. de Boer and R. B. Bird, *Physica*, **20**, 185 (1954).
† E. A. Mason and W. E. Rice, *J. Chem. Phys.* **22**, 522 (1954).
‡ H. S. W. Massey and C. B. O. Mohr, *Proc. R. Soc.* A, **141**, 434 (1933).

temperature approaches the absolute zero, the apparent radius for viscosity is four times the actual radius. In Table 29 values calculated by the classical and quantum theories for rigid elastic spheres with suitably chosen radii are compared with the experimental values for helium.* The improved agreement with experiment at low temperatures as compared with classical theory is striking. However, this cannot be regarded as justifying the use of the rigid spherical model; the decrease in apparent molecular radius inferred from experiment on classical theory continues to high temperatures, where quantum effects are inappreciable.

Several alternative comparisons of quantum results with experiment have been made for helium, using the 12,6 and exp;6 models, and the full Buckingham (exp;8,6) model.† These refer both to normal helium (^{4}He) and to the ^{3}He isotope. On classical theory, assuming the same force-law for molecular interactions in the two, the viscosity of ^{3}He should be $\sqrt{(3/4)}$ times that of ^{4}He, the molecular masses being in the ratio 3/4. Quantum theory predicts deviations from this simple law, both because the smaller mass makes quantum effects in ^{3}He more important than in ^{4}He, and because ^{3}He and ^{4}He are respectively Fermi–Dirac and Bose–Einstein particles.

Let α_e and α_o denote the values of α_1 given by (17.3, 13), when the sum is taken over even and odd values of n respectively. Since ^{4}He consists of Bose–Einstein molecules without nuclear spin, all its molecules are identical, and for it $\alpha_1 = \alpha_e$. On the other hand, ^{3}He consists of Fermi–Dirac molecules with nuclear spin $\pm\frac{1}{2}$, and half of the encounters in it are between molecules of opposite spin; for these (17.3, 11) must be used, corresponding to $\alpha_1 = \frac{1}{2}(\alpha_e+\alpha_o)$. Hence the average value of α_1 for all types of encounter in ^{3}He is

$$\bar{\alpha}_1 = \tfrac{1}{2}\{\alpha_o+\tfrac{1}{2}(\alpha_e+\alpha_o)\} = \tfrac{3}{4}\alpha_o+\tfrac{1}{4}\alpha_e \qquad (17.4, 1)$$

with a corresponding result for the effective molecular cross-section.

The calculated viscosities of ^{3}He and ^{4}He at very low temperatures are compared with the experimental values in Fig. 13 and in the last column of Table 29. The calculated values are those of Keller‡ based on the 12,6 model, using values of the constants ϵ, σ derived by the virial method for more normal temperatures. Similar results are obtained using the exp;6 and Buckingham models. As Fig. 13 shows, the agreement with experiment is

* E. A. Uehling (*Phys. Rev.* **46**, 917 (1934)) recalculated these values and found a considerable divergence from the results of Massey and Mohr.

† H. S. W. Massey and C. B. O. Mohr, *Proc. R. Soc.* A, **144**, 188 (1934); R. A. Buckingham, J. Hamilton and H. S. W. Massey, *Proc. R. Soc.* A, **179**, 103 (1941); J. de Boer, *Physica*, **10**, 348 (1943); J. de Boer and E. G. D. Cohen, *Physica*, **17**, 993 (1951); O. Halpern and R. A. Buckingham, *Phys. Rev.* **98**, 1626 (1955); W. E. Keller, *Phys. Rev.* **105**, 41 (1956).

‡ W. E. Keller, *Phys. Rev.* **105**, 41 (1956). The values in the last column of Table 29 are obtained from Keller's tables by interpolation. At 40 °K., the highest temperature considered by Keller, the quantum values of the viscosity exceed the classical by less than 4 per cent.

Table 29. *Viscosity of helium on the quantum theory*

Absolute temperature	$\mu \times 10^7$ (obs.)	$\mu \times 10^7$ (calculated): Rigid elastic spheres, Classical theory	Rigid elastic spheres, Quantum theory	12,6 model (Keller, quantum theory)
294·5	1994	2000	1850	—
273·1	1870	1930	1770	—
250·3	1788	1840	1690	—
203·1	1564	1670	1500	—
170·5	1392	1520	1350	—
89·8	918	1100	920	—
75·1	815	1010	815	—
20·2	350·3	520	355	354
17·04	310 (1)	—	—	317
15	294·6	450	300	291
14·12	280 (1)	—	—	279·5
4·2	107·8 (2)	—	—	108·6
2·68	66·0 (2)	—	—	66·0
1·29	34·0 (2)	—	—	37·0

Experimental references: (1) E. W. Becker and R. Misenta, *Z. Phys.* **140**, 535 (1955); (2) E. W. Becker, R. Misenta and F. Schmeissner, *Z. Phys.* **137**, 126 (1954); the other values are, as in Table 13, from H. K. Onnes and S. Weber, *Proc. Sect. Sci. K. ned. Akad. Wet.* **21**, 1385 (1913).

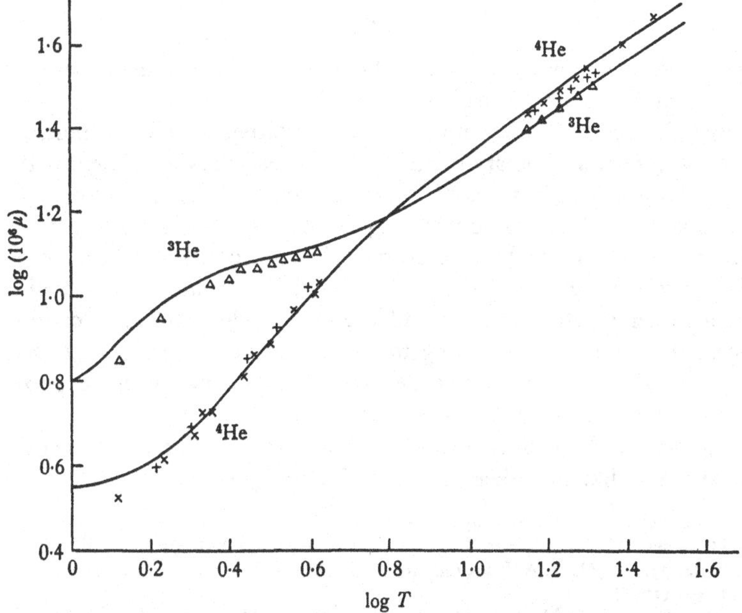

Fig. 13. Keller's quantum calculations of μ for ^{3}He and ^{4}He, based on the 12,6 model, compared with observation. Continuous curves, theory; $\times$, $\triangle$, observed values for ^{4}He and ^{3}He respectively; $+$, values calculated for ^{4}He from observed values of λ, using $\lambda = \frac{5}{2}\mu c_v$. Observed values of μ from Becker and Misenta and from Becker, Misenta and Schmeissner; observed values of λ from J. B. Ubbink and W. J. de Haas, *Physica*, **10**, 465 (1943).

remarkably good considering the wide temperature-range over which the force-law is extrapolated. Contrary to classical theory, the viscosity of ^{3}He actually exceeds that of ^{4}He at very low temperatures. Close agreement between theory and experiment can be obtained for ^{3}He only when the combination rule (17.4, 1) is used in determining the effective cross-section.

A comparison of theory and experiment for the coefficient of diffusion in mixtures of ^{3}He and ^{4}He at low temperatures has been made by Bendt,* likewise with satisfactory results.

17.41. Hydrogen at low temperatures

The theory of this chapter, though strictly applicable only to spherical molecules, may be expected also to apply moderately well to hydrogen at low temperatures, at which the rotational degrees of freedom are not excited (cf. 17.63). However, hydrogen exists in two forms, para- and orthohydrogen, which have to be regarded as independent gases; in normal hydrogen these are present in the ratio 3:1. A hydrogen molecule is a Bose–Einstein particle; however, because of nuclear spins orthohydrogen has nine independent quantum states, even at low temperatures. In consequence the relation, similar to (17.4, 1), giving the mean value of α_1 for orthohydrogen is

$$\bar{\alpha}_1 = \tfrac{1}{9}\alpha_e + \tfrac{8}{9}[\tfrac{1}{2}(\alpha_e+\alpha_o)] = \tfrac{5}{9}\alpha_e + \tfrac{4}{9}\alpha_o. \qquad (17.41, 1)$$

Parahydrogen has only one quantum state at low temperatures, and $\alpha_1 = \alpha_e$ in it; for encounters between molecules of different modifications,

$$\alpha_{12} = \tfrac{1}{2}(\alpha_e+\alpha_o).$$

The quantities α_e, α_o may conceivably differ for para- and orthohydrogen, but this possibility is not normally considered.

Calculated viscosities for para- and orthohydrogen, and for normal hydrogen, regarded as a binary mixture, show satisfactory agreement with experiment at low temperatures.† The calculations were made on the 12,6 and Buckingham models, using values of the force-constants derived from experiments at more normal temperatures. Theory predicts that, because of the difference between the quantum values of $\bar{\alpha}_1$, parahydrogen should possess a rather greater viscosity and thermal conductivity than orthohydrogen, the difference increasing to about 1 per cent near $T = 10\,^\circ$K. This is in rough agreement with experiment,‡ though part of the experimental difference may be due to other effects.

A comparison of H_2 with HD and D_2 is not straightforward, because the HD molecule, unlike the others, is a Fermi–Dirac particle; also para- and

* P. J. Bendt, *Phys. Rev.* **110**, 85 (1958).

† E. G. D. Cohen, M. J. Offerhaus, J. M. J. van Leeuwen, B. W. Roos and J. de Boer, *Physica*, **22**, 791 (1956); R. A. Buckingham, A. R. Davies and D. C. Gilles, *Proc. Phys. Soc.* **71**, 457 (1958).

‡ E. W. Becker and O. Stehl, *Z. Phys.* **133**, 615 (1952); K. Heinzinger, A. Klemm and L. Waldmann, *Z. Naturf.* **16a**, 1338 (1961).

orthodeuterium, differing from the corresponding H_2 modifications in nuclear spin, have in place of (17.41, 1)

$$\bar{\alpha}_1 = \tfrac{7}{12}\alpha_e + \tfrac{5}{12}\alpha_o \text{ (ortho)}, \quad \bar{\alpha}_1 = \tfrac{5}{9}\alpha_e + \tfrac{4}{9}\alpha_o \text{ (para)}. \qquad (17.41, 2)$$

We may note, however, that on classical theory, if the force-law is the same for D_2 and H_2, their viscosities should be in the ratio $\sqrt{2}$, or 1·414. In fact the ratio, though not far from $\sqrt{2}$ at ordinary temperatures, falls to 1·37 at 71·5 °K., and 1·25 at 14·4 °K.;* this can fairly be attributed to quantum effects.

17.5. Degeneracy for Fermi–Dirac particles

As noted in 17.1, a state of congestion results when the mean distance between neighbouring molecules is comparable with the size of the quantum wave fields with which molecules are surrounded. Consider, for example, a gas composed of electrons or other Fermi–Dirac particles. According to quantum theory, not more than $V\beta\, d\mathbf{c}(m/h)^3$ such particles in a volume V can possess velocities in the range $d\mathbf{c}$, where m and β are the mass and the 'statistical weight' of a particle,† and h is Planck's constant. Hence if in the velocity-range there is already this number of particles, and the result of a collision would be that a further particle would enter the range, the collision cannot happen. More generally, if $Vf(\mathbf{c})d\mathbf{c}$ particles have velocities in this range, the probability of a collision which would result in a particle entering this range is reduced in the ratio $1-f(\mathbf{c})h^3/m^3\beta$. Accordingly (3.52, 9) is to be replaced by

$$\left(\frac{\partial_e f_1}{\partial t}\right)_2 = \iiint \left\{ f_1' f_2' \left(1 - \frac{f_1 h^3}{m_1^3 \beta_1}\right)\left(1 - \frac{f_2 h^3}{m_2^3 \beta_2}\right) - f_1 f_2 \left(1 - \frac{f_1' h^3}{m_1^3 \beta_1}\right)\left(1 - \frac{f_2' h^3}{m_2^3 \beta_2}\right)\right\}$$
$$\times g\alpha_{12}(g,\chi)\sin\chi\, d\chi\, d\epsilon\, d\mathbf{c}_2. \qquad (17.5, 1)$$

The uniform steady state for a gas-mixture is given by

$$f_1' f_2' \Big/ \left(1 - \frac{f_1' h^3}{m_1^3 \beta_1}\right)\left(1 - \frac{f_2' h^3}{m_2^3 \beta_2}\right) = f_1 f_2 \Big/ \left(1 - \frac{f_1 h^3}{m_1^3 \beta_1}\right)\left(1 - \frac{f_2 h^3}{m_2^3 \beta_2}\right)$$

and similar equations, whence it follows that

$$\ln\left\{ f_1 \Big/ \left(1 - \frac{f_1 h^3}{m_1^3 \beta_1}\right)\right\}, \quad \ln\left\{ f_2 \Big/ \left(1 - \frac{f_2 h^3}{m_2^3 \beta_2}\right)\right\}$$

* A. O. Rietveld, A. van Itterbeek and C. A. Velds, *Physica*, **25**, 205 (1959). A small difference in the molecular force-law between D_2 and H_2 is probable; see, e.g., D. E. Diller and E. A. Mason, *J. Chem. Phys.* **44**, 2604 (1966).

† The statistical weight is the number of independent quantum states in which the particle can possess the same internal energy: for an electron it is 2, corresponding to the two possible values of the spin. According to quantum mechanics, not more than one electron in an atom can occupy a given quantum state. It is also a fundamental assumption of statistical mechanics that the *a priori* probability that an electron should occupy any one quantum state is the same as the probability that it should occupy a volume h^3 of the six-dimensional space whose coordinates are the space-coordinates and components of momentum of the electron. Thus not more than one electron with a given spin, and not more than two in all, can occupy the volume h^3. The result is generalized to apply to all Fermi–Dirac particles.

are summational invariants for encounters. This gives

$$f_1 = m_1^3\beta_1/h^3(1 + A_1 e^{\alpha m_1 C_1^2}), \quad f_2 = m_2^3\beta_2/h^3(1 + A_2 e^{\alpha m_2 C_2^2}), \qquad (17.5, 2)$$

where A_1, A_2, α are related to the number-densities and temperature of the gases, and $\mathbf{C}_1$, $\mathbf{C}_2$ are, as before, the velocities of the particles relative to the mean motion. Large values of A_1, A_2 correspond, for a given α, to small number-densities of the gas, and vice versa.

If A_1, A_2 are large, (17.5, 2) reduces to the usual (Maxwellian) form

$$f_1 = \frac{m_1^3\beta_1}{h^3 A_1} e^{-\alpha m_1 C_1^2}, \quad f_2 = \frac{m_2^3\beta_2}{h^3 A_2} e^{-\alpha m_2 C_2^2},$$

and so

$$\alpha = \frac{1}{2kT}. \qquad (17.5, 3)$$

If the velocity-distribution of one of the gases is independent of the density of the other, as is true in the classical theory, α will have the same value if A_1 is large and A_2 is not large, or, again, by the same argument, if neither A_1 nor A_2 is large. Statistical mechanics shows that this is actually the case: thus (17.5, 3) is valid for all densities.

The relations connecting the quantities A_1, A_2 with the corresponding number-densities n_1, n_2 are of the same form as for a simple gas, whose velocity-distribution function is

$$f = m^3\beta/h^3(1 + A e^{mC^2/2kT}). \qquad (17.5, 4)$$

If A is large compared with unity, f approximates to the Maxwellian form with n given by

$$\frac{m^3\beta}{h^3 A} = n\left(\frac{m}{2\pi kT}\right)^{\frac{3}{2}}.$$

Hence the condition ($A \gg 1$) that there is no appreciable departure from the Maxwellian distribution implies that

$$h^3 n/\beta(2\pi mkT)^{\frac{3}{2}} \ll 1. \qquad (17.5, 5)$$

When this condition is satisfied, the gas is said to be non-degenerate.

If A is small compared with unity, f is nearly equal to $m^3\beta/h^3$ when C is less than the quantity w defined by

$$\frac{mw^2}{2kT} = \ln\left(\frac{1}{A}\right),$$

but falls off very rapidly as C increases beyond this value. In this case the gas is said to be degenerate. To a first approximation we can write

$$f = \frac{m^3\beta}{h^3} \quad \text{if} \quad C < w,$$

$$f = 0 \quad \text{if} \quad C > w.$$

This gives

$$n = \frac{m^3\beta}{h^3}\int_0^w 4\pi C^2 dC$$

$$= \frac{4\pi}{3}\frac{m^3\beta}{h^3}\left(\frac{2kT\ln(1/A)}{m}\right)^{\frac{3}{2}}. \qquad (17.5, 6)$$

Also, using the same approximation, we find

$$n\overline{C^2} = \frac{4\pi}{5}\frac{m^3\beta}{h^3}\left(\frac{2kT\ln(1/A)}{m}\right)^{\frac{5}{2}}$$

$$= \frac{3}{5}\left(\frac{3}{4\pi\beta}\right)^{\frac{2}{3}}\frac{h^2}{m^2}n^{\frac{5}{3}}. \qquad (17.5, 7)$$

Hence the energy of the molecules is, to this approximation, independent of the temperature. The pressure of the gas, which is given by

$$p = \tfrac{1}{3}nm\overline{C^2} = \frac{1}{5}\left(\frac{3}{4\pi\beta}\right)^{\frac{2}{3}}\frac{h^2}{m}n^{\frac{5}{3}}, \qquad (17.5, 8)$$

is also, to this approximation, independent of the temperature.

Using the fact that A is small compared with unity, in conjunction with equation (17.5, 6), we obtain as the condition for degeneracy

$$h^3n/\beta(2\pi mkT)^{\frac{3}{2}} \gg 1. \qquad (17.5, 9)$$

It follows from this condition that the mean translational energy of the molecules, and the pressure, are large compared with the values $3k/2m$ and knT which they would have in the absence of degeneracy. The relations (17.5, 5) and (17.5, 9) constitute Sommerfeld's degeneracy criterion.* It is clear from them that degeneracy is most probable if m is small: thus it is most likely to occur in an electron gas. For other gases it is most likely to occur at great densities and low temperatures. It is found, however, that even for the lightest gases at temperatures as low as 15 °K. the congestion produced by degeneracy is less than that due to the finite size of the molecules, considered in Chapter 16, and so the effect of degeneracy can normally be neglected, save for an electron-gas (cf. p. 336, second footnote).†

17.51. Degeneracy for Bose–Einstein particles

Rather different results apply to Bose–Einstein particles. For these the presence of a like particle in the velocity-range $d\mathbf{c}$ increases the probability that a particle will enter that range; the presence of $f(\mathbf{c})\,d\mathbf{c}$ particles per unit volume increases this probability in the ratio $1+f(\mathbf{c})\,h^3/m^3\beta$. The analysis in this case is similar to that of 17.5, save that the sign of $h^3/m^3\beta$ is changed: in the uniform steady state it is found that

$$f = m^3\beta/h^3(A\,e^{mC^2/2kT}-1). \qquad (17.51, 1)$$

The constant A, which is related to the number-density, is in this case always greater than unity; the limiting case of extreme degeneracy is approached as $A \to 1$. The molecular energy and pressure corresponding to (17.51, 1) are less than those given by the classical formulae. It is, however, unnecessary to go into details, since the results for ordinary gases are not appreciably affected by degeneracy.

* A Sommerfeld, *Z. Phys.* **47**, 1, 43 (1928).

† In a dense star, because of the increased electron energies due to degeneracy, relativistic effects often are important.

17.52. Transport phenomena in a degenerate gas

The equation (17.5, 1) and the corresponding equation for Bose–Einstein particles were used by Uehling and Uhlenbeck* and later authors to construct a theory of the transport phenomena in a degenerate gas. However, (17.5, 1) is at most only approximately valid; it is based on the assumption of binary encounters, which is not valid in a degenerate gas. In fact, since the wave-length $h/\pi mg$ of the quantum wave field surrounding a pair of molecules is comparable with $h/2\pi(mkT)^{\frac{1}{2}}$, the condition (17.5, 9) for degeneracy implies that this wave-length considerably exceeds the mean distance $n^{-\frac{1}{3}}$ between neighbouring particles.

Transport phenomena in ordinary gases are not appreciably affected by degeneracy, even at low temperatures.† The situation is different for an electron-gas; however, an exact discussion for such a gas needs to take into account the interaction of electron-waves with the whole of the neighbouring particles, rather than binary encounters. An elementary discussion in terms of free path ideas was given by Sommerfeld;‡ he was able to account for some of the anomalies encountered in the classical theory of an electron-gas. However, a full discussion of (say) the electron-gas in a metal demands far more sophisticated concepts than those embodied in (17.5, 1).

INTERNAL ENERGY

17.6. Quantized internal energies

In discussing molecules with internal energy, a simple semi-classical approach is here employed. The translatory motions are described classically, the rotations and internal vibrations in terms of quantum states: for simplicity these states are assumed to be non-degenerate, so that to each energy level corresponds one and only one quantum state. A correspondence principle is invoked to derive quantum results from the corresponding classical ones. § Such an approach is simpler than a full quantum treatment but it is incomplete; in particular, it is unable to take into account the

* E. A. Uehling and G. E. Uhlenbeck, *Phys. Rev.* **43**, 552 (1933).

† E. A. Uehling, *Phys. Rev.* **46**, 917 (1934). The onset of degeneracy should be accompanied by a considerable variation of viscosity with pressure. The observed variation for helium at very low temperatures is much more modest—only $\frac{1}{2}$ per cent per atmosphere (J. M. J. Coremans, J. J. M. Beenakker, A. van Itterbeek and P. Zandbergen, *Bull. Inst. Int. Froid*, Annexe 1958–1, p. 289).

‡ A. Sommerfeld, *Z. Phys.* **47**, 1, 43 (1928).

§ The approach is essentially that of C. S. Wang-Chang and G. E. Uhlenbeck, *Univ. of Michigan report* CM-681, 1951; see also C. S. Wang-Chang, G. E. Uhlenbeck and J. de Boer, in *Studies in Statistical Mechanics* (ed. J. de Boer and G. E. Uhlenbeck), vol. 2, 249 (North-Holland, 1964). Its defects have been discussed by L. Waldmann, who has attempted a more exact discussion in a special case (*Handbuch der Physik*, vol. 12, 469–93 (Springer, 1958); *Proc. Int. Seminar on Transport Properties of Gases*, pp. 59–80 (Brown University, Providence, R.I., 1964).

correlation between linear and angular momentum indicated by classical theory (in the second and third terms on the right of (11.41, 3), etc.).

The state of a molecule m_s is specified by the velocity $\boldsymbol{c}_s$ of its mass-centre, and by a set of internal quantum numbers denoted collectively by K_s. A separate velocity-distribution function $f_s(\boldsymbol{c}_s, K_s, \boldsymbol{r}, t)$ has to be defined for molecules in each of the quantum states K_s; it is, as usual, such that the probable number of molecules m_s in the state K_s at time t, with positions and velocities in the ranges $\boldsymbol{r}$, $d\boldsymbol{r}$ and $\boldsymbol{c}_s$, $d\boldsymbol{c}_s$, is $f_s(\boldsymbol{c}_s, K_s, \boldsymbol{r}, t)\,d\boldsymbol{r}\,d\boldsymbol{c}_s$. Each internal state K_s corresponds to an elementary volume h^{R-3} of the internal phase-space of Chapter 11 ($R-3$ is the number of internal coordinates);* an integration over the internal phase-space is to be replaced by a sum over all values of K_s. For brevity, $f_s(\boldsymbol{c}_s, K_s, \boldsymbol{r}, t)$ is written f_{sK}; likewise $E_{sK}^{(i)}$ is used to denote the internal energy of a molecule m_s in the state K_s (measured from the state of least internal energy as zero).

17.61. Encounter probabilities

An encounter between a molecule m_s and a molecule m_t may alter their internal states as well as their translatory velocities. Let $E_{sK'}^{(i)}$, $E_{tL'}^{(i)}$, $\boldsymbol{g}'_{ts}$ and $E_{sK}^{(i)}$, $E_{tL}^{(i)}$, $\boldsymbol{g}_{ts}$ denote the internal energies and relative velocities *before* and *after* the encounter; we write $\boldsymbol{g}'_{ts} = g'\mathbf{e}'$, $\boldsymbol{g}_{ts} = g\mathbf{e}$, where $\mathbf{e}$, $\mathbf{e}'$ are unit vectors. From the conservation of energy relative to the mass-centre of the two molecules,

$$\frac{1}{2}\frac{m_s m_t}{m_s+m_t}(g'^2-g^2) = \Delta(E^{(i)}), \tag{17.61, 1}$$

where $\Delta(E^{(i)})$ denotes the increase of internal energy at the encounter, i.e.

$$\Delta(E^{(i)}) = E_{sK}^{(i)}+E_{tL}^{(i)}-E_{sK'}^{(i)}-E_{tL'}^{(i)}. \tag{17.61, 2}$$

When the values of K_s, L_t, K'_s, L'_t are assigned, equation (17.61, 1) gives either of g and g' in terms of the other. In order to specify an encounter completely, one needs to know the initial states $\boldsymbol{c}'_s$, K'_s and $\boldsymbol{c}'_t$, L'_t, the final internal states K_s, L_t, and the final direction $\mathbf{e}$ (or alternatively $\boldsymbol{c}_s$, K_s, $\boldsymbol{c}_t$, L_t K'_s, L'_t and $\mathbf{e}'$).

The number of encounters per unit volume and time between pairs of molecules m_s, m_t in the sets $\boldsymbol{c}'_s$, $d\boldsymbol{c}'_s$, K'_s and $\boldsymbol{c}'_t$, $d\boldsymbol{c}'_t$, L'_t which result in the molecules entering the internal states K_s, L_t, and having the direction $\mathbf{e}$ of the final relative velocity in the solid angle $d\mathbf{e}$, is proportional to $d\mathbf{e}$ and to the numbers $f'_{sK'}d\boldsymbol{c}'_s$ and $f'_{tL'}d\boldsymbol{c}'_t$ of molecules involved. We write it as

$$f'_{sK'}f'_{tL'}\,\alpha_{KL}^{K'L'}(\boldsymbol{g}'_{ts}, \boldsymbol{g}_{ts})\,g\,d\mathbf{e}\,d\boldsymbol{c}'_s d\boldsymbol{c}'_t. \tag{17.61, 3}$$

In this $\alpha_{KL}^{K'L'}d\mathbf{e}$, like the earlier $\alpha_{12}d\mathbf{e}'$, denotes an elementary collision cross-section. If $\boldsymbol{G}$ is the velocity of the mass-centre of the pair of molecules, by (3.52, 5),

$$\begin{aligned} g\,d\mathbf{e}\,d\boldsymbol{c}'_s d\boldsymbol{c}'_t &= g\,d\mathbf{e}\,d\boldsymbol{G}\,d\boldsymbol{g}'_{ts} = g\,d\mathbf{e}\,d\boldsymbol{G}\,g'^2 dg'\,d\mathbf{e}' \\ &= g'\,d\mathbf{e}'\,d\boldsymbol{G}\,d\boldsymbol{g}_{ts} = g'\,d\mathbf{e}'\,d\boldsymbol{c}_s d\boldsymbol{c}_t, \end{aligned} \tag{17.61, 4}$$

* The function $f_s(\boldsymbol{c}_s, K_s, \boldsymbol{r}, t)$ does not correspond exactly to the function $f_s(\boldsymbol{Q}_s, \boldsymbol{P}_s, t)$ of Chapter 11, but to $h^{R-3}f_s(\boldsymbol{Q}_s, \boldsymbol{P}_s, t)$.

since, by (17.61, 1), $g'\,dg' = g\,dg$. Hence (17.61, 3) can also be written

$$f'_{sK'} f'_{tL'} \alpha^{K'L'}_{KL}(\mathbf{g}'_{ts}, \mathbf{g}_{ts}) g'\,d\mathbf{e}'\,d\mathbf{c}_s\,d\mathbf{c}_t. \qquad (17.61, 5)$$

17.62. The Boltzmann equation

The state of a molecule m_s is assumed to change only at binary encounters. Then f_{sK} satisfies a Boltzmann equation of the type

$$\frac{\partial f_{sK}}{\partial t} + \mathbf{c}_s \cdot \frac{\partial f_{sK}}{\partial \mathbf{r}} + \mathbf{F}_s \cdot \frac{\partial f_{sK}}{\partial \mathbf{c}_s} = \frac{\partial_e f_{sK}}{\partial t}, \qquad (17.62, 1)$$

where $\partial_e/\partial t$ denotes a rate of change due to encounters.

The total number of molecules m_s entering the velocity-range $\mathbf{c}_s$, $d\mathbf{c}_s$ and the internal state K_s as a result of encounters per unit volume and time is, by (17.61, 5),

$$d\mathbf{c}_s \sum_t \sum_{K'_s} \sum_{L'_t} \sum_{L_t} \iint f'_{sK'} f'_{tL'} \alpha^{K'L'}_{KL}(\mathbf{g}'_{ts}, \mathbf{g}_{ts}) g'\,d\mathbf{e}'\,d\mathbf{c}_t. \qquad (17.62, 2)$$

In this, $\mathbf{c}'_s$, $\mathbf{c}'_t$ are determined, for given $\mathbf{c}_s$, $\mathbf{c}_t$, $\mathbf{e}'$, from the vectors $\mathbf{G}$, $\mathbf{g}'_{ts}$; $\mathbf{G}$ is known directly from $\mathbf{c}_s$, $\mathbf{c}_t$, and $\mathbf{g}'_{ts}$ is $g'\mathbf{e}'$, where g' is given by (17.61, 1). Similarly the total number of molecules m_s leaving the state $\mathbf{c}_s$, $d\mathbf{c}_s$, K_s per unit volume and time is (cf. (17.61, 3))

$$d\mathbf{c}_s \sum_t \sum_{K''_s} \sum_{L''_t} \sum_{L_t} \iint f_{sK} f_{tL} \alpha^{KL}_{K''L''}(\mathbf{g}_{ts}, \mathbf{g}''_{ts}) g''\,d\mathbf{e}''\,d\mathbf{c}_t, \qquad (17.62, 3)$$

in which K''_s, L''_t, $\mathbf{g}''_{ts}$, g'', $\mathbf{e}''$ refer to the state of the molecules after encounter. Taking the difference of (17.62, 2) and (17.62, 3), and dividing by $d\mathbf{c}_s$, we get

$$\frac{\partial_e f_{sK}}{\partial t} = \sum_t \sum_{L_t} \int \left\{ \sum_{K'_s} \sum_{L'_t} \int f'_{sK'} f'_{tL'} \alpha^{K'L'}_{KL}(\mathbf{g}'_{ts}, \mathbf{g}_{ts}) g'\,d\mathbf{e}' - \sum_{K''_s} \sum_{L''_t} \int f_{sK} f_{tL} \alpha^{KL}_{K''L''}(\mathbf{g}_{ts}, \mathbf{g}''_{ts}) g''\,d\mathbf{e}'' \right\} d\mathbf{c}_t. \qquad (17.62, 4)$$

Equation (17.62, 4) differs in form from the classical expression for $\partial_e f_s/\partial t$, which involves only the combination $f'_s f'_t - f_s f_t$. For (17.62, 4) to be reducible to the classical form it is necessary that

$$\sum_{K''_s} \sum_{L''_t} \int \alpha^{KL}_{K''L''}(\mathbf{g}_{ts}, \mathbf{g}''_{ts}) g''\,d\mathbf{e}'' = \sum_{K'_s} \sum_{L'_t} \int \alpha^{K'L'}_{KL}(\mathbf{g}'_{ts}, \mathbf{g}_{ts}) g'\,d\mathbf{e}'. \qquad (17.62, 5)$$

This condition can be shown to be satisfied;* hence

$$\frac{\partial_e f_{sK}}{\partial t} = \sum_t \sum_{L_t} \sum_{K'_s} \sum_{L'_t} \iint (f'_{sK'} f'_{tL'} - f_{sK} f_{tL}) \alpha^{K'L'}_{KL} g'\,d\mathbf{e}'\,d\mathbf{c}_t. \qquad (17.62, 6)$$

* In order to justify (17.62, 5) by the correspondence principle, one seeks the quantum equivalent of (11.22, 1), i.e.

$$g'\,dt\,b'\,db'\,d\mathbf{e}'\,d\mathbf{r}\,d\mathbf{c}'_s\,d\mathbf{c}'_t\,d\Omega'_s\,d\Omega'_t = g\,dt\,b\,db\,d\mathbf{e}\,d\mathbf{r}\,d\mathbf{c}_s\,d\mathbf{c}_t\,d\Omega_s\,d\Omega_t.$$

This equation implies that, for each phase-element occupied by molecules before encounters occurring in $d\mathbf{r}$ during dt, and carrying them into ranges $d\mathbf{c}_s d\Omega_s$ and $d\mathbf{c}_t d\Omega_t$, there is an equal phase-element occupied by molecules before encounters which carry them out of these ranges. In replacing the internal phase elements by quantum states,

17.63. The uniform steady state

The derivation of the velocity-distribution function in the uniform steady state follows much the same lines as in 11.3 and 11.32. One finds that $\ln f_{sK}$ is a summational invariant; it follows that (cf. (11.33, 1))

$$f_{sK} = n_s Z_s^{-1} \exp(-E_{sK}/kT) \qquad (17.63,\ 1)$$

$$= n_s Z_s^{-1} \exp\{-(\tfrac{1}{2}m_s C_s^2 + E_{sK}^{(i)})/kT\}, \qquad (17.63,\ 2)$$

where E_{sK} is the total heat-energy of a molecule (energy relative to the mass-velocity $\boldsymbol{c}_0$). Equation (17.63, 1) expresses the quantum form of the Boltzmann distribution. The equation giving the partition function Z_s is now

$$Z_s = \sum_{K_s} \int \exp(-E_{sK}/kT)\, d\boldsymbol{c}_s$$

$$= \int \exp(-m_s C_s^2/2kT)\, d\boldsymbol{c}_s \times \sum_{K_s} \exp(-E_{sK}^{(i)}/kT)$$

$$= (2\pi kT/m_s)^{\frac{3}{2}} Z_s^{(i)}, \qquad (17.63,\ 3)$$

where

$$Z_s^{(i)} = \sum_{K_s} \exp(-E_{sK}^{(i)}/kT). \qquad (17.63,\ 4)$$

The sum $Z_s^{(i)}$ is a sum over all possible internal states. To a crude first approximation, $E_{sK}^{(i)}$ can often be divided into independent parts

$$E_{sK}^{(i)} = E_{sR} + E_{sV} + E_{sV'} + \ldots, \qquad (17.63,\ 5)$$

where E_{sR} is the energy of rotation and E_{sV}, $E_{sV'}$, ... are the energies of the various internal vibrations. To the same approximation,

$$Z_s^{(i)} = \Big(\sum_{R_s} e^{-E_{sR}/kT}\Big)\Big(\sum_{V_s} e^{-E_{sV}/kT}\Big)(\ldots)$$

$$\equiv Z_{sR} Z_{sV} \ldots. \qquad (17.63,\ 6)$$

For actual molecules, the rotational states are degenerate, i.e. there is a number w_{sR} of such states all possessing the same rotational energy E_{sR}. This degeneracy is the quantum equivalent of the fact that in classical mechanics the angular momentum can vary in direction as well as magnitude. The analysis of the preceding sections is strictly not applicable to such degenerate states. However, the formulae of this section are valid even when there is degeneracy; one may then write

$$Z_{sR} = \sum_{E_{sR}} w_{sR} e^{-E_{sR}/kT}, \qquad (17.63,\ 7)$$

$d\boldsymbol{\Omega}'_s$ and $d\boldsymbol{\Omega}_s$ have to be replaced by equal powers of h, and similarly for $d\boldsymbol{\Omega}'_t$ and $d\boldsymbol{\Omega}_t$. Also $g'b'db'd\epsilon'$ and $gb\,db\,d\epsilon$ correspond to

$$\sum_{K_s'} \sum_{L_t'} \alpha_{KL}^{K'L'}(\boldsymbol{g}'_{ts}, \boldsymbol{g}_{ts})\, g\, d\mathbf{e} \quad \text{and} \quad \sum_{K_s''} \sum_{L_t''} \alpha_{K''L''}^{KL}(\boldsymbol{g}_{ts}, \boldsymbol{g}''_{ts})\, g''\, d\mathbf{e}'',$$

the summations being introduced to allow for the microscopic indeterminacy inherent in quantum mechanics. Hence, on using (17.61, 4), and integrating over all $d\mathbf{e}'$ and $d\mathbf{e}''$, we recover equation (17.62, 5).

the sum being over all possible rotational energies E_{sR}; there is a similar formula for $Z_s^{(i)}$. The number w_{sR} is the statistical weight attached to the rotational energy E_{sR}.

The mean thermal energy of a molecule satisfies (11.33, 5), i.e.

$$\bar{E}_s = kT^2 \frac{d}{dT}(\ln Z_s)$$

$$= \tfrac{3}{2}kT + (Z_s^{(i)})^{-1} \sum_{K_s} E_{sK}^{(i)} \exp(-E_{sK}^{(i)}/kT). \qquad (17.63, 8)$$

In the approximation (17.63, 5) this becomes

$$\bar{E}_s = \tfrac{3}{2}kT + (Z_{sR})^{-1} \sum_{R_s} E_{sR}\, e^{-E_{sR}/kT}$$

$$+ (Z_{sV})^{-1} \sum_{V_s} E_{sV}\, e^{-E_{sV}/kT} + \ldots. \qquad (17.63, 9)$$

The first term on the right of (17.63, 9) gives the mean translational energy; the second and later terms give the mean energies of rotation and of the various internal vibrations.

If any of the sums in (17.63, 9) contains a moderately large number of terms such that the partial energy (E_{sR} or E_{sV}, etc.) is smaller than or comparable with kT, the sum approximates to an integral, and the corresponding contribution to $\bar{E}_s$ approximates to the equipartition value found in Chapter 11. However, as can readily be shown, the contributions of any of the sums to $\bar{E}_s$, and to the specific heat per molecule $d\bar{E}_s/dT$, are both small when kT is small compared with all the partial excitation energies (the non-zero values of E_{sR}, E_{sV}, etc.) appearing in that sum. In this case the degrees of freedom corresponding to that sum are effectively not excited.

For example, the possible excitation energies of molecular rotation are multiples of $h^2/4\pi^2 I$, where I is the moment of inertia involved; thus equipartition obtains if $4\pi^2 IkT/h^2$ is large, but the molecules effectively do not rotate if it is small. Moments of inertia arising from the mass of electrons alone are very small; because of this, molecular rotations in a monatomic gas are not excited, and the same is true of the rotation of a diatomic molecule about the line joining the two atomic nuclei (cf. 11.34). For molecules of H_2 rotating about a line perpendicular to that joining the nuclei, I is such that the rotational contribution to the specific heat per molecule has nearly its equipartition value k when $T > 300$ °K. but is negligible when $T < 50$ °K. For the hydrogen isotopes HD and D_2, equipartition prevails above about 150 °K. For other diatomic and polyatomic gases I is much larger, and at all significant temperatures the rotational specific heat per molecule has nearly its equipartition value (k for diatomic gases, $\tfrac{3}{2}k$ for polyatomic).

However, equipartition is normally far from attainment for internal vibrations. When it comes near to attainment, the internal motions are becoming so violent that dissociation of the molecules is not far off. The vibrational contribution to the specific heat is of most importance for polyatomic gases, e.g. carbon dioxide; the only common diatomic gas for which it is important at ordinary temperatures is chlorine.

17.64. Internal energy and the transport phenomena

The quantum discussion for a non-uniform gas also follows much the same lines as the classical discussion in Chapter 11. The velocity-distribution function f_{sK} is, to a second approximation, given by

$$f_{sK} = f_{sK}^{(0)}(1+\Phi_{sK}^{(1)}), \tag{17.64, 1}$$

where $f_{sK}^{(0)}$ is the Boltzmann distribution function of (17.63, 1), and $\Phi_{sK}^{(1)}$ is small. In a binary mixture $\Phi_{1K}^{(1)}$, $\Phi_{2K}^{(1)}$ satisfy (cf. 11.4, 8)

$$-J_{1K}(\Phi^{(1)}) = f_{1K}^{(0)}\Big\{x_1^{-1}\boldsymbol{C}_1\cdot\boldsymbol{d}_{12}+(\mathscr{E}_{1K}-\bar{\mathscr{E}}_1-1)\boldsymbol{C}_1\cdot\nabla\ln T$$
$$+2\overset{\circ}{\mathscr{C}_1}\mathscr{C}_1:\nabla\boldsymbol{c}_0+\Big[\tfrac{2}{3}(\mathscr{C}_1^2-\tfrac{3}{2})-\frac{2}{N}(\mathscr{E}_s-\bar{\mathscr{E}}_s)\Big]\nabla\cdot\boldsymbol{c}_0\Big\} \tag{17.64, 2}$$

and a similar equation. In these, as before,

$$\mathscr{C}_1 = \boldsymbol{C}_1(m_1/2kT)^{\frac{1}{2}}, \quad \mathscr{E}_{1K} = E_{1K}/kT,$$

and $\frac{1}{2}Nk$ is the specific heat per molecule; also

$$J_{1K}(\Phi^{(1)}) = \sum_t\sum_{L_t}\sum_{K_1'}\sum_{L_t'}\iint f_{1K}^{(0)}f_{tL}^{(0)}(\Phi_{1K}^{(1)}+\Phi_{tL}^{(1)}-\Phi_{1K'}^{(1)\prime}-\Phi_{tL'}^{(1)\prime})\,\alpha_{KL}^{K'L'}g'\,de'\,d\boldsymbol{c}_t. \tag{17.64, 3}$$

The expression for $\Phi_{1K}^{(1)}$ is of the form (cf. (11.41, 1))

$$\Phi_{1K}^{(1)} = -\boldsymbol{A}_{1K}\cdot\nabla\ln T-\boldsymbol{D}_{1K}\cdot\boldsymbol{d}_{12}-2\mathsf{B}_{1K}:\nabla\boldsymbol{c}_0-2B_{1K}\nabla\cdot\boldsymbol{c}_0. \tag{17.64, 4}$$

From these relations the equations for $J_{1K}(\boldsymbol{A})$, $J_{1K}(\boldsymbol{D})$, etc., can readily be obtained.

First approximations to the transport coefficients are deduced as in Chapter 11. For example, a first approximation to $\boldsymbol{A}_K$ for a simple gas is taken to be

$$n\boldsymbol{A}_K = a_1(\mathscr{C}^2-\tfrac{5}{2})\boldsymbol{C}+a_2(\mathscr{E}_K^{(i)}-\overline{\mathscr{E}^{(i)}})\boldsymbol{C}, \tag{17.64, 5}$$

where the coefficients a_1, a_2 are independent of the internal quantum state K; they are determined from equations derived by multiplying the equation for $J_K(\boldsymbol{A})$ in turn by $(\mathscr{C}^2-\frac{5}{2})\boldsymbol{C}$ and $(\mathscr{E}_K^{(i)}-\overline{\mathscr{E}^{(i)}})\boldsymbol{C}$, summing over all K, and integrating with respect to $\boldsymbol{c}$. Similar modifications in the argument of Chapter 11 are required in discussing the quantities $\boldsymbol{A}$, B, B.

The final formulae for μ, λ, ϖ and D_{12} derived in this manner have the same form as those of Chapter 11, in terms of quantities a_{11}, a_{22}, etc., differing only trivially from those of that chapter. The difference is essentially that the element $gb\,db\,d\epsilon$ is replaced by $g'\alpha_{KL}^{K'L'}de'$, and the integrations over the internal phase spaces are replaced by summations; also the value of Z is now given by (17.63, 4). Thus, for example, the a of (11.5, 13) becomes

$$a = \frac{\pi^{\frac{3}{2}}}{Z^2}\left(\frac{kT}{m}\right)^3 \sum_K \sum_L \sum_{K'} \sum_{L'} \iint \exp\left(-\mathscr{g}^2 - \mathscr{E}_K^{(i)} - \mathscr{E}_L^{(i)}\right)(\mathscr{g}^2 - \mathscr{g}'^2)^2 \mathscr{g}' \, \alpha_{KL}^{K'L'} \, d\mathbf{e}' \, d\mathscr{g}. \tag{17.64, 6}$$

The changes in (11.5, 14, 15), and (11.52, 5) are similar.

The general conclusions regarding volume viscosity and heat conduction remain as in Chapter 11; the only serious modification is through the quantum effect on the specific heat. In particular, the discussion of the modified Eucken theory and of the Mason–Monchick formula (cf. 11.8, 11.81) remains unaffected.*

Once more it is necessary to emphasize that the theory outlined above is strictly not applicable to molecular rotations, because it ignores the degeneracy of the rotational states. As Waldmann† has stressed, the theory is in consequence unable to take proper account of the vector properties of the molecular angular momentum. The quantum equivalent of these vector properties is that a molecule cannot be regarded as in a pure state, its internal wave function between collisions being a linear combination of all the wave functions corresponding to its rotational energy and total scalar angular momentum. For a full discussion involving the vector properties of the angular momentum it is not sufficient simply to introduce cross-sections $\alpha_{KL}^{K'L'}$ for transitions between pure states.

The vector properties of the angular momentum are represented only by terms like those involving $\boldsymbol{h} \wedge \boldsymbol{C}$ and $\boldsymbol{h} \wedge (\boldsymbol{h} \wedge \boldsymbol{C})$ in (11.41, 3). Since the expression finally assumed for $\boldsymbol{A}$, in (11.5, 4) as well as in equation (17.64, 5), includes no terms of these types, one may expect that the first approximations of this section will not be affected by the neglect of degeneracy. However, as noted in 11.61, the neglect of such terms in higher approximations may lead to appreciable errors in the transport coefficients.

In principle it should be possible to complete the evaluation of the transport coefficients by a quantum determination of $\alpha_{KL}^{K'L'}$, followed by an evaluation of integrals like (17.64, 6). In practice this has proved very laborious, and no detailed quantum discussion has so far been given; however, some progress has been made with the theory of diatomic molecules.‡

* The work of Mason and Monchick was, in fact, originally based on the formulae of Wang-Chang and Uhlenbeck.

† L. Waldmann, *Proc. Int. Seminar on Transport Properties of Gases*, pp. 59–80 (Brown University, Providence, R.I., 1964).

‡ A. M. Arthurs and A. Dalgarno, *Proc. R. Soc.* A, **256**, 540 (1960); R. B. Bernstein, A. Dalgarno, H. S. W. Massey and I. C. Percival, *Proc. R. Soc.* A, **274**, 427 (1963). See also K. Takayanagi, *Prog. Theor. Phys., Osaka*, **11**, 557 (1954).

18

MULTIPLE GAS MIXTURES

18.1. Mixtures of several constituents

The methods of Chapter 8 can be generalized to apply to a mixture of three or more constituent gases.* Equations of type (8.1, 1, 3, 4, 7, 9) hold for each constituent, and the equations of continuity, momentum and energy of the gas as a whole are given by (8.1, 5, 6, 8, 10), with the obvious modifications required to take account of the additional constituents.

Enskog's method of solution of the Boltzmann equation is evidently still applicable. We discuss only the second approximation to the velocity-distribution function. Formal results are given for a K-constituent mixture; detailed applications are, however, limited to the case $K = 3$.

18.2. The second approximation

The first approximation to f_s is the Maxwellian function

$$f_s^{(0)} = n_s\left(\frac{m_s}{2\pi kT}\right)^{\frac{3}{2}}\exp(-\mathscr{C}_s^2), \quad \mathscr{C}_s = \left(\frac{m_s}{2kT}\right)^{\frac{1}{2}}\boldsymbol{C}_s. \qquad (18.2, 1)$$

The second approximation is $f_s^{(0)}(1+\Phi_s)$, where Φ_s satisfies the equation

$$\frac{\partial_0 f_s^{(0)}}{\partial t}+\boldsymbol{c}_s.\frac{\partial f_s^{(0)}}{\partial \boldsymbol{r}}+\boldsymbol{F}_s.\frac{\partial f_s^{(0)}}{\partial \boldsymbol{c}_s} = -\sum_t n_s n_t I_{st}(\Phi_s+\Phi_t) \qquad (18.2, 2)$$

(cf. (8.3, 1, 2)). In the sum on the right of (18.2, 2) it is to be understood that $I_{st}(\Phi_s+\Phi_t)$ reduces to $I_s(\Phi_s)$ when $t = s$.

The left-hand side of (18.2, 2) can be transformed as in 8.3, using generalized forms of (8.21, 2, 4, 6). The result (cf. (8.3, 8, 9)) is

$$f_s^{(0)}[\boldsymbol{C}_s.(\mathrm{x}_s^{-1}\boldsymbol{d}_s+(\mathscr{C}_s^2-\tfrac{5}{2})\nabla\ln T)+2\overset{\circ}{\mathscr{C}_s\mathscr{C}_s}:\nabla\boldsymbol{c}_0], \qquad (18.2, 3)$$

where $\mathrm{x}_s = n_s/n$, and

$$\boldsymbol{d}_s = \frac{1}{p}\left\{\nabla p_s-\rho_s\left(\boldsymbol{F}_s-\frac{D_0\boldsymbol{c}_0}{Dt}\right)\right\} \qquad (18.2, 4)$$

$$= \frac{1}{p}\left\{\nabla p_s-\rho_s\boldsymbol{F}_s-\frac{\rho_s}{\rho}(\nabla p-\sum_t\rho_t\boldsymbol{F}_t)\right\}. \qquad (18.2, 5)$$

* For early discussions, see E. J. Hellund, *Phys. Rev.* **57**, 319, 328, 737, 743 (1940); L. Waldmann, *Z. Phys.* **124**, 175 (1947); C. F. Curtiss and J. O. Hirschfelder, *J. Chem. Phys.* **17**, 550 (1949). An elegant account of the theory, given by L. Waldmann in *Handbuch der Physik*, vol. 12, 402–26 (1958), is largely followed here.

This may also be written (cf. (8.3, 12))

$$\boldsymbol{d}_s = \nabla \mathrm{x}_s + (\mathrm{x}_s - \rho_s/\rho)\nabla \ln p - (\rho_s/\rho p)(\rho \boldsymbol{F}_s - \sum_t \rho_t \boldsymbol{F}_t). \quad (18.2, 6)$$

Clearly, from (18.2, 5 or 6) $\quad \sum_s \boldsymbol{d}_s = 0. \quad (18.2, 7)$

From (18.2, 4), if the gas is at rest $\boldsymbol{d}_s$ reduces to $(\nabla p_s - \rho_s \boldsymbol{F}_s)/p$. However, in general (18.2, 7) does not imply any relation between the vectors $\nabla p_s - \rho_s \boldsymbol{F}_s$, which are linearly independent.

As in 8.31, the functions Φ_s must be linear in $\nabla \ln T$, $\nabla \boldsymbol{c}_0$ and the vectors $\boldsymbol{d}_t$. We write

$$\Phi_s = -\boldsymbol{A}_s . \nabla \ln T - 2\mathsf{B}_s : \nabla \boldsymbol{c}_0 - \sum_t \boldsymbol{D}_s^{(t)} . \boldsymbol{d}_t, \quad (18.2, 8)$$

where $\boldsymbol{A}_s$, B_s, $\boldsymbol{D}_s^{(t)}$ are respectively of the forms

$$\boldsymbol{C}_s A_s(C_s), \quad \overset{\circ}{\boldsymbol{C}_s \boldsymbol{C}_s} B_s(C_s), \quad \boldsymbol{C}_s D_s(C_s).$$

In view of (18.2, 7), the vectors $\boldsymbol{D}_s^{(t)}$ (for given s) possess a degree of indeterminacy; this is removed by imposing the further condition

$$\sum_t \rho_t \boldsymbol{D}_s^{(t)} = 0. \quad (18.2, 9)$$

Then, from (18.2, 4),

$$\sum_t \boldsymbol{D}_s^{(t)} . \boldsymbol{d}_t = p^{-1} \sum_t \boldsymbol{D}_s^{(t)} . (\nabla p_t - \rho_t \boldsymbol{F}_t),$$

so that (18.2, 8) becomes

$$\Phi_s = -\boldsymbol{A}_s . \nabla \ln T - 2\mathsf{B}_s : \nabla \boldsymbol{c}_0 - p^{-1} \sum_t \boldsymbol{D}_s^{(t)} . (\nabla p_t - \rho_t \boldsymbol{F}_t). \quad (18.2, 10)$$

The equations satisfied by $\boldsymbol{A}_s$, B_s and $\boldsymbol{D}_s^{(t)}$ are found by substituting from (18.2, 3, 10) into (18.2, 2) and equating coefficients of $\nabla \ln T$, $\nabla \boldsymbol{c}_0$ and $\nabla p_t - \rho_t \boldsymbol{F}_t$. This gives

$$\sum_t n_s n_t I_{st}(\boldsymbol{A}_s + \boldsymbol{A}_t) = f_s^{(0)}(\mathscr{C}_s^2 - \tfrac{5}{2}) \boldsymbol{C}_s, \quad (18.2, 11)$$

$$\sum_t n_s n_t I_{st}(\mathsf{B}_s + \mathsf{B}_t) = f_s^{(0)} \overset{\circ}{\mathscr{C}_s \mathscr{C}_s}, \quad (18.2, 12)$$

$$\sum_t n_s n_t I_{st}(\boldsymbol{D}_s^{(u)} + \boldsymbol{D}_t^{(u)}) = \mathrm{x}_s^{-1}(\delta_{su} - \rho_s/\rho) f_s^{(0)} \boldsymbol{C}_s, \quad (18.2, 13)$$

where δ_{su} is the Kronecker delta. Equations (18.2, 13) and (18.2, 9) are consistent, as can be seen by multiplying (18.2, 13) by ρ_u and summing over all u.

18.3. Diffusion

The velocity of diffusion $\overline{\boldsymbol{C}}_s$ of the sth constituent relative to the mass motion is given by

$$n_s \overline{\boldsymbol{C}}_s = \int f_s^{(0)} \Phi_s \boldsymbol{C}_s d\boldsymbol{c}_s. \quad (18.3, 1)$$

On substituting for Φ_s from (18.2, 10) this becomes

$$\overline{\boldsymbol{C}}_s = -\sum_t \Delta_{st} p^{-1}(\nabla p_t - \rho_t \boldsymbol{F}_t) - D_{Ts}\nabla \ln T, \qquad (18.3, 2)$$

where $$\Delta_{st} = \frac{1}{3n_s}\int f_s^{(0)} \boldsymbol{D}_s^{(t)} \cdot \boldsymbol{C}_s d\boldsymbol{c}_s, \quad D_{Ts} = \frac{1}{3n_s}\int f_s^{(0)} \boldsymbol{A}_s \cdot \boldsymbol{C}_s d\boldsymbol{c}_s. \qquad (18.3, 3)$$

Here D_{Ts} is a thermal diffusion coefficient for the sth constituent; Δ_{st} is a generalized diffusion coefficient for the sth constituent, measuring the 'induced' diffusion due to the forces acting on the tth constituent.*

The diffusion coefficients are not all independent. From (18.3, 3) and (18.2, 9),

$$\sum_t \rho_t \Delta_{st} = 0. \qquad (18.3, 4)$$

Again, the solutions of equations (18.2, 11, 13) (which are not unique, in that multiples $\alpha m_s \boldsymbol{C}_s$, $\delta_u m_s \boldsymbol{C}_s$ of the summational invariants $m_s \boldsymbol{C}_s$ can be added to each $\boldsymbol{A}_s$ or $\boldsymbol{D}_s^{(u)}$) are to be fixed by imposing the further condition $\sum_s \rho_s \overline{\boldsymbol{C}}_s = 0$ (cf. (8.31, 7, 8)). It follows that

$$\sum_s \rho_s \Delta_{st} = 0, \quad \sum_s \rho_s D_{Ts} = 0. \qquad (18.3, 5)$$

In virtue of (18.3, 4) and (18.2, 4), equation (18.3, 2) can be written in the form

$$\overline{\boldsymbol{C}}_s = -\sum_t \Delta_{st} \boldsymbol{d}_t - D_{Ts}\nabla \ln T. \qquad (18.3, 6)$$

Multiply (18.2, 13) in turn by $\boldsymbol{D}_s^{(v)}$ and $\boldsymbol{A}_s$, integrate with respect to $\boldsymbol{c}_s$, and sum over all values of s. Then, using (18.3, 3) and (18.3, 5),

$$\Delta_{uv} = \tfrac{1}{3}n\{\boldsymbol{D}^{(u)}, \boldsymbol{D}^{(v)}\}, \qquad (18.3, 7)$$

$$D_{Tu} = \tfrac{1}{3}n\{\boldsymbol{D}^{(u)}, \boldsymbol{A}\}, \qquad (18.3, 8)$$

the bracket expressions being defined by obvious generalizations of (4.4, 12). From (18.3, 7) follows the symmetry relation $\Delta_{uv} = \Delta_{vu}$, so that the set of equations (18.3, 4) are identical with the first set of (18.3, 5). Either set may be used to express the coefficients Δ_{ss} in terms of the coefficients Δ_{st} $(s \neq t)$. In view of this and the symmetry relations, there are only $\frac{1}{2}K(K-1)$ independent coefficients Δ_{st}; similarly there are only $K-1$ independent thermal diffusion coefficients D_{Ts}.

An alternative way of writing (18.3, 6) is

$$\overline{\boldsymbol{C}}_s = -\sum_t \Delta_{st}(\boldsymbol{d}_t + \mathrm{k}_{Tt}\nabla \ln T). \qquad (18.3, 9)$$

Here the coefficients k_{Tt} are defined by

$$D_{Ts} = \sum_t \Delta_{st}\mathrm{k}_{Tt}, \qquad (18.3, 10)$$

$$\sum_t \mathrm{k}_{Tt} = 0, \qquad (18.3, 11)$$

* Note that Δ_{st} does not reduce to the usual mutual diffusion coefficient for a binary mixture. The above discussion differs from that given by Waldmann (*Handbuch der Physik*, vol. 12, 406–11) in that he used diffusion velocities relative to the mean molecular velocity $\bar{\boldsymbol{c}}$ instead of those relative to $\boldsymbol{c}_0$.

the supplementary equation (18.3, 11) being required because (cf. (18.3, 5)) only $K-1$ of the set (18.3, 10) are independent relations. Together with (18.2, 7), this choice of the supplementary equation ensures that

$$\sum_s (\boldsymbol{d}_s + \mathrm{k}_{Ts} \nabla \ln T) = 0. \tag{18.3, 12}$$

Equations (18.3, 6, 9) again imply that relative diffusion of two of the constituent gases may be induced by factors not directly affecting these gases, such as forces acting on the molecules of another constituent gas.

Using (18.3, 12) together with (18.3, 9) (which also is equivalent to only $K-1$ independent equations, by (18.3, 5)), we can express the vectors $\boldsymbol{d}_s + \mathrm{k}_{Ts} \nabla \ln T$ in terms of the diffusion velocities, in the form*

$$\boldsymbol{d}_s + \mathrm{k}_{Ts} \nabla \ln T = -\sum_t \mathrm{x}_s \mathrm{x}_t (\overline{\boldsymbol{C}}_s - \overline{\boldsymbol{C}}_t)/D_{st}, \tag{18.3, 13}$$

where $D_{st} = D_{ts}$. In (18.3, 13), D_{st} is closely analogous to the binary diffusion coefficient of Chapter 8, to which it reduces for a binary gas ($K = 2$); also (see (18.41), below) the first approximation to D_{st} is equal to the binary gas $[D_{st}]_1$ of 9.81. For these reasons, and because it does not involve redundant diffusion coefficients, (18.3, 13) is usually preferable to (18.3, 6) as the basic diffusion equation.

If all the diffusion velocities vanish,

$$\boldsymbol{d}_s + \mathrm{k}_{Ts} \nabla \ln T = 0 \tag{18.3, 14}$$

for all s. In a gas at rest under no forces, this becomes

$$\nabla \mathrm{x}_s + \mathrm{k}_{Ts} \nabla \ln T = 0. \tag{18.3, 15}$$

Hence k_{Ts}, like k_T in a binary mixture, measures changes in equilibrium concentrations produced by a temperature gradient.

18.31. Heat conduction

The heat flow $\boldsymbol{q}$ is given by

$$\boldsymbol{q}/kT = \tfrac{5}{2} \sum_s n_s \overline{\boldsymbol{C}}_s + \sum_s \int f_s^{(0)} \Phi_s (\mathscr{C}_s^2 - \tfrac{5}{2}) \boldsymbol{C}_s d\boldsymbol{c}_s$$

* The coefficient $\mathrm{x}_s \mathrm{x}_t / D_{st}$ of $\overline{\boldsymbol{C}}_s - \overline{\boldsymbol{C}}_t$ in (18.3, 13) can be shown to be $-\mathscr{D}_{st}/\mathscr{D}$, where $\mathscr{D}$ is the determinant of $K+1$ rows and columns, obtained by bordering the determinant whose elements are Δ_{st} with an extra row and column whose common element is zero, every other element being unity; $\mathscr{D}_{st}$ is the cofactor of Δ_{st} in $\mathscr{D}$. Since $\sum_t \mathscr{D}_{st} = 0$, (18.3, 13) is equivalent to

$$\boldsymbol{d}_s + \mathrm{k}_{Ts} \nabla \ln T = \sum_t \mathscr{D}_{st} \overline{\boldsymbol{C}}_t / \mathscr{D} = \sum_t \mathrm{x}_s \mathrm{x}_t \overline{\boldsymbol{C}}_t / D_{st},$$

where D_{ss} is defined as such that $\sum_t \mathrm{x}_t / D_{st} = 0$. Similarly (18.3, 9) can be derived by solving for $\overline{\boldsymbol{C}}_s$ equations (18.3, 13) together with (18.2, 9). One finds

$$\Delta_{st} = -\mathscr{D}'_{st}/\mathscr{D}',$$

where $\mathscr{D}'$ is the determinant obtained by bordering that whose elements are $\mathrm{x}_s \mathrm{x}_t / D_{st}$ with an extra row and column whose common element is zero, the other elements being $\rho_1, \rho_2, \ldots, \rho_K$; $\mathscr{D}'_{st}$ is the cofactor of $\mathrm{x}_s \mathrm{x}_t / D_{st}$ in $\mathscr{D}'$.

(cf. (8.41, 2)). On substituting for Φ_s from (18.2, 8) this becomes

$$\boldsymbol{q}/kT = \tfrac{5}{2}\sum_s n_s\overline{\boldsymbol{C}}_s - \tfrac{1}{3}\nabla \ln T \sum_s \int f_s^{(0)}(\mathscr{C}_s^2 - \tfrac{5}{2})\boldsymbol{A}_s.\boldsymbol{C}_s d\boldsymbol{c}_s$$
$$-\tfrac{1}{3}\sum_t \boldsymbol{d}_t \sum_s \int f_s^{(0)}(\mathscr{C}_s^2 - \tfrac{5}{2})\boldsymbol{D}_s^{(t)}.\boldsymbol{C}_s d\boldsymbol{c}_s. \quad (18.31, 1)$$

Let $\boldsymbol{a}_s$ be an arbitrary vector function of $\boldsymbol{c}_s$; then, on multiplying (18.2, 11) by $\boldsymbol{a}_s$, integrating with respect to $\boldsymbol{c}_s$, and summing over all values of s,

$$\sum_s \int f_s^{(0)}(\mathscr{C}_s^2 - \tfrac{5}{2})\boldsymbol{a}_s.\boldsymbol{C}_s d\boldsymbol{c}_s = n^2\{\boldsymbol{a}, \boldsymbol{A}\}. \quad (18.31, 2)$$

Thus, using (18.3, 8),

$$\boldsymbol{q}/kT = \tfrac{5}{2}\sum_s n_s\overline{\boldsymbol{C}}_s - \tfrac{1}{3}n^2\{\boldsymbol{A}, \boldsymbol{A}\}\nabla \ln T - n\sum_s D_{Ts}\boldsymbol{d}_s. \quad (18.31, 3)$$

The thermal conductivity λ, as usually defined, gives the heat flow in the absence of diffusion. Since the vectors $\boldsymbol{d}_s + \mathrm{k}_{Ts}\nabla \ln T$ all vanish in the absence of diffusion, we write (18.31, 3) in the form

$$\boldsymbol{q} = \tfrac{5}{2}kT\sum_s n_s\overline{\boldsymbol{C}}_s - p\sum_s D_{Ts}(\boldsymbol{d}_s + \mathrm{k}_{Ts}\nabla \ln T) - \lambda\nabla T, \quad (18.31, 4)$$

where

$$\lambda = \tfrac{1}{3}kn^2\{\boldsymbol{A}, \boldsymbol{A}\} - kn\sum_s \mathrm{k}_{Ts}D_{Ts}. \quad (18.31, 5)$$

In (18.31, 4) the first term on the right represents the difference between the heat flows relative to the mass velocity $\boldsymbol{c}_0$ and the mean velocity $\overline{\boldsymbol{c}}$ of the molecules. The second term is the heat flow, relative to $\overline{\boldsymbol{c}}$, due to diffusion; by (18.3, 9, 10), it is equal to

$$-p\sum_t \mathrm{k}_{Tt}\sum_s \Delta_{st}(\boldsymbol{d}_s + \mathrm{k}_{Ts}\nabla \ln T) = p\Sigma \mathrm{k}_{Tt}\overline{\boldsymbol{C}}_t. \quad (18.31, 6)$$

The last term gives the heat flow due to the thermal conductivity λ.

An alternative expression for λ is obtained by writing

$$\tilde{\boldsymbol{A}}_s = \boldsymbol{A}_s - \sum_t \mathrm{k}_{Tt}\boldsymbol{D}_s^{(t)}. \quad (18.31, 7)$$

Then by (18.31, 5) and (18.3, 8),

$$\lambda = \tfrac{1}{3}kn^2\{\boldsymbol{A}, \tilde{\boldsymbol{A}}\}. \quad (18.31, 8)$$

Also by (18.3, 7, 8, 10),

$$\{\tilde{\boldsymbol{A}}, \boldsymbol{D}^{(u)}\} = (3/n)(D_{Tu} - \sum_t \Delta_{ut}\mathrm{k}_{Tt})$$
$$= 0. \quad (18.31, 9)$$

Hence

$$\lambda = \tfrac{1}{3}kn^2\{\tilde{\boldsymbol{A}}, \tilde{\boldsymbol{A}}\}. \quad (18.31, 10)$$

The equality of the coefficients k_{Ts} appearing in (18.31, 6) and (18.3, 9) (more correctly, the equality of the coefficients D_{Ts} in (18.31, 4) and (18.3, 6)) is an example of the symmetry relations met in the thermodynamics of irreversible processes; further examples are the symmetry conditions $\Delta_{st} = \Delta_{ts}$, $D_{st} = D_{ts}$ between the generalized diffusion coefficients in (18.3, 2, 13).

18.32. Viscosity

The deviation $\mathsf{p}^{(1)}$ of the pressure tensor from the hydrostatic pressure is given by

$$\mathsf{p}^{(1)} = \sum_s \int f_s^{(0)} m_s \boldsymbol{C}_s \boldsymbol{C}_s \Phi_s d\boldsymbol{c}_s.$$

As in 8.42 it follows from (18.2, 8) that

$$\mathsf{p}^{(1)} = -2\mu \overline{\overline{\overset{\circ}{\nabla} \boldsymbol{c}_0}},$$

where
$$\mu = \tfrac{2}{5} kT \sum_s \int f_s^{(0)} \overset{\circ}{\mathscr{C}_s \mathscr{C}_s} : \mathsf{B}_s \, d\boldsymbol{c}_s, \qquad (18.32, 1)$$

or, by (18.2, 12),
$$\mu = \tfrac{2}{5} kn^2 T \{\mathsf{B}, \mathsf{B}\}. \qquad (18.32, 2)$$

18.4. Expressions for the gas coefficients

Explicit expressions for the gas coefficients introduced in the last three sections are obtained by expanding $\boldsymbol{A}_s$, $\boldsymbol{D}_s^{(t)}$ and B_s in series of terms

$$\boldsymbol{C}_s S_{\frac{3}{2}}^{(p)}(\mathscr{C}_s^2), \quad \overset{\circ}{\boldsymbol{C}_s \boldsymbol{C}_s} S_{\frac{5}{2}}^{(p)}(\mathscr{C}_s^2).$$

Equations to determine the coefficients are obtained by multiplying (18.2, 11, 13) by $\boldsymbol{C}_s S_{\frac{3}{2}}^{(q)}(\mathscr{C}_s^2)$, or (18.2, 12) by $\overset{\circ}{\boldsymbol{C}_s \boldsymbol{C}_s} S_{\frac{5}{2}}^{(q)}(\mathscr{C}_s^2)$, and integrating with respect to $\boldsymbol{c}_s$. The mth order approximation is found by restricting p and q to their first m permissible values. The mth approximations to λ, μ and Δ_{ss} increase with increasing m; and they are best possible approximations in the sense used in the first footnote to p. 131.*

The mth approximation in general involves the solution of mK simultaneous linear equations. In view of the complexity of the solution, and of the fact that the first or second approximations for a simple or a binary gas usually differ little from the true values, attention is here confined to the lowest non-vanishing approximation in each case.

18.41. The diffusion coefficients

In the absence of temperature and velocity gradients, the equation for Φ_s becomes

$$f_s^{(0)} \boldsymbol{C}_s . \boldsymbol{d}_s = -x_s \sum_t n_s n_t I_{st}(\Phi_s + \Phi_t) \qquad (18.41, 1)$$

(cf. (18.2, 2, 3)). The first approximation to Φ_s in this case is taken to be

$$\Phi_s = m_s \boldsymbol{\alpha}_s . \boldsymbol{C}_s / kT, \qquad (18.41, 2)$$

where $\boldsymbol{\alpha}_s$ is a vector independent of $\boldsymbol{c}_s$.

To determine $\boldsymbol{\alpha}_s$, multiply (18.41, 1) by $m_s \boldsymbol{C}_s$ and integrate over all values of $\boldsymbol{c}_s$; this gives

$$nkT \boldsymbol{d}_s = -n_s^2 [m_s \boldsymbol{C}_s, \Phi_s]_s - \sum_{t \neq s} n_s n_t [m_s \boldsymbol{C}_s, \Phi_s + \Phi_t]_{st},$$

* E. J. Hellund, *Phys. Rev.* **57**, 319 (1940).

where the bracket expressions are defined as in (4.4, 7, 9). The first term on the right of this is zero, since $m_s \boldsymbol{C}_s$ is a summational invariant. Thus, substituting from (18.41, 2),

$$nk^2T^2\boldsymbol{d}_s = -\tfrac{1}{3}\sum_{t \neq s} m_s n_s n_t \{\boldsymbol{\alpha}_s[\boldsymbol{C}_s, m_s \boldsymbol{C}_s]_{st} + \boldsymbol{\alpha}_t[\boldsymbol{C}_s, m_t \boldsymbol{C}_t]_{st}\}.$$

Using equations similar to (9.6, 1, 4) and (9.81, 1), this becomes

$$\boldsymbol{d}_s = -\frac{16n}{3kT}\sum_t \frac{m_s m_t}{m_s + m_t} \mathrm{x}_s \mathrm{x}_t (\boldsymbol{\alpha}_s - \boldsymbol{\alpha}_t)\,\Omega_{st}^{(1)}(1)$$

$$= -\sum_t \mathrm{x}_s \mathrm{x}_t (\boldsymbol{\alpha}_s - \boldsymbol{\alpha}_t)/[D_{st}]_1.$$

Here $[D_{st}]_1$ denotes the first approximation to the diffusion coefficient in a binary mixture of the sth and tth gases, at the pressure and temperature of the actual multiple mixture. Also (cf. (18.3, 1)) the diffusion velocity $\overline{\boldsymbol{C}}_s$ corresponding to (18.41, 2) is $\boldsymbol{\alpha}_s$; hence to the present first approximation

$$\boldsymbol{d}_s = -\sum_t \mathrm{x}_s \mathrm{x}_t (\overline{\boldsymbol{C}}_s - \overline{\boldsymbol{C}}_t)/[D_{st}]_1. \tag{18.41, 3}$$

Equation (18.41, 3) shows that the generalized diffusion coefficients D_{st} of (18.3, 13) are identical, to a first approximation, with binary diffusion coefficients at the same temperature and pressure. An interpretation of (18.41, 3) as an (approximate) equation of motion of the sth gas was given in 6.63.

Equation (18.41, 3) is equivalent to $K-1$ independent equations, from which the $K-1$ velocities of diffusion of the other constituents relative to a given constituent can be calculated. For example, for a ternary mixture, from the first two of equations (18.41, 3),

$$\overline{\boldsymbol{C}}_1 - \overline{\boldsymbol{C}}_2 = \frac{[D_{12}]_1([D_{23}]_1 \mathrm{x}_1 \boldsymbol{d}_2 - [D_{13}]_1 \mathrm{x}_2 \boldsymbol{d}_1)}{\mathrm{x}_1 \mathrm{x}_2(\mathrm{x}_1[D_{23}]_1 + \mathrm{x}_2[D_{31}]_1 + \mathrm{x}_3[D_{12}]_1)}. \tag{18.41, 4}$$

Similar expressions, in terms of ratios of determinants, can be given for general values of K (cf. 18.3, footnote to p. 346). However, in view of the difficulty of evaluating the determinants, it is normally simpler to retain the equations (18.41, 3).

18.42. Thermal conductivity

By (18.31, 2, 8) the thermal conductivity λ is given by $\sum_s \lambda_{(s)}$, where

$$\lambda_{(s)} = \tfrac{1}{3}k\int f_s^{(0)}(\mathscr{C}_s^2 - \tfrac{5}{2})\,\boldsymbol{C}_s . \boldsymbol{A}_s\, d\boldsymbol{c}_s. \tag{18.42, 1}$$

We may regard $\lambda_{(s)}$ as the contribution of the sth gas to the total conductivity. From (18.2, 11, 13) and (18.31, 7), the equation satisfied by $\boldsymbol{A}_s$ is

$$\sum_t n_s n_t I_{st}(\boldsymbol{A}_s + \boldsymbol{A}_t) = f_s^{(0)} \boldsymbol{C}_s(\mathscr{C}_s^2 - \tfrac{5}{2} - x_s^{-1}\mathrm{k}_{Ts}), \tag{18.42, 2}$$

since $\sum_t \mathrm{k}_{Tt} = 0$.

Equation (18.42, 2) can be used to determine λ without first knowing k_{Ts}. From the definitions of $\tilde{\boldsymbol{A}}_s$ and k_{Ts}, using (18.3, 3),

$$\int f_s^{(0)} \tilde{\boldsymbol{A}}_s . \boldsymbol{C}_s d\boldsymbol{c}_s = 3n_s(D_{Ts} - \sum_t k_{Tt}\Delta_{st}) = 0. \qquad (18.42, 3)$$

Thus $\tilde{\boldsymbol{A}}_s$ does not contribute to the diffusion velocity $\overline{\boldsymbol{C}}_s$ (as is natural, because $\tilde{\boldsymbol{A}}_s$ was introduced to represent heat conduction in the absence of diffusion). Hence when $\tilde{\boldsymbol{A}}_s$ is expressed in terms of the vectors $S_{\frac{3}{2}}^{(p)}(\mathscr{C}_s^2)\,\boldsymbol{C}_s$ the term $p = 0$, which alone contributes to $\overline{\boldsymbol{C}}_s$, must be absent. The coefficients of the vectors for $p > 0$ can be determined by multiplying (18.42, 2) by the vectors $S_{\frac{3}{2}}^{(q)}(\mathscr{C}_s^2)\,\boldsymbol{C}_s$, where $q > 0$, and integrating with respect to $\boldsymbol{c}_s$; when this is done, the term involving k_{Ts} is found to disappear.

The first approximation to $\tilde{\boldsymbol{A}}_s$ can be written

$$\tilde{\boldsymbol{A}}_s = -(2m_s/5k^2Tn_s)\,a_s S_{\frac{3}{2}}^{(1)}(\mathscr{C}_s^2)\,\boldsymbol{C}_s; \qquad (18.42, 4)$$

the factor multiplying a_s is introduced in order to make a_s equal to the first approximation to $\lambda_{(s)}$. To determine a_s we multiply (18.42, 2) by $-S_{\frac{3}{2}}^{(1)}(\mathscr{C}_s^2)\,\boldsymbol{C}_s$ and integrate with respect to $\boldsymbol{c}_s$. Then on substituting for $\tilde{\boldsymbol{A}}_s$ from (18.42, 4) we get

$$a_s\{n_s[S_{\frac{3}{2}}^{(1)}(\mathscr{C}_s^2)\,\mathscr{C}_s, S_{\frac{3}{2}}^{(1)}(\mathscr{C}_s^2)\,\mathscr{C}_s]_s + \sum_{t \neq s} n_t[S_{\frac{3}{2}}^{(1)}(\mathscr{C}_s^2)\,\mathscr{C}_s, S_{\frac{3}{2}}^{(1)}(\mathscr{C}_s^2)\,\mathscr{C}_s]_{st}\}$$
$$+ n_s \sum_{t \neq s} a_t(m_t/m_s)^{\frac{1}{2}}\,[S_{\frac{3}{2}}^{(1)}(\mathscr{C}_s^2)\,\mathscr{C}_s, S_{\frac{3}{2}}^{(1)}(\mathscr{C}_t^2)\,\mathscr{C}_t]_{st} = 75k^2Tn_s/8m_s. \qquad (18.42, 5)$$

The bracket integrals are given by equations similar to (9.6, 3, 6, 8); the expressions for them can be simplified by using equations similar to (9.7, 1), (9.8, 7, 8) and (9.81, 1). Using these equations, (18.42, 5) can be reduced to the form

$$a_s\{x_s/[\lambda_s]_1 + \sum_{t \neq s}(Tx_t/5p[D_{st}]_1)(6M_{ts}^2 + (5 - 4B_{st})\,M_{st}^2 + 8M_{st}\,M_{ts}\,A_{st})\}$$
$$- x_s \sum_{t \neq s} a_t(TM_{st}M_{ts}/5p[D_{st}]_1)(11 - 4B_{st} - 8A_{st}) = x_s. \qquad (18.42, 6)$$

Here $[\lambda_s]_1$ and $[D_{st}]_1$ denote, as before, first approximations to the thermal conductivity of the sth gas (when pure) and the diffusion coefficient in a binary s, t mixture, at pressure p and temperature T; A_{st}, B_{st} are defined in terms of collision-integrals for collisions of molecules m_s, m_t in the same way as the A, B of (9.8, 7); and

$$M_{st} = m_s/(m_s + m_t), \quad M_{ts} = m_t/(m_s + m_t). \qquad (18.42, 7)$$

The first approximation $[D_{st}]_1$ is used instead of E_{st} because of its closer relation to experimental quantities.

In any actual calculation, the quantities a_s are best determined by solving the simultaneous equations (18.42, 6) by numerical methods. The thermal conductivity is then given, to a first approximation, by

$$[\lambda]_1 = \sum_s [\lambda_{(s)}]_1 = \sum_s a_s. \qquad (18.42, 8)$$

The expression for $[\lambda]_1$ given by (18.42, 6, 8) may be compared with the Wassiljewa expression given in 6.63. Whereas the latter expression implies that the effect of collisions is solely to oppose the conduction of heat, (18.42, 6) implies that there is also a secondary effect, the transport of heat $-\lambda_{(t)}\nabla T$ by the tth gas being in part converted into heat transport by the sth gas at collisions between molecules m_s and m_t.*

18.43. Thermal diffusion

We here consider the thermal diffusion ratio k_{Ts} rather than D_{Ts}, because of the physical significance of k_{Ts} in relations like (18.3, 15) and (18.31, 6). In order to determine k_{Ts} we multiply (18.42, 2) by $m_s \boldsymbol{C}_s$ and integrate with respect to $\boldsymbol{c}_s$. Then the term arising from $\mathscr{C}_s^2 - \frac{5}{2}$ on the right vanishes, and we find

$$-3p\mathrm{k}_{Ts} = n_s^2[m_s \boldsymbol{C}_s, \boldsymbol{\bar{A}}_s]_s + \sum_{t \neq s} n_s n_t [m_s \boldsymbol{C}_s, \boldsymbol{\bar{A}}_s + \boldsymbol{\bar{A}}_t]_{st}. \qquad (18.43, 1)$$

The first term on the right vanishes because $m_s \boldsymbol{C}_s$ is a summational invariant. In the second term we substitute for $\boldsymbol{\bar{A}}_s$ the approximate form given by (18.42, 4). Since $[m_s \boldsymbol{C}_s + m_t \boldsymbol{C}_t, \boldsymbol{\bar{A}}_s]_{st} = 0$, this gives

$$15kp[\mathrm{k}_{Ts}]_1 = \sum_{t \neq s} 4(m_s m_t)^{\frac{1}{2}} \{n_s a_t [\boldsymbol{\mathscr{C}}_s, S_{\frac{3}{2}}^{(1)}(\mathscr{C}_t^2)\boldsymbol{\mathscr{C}}_t]_{st} - n_t a_s [\boldsymbol{\mathscr{C}}_t, S_{\frac{3}{2}}^{(1)}(\mathscr{C}_s^2)\boldsymbol{\mathscr{C}}_s]_{st}\}. \qquad (18.43, 2)$$

The bracket integrals in (18.43, 2) are given by formulae like (9.6, 2). On substituting their values, and using equations similar to (9.81, 1), we get

$$[\mathrm{k}_{Ts}]_1 = \sum_t (T\mathrm{c}_{st}/p[D_{st}]_1)(\mathrm{x}_s a_t M_{st} - \mathrm{x}_t a_s M_{ts}). \qquad (18.43, 3)$$

Here $[D_{st}]_1$, M_{st}, M_{ts} are as defined in 18.42, and c_{st} is the equivalent, for encounters of molecules m_s, m_t, of the c defined in (9.8, 7). The value of $[\mathrm{k}_{Ts}]_1$ is now obtained by substituting for a_s, a_t their values found by solving (18.42, 6). Since a_s, being the first approximation to $\lambda_{(s)}$, has x_s as a factor, we can write in generalization of (8.4, 8)

$$[\mathrm{k}_{Ts}]_1 = \sum_t \mathrm{x}_s \mathrm{x}_t \alpha_{st}, \qquad (18.43, 4)$$

where $\alpha_{st} = -\alpha_{ts}$.

Equation (18.43, 3) indicates a close dependence of k_{Ts} on the parts $\lambda_{(t)}$ of the thermal conductivity λ. However, whereas λ is appreciably affected by the internal energy of the molecules, it is usually assumed that the internal energy hardly affects the value of k_{Ts}. Thus (18.43, 3) should apply approximately to molecules possessing internal energy, provided that in (18.42, 6) $[\lambda_s]_1$ is given its monatomic value $15k[\mu_s]_1/4m_s$.

* T. G. Cowling, P. Gray and P. G. Wright, *Proc. R. Soc.* A, **276**, 69 (1963). See also J. M. Yos, *AVCO Report*, AVSSD-0112-67-RM (1967).

18.44. The viscosity

The viscosity is given by (18.32, 1), i.e.

$$\mu = \Sigma\mu_{(s)}, \tag{18.44, 1}$$

where $\mu_{(s)}$, the contribution of the sth gas to μ, is given by

$$\mu_{(s)} = \tfrac{2}{5}kT\int f_s^{(0)}\overset{\circ}{\mathscr{C}_s\mathscr{C}_s} : \mathsf{B}_s\, d\boldsymbol{c}_s \tag{18.44, 2}$$

and B_s satisfies (18.2, 12). The first approximation to μ is derived by using the approximate forms

$$\mathsf{B}_s = (b_s/n_s kT)\overset{\circ}{\mathscr{C}_s\mathscr{C}_s}; \tag{18.44, 3}$$

the factor $n_s kT$ is inserted in (18.44, 3) so as to make b_s equal to the first approximation to $\mu_{(s)}$. To determine b_s we multiply (18.2, 12) by $\overset{\circ}{\mathscr{C}_s\mathscr{C}_s}$ and integrate with respect to $\boldsymbol{c}_s$. This gives, after substituting for B_s from (18.44, 3),

$$\tfrac{5}{2}n_s kT = n_s b_s[\overset{\circ}{\mathscr{C}_s\mathscr{C}_s}, \overset{\circ}{\mathscr{C}_s\mathscr{C}_s}]_s + \sum_{t\neq s}(n_t b_s[\overset{\circ}{\mathscr{C}_s\mathscr{C}_s}, \overset{\circ}{\mathscr{C}_s\mathscr{C}_s}]_{st} + n_s b_t[\overset{\circ}{\mathscr{C}_s\mathscr{C}_s}, \overset{\circ}{\mathscr{C}_t\mathscr{C}_t}]_{st}). \tag{18.44, 4}$$

The bracket integrals in (18.44, 4) are determined from equations similar to (9.6, 14–16), (9.7, 1), (9.8, 7) and (9.81, 1). Using these, we find from (18.44, 4)

$$x_s = b_s\left\{\frac{x_s}{[\mu_s]_1} + \sum_{t\neq s}\frac{3x_t}{(\rho'_s+\rho'_t)[D_{st}]_1}\left(\frac{2}{3}+\frac{m_t}{m_s}A_{st}\right)\right\} - x_s\sum_{t\neq s}\frac{3b_t}{(\rho'_s+\rho'_t)[D_{st}]_1}(\tfrac{2}{3}-A_{st}). \tag{18.44, 5}$$

In (18.44, 5), $[\mu_s]_1$, $[D_{st}]_1$, A_{st} are as defined earlier; ρ'_s, ρ'_t denote the densities of the sth and tth gases when pure, at the pressure and temperature of the actual gas. When the quantities b_s have been found from (18.44, 5) the first approximation to μ is given by

$$[\mu]_1 = \sum_s[\mu_{(s)}]_1 = \sum_s b_s. \tag{18.44, 6}$$

APPROXIMATE VALUES IN SPECIAL CASES

18.5. Isotopic mixtures

Consider a mixture of several isotopes of the same gas. The mean mass $\overline{m}$ of molecules in the mixture is defined by

$$n\overline{m} = \sum_i n_i m_i, \tag{18.5, 1}$$

and the quantities $(m_i-\overline{m})/\overline{m}$ are supposed to be small enough for their squares and products to be neglected. Then, assuming that the same force-law applies to all types of encounter in the mixture, the thermal conductivity and viscosity are found to be those corresponding to a simple gas of molecular mass $\overline{m}$. The constants A, B, C are the same for each pair of isotopes; the thermal diffusion ratio k_{Ts} is such that*

$$[k_{Ts}]_1 = \frac{15C(1+A)x_s}{2A(11-4B+8A)} \frac{m_s-\overline{m}}{\overline{m}}. \tag{18.5, 2}$$

The Kihara approximation is found by replacing B by $\frac{3}{4}$.

In the steady state in which thermal diffusion is balanced by diffusion due to inhomogeneity, $\nabla x_s = -k_{Ts} \nabla \ln T$. Thus, to a first approximation,

$$\nabla \ln\left(\frac{x_s}{x_t}\right) = -\nabla \ln T \frac{15C(1+A)}{2A(11-4B+8A)} \frac{m_s-m_t}{\overline{m}}. \tag{18.5, 3}$$

Since this is valid only to the first order in $(m_s-m_t)/\overline{m}$, the quantity $\overline{m}$ in the denominator could be replaced by any other mean mass. Hence (18.5, 3) shows that the relative concentration of molecules m_s, m_t has the same spatial variation as if the other isotopes were absent.

Equations (18.5, 2) and (18.5, 3) indicate that results for a mixture of several isotopes can be used to determine the thermal diffusion ratio for a pair of isotopes, and vice versa. A similar remark also applies to the diffusion thermo-effect.†

The thermal separation of two isotopic gases (m_1, m_2) for which C_{12} is small can be greatly increased by the introduction, in excess, of a third gas (m_3) such that C_{13}, C_{23} are large. Suppose that the properties of the third gas are otherwise intermediate between those of the two isotopes, so that in (18.43, 3)

$$a_1/x_1 m_1 > a_3/x_3 m_3 > a_2/x_2 m_2.$$

Then in the steady state in which thermal diffusion is balanced by that due to inhomogeneity

$$\nabla \ln\left(\frac{x_1}{x_2}\right) = -\nabla T\left[\frac{(x_1+x_2)m_1 m_2 C_{12}}{p(m_1+m_2)[D_{12}]_1}\left(\frac{a_1}{x_1 m_1}-\frac{a_2}{x_2 m_2}\right)\right.$$
$$\left.+\frac{x_3 m_3}{p}\left\{\frac{M_{13}C_{13}}{[D_{13}]_1}\left(\frac{a_1}{x_1 m_1}-\frac{a_3}{x_3 m_3}\right)+\frac{M_{23}C_{23}}{[D_{23}]_1}\left(\frac{a_3}{x_3 m_3}-\frac{a_2}{x_2 m_2}\right)\right\}\right]. \tag{18.5, 4}$$

Because the properties of molecules m_3 are intermediate between those of the two isotopes, the three quantities

$$\frac{m_1 m_2}{(m_1+m_2)[D_{12}]_1}, \quad \frac{m_1 m_3}{(m_1+m_3)[D_{13}]_1}, \quad \frac{m_2 m_3}{(m_2+m_3)[D_{23}]_1}$$

are nearly equal. Also C_{13}, C_{23} are nearly equal; hence (18.5, 4) approximates to

$$\nabla \ln\left(\frac{x_1}{x_2}\right) = -\frac{\{(1-x_3)C_{12}+x_3 C_{13}\}}{p(m_1+m_2)[D_{12}]_1} m_1 m_2 \left(\frac{a_1}{x_1 m_1}-\frac{a_2}{x_2 m_2}\right)\nabla T. \tag{18.5, 5}$$

* R. C. Jones, *Phys. Rev.* **59**, 1019 (1941).

† L. Waldmann, *Z. Naturforschung*, **1**, 12 (1946).

Suppose that a_1/x_1, a_2/x_2 (the contributions of molecules m_1, m_2 to the thermal conductivity, in proportion to their numbers) are not greatly altered by the presence of the third gas. Then the effect of its presence is to increase the expression on the right of (18.5, 5) in the ratio

$$\{(1-x_3)c_{12}+x_3c_{13}\}/c_{12} = 1+x_3(c_{13}/c_{12}-1).$$

This is large if c_{13}/c_{12} is large and x_3 is not small.

Clusius and his associates have used the enhanced separation due to an added gas in separating isotopes with the thermal diffusion column. A nearly steady state is eventually reached in which the two isotopes, each containing a small amount of the added gas, are separated by a mass of that gas in an approximately pure state. The trace of the added gas is then removed from the isotopes by a chemical process.*

18.51. One gas present as a trace

Consider a ternary mixture, with molecules m_3 present in trace only. The viscosity and thermal conductivity are normally those of a binary gas of molecules m_1, m_2, as if the third gas were absent; the relative diffusion of molecules m_1, m_2 is likewise not appreciably affected by the molecules m_3.

The diffusion of molecules m_3 through an otherwise homogeneous gas of molecules m_1, m_2 at rest under no forces is given, to the first approximation, by

$$\nabla x_3 = -x_3\overline{C}_3(x_1/[D_{13}]_1+x_2/[D_{23}]_1) \qquad (18.51, 1)$$

(cf. (18.41, 3)). This is the same as for diffusion through a single gas, with coefficient of diffusion $D_{12,3}$, where

$$1/[D_{12,3}]_1 = x_1/[D_{13}]_1+x_2/[D_{23}]_1. \qquad (18.51, 2)$$

Equation (18.51, 2) was verified experimentally by Walker, de Haas and Westenberg in the diffusion of CO_2 through mixtures of He and N_2.† It should be noted, however, that the linear dependence indicated by (18.51, 2) is correct to a first approximation only.‡

18.52. Ternary mixture, including electrons

Consider a ternary mixture in which the molecules m_3 are electrons, so that m_3/m_1 and m_3/m_2 are small. We shall give approximate expressions, regarding $(m_3/m_1)^{\frac{1}{2}}$ and $(m_3/m_2)^{\frac{1}{2}}$ as negligible, and assuming that x_3 is not small.

The quantities λ_3, D_{13}, D_{23} in (18.42, 6) are all proportional to $m_3^{-\frac{1}{2}}$, and so are large. Hence a_3 is also proportional to $m_3^{-\frac{1}{2}}$, and is large compared with

* See, for example, K. Clusius and E. Schumacher, *Helv. Chim. Acta*, **36**, 969 (1953), for the separation of ^{36}A, ^{38}A, ^{40}A using H^{35}Cl and H^{37}Cl or D^{35}Cl and D^{37}Cl as auxiliary gases.

† R. E. Walker, N. de Haas and A. A. Westenberg, *J. Chem. Phys.* **32**, 1314 (1960); see also B. N. Srivastava and R. Paul, *Physica*, **28**, 646 (1962); R. Paul, *Ind. J. Phys.* **36**, 464 (1962).

‡ S. I. Sandler and E. A. Mason, *J. Chem. Phys.* **48**, 2873 (1968).

a_1 and a_2. This implies that the thermal conductivity is, to the present approximation, due wholly to the electrons, being given by

$$[\lambda]_1 = a_3 = x_3 \bigg/ \left\{\frac{x_3}{[\lambda_3]_1} + \frac{T}{5p}\left[\frac{x_1(5-4B_{13})}{[D_{13}]_1} + \frac{x_2(5-4B_{23})}{[D_{23}]_1}\right]\right\}. \quad (18.52, 1)$$

The error involved in neglecting $(m_3/m_1)^{\frac{1}{2}}$ and $(m_3/m_2)^{\frac{1}{2}}$ may be 2 or 3 per cent. This is likely to be smaller than that involved in proceeding only to a first approximation $[\lambda]_1$, especially when the molecules m_1 amd m_2 are both ionized. If further approximations to λ are required (still regarding $(m_3/m_1)^{\frac{1}{2}}$ and $(m_3/m_2)^{\frac{1}{2}}$ as negligible), they can best be calculated by regarding the molecules m_1 and m_2 as inert targets, but taking into account the encounters of electrons with other electrons as well as with these targets. In a wholly ionized gas the thermal conductivity then is that of a binary electron-ion gas, with a number-density n_1+n_2 of heavy particles each possessing charge e_{12}, where

$$(n_1+n_2)\,e_{12}^2 = n_1 e_1^2 + n_2 e_2^2. \quad (18.52, 2)$$

In the viscosity formulae, $[\mu_3]_1$, $m_3[D_{13}]_1$ and $m_3[D_{23}]_1$ are each small, of order $m_3^{\frac{1}{2}}$; thus from (18.44, 5) b_3 is also of order $m_3^{\frac{1}{2}}$, and may be neglected. Hence to the present order of approximation the viscosity is that of the two heavy gases, unaffected by the presence of electrons.

In order to calculate the thermal diffusion ratios it is necessary to calculate a_1 and a_2; these have, to the present approximation, the same values as if electrons were absent. In terms of a_1, a_2, a_3, the thermal diffusion ratios are given by

$$[k_{T3}]_1 = -\frac{a_3 T}{p}\left(\frac{C_{13}\,x_1}{[D_{13}]_1} + \frac{C_{23}\,x_2}{[D_{23}]_1}\right), \quad (18.52, 3)$$

$$[k_{T1}]_1 = \frac{T C_{12}}{p[D_{12}]_1}(x_1 a_2 M_{12} - x_2 a_1 M_{21}) + \frac{T C_{13} x_1 a_3}{p[D_{13}]_1}, \quad (18.52, 4)$$

with a similar relation for $[k_{T2}]_1$. Since a_3, $[D_{13}]_1$, $[D_{23}]_1$ are all proportional to $m_3^{-\frac{1}{2}}$, all the terms in (18.52, 3, 4) are similar in order of magnitude.

The diffusion velocities are given, as in (18.3, 13), by

$$\boldsymbol{d}'_s \equiv \boldsymbol{d}_s + k_{Ts}\nabla \ln T = -\sum_t x_s x_t(\overline{\boldsymbol{C}}_s - \overline{\boldsymbol{C}}_t)/D_{st}. \quad (18.52, 5)$$

In this the coefficients k_{Ts} are similar in order of magnitude for all s; the same may be assumed for $\boldsymbol{d}_s$, since in (18.2, 4) the various pressure-gradients ∇p_s are similar in order of magnitude, and the volume force $\rho_s \boldsymbol{F}_s$ due to an electric field is comparable in magnitude for electrons and heavy ions. Solving (18.52, 5) for the diffusion velocities as in (18.41, 4) we get

$$\overline{\boldsymbol{C}}_1 - \overline{\boldsymbol{C}}_2 = \frac{D_{12}(D_{23} x_1 \boldsymbol{d}'_2 - D_{13} x_2 \boldsymbol{d}'_1)}{x_1 x_2(x_1 D_{23} + x_2 D_{31} + x_3 D_{12})} \quad (18.52, 6)$$

and two similar relations. Thus, in view of the smallness of D_{12} compared with D_{13} and D_{23}, the diffusion process falls into two parts. The first and major part is the diffusion of electrons relative to heavy molecules, with velocity

$$\overline{C}_3-\overline{C}_1 \doteqdot \overline{C}_3-\overline{C}_2 \doteqdot -D_{13}D_{23}\boldsymbol{d}_3'/x_3(x_1D_{23}+x_2D_{13}). \quad (18.52, 7)$$

This is the same as the velocity of diffusion of electrons relative to a single heavy gas of number-density n_1+n_2, with diffusion coefficient $D_{12,3}$, where

$$(x_1+x_2)/D_{12,3} = x_1/D_{13}+x_2/D_{23}. \quad (18.52, 8)$$

The second part consists of the (much smaller) relative diffusion of the two heavy gases, given by

$$\overline{C}_1-\overline{C}_2 = \frac{D_{12}(D_{23}x_1\boldsymbol{d}_2'-D_{13}x_2\boldsymbol{d}_1')}{x_1x_2(x_1D_{23}+x_2D_{13})}. \quad (18.52, 9)$$

Let $\boldsymbol{d}_1'$, $\boldsymbol{d}_2'$ be expressed in terms of $\boldsymbol{d}_3'$ $(\equiv -\boldsymbol{d}_1'-\boldsymbol{d}_2')$ and $\boldsymbol{d}_{12}'$, defined by

$$\boldsymbol{d}_{12}' = (\rho_2\boldsymbol{d}_1'-\rho_1\boldsymbol{d}_2')/(\rho_1+\rho_2); \quad (18.52, 10)$$

for a binary gas in the absence of a temperature gradient $\boldsymbol{d}_{12}'$ reduces to the $\boldsymbol{d}_{12}$ of 8.3. Then (18.52, 9) becomes

$$x_1x_2(\overline{C}_1-\overline{C}_2) = -D_{12}\boldsymbol{d}_{12}'-\frac{D_{12}(\rho_2D_{23}x_1-\rho_1D_{13}x_2)\boldsymbol{d}_3'}{(\rho_1+\rho_2)(x_1D_{23}+x_2D_{13})}. \quad (18.52, 11)$$

This indicates the presence of 'induced' diffusion due to $\boldsymbol{d}_3'$, as well as the usual diffusion due to $\boldsymbol{d}_{12}'$.

Equations (18.52, 7, 9, 11) do not depend on any particular order of approximation to D_{12}, D_{13} and D_{23}. In general it is sufficient to replace D_{12} by its first approximation $[D_{12}]_1$; however, as with λ it may be desirable, especially with a fully ionized gas, to proceed to higher approximations in calculating D_{13} and D_{23}. Electron diffusion in a fully ionized gas can be calculated from binary-gas formulae, using (18.52, 2).

If we are concerned only with thermal diffusion, $\boldsymbol{d}_s' = k_{Ts}\nabla \ln T$. Thus, using (18.52, 4) and a similar equation, we find from (18.52, 9) that, to a first approximation,

$$\overline{C}_1-\overline{C}_2 = p^{-1}\nabla T\{(x_1x_2)^{-1}c_{12}(x_2a_1M_{21}-x_1a_2M_{12}) + (c_{23}-c_{13})a_3D_{12}/(x_1D_{23}+x_2D_{13})\}. \quad (18.52, 12)$$

Since a_1, a_2, a_3 denote first approximations to the contributions of the three gases to the thermal conductivity, the a_3 term in (18.52, 12) represents an induced thermal diffusion due to the presence of the electrons; it is normally similar in order of magnitude to the other terms, but vanishes if $c_{13} = c_{23}$, as is the case for example in a fully ionized gas ($c_{13} = c_{23} = c_{12} = -0{\cdot}6$). Thermal diffusion is especially important in a fully ionized gas, because for such a gas the quantities c_{st} are several times as large as for a neutral gas.

A case of special interest is when molecules m_1 are protons, and molecules m_2, which are few in number, are ions of much greater mass and charge. In this case M_{12} is small and $M_{21} \doteqdot 1$; also $x_1 a_2/x_2 a_1$ is of order $(M_{12})^{\frac{1}{2}}$ and a_1 is approximately $[\lambda_1]_1$. Hence (18.52, 12) becomes approximately

$$\overline{C}_1 - \overline{C}_2 = p^{-1} x_1^{-1} c_{12} [\lambda_1]_1 \nabla T. \tag{18.52, 13}$$

This does not depend on the charge e_2 on the heavy ions, whereas the coefficient of diffusion D_{12} is proportional to $1/e_2^2$. Thus if e_2 is large, thermal diffusion may be able to produce a greatly enhanced separation of the two kinds of positive ion before being balanced by inhomogeneity diffusion,* the heavy ions travelling towards the hotter regions.

* S. Chapman, *Proc. Phys. Soc.* **72**, 353 (1958). A more exact discussion was given by J. M. Burgers, in *Plasma Physics* (ed. F. H. Clauser), pp. 260–4 (Pergamon Press, 1960). This took into account electric fields of thermal origin, and obtained a rather larger diffusion velocity.

19

ELECTROMAGNETIC PHENOMENA IN IONIZED GASES

19.1. Convection currents and conduction currents

In ordinary gases the molecules are electrically neutral; when some of them carry electric charges, the gas is said to be ionized. In an ionized gas the different types of particle (to all of which the general term 'molecule' will be applied) may include neutral molecules or atoms, atomic or molecular ions of either sign, and electrons. The charged particles will carry an integral multiple (positive or negative) of the charge e, where e denotes the numerical value of the electronic charge.

Let n_s, m_s, e_s denote the number-density, mass, and charge of the sth type of molecule in an ionized gas ($s = 1, 2, \ldots$). It is convenient to enumerate first the various kinds of neutral molecules, so that if, for example, there is only one kind of neutral molecule present, $s = 1$ will refer to this.

The volume-density of charge, or charge per unit volume, may be denoted by ρ_e; clearly

$$\rho_e = \Sigma n_s e_s. \tag{19.1, 1}$$

The current-intensity is defined as the vector in the direction of current-flow at any point, with magnitude equal to the current per unit cross-section normal to this direction: it is equal to

$$\begin{aligned}\Sigma n_s e_s \bar{\boldsymbol{c}}_s &= \Sigma n_s e_s \boldsymbol{c}_0 + \Sigma n_s e_s \overline{\boldsymbol{C}}_s \\ &= \rho_e \boldsymbol{c}_0 + \boldsymbol{j},\end{aligned} \tag{19.1, 2}$$

where

$$\boldsymbol{j} \equiv \Sigma n_s e_s \overline{\boldsymbol{C}}_s. \tag{19.1, 3}$$

Clearly $\boldsymbol{j}$ is the *charge* flux-vector of 2.3 or 2.5, corresponding to $\phi_s = e_s$.

In (19.1, 2) the first term is due to transport of the volume-charge with velocity $\boldsymbol{c}_0$; this part of the current will be called the *convection current*, and the remaining part, $\boldsymbol{j}$, will be called the *conduction current*. The division here made between the two is to some extent arbitrary, because, so far as electrical considerations are concerned, we might equally well have written $\boldsymbol{c}_s = \bar{\boldsymbol{c}} + \boldsymbol{C}_s'$, where $\bar{\boldsymbol{c}}$ denotes the mean molecular velocity (not the *mass*-velocity of (2.5, 7)); in that case we might regard $\rho_e \bar{\boldsymbol{c}}$ as the convection current, and $\Sigma n_s e_s \overline{\boldsymbol{C}_s'}$ as the conduction current. The choice actually made here, however, is that indicated by (19.1, 2).

If either $\rho_e = 0$, or $\boldsymbol{c}_0 = 0$, there is no convection current. In many ionized gases $\rho_e = 0$ to a high degree of approximation; when this is not so, an electric field is set up which tends to dissipate the volume charge; the gas retains an appreciable charge only if external agencies maintain it.

19.11. The electric current in a binary mixture

In the case of a binary mixture, (19.1, 3) reduces to

$$\begin{aligned} \boldsymbol{j} &= n_1 e_1 \overline{\boldsymbol{C}}_1 + n_2 e_2 \overline{\boldsymbol{C}}_2 \\ &= n_1(e_1 - e_2 m_1/m_2)\overline{\boldsymbol{C}}_1 \\ &= (n_1 n_2/\rho)(e_1 m_2 - e_2 m_1)(\overline{\boldsymbol{C}}_1 - \overline{\boldsymbol{C}}_2). \end{aligned} \qquad (19.11, 1)$$

In this we substitute for $\boldsymbol{C}_1 - \boldsymbol{C}_2$ from (14.1, 1), giving

$$\boldsymbol{j} = -\frac{n^2}{\rho}(e_1 m_2 - e_2 m_1) D_{12} \left\{ \nabla x_1 + \frac{n_1 n_2 (m_2 - m_1)}{n\rho} \nabla \ln p \right. \\ \left. - \frac{\rho_1 \rho_2}{p\rho}(\boldsymbol{F}_1 - \boldsymbol{F}_2) + k_T \nabla \ln T \right\}. \qquad (19.11, 2)$$

If the forces $m_1 \boldsymbol{F}_1$ and $m_2 \boldsymbol{F}_2$ acting on the two types of molecule are due to the presence of an electric field $\boldsymbol{E}$, or include a part due to this cause, the corresponding part of $\boldsymbol{j}$ (the ohmic current) is

$$\frac{n_1 n_2 n^2}{p\rho^2}(e_1 m_2 - e_2 m_1)^2 D_{12} \boldsymbol{E} = \vartheta \boldsymbol{E}, \qquad (19.11, 3)$$

where

$$\vartheta \equiv \frac{n_1 n_2 n}{\rho^2 kT}(e_1 m_2 - e_2 m_1)^2 D_{12}. \qquad (19.11, 4)$$

The coefficient ϑ, which is essentially positive, is called the *electrical conductivity* of the gas.

If the molecules m_1 are either ions or neutral molecules, and the others are electrons, the ratio m_2/m_1 is so small that to a high degree of approximation $\rho = n_1 m_1$, and (19.11, 4) is equivalent to

$$\vartheta = \frac{n_2 n e_2^2}{k n_1 T} D_{12}. \qquad (19.11, 5)$$

Since this involves e_2, and not also e_1, it shows that the conductivity is nearly all due to the electrons. If the gas is electrically neutral, so that $n_1 e_1 = -n_2 e_2$, the convection current vanishes, and (19.11, 5) may also be written in the more symmetric form

$$\vartheta = -\frac{n e_1 e_2}{kT} D_{12}; \qquad (19.11, 6)$$

this expression for ϑ is still positive, because on the present hypothesis e_1 and e_2 are opposite in sign. If the ions are singly charged, (19.11, 6) reduces to

$$\vartheta = \frac{n e^2}{kT} D_{12}. \qquad (19.11, 7)$$

Besides the conduction current due to an electric field, contributions to the total conduction current also arise from gradients of concentration, pressure, and temperature, and also from any non-electrical forces which tend to produce unequal accelerations of the two types of molecule; gravitational or centrifugal forces tend to accelerate all particles equally, so that they do not contribute directly to the electric current, though they may contribute indirectly, by setting up a pressure or concentration gradient.

19.12. Electrical conductivity in a slightly ionized gas

Consider a gas, the great majority of whose molecules are neutral. If there is only one kind of neutral molecule ($e_1 = 0$) and one kind of charged molecule ($e_2 \neq 0$), the electric conductivity ϑ is, by (19.11, 4) (since n and ρ are approximately n_1 and $n_1 m_1$),

$$\vartheta = \frac{n_2 e_2^2}{kT} D_{12},$$

as if each charged particle contributed an amount

$$\vartheta / n_2 = \frac{e_2^2}{kT} D_{12}$$

to ϑ.

In this case, by (19.11, 1), using similar approximations,

$$\begin{aligned} \bar{c}_2 - \bar{c}_1 = \overline{C}_2 - \overline{C}_1 &= j / n_2 e_2 \\ &= (\vartheta / n_2 e_2) E = (e_2 D_{12} / kT) E. \end{aligned}$$

This gives the mean velocity of the charged particles relative to the neutral particles. When the gas is at atmospheric pressure, the coefficient

$$\frac{e_2 D_{12}}{kT}$$

in this equation is called the *mobility* of the charged particles in the neutral gas.

If more than one kind of charged particle is present, each in numbers so small that its influence on the mean velocities of the others is negligible, we have equations such as

$$\overline{C}_s - \overline{C}_1 = \frac{e_s D_{1s}}{kT} E$$

for each type s. In this case the total current-intensity is

$$\begin{aligned} \Sigma n_s e_s \bar{c}_s &= \rho_e c_0 + \Sigma n_s e_s \overline{C}_s \\ &= \rho_e (c_0 + \overline{C}_1) + \Sigma n_s e_s (\overline{C}_s - \overline{C}_1) \\ &= \rho_e \bar{c}_1 + \Sigma \frac{n_s e_s^2 D_{1s}}{kT} E. \end{aligned}$$

If $\rho_e = 0$, the current is ϑE, where now

$$\vartheta = (\Sigma n_s e_s^2 D_{1s}) / kT;$$

thus each charged particle m_s makes an average contribution to ϑ equal to

$$\frac{e_s^2 D_{1s}}{kT}.$$

19.13. Electrical conduction in a multiple mixture

If more than one kind of charged particle is present in the gas, and the degree of ionization is not small, the electrical conductivity must be calculated in terms of the generalized diffusion coefficients Δ_{st} or D_s, of 18.3. If diffusion is due wholly to an electric field, and the gas as a whole is electrically neutral, by (18.3, 2),

$$\boldsymbol{j} = \sum_s n_s e_s \overline{\boldsymbol{C}}_s = \sum_s \sum_t n_s n_t e_s e_t \Delta_{st} \boldsymbol{E}/p.$$

Using the relation $\sum_t \rho_t \Delta_{st} = 0$, one finds that $\boldsymbol{j} = \vartheta \boldsymbol{E}$, where

$$\vartheta = -\frac{1}{p} \sum_{s \neq t} \frac{n_s n_t}{m_s m_t} (e_s m_t - e_t m_s)^2 \Delta_{st}. \qquad (19.13, 1)$$

The sum in (19.13, 1) is over all unequal *pairs* s, t, each pair being taken once only; ϑ is positive,* because negative values of Δ_{st} predominate if $s \neq t$. This gives the appropriate generalization of (19.11, 4).

In the important case when electrons are present in numbers comparable with those of the other charged particles, the contribution of the heavy particles to ϑ is negligible compared with that of the electrons (cf. 18.52). In this case, if the suffix 2 refers to electrons, by (18.3, 13), approximately

$$\boldsymbol{j} = n_2 e_2 \overline{\boldsymbol{C}}_2 = -\boldsymbol{d}_2 e_2 n / \Big(\sum_{t \neq 2} x_t / D_{2t}\Big)$$

$$= \boldsymbol{E} e_2^2 n n_2 / p \Big(\sum_{t \neq 2} x_t / D_{2t}\Big).$$

Hence

$$\vartheta = n_2 e_2^2 / kT \Big(\sum_{t \neq 2} x_t / D_{2t}\Big), \qquad (19.13, 2)$$

indicating that the other gases in the mixture contribute additively to the electrical *resistance*.

MAGNETIC FIELDS

19.2. Boltzmann's equation for an ionized gas in the presence of a magnetic field

All the work of previous chapters has been based on the assumption that the applied forces acting on the molecules are independent of the molecular velocities. In the presence of a magnetic field of intensity $\boldsymbol{H}$, the force on a particle of mass m and charge e *electromagnetic* units, moving with velocity $\boldsymbol{c}$, includes a term $e\boldsymbol{c} \wedge \boldsymbol{H}$ (cf. p. 11 for this notation). The acceleration of a molecule can therefore be divided into two parts, one, denoted by $\boldsymbol{F}$, independent of $\boldsymbol{c}$, and the other, due to the magnetic field, equal to $(e/m)\boldsymbol{c} \wedge \boldsymbol{H}$.

* By (18.3, 7),

$$\vartheta = (3kT)^{-1} \Big\{\sum_s n_s e_s \boldsymbol{D}^{(s)}, \sum_s n_s e_s \boldsymbol{D}^{(s)}\Big\} > 0.$$

The occurrence of the second term necessitates a re-examination of the argument of this book, from 3.1 onwards.*

The first modification comes in the proof of Boltzmann's equation. In 3.1 it was taken as obvious that molecules in the velocity-range $\boldsymbol{c}$, $d\boldsymbol{c}$ at the beginning of an interval dt must occupy an equal velocity-range at the end of this time; this is no longer self-evident. Since the molecular velocity increases during dt from $\boldsymbol{c}$ to $\boldsymbol{c}+\boldsymbol{F}dt+(e/m)\,\boldsymbol{c}\wedge\boldsymbol{H}\,dt$, the range occupied by these molecules at the end of dt is now of magnitude†

$$d\boldsymbol{c}\,\frac{\partial[\boldsymbol{c}+\{\boldsymbol{F}+(e/m)\,\boldsymbol{c}\wedge\boldsymbol{H}\}dt]}{\partial(\boldsymbol{c})} = d\boldsymbol{c}\begin{vmatrix} 1, & \frac{e}{m}H_z dt, & -\frac{e}{m}H_y dt \\ -\frac{e}{m}H_z dt, & 1, & \frac{e}{m}H_x dt \\ \frac{e}{m}H_y dt, & -\frac{e}{m}H_x dt, & 1 \end{vmatrix}$$

which equals $d\boldsymbol{c}$, since dt^2 is to be neglected. The velocity-range is therefore still invariant in magnitude; the derivation of Boltzmann's equation proceeds as before, and yields the result

$$\frac{\partial_e f}{\partial t} = \frac{\partial f}{\partial t}+\boldsymbol{c}.\frac{\partial f}{\partial \boldsymbol{r}}+\left(\boldsymbol{F}+\frac{e}{m}\boldsymbol{c}\wedge\boldsymbol{H}\right).\frac{\partial f}{\partial \boldsymbol{c}}. \tag{19.2, 1}$$

As was noted in 10.34, a simple ionized gas (consisting of molecules with charges all of one sign) does not exist in nature. However, in a gas-mixture the Boltzmann equation for each f_s is identical in form with (19.2, 1). If f_s is regarded as a function of $\boldsymbol{C}_s$, $\boldsymbol{r}$, t instead of $\boldsymbol{c}_s$, $\boldsymbol{r}$, t, the equation becomes

$$\frac{\partial_e f_s}{\partial t} = \frac{Df_s}{Dt}+\boldsymbol{C}_s.\frac{\partial f_s}{\partial \boldsymbol{r}}+\left(\boldsymbol{F}_s+\frac{e_s}{m_s}\boldsymbol{c}_0\wedge\boldsymbol{H}-\frac{D_0\boldsymbol{c}_0}{Dt}\right).\frac{\partial f_s}{\partial \boldsymbol{C}_s}$$
$$+\frac{e_s}{m_s}(\boldsymbol{C}_s\wedge\boldsymbol{H}).\frac{\partial f_s}{\partial \boldsymbol{C}_s}-\frac{\partial f_s}{\partial \boldsymbol{C}_s}\boldsymbol{C}_s:\frac{\partial}{\partial \boldsymbol{r}}\boldsymbol{c}_0. \tag{19.2, 2}$$

* The first theoretical work on the effect of magnetic fields on gas phenomena seems to have been done by R. Gans (*Annln Phys.* **20**, 203 (1906)), who considered the effect of a magnetic field on the flow of heat and electricity in a Lorentzian gas. Further work on a Lorentzian gas was done by N. Bohr (thesis). The general theory of the transport phenomena in an ionized gas subject to a magnetic field was first discussed in the first edition of this book; for a special case of electrical conductivity, see T. G. Cowling, *Mon. Not. R. astr. Soc.* **93**, 90 (1932). The subject has, however, frequently been considered by collision-interval methods, following J. S. Townsend (*Proc. R. Soc.* A, **86**, 571 (1912), and *Electricity in Gases*, §§89–92). It is of great importance in the theory of radio propagation, in which an important part is played by the ionized layers of the upper atmosphere; for early work on this, see P. O. Pedersen, *The Propagation of Radio Waves* (Copenhagen, 1927), chapters 6 and 7. The theory is also of importance for solar physics, owing to the sun's general magnetic field and the much more intense magnetic fields of sunspots; and for the physics of interplanetary and interstellar gas.

† The notation for Jacobians is that of 1.411.

The equation of change of a molecular property ϕ_s can be deduced in the form

$$n_s \Delta\bar{\phi}_s = \frac{D(n_s\bar{\phi}_s)}{Dt} + n_s\bar{\phi}_s \frac{\partial}{\partial \boldsymbol{r}}.\boldsymbol{c}_0 + \frac{\partial}{\partial \boldsymbol{r}}.n_s\overline{\phi_s \boldsymbol{C}_s} - n_s\left\{\overline{\frac{D\phi_s}{Dt}} + \overline{\boldsymbol{C}_s.\frac{\partial\phi_s}{\partial \boldsymbol{r}}}\right.$$
$$\left. + \left(\boldsymbol{F}_s + \frac{e_s}{m_s}\boldsymbol{c}_0 \wedge \boldsymbol{H} - \frac{D\boldsymbol{c}_0}{Dt}\right).\overline{\frac{\partial \phi_s}{\partial \boldsymbol{C}_s}} + \frac{e_s}{m_s}\overline{(\boldsymbol{C}_s \wedge \boldsymbol{H}).\frac{\partial\phi_s}{\partial \boldsymbol{C}_s}} - \overline{\frac{\partial \phi_s}{\partial \boldsymbol{C}_s}\boldsymbol{C}_s}:\frac{\partial}{\partial \boldsymbol{r}}\boldsymbol{c}_0\right\}. \tag{19.2, 3}$$

The equations of continuity, mass motion and energy are obtained by giving ϕ_s appropriate values in (19.2, 3) (cf. 8.1). They are

$$\frac{Dn_s}{Dt} + n_s\frac{\partial}{\partial \boldsymbol{r}}.\boldsymbol{c}_0 + \frac{\partial}{\partial \boldsymbol{r}}.(n_s\overline{\boldsymbol{C}}_s) = 0, \quad \frac{D\rho}{Dt} + \rho\frac{\partial}{\partial \boldsymbol{r}}.\boldsymbol{c}_0 = 0, \tag{19.2, 4}$$

$$\rho\frac{D\boldsymbol{c}_0}{Dt} = \sum_s \rho_s \boldsymbol{F}_s + (\sum_s n_s e_s)\boldsymbol{c}_0 \wedge \boldsymbol{H} + \boldsymbol{j} \wedge \boldsymbol{H} - \frac{\partial}{\partial \boldsymbol{r}}.\mathsf{p}, \tag{19.2, 5}$$

$$\tfrac{3}{2}kn\frac{DT}{Dt} = \tfrac{3}{2}kT\frac{\partial}{\partial \boldsymbol{r}}.\sum_s n_s\overline{\boldsymbol{C}}_s + \sum_s \rho_s \boldsymbol{F}_s.\overline{\boldsymbol{C}}_s + \boldsymbol{j}.(\boldsymbol{c}_0 \wedge \boldsymbol{H}) - \mathsf{p}:\frac{\partial}{\partial \boldsymbol{r}}\boldsymbol{c}_0 - \frac{\partial}{\partial \boldsymbol{r}}.\boldsymbol{q}, \tag{19.2, 6}$$

where $\boldsymbol{j}$ is given by (19.1, 3).

It may be proved by the methods of Chapter 4 that the distribution of velocities in a stationary gas in a uniform steady state is of Maxwellian form, provided only that the forces $\boldsymbol{F}_s$ are derived from potential functions Ψ_s. The density distribution is unaffected by the presence of the magnetic field, being given by the equation

$$n_s = (n_s)_0 e^{-m_s\Psi_s/kT}$$

as in (4.14, 7).

Maxwell's formula is also valid for a gas rotating as if rigid, with angular velocity ω, in the presence of a magnetic field whose lines of force are in planes passing through the axis of rotation, and of fields of force whose potentials Ψ_s are symmetric about this axis. Let Oz be the axis of rotation, and let the vector potential of the magnetic field have components $-yA$, xA, 0. Then (cf. (4.14, 11)) it is found that the variation of density of the sth gas is the same as if it were at rest in a field of force of potential

$$\Psi_s - \omega^2(x^2+y^2) - (e_s/m_s)\,\omega(x^2+y^2)A.$$

19.3. The motion of a charged particle in a magnetic field

As a preliminary to considering an elementary theory of the transport phenomena in a magnetic field we first consider in detail the motion of a single molecule of mass m_1 and charge e_1 during a free path. The fields of force—magnetic and otherwise—which act on the particle are supposed uniform. Let Ox be taken parallel to the magnetic intensity $\boldsymbol{H}$, and let Oz be taken in the direction of the component of $\boldsymbol{F}_1$ which is perpendicular to $\boldsymbol{H}$. Then if the components of $\boldsymbol{F}_1$ are X_1, 0, Z_1, the equations of motion of the molecule are

$$m_1\ddot{x} = m_1 X_1, \quad m_1\ddot{y} = e_1 H\dot{z}, \quad m_1\ddot{z} = m_1 Z_1 - e_1 H\dot{y}.$$

Let
$$\omega_1 \equiv e_1 H/m_1. \tag{19.3, 1}$$

Then these equations become

$$\ddot{x} = X_1, \quad \ddot{y} = \omega_1\dot{z}, \quad \ddot{z} = Z_1 - \omega_1\dot{y}. \tag{19.3, 2}$$

Integrating, we obtain the following equations for the position $\boldsymbol{r}$ and velocity $\boldsymbol{c}_1$ of the molecule at any time t, in terms of the position $\boldsymbol{r}'$ and velocity $\boldsymbol{c}_1'$ at time $t = 0$:

$$\left.\begin{aligned} u_1 &= u_1' + X_1 t, \\ v_1 &= (v_1' - Z_1/\omega_1)\cos\omega_1 t + w_1'\sin\omega_1 t + Z_1/\omega_1, \\ w_1 &= w_1'\cos\omega_1 t - (v_1' - Z_1/\omega_1)\sin\omega_1 t, \end{aligned}\right\} \tag{19.3, 3}$$

$$\left.\begin{aligned} x &= x' + u_1' t + \tfrac{1}{2}X_1 t^2, \\ y &= y' + \frac{1}{\omega_1}\left\{\left(v_1' - \frac{Z_1}{\omega_1}\right)\sin\omega_1 t + w_1'(1 - \cos\omega_1 t) + Z_1 t\right\}, \\ z &= z' + \frac{1}{\omega_1}\left\{w_1'\sin\omega_1 t - \left(v_1' - \frac{Z_1}{\omega_1}\right)(1 - \cos\omega_1 t)\right\}. \end{aligned}\right\} \tag{19.3, 4}$$

The motion parallel to $\boldsymbol{H}$ or Ox is clearly unaffected by the magnetic field. If there is no non-magnetic field, the molecule moves in this direction with uniform velocity; in the transverse direction the motion is circular, so that the resultant motion is spiral, the spiral-axes being lines of magnetic force; the *spiral-frequency* is $\omega_1/2\pi$. When non-magnetic forces act, the motion transverse to $\boldsymbol{H}$ includes, in addition to the circular motion, a mean drift in the y-direction, i.e. perpendicular to $\boldsymbol{H}$ and $\boldsymbol{F}_1$; the drift velocity is given vectorially by $m_1\boldsymbol{F}_1 \wedge \boldsymbol{H}/e_1 H^2$. The resultant transverse motion is trochoidal.*

Because motion parallel to the lines of force is unaffected by the presence of the magnetic field, it is natural to expect that the presence of a magnetic field will not affect diffusion and heat conduction in the direction of the field. The exact theory confirms that this is so.

19.31. Approximate theory of diffusion in a magnetic field

An approximate theory can be given for the transport phenomena in an ionized gas in a magnetic field,† using a relaxation approach similar to that of 6.6–6.63. We first consider diffusion. Since the relaxation approach is

* Here appears one of the interesting paradoxes of the kinetic theory. If the interval between two consecutive encounters is very long, every molecule m_1 will possess an average velocity Z_1/ω_1 perpendicular to $\boldsymbol{F}_1$ and $\boldsymbol{H}$; yet a uniform steady state is possible, such that the gas at any point is at rest. A similar remark applies when $F_1 = 0$ and the magnetic field varies from point to point. For a discussion of the paradox, see T. G. Cowling, *Mon. Not. R. astr. Soc.* **90**, 140 (1929); **92**, 407 (1932).

† The theory given here is essentially equivalent to that given in earlier editions of this book, in spite of considerable apparent differences.

unable to give a theory of thermal diffusion, we assume the temperature to be uniform; velocity-gradients are also ignored.

The relaxation theory of diffusion (cf. (6.63, 4)) was based on a consideration of the equation of mass motion of the sth constituent gas in a mixture, under the influence of its own partial pressure gradient, of body forces, and of drags exerted on it by the other gases. The effect of the magnetic field is simply to introduce an extra body force $n_s e_s \bar{\boldsymbol{c}}_s \wedge \boldsymbol{H}$ per unit volume. Thus assuming that the acceleration of the sth gas can be taken to equal $D\boldsymbol{c}_0/Dt$, the acceleration of the mass as a whole, the equation of motion of this gas becomes

$$\rho_s \frac{D\boldsymbol{c}_0}{Dt} = \rho_s \boldsymbol{F}_s + n_s e_s \bar{\boldsymbol{c}}_s \wedge \boldsymbol{H} - \nabla p_s + \sum_t \frac{p_s p_t}{pD_{st}} (\bar{\boldsymbol{c}}_t - \bar{\boldsymbol{c}}_s), \quad (19.31, 1)$$

where D_{st} is the coefficient of diffusion in a binary s, t mixture at the pressure and temperature of the actual mixture. The set of equations (19.31, 1) gives the diffusion velocities $\bar{\boldsymbol{c}}_t - \bar{\boldsymbol{c}}_s$, and also (on summation over all s) the equation of motion of the gas as a whole. The term involving $\boldsymbol{H}$ does not affect the component of (19.31, 1) parallel to $\boldsymbol{H}$; hence (in agreement with 19.3) diffusion parallel to the magnetic field is unaffected by the field.

Consider now diffusion perpendicular to the field $\boldsymbol{H}$. We consider first a slightly ionized gas, in which encounters between pairs of charged particles are negligible in importance. The gas as a whole is supposed to be at rest relative to the axes used; thus $\boldsymbol{c}_0 = 0$, $D\boldsymbol{c}_0/Dt = 0$. Because of the preponderance of neutral molecules, their mean velocity $\bar{\boldsymbol{c}}_1$ is effectively $\boldsymbol{c}_0$, i.e. zero. Hence for any of the charged constituents, (19.31, 1) reduces to

$$0 = \rho_s \boldsymbol{F}_s - \nabla p_s + n_s e_s \bar{\boldsymbol{c}}_s \wedge \boldsymbol{H} - n_s m_s \bar{\boldsymbol{c}}_s \nabla/\tau_s, \quad (19.31, 2)$$

where τ_s is defined by

$$p_1 p_s / pD_{1s} = n_s m_s/\tau_s. \quad (19.31, 3)$$

Thus (cf. (6.62, 7)) τ_s is $(m_1+m_s)/m_1$ times the effective collision-interval for a molecule m_s in collisions with molecules m_1; if m_s refers to electrons, m_s/m_1 is negligible, and τ_s is the actual collision-interval.

As in 19.3, the field $\boldsymbol{H}$ is assumed parallel to Ox; in addition it is assumed that $\boldsymbol{F}_s$ and ∇p_s reduce to their components Z_s and $\partial p_s/\partial z$ in the z-direction. Then, taking the components of (19.31, 2) in the y and z directions,

$$0 = \omega_s \bar{w}_s - \bar{v}_s/\tau_s, \quad 0 = Z_s - \frac{1}{\rho_s}\frac{\partial p_s}{\partial z} - \omega_s \bar{v}_s - \bar{w}_s/\tau_s,$$

where $\omega_s = e_s H/m_s$, as before. Thus

$$\bar{v}_s = \frac{\omega_s \tau_s^2}{1+\omega_s^2\tau_s^2}\left(Z_s - \frac{1}{\rho_s}\frac{\partial p_s}{\partial z}\right), \quad \bar{w}_s = \frac{\tau_s}{1+\omega_s^2\tau_s^2}\left(Z_s - \frac{1}{\rho_s}\frac{\partial p_s}{\partial z}\right). \quad (19.31, 4)$$

The velocity of diffusion in the absence of a magnetic field is found by putting $\omega_s = 0$. The presence of the field accordingly reduces the velocity of diffusion in the x-direction in the ratio $1/(1+\omega_s^2\tau_s^2)$. It also leads to a flow of molecules perpendicular both to $\boldsymbol{H}$ and to the direction of the ordinary diffusion; this may be called *transverse* diffusion.

The case when the diffusion is due to an electric force $\boldsymbol{E}$ is of special interest. It then appears that the electric current set up by an electric force transverse to a magnetic field is not altogether in the direction of the electric force; a component perpendicular to both $\boldsymbol{H}$ and $\boldsymbol{E}$ also exists, which is known as the Hall current. Also the flow in the direction of the electric force is less than it would be in the absence of the magnetic field, and so we can regard the electrical conductivity in this direction as reduced by the magnetic field. The currents in the x and y directions are given by $j_x = \vartheta^{\text{I}}E, j_y = \vartheta^{\text{II}}E$, where

$$\vartheta^{\text{I}} = \sum_s \frac{n_s e_s^2 \tau_s}{m_s(1+\omega_s^2\tau_s^2)}, \quad \vartheta^{\text{II}} = \sum_s \frac{n_s e_s^2 \omega_s \tau_s^2}{m_s(1+\omega_s^2\tau_s^2)}. \tag{19.31, 5}$$

The quantities ϑ^{I}, ϑ^{II} may be termed the direct (Pedersen) and transverse (Hall) conductivities.

Equation (19.31, 5) implies that the direct conductivity is much reduced when the gas is so rare, or the magnetic field so strong, that the collision-frequencies τ_s^{-1} are not large compared with the spiral-frequencies ω_s; under these conditions the transverse current may be comparable with the direct current. In a gas containing electrons and positive ions, when H is not too great, the electric current is mainly due to the diffusion of electrons; but owing to the dependence of ω_s on m_s, when the magnetic field is so strong that $\omega_s\tau_s$ is large compared with unity for both electrons and ions, the current is due mainly to the motion of ions. This can occur only for very intense magnetic fields or very low densities. For example, in a gas of atmospheric composition and temperature, $\omega_s\tau_s$ is large for electrons only when $H/p > 10^5$, and for ions only when $H/p > 10^7$, if H is measured in gauss and p in atmospheres.

In metals, when the direction of the Hall current is perpendicular to the boundary of the metal, an electric field is set up which prevents any further flow of the Hall current. If the Hall current is similarly inhibited in a gas, then, in addition to the applied electric field E in the x-direction, there is a 'polarization' field E' in the negative y-direction, such that

$$\vartheta^{\text{I}}E' = \vartheta^{\text{II}}E.$$

The total current-density in the direction of E is now $\vartheta^{\text{I}}E + \vartheta^{\text{II}}E' = \vartheta^{\text{III}}E$, where

$$\vartheta^{\text{III}} \equiv \vartheta^{\text{I}} + (\vartheta^{\text{II}})^2/\vartheta^{\text{I}}. \tag{19.31, 6}$$

Thus ϑ^{III} (the Cowling conductivity*) is the conductivity effective in this case. It is also in general the conductivity effective in the dissipation of the

* The conductivities ϑ^{I} and ϑ^{III}, important in ionospheric theory, were first named the 'Pedersen' and 'Cowling' conductivities by S. Chapman, *Nuovo Cim.* x, 4 (Supp. 4), 1385 (1956). For their introduction, see the references quoted in the footnote to p. 362.

energy of currents. When an electric field E produces direct and transverse currents $\vartheta^{\mathrm{I}}E$, $\vartheta^{\mathrm{II}}E$, the rate of supply or dissipation of energy per unit volume is $\boldsymbol{j}.\boldsymbol{E}$, i.e. $\vartheta^{\mathrm{I}}E^2$; this is also expressible as $j^2/\vartheta^{\mathrm{III}}$.

As a second application of (19.31, 1), consider a fully ionized gas composed of electrons of mass m_e and charge $-e$ and an equal number of positive ions of mass m_i and charge e (the suffixes e, i here refer throughout to the electronic and ionic constituents). Because of the smallness of m_e, the mean velocity of the ions is effectively the mass-velocity $\boldsymbol{c}_0$ of the gas, and the inertia term $\rho_e D\boldsymbol{c}_0/Dt$ can be omitted* from the equation (19.31, 1) for electrons. Thus (cf. (19.31, 3)) this equation becomes

$$0 = \rho_e \boldsymbol{F}_e - n_e e\boldsymbol{c}_0 \wedge \boldsymbol{H} - \nabla p_e - n_e(m_e \overline{\boldsymbol{C}}_e/\tau_e + e\overline{\boldsymbol{C}}_e \wedge \boldsymbol{H}), \quad (19.31, 7)$$

where τ_e is the effective mean collision-interval for an electron colliding with ions. In place of the ion equation we may use the equation of motion of the gas as a whole,

$$\rho \frac{D\boldsymbol{c}_0}{Dt} = \sum_s \rho_s \boldsymbol{F}_s + \boldsymbol{j} \wedge \boldsymbol{H} - \nabla p, \quad (19.31, 8)$$

where now

$$\boldsymbol{j} = -n_e e \overline{\boldsymbol{C}}_e. \quad (19.31, 9)$$

Equation (19.31, 7) is similar in form to (19.31, 2), save for the extra term $-n_e e\boldsymbol{c}_0 \wedge \boldsymbol{H}$ due to the electric field $\boldsymbol{c}_0 \wedge \boldsymbol{H}$ acting on moving matter. Hence the diffusion consists again of direct and transverse components, whose magnitudes are in the ratio $1/\omega_e\tau_e$, where $\omega_e = eH/m_e$. One might expect also to be able to introduce direct and transverse conductivities ϑ^{I}, ϑ^{II}, related to the conductivity $\vartheta_0 \equiv n_e e^2 \tau_e/m_e$ in the absence of the field $\boldsymbol{H}$ by

$$\vartheta^{\mathrm{I}} = \vartheta_0/(1+\omega_e^2\tau_e^2), \quad \vartheta^{\mathrm{II}} = \vartheta_0\omega_e\tau_e/(1+\omega_e^2\tau_e^2). \quad (19.31, 10)$$

The third conductivity ϑ^{III} would in this case be

$$\vartheta^{\mathrm{III}} \equiv \vartheta^{\mathrm{I}} + (\vartheta^{\mathrm{II}})^2/\vartheta^{\mathrm{I}} = \vartheta_0,$$

so that the magnetic field would neither increase the dissipation of currents, nor alter the effective conductivity when the Hall current is prevented from flowing.

It is, however, questionable whether the concept of conductivity has any real significance when $\omega_e\tau_e$ is large. If diffusion is due wholly to electric fields, and $(\omega_e\tau_e)^{-1}$ is negligible, (19.31, 7) approximates to

$$0 = -n_e e(\boldsymbol{E} + \bar{\boldsymbol{c}}_e \wedge \boldsymbol{H}). \quad (19.31, 11)$$

Hence the electrons move with a mean velocity $\bar{\boldsymbol{c}}_e$ such that no resultant electric force acts on them; this velocity is just the drift velocity found

* This omission is not justified in the case of very rapid (plasma) oscillations, for which, in any case, the acceleration of the electron gas cannot be taken as $D\boldsymbol{c}_0/Dt$.

in 19.3. The ions move with velocity $\mathbf{c}_0$, where, by (19.31, 8, 9),

$$n_e m_i \frac{D\mathbf{c}_0}{Dt} = \mathbf{j} \wedge \mathbf{H} = -n_e e \overline{\mathbf{C}}_e \wedge \mathbf{H}$$

or

$$m_i \frac{D(\mathbf{c}_0 - \overline{\mathbf{c}}_e)}{Dt} = e(\mathbf{c}_0 - \overline{\mathbf{c}}_e) \wedge \mathbf{H}.$$

Thus the motion of the ions relative to $\overline{\mathbf{c}}_e$ is the circular motion, with angular velocity $\omega_i \equiv eH/m_i$, of a single ion in the field H; this again agrees with the results of 19.3. However, such a rapid rotation of $\mathbf{c}_0 - \overline{\mathbf{c}}_e$, and so of $\mathbf{j}$ $(\equiv n_e e(\mathbf{c}_0 - \overline{\mathbf{c}}_e))$, is not normally contemplated in considering the conductivity. When collisions are not completely negligible, $\mathbf{c}_0 - \overline{\mathbf{c}}_e$ is found to tend to zero; thus ultimately the ion gas and the electron gas move with a common mean velocity, such that no resultant electric force acts on either. In this case there are no electric currents. In the more general case in which pressure gradients exist, electric currents can persist; but these can hardly be regarded as due to a conductivity, in the normal sense.

The general problem of currents in a ternary mixture of neutral molecules, ions and electrons has been considered by a number of authors.* A discussion similar to those for the cases treated above can be given, and conductivities ϑ^{I}, ϑ^{II}, ϑ^{III} can be introduced as before. The main new feature is the possibility, especially when collisions are few, of *ambipolar* diffusion, i.e. diffusion in which the relative motion of ions and electrons is small, the main diffusion velocity being a common velocity of all the charged particles relative to the neutral molecules.

Thermal diffusion is affected by the magnetic field in much the same way as ordinary diffusion, as the general theory shows. The production of a transverse electric current in this case is called the Nernst effect.

19.32. Approximate theory of heat conduction and viscosity

The relaxation theory of heat conduction and viscosity in a magnetic field involves fewer difficulties of interpretation than that of diffusion.

We assume that the gas is in a steady state, uniform in composition, and subject to no non-magnetic forces, and that $\mathbf{c}_0$ (but not necessarily its gradient) is zero at the point considered. Then the Boltzmann equation for f_s in a gas-mixture is

$$\frac{\partial_e f_s}{\partial t} = \mathbf{c}_s \cdot \frac{\partial f_s}{\partial \mathbf{r}} + \frac{e_s}{m_s}(\mathbf{c}_s \wedge \mathbf{H}) \cdot \frac{\partial f_s}{\partial \mathbf{c}_s}.$$

The relaxation approximation replaces $\partial_e f_s/\partial t$ on the left by $-(f_s - f_s^{(0)})/\tau_s$. Also in the first term on the right f_s is replaced by $f_s^{(0)}$; in the second term f_s can be replaced by $f_s - f_s^{(0)}$ without altering its value. Hence the equation becomes

$$\frac{f_s - f_s^{(0)}}{\tau_s} + \frac{e_s}{m_s}(\mathbf{c}_s \wedge \mathbf{H}) \cdot \frac{\partial (f_s - f_s^{(0)})}{\partial \mathbf{c}_s} = -\mathbf{c}_s \cdot \frac{\partial f_s^{(0)}}{\partial \mathbf{r}}. \qquad (19.32, 1)$$

* A. Schlüter, *Z. Naturforschung*, **6a**, 73 (1951); J. H. Piddington, *Mon. Not. R. astr. Soc.* **114**, 638, 651 (1955); T. G. Cowling, *Mon. Not. R. astr. Soc.* **116**, 114 (1956).

We first consider heat conduction perpendicular to $\boldsymbol{H}$ in a gas at rest. In a steady state the pressure must be uniform, and so $n_s \propto 1/T$. Hence in (19.32, 1)

$$\frac{\partial f_s^{(0)}}{\partial \boldsymbol{r}} = f_s^{(0)}(\mathscr{C}_s^2 - \tfrac{5}{2})\frac{1}{T}\frac{\partial T}{\partial \boldsymbol{r}}.$$

Suppose that $\boldsymbol{H}$ is parallel to Ox, and that the temperature is a function of z alone. Then (19.32, 1) becomes

$$f_s - f_s^{(0)} - \omega_s \tau_s \left(v_s \frac{\partial}{\partial w_s} - w_s \frac{\partial}{\partial v_s}\right)(f_s - f_s^{(0)}) = -f_s^{(0)}(\mathscr{C}_s^2 - \tfrac{5}{2}) w_s \frac{\tau_s}{T}\frac{\partial T}{\partial z},$$

the solution of which is

$$f_s - f_s^{(0)} = -\frac{\tau_s f_s^{(0)}(\mathscr{C}_s^2 - \frac{5}{2})}{1 + \omega_s^2 \tau_s^2}(w_s + \omega_s \tau_s v_s)\frac{1}{T}\frac{\partial T}{\partial z}.$$

The corresponding heat flux $\frac{1}{2}\rho_s \overline{c_s^2 \boldsymbol{c}_s}$ carried by the sth gas has components

$$\tfrac{1}{2}\rho_s \overline{c_s^2 w_s} = -\frac{5}{2}\frac{n_s k^2 T}{m_s}\frac{\tau_s}{1 + \omega_s^2 \tau_s^2}\frac{\partial T}{\partial z} \tag{19.32, 2}$$

in the direction of the temperature gradient, and

$$\tfrac{1}{2}\rho_s \overline{c_s^2 v_s} = -\frac{5}{2}\frac{n_s k^2 T}{m_s}\frac{\omega_s \tau_s^2}{1 + \omega_s^2 \tau_s^2}\frac{\partial T}{\partial z} \tag{19.32, 3}$$

transverse to that direction. Thus when the temperature gradient is perpendicular to $\boldsymbol{H}$ the thermal flux down the temperature gradient is reduced, and in addition there is a flux perpendicular to both $\boldsymbol{H}$ and ∇T (the Righi–Leduc effect).

In discussing viscosity we consider a gas of uniform temperature; the term $\partial f_s^{(0)}/\partial \boldsymbol{r}$ depends on the gradient of $\boldsymbol{c}_0$, which does not vanish at the point considered, though $\boldsymbol{c}_0$ itself does. We find

$$\boldsymbol{C}_s \cdot \frac{\partial f_s^{(0)}}{\partial \boldsymbol{r}} = f_s^{(0)} \frac{m_s}{kT}\boldsymbol{C}_s \boldsymbol{C}_s : \mathsf{e},$$

where e denotes the tensor $\overline{\overline{\nabla \boldsymbol{c}_0}}$, as before.

Suppose that $\boldsymbol{H}$ is parallel to Ox, and that either $\boldsymbol{c}_0$ is parallel to Ox and depends only on z, or it is parallel to Oz and depends only on x; in either case e reduces to the pair of equal components e_{xz} and e_{zx}. Then (19.32, 1) becomes

$$f_s - f_s^{(0)} - \omega_s \tau_s \left(v_s \frac{\partial}{\partial w_s} - w_s \frac{\partial}{\partial v_s}\right)(f_s - f_s^{(0)}) = -f_s^{(0)}\frac{2\tau_s m_s u_s w_s}{kT} e_{xz},$$

whose solution is

$$f_s - f_s^{(0)} = -f_s^{(0)} \frac{2\tau_s m_s u_s e_{xz}}{kT(1 + \omega_s^2 \tau_s^2)}(w_s + \omega_s \tau_s v_s).$$

Thus the sth gas contributes to the components p_{xz}, p_{xy} of the pressure tensor the amounts given by

$$\rho_s \overline{u_s w_s} = -\frac{2p_s \tau_s e_{xz}}{1+\omega_s^2 \tau_s^2}, \tag{19.32, 4}$$

$$\rho_s \overline{u_s v_s} = -\frac{2p_s \omega_s \tau_s^2 e_{xz}}{1+\omega_s^2 \tau_s^2}. \tag{19.32, 5}$$

The magnetic field accordingly reduces the contribution of the sth gas to the viscosity, and produces a cross-stress perpendicular to the usual viscous stress and to $\boldsymbol{H}$.

Similarly if $\mathbf{c}_0$ is parallel to Oy and depends only on Oz, so that $\mathbf{e}$ reduces to the pair of equal components e_{yz} and e_{zy}, the sth gas makes viscous contributions to p_{yz}, p_{yy} and p_{zz} given respectively by

$$-\frac{2p_s \tau_s e_{yz}}{1+4\omega_s^2 \tau_s^2}, \quad -\frac{4p_s \omega_s \tau_s^2 e_{yz}}{1+4\omega_s^2 \tau_s^2}, \quad \frac{4p_s \omega_s \tau_s^2 e_{yz}}{1+4\omega_s^2 \tau_s^2}. \tag{19.32, 6}$$

As in 6.6, the collision-intervals τ_s effective in heat conduction and viscosity need not be identical.

19.4. Boltzmann's equation: the second approximation to f for an ionized gas

We now return to the general theory. The solution of Boltzmann's equation by successive approximation will be given for a binary gas-mixture. Boltzmann's equation for the first gas is

$$\frac{Df_1}{Dt} + \boldsymbol{C}_1 \cdot \frac{\partial f_1}{\partial \boldsymbol{r}} + \left(\boldsymbol{F}_1 + \frac{e_1}{m_1}\mathbf{c}_0 \wedge \boldsymbol{H} - \frac{D\mathbf{c}_0}{Dt}\right) \cdot \frac{\partial f_1}{\partial \boldsymbol{C}_1} + \frac{e_1}{m_1}(\boldsymbol{C}_1 \wedge \boldsymbol{H}) \cdot \frac{\partial f_1}{\partial \boldsymbol{C}_1} - \frac{\partial f_1}{\partial \boldsymbol{C}_1}\boldsymbol{C}_1 : \frac{\partial}{\partial \boldsymbol{r}}\mathbf{c}_0$$
$$= -J_1(ff_1) - J_{12}(f_1 f_2), \tag{19.4, 1}$$

where the integrals J are as defined in (8.2, 2, 3). Because of the large molecular velocities, the term on the left which involves $\boldsymbol{C}_1 \wedge \boldsymbol{H}$ is in general much larger than the other terms on this side; the first approximation to (19.4, 1) is accordingly taken to be

$$\frac{e_1}{m_1}(\boldsymbol{C}_1 \wedge \boldsymbol{H}) \cdot \frac{\partial f_1^{(0)}}{\partial \boldsymbol{C}_1} = -J_1(ff_1) - J_{12}(f_1 f_2), \tag{19.4, 2}$$

and the second approximation is

$$\frac{D_0 f_1^{(0)}}{Dt} + \boldsymbol{C}_1 \cdot \frac{\partial f_1^{(0)}}{\partial \boldsymbol{r}} + \left(\boldsymbol{F}_1 + \frac{e_1}{m_1}\mathbf{c}_0 \wedge \boldsymbol{H} - \frac{D_0 \mathbf{c}_0}{Dt}\right) \cdot \frac{\partial f_1^{(0)}}{\partial \boldsymbol{C}_1}$$
$$+ \frac{e_1}{m_1}(\boldsymbol{C}_1 \wedge \boldsymbol{H}) \cdot \frac{\partial f_1^{(1)}}{\partial \boldsymbol{C}_1} - \frac{\partial f_1^{(0)}}{\partial \boldsymbol{C}_1}\boldsymbol{C}_1 : \frac{\partial}{\partial \boldsymbol{r}}\mathbf{c}_0$$
$$= -J_1(f^{(0)}f_1^{(1)}) - J_1(f^{(1)}f_1^{(0)}) - J_{12}(f_1^{(0)}f_2^{(1)}) - J_{12}(f_1^{(1)}f_2^{(0)}). \tag{19.4, 3}$$

The novel feature is the retention of terms involving $f_1^{(0)}$ and $f_1^{(1)}$ respectively in the differential parts of (19.4, 2, 3).

The general solution of (19.4, 2) is of the form

$$f_1^{(0)} = n_1\left(\frac{m_1}{2\pi kT}\right)^{\frac{3}{2}} \exp(-m_1 C_1^2/2kT), \qquad (19.4, 4)$$

as can be shown by a generalized form of the H-theorem; n_1 and T, which are arbitrary constants in the solution, are taken to be identical with the number-density of the first gas and the temperature. Writing

$$f_1^{(1)} = f_1^{(0)}\Phi_1^{(1)}, \quad f_2^{(1)} = f_2^{(0)}\Phi_2^{(1)}, \qquad (19.4, 5)$$

we can put (19.4, 3) in the form

$$\frac{D_0 f_1^{(0)}}{Dt} + \boldsymbol{C}_1 . \frac{\partial f_1^{(0)}}{\partial \boldsymbol{r}} + \left(\boldsymbol{F}_1 + \frac{e_1}{m_1}\boldsymbol{c}_0 \wedge \boldsymbol{H} - \frac{D_0 \boldsymbol{c}_0}{Dt}\right) . \frac{\partial f_1^{(0)}}{\partial \boldsymbol{C}_1} - \frac{\partial f_1^{(0)}}{\partial \boldsymbol{C}_1}\boldsymbol{C}_1 : \frac{\partial}{\partial \boldsymbol{r}}\boldsymbol{c}_0$$

$$= -\frac{e_1}{m_1} f_1^{(0)}(\boldsymbol{C}_1 \wedge \boldsymbol{H}) . \frac{\partial \Phi_1^{(1)}}{\partial \boldsymbol{C}_1} - n_1^2 I_1(\Phi_1^{(1)}) - n_1 n_2 I_{12}(\Phi_1^{(1)} + \Phi_2^{(1)}), \qquad (19.4, 6)$$

the integrals I being as defined in (4.4, 3, 4).

The term involving $\boldsymbol{j}$ in (19.2, 5) corresponds to the term involving $\boldsymbol{C}_1 \wedge \boldsymbol{H}$ in (19.2, 3): hence, corresponding to the retention of $f^{(1)}$ in the differential part of (19.4, 3) we must substitute $\boldsymbol{j}^{(1)}$ for $\boldsymbol{j}$ when obtaining a first approximation to $D_0\boldsymbol{c}_0/Dt$ from (19.2, 5). On the other hand, the term involving $\boldsymbol{j}$ in (19.2, 6) corresponds to the term involving $\boldsymbol{c}_0 \wedge \boldsymbol{H}$ in (19.2, 3), and so in obtaining the first approximation to DT/Dt from (19.2, 6) we put $\boldsymbol{j} = \boldsymbol{j}^{(0)} = 0$.* The first approximations to equations (19.2, 4–6) are thus

$$\frac{D_0 n_1}{Dt} + n_1 \frac{\partial}{\partial \boldsymbol{r}} . \boldsymbol{c}_0 = 0, \quad \frac{D_0 n_2}{Dt} + n_2 \frac{\partial}{\partial \boldsymbol{r}} . \boldsymbol{c}_0 = 0, \qquad (19.4, 7)$$

$$\rho \frac{D_0 \boldsymbol{c}_0}{Dt} = \rho_1 \boldsymbol{F}_1 + \rho_2 \boldsymbol{F}_2 + (n_1 e_1 + n_2 e_2)\boldsymbol{c}_0 \wedge \boldsymbol{H} + \boldsymbol{j}^{(1)} \wedge \boldsymbol{H} - \frac{\partial p}{\partial \boldsymbol{r}}, \qquad (19.4, 8)$$

$$\tfrac{3}{2}kn \frac{D_0 T}{Dt} = -p \frac{\partial}{\partial \boldsymbol{r}} . \boldsymbol{c}_0 = -knT \frac{\partial}{\partial \boldsymbol{r}} . \boldsymbol{c}_0. \qquad (19.4, 9)$$

Using these relations, and substituting for $f_1^{(0)}$ from (19.4, 4), equation (19.4, 6) can be put in the form

$$f_1^{(0)}\{(\mathscr{C}_1^2 - \tfrac{5}{2})\boldsymbol{C}_1 . \nabla \ln T + x_1^{-1}\boldsymbol{d}_{12} . \boldsymbol{C}_1 + 2\overset{\circ}{\mathscr{C}_1\mathscr{C}_1} : \nabla \boldsymbol{c}_0\}$$

$$= -f_1^{(0)}\left\{\frac{m_1}{\rho kT}\boldsymbol{C}_1 . (\boldsymbol{j}^{(1)} \wedge \boldsymbol{H}) + \frac{e_1}{m_1}(\boldsymbol{C}_1 \wedge \boldsymbol{H}) . \frac{\partial \Phi_1^{(1)}}{\partial \boldsymbol{C}_1}\right\}$$

$$- n_1^2 I_1(\Phi_1^{(1)}) - n_1 n_2 I_{12}(\Phi_1^{(1)} + \Phi_2^{(1)}), \qquad (19.4, 10)$$

* It may readily be verified that (19.4, 7–9) can be deduced by multiplying (19.4, 6) in turn by 1, $m_1 \boldsymbol{C}_1$, and $\frac{1}{2}m_1 C_1^2$, integrating over all values of $\boldsymbol{c}_1$, and adding to the corresponding equation for the second gas.

where in the present problem

$$d_{12} = \nabla x_1 + \frac{n_1 n_2(m_2 - m_1)}{n\rho p}\nabla p - \frac{\rho_1\rho_2(F_1 - F_2)}{\rho p} - \frac{n_1 n_2}{\rho p}(m_2 e_1 - m_1 e_2)\, c_0 \wedge H. \tag{19.4, 11}$$

Similarly from Boltzmann's equation for the second gas we obtain the equation

$$f_2^{(0)}\{(\mathscr{C}_2^2 - \tfrac{5}{2}) C_2 . \nabla \ln T - x_2^{-1} d_{12} . C_2 + 2\overset{\circ}{\mathscr{C}_2\mathscr{C}_2} : \nabla c_0\}$$

$$= -f_2^{(0)}\left\{\frac{m_2}{\rho kT} C_2 . (j^{(1)} \wedge H) + \frac{e_2}{m_2}(C_2 \wedge H) . \frac{\partial \Phi_2^{(1)}}{\partial C_2}\right\}$$

$$- n_2^2 I_2(\Phi_2^{(1)}) - n_1 n_2 I_{21}(\Phi_2^{(1)} + \Phi_1^{(1)}). \tag{19.4, 12}$$

Since

$$j^{(1)} = n_1 e_1 \overline{C}_1^{(1)} + n_2 e_2 \overline{C}_2^{(1)}$$

$$= \int f_1^{(0)} \Phi_1^{(1)} e_1 C_1 dc_1 + \int f_2^{(0)} \Phi_2^{(1)} e_2 C_2 dc_2, \tag{19.4, 13}$$

each term on the right-hand sides of (19.4, 10, 12) involves the unknown functions $\Phi_1^{(1)}$ and $\Phi_2^{(1)}$ linearly. The left-hand sides are linear functions of ∇T, ∇c_0 and d_{12}; hence

$$\left.\begin{aligned} \Phi_1^{(1)} &= -A_1 . \nabla \ln T - 2\mathsf{B}_1 : \nabla c_0 - D_1 . d_{12}, \\ \Phi_2^{(1)} &= -A_2 . \nabla \ln T - 2\mathsf{B}_2 : \nabla c_0 - D_2 . d_{12}, \end{aligned}\right\} \tag{19.4, 14}$$

where the quantities A and D are vectors and the quantities B are tensors, which involve only the vectors C_1, C_2, H and scalars. Now H is a rotation-vector in the sense of 1.11; in mechanical equations of translation it appears only through its vector products with other vectors. Thus the only true vectors (not rotation-vectors) on which A_1 can depend are scalar multiples of C_1, $C_1 \wedge H$, $(C_1 \wedge H) \wedge H$, etc.; only the first three of this series need be considered, since

$$[(C_1 \wedge H) \wedge H] \wedge H = -H^2 C_1 \wedge H.$$

Hence we can write

$$A_1 = A_1^{\mathrm{I}}(C_1, H)\, C_1 + A_1^{\mathrm{II}}(C_1, H)\, C_1 \wedge H + A_1^{\mathrm{III}}(C_1, H)(C_1 \wedge H) \wedge H, \tag{19.4, 15}$$

with a similar equation for A_2. Similarly

$$D_1 = D_1^{\mathrm{I}}(C_1, H)\, C_1 + D_1^{\mathrm{II}}(C_1, H)\, C_1 \wedge H + D_1^{\mathrm{III}}(C_1, H)(C_1 \wedge H) \wedge H. \tag{19.4, 16}$$

The corresponding equations for B_1, B_2 involve six independent non-divergent symmetrical tensors, formed by combining the three vectors $C_1, C_1 \wedge H, (C_1 \wedge H) \wedge H$.

19.41. Direct and transverse diffusion

Suppose for the moment that ∇T and ∇c_0 vanish. Then to the present approximation

$$\begin{aligned} n_1\overline{C}_1 &= \int f_1^{(0)}\Phi_1^{(1)}C_1\,dc_1 \\ &= -\int f_1^{(0)}C_1\{D_1^{\mathrm{I}}C_1+D_1^{\mathrm{II}}C_1\wedge H+D_1^{\mathrm{III}}(C_1\wedge H)\wedge H\}.d_{12}\,dc_1 \\ &= -\int f_1^{(0)}C_1C_1.\{D_1^{\mathrm{I}}d_{12}+D_1^{\mathrm{II}}H\wedge d_{12}+D_1^{\mathrm{III}}H\wedge(H\wedge d_{12})\}\,dc_1 \\ &= -\tfrac{1}{3}\int f_1^{(0)}C_1^2\{D_1^{\mathrm{I}}d_{12}+D_1^{\mathrm{II}}H\wedge d_{12}+D_1^{\mathrm{III}}H\wedge(H\wedge d_{12})\}\,dc_1. \end{aligned} \quad (19.41, 1)$$

Thus the velocity of diffusion has components, not only in the direction of d_{12}, but also in the directions of $H\wedge d_{12}$ and $H\wedge(H\wedge d_{12})$. To see the meaning of this result more clearly we consider certain special cases.

Suppose first that d_{12} is parallel to H; then $H\wedge d_{12}=0$, and so

$$\Phi_1^{(1)} = -D_1^{\mathrm{I}}C_1.d_{12},$$

the terms in D_1^{II}, D_1^{III} vanishing. Thus $\overline{C}_1$, $\overline{C}_2$ and in consequence also $j^{(1)}$ are parallel to d_{12} and H, and

$$j^{(1)}\wedge H = 0, \quad (C_1\wedge H).\frac{\partial\Phi_1^{(1)}}{\partial C_1} = -(C_1\wedge H).D_1^{\mathrm{I}}d_{12} = 0.$$

Hence equation (19.4, 10) reduces to

$$x_1^{-1}f_1^{(0)}d_{12}.C_1 = -n_1^2 I_1(\Phi_1^{(1)}) - n_1n_2I_{12}(\Phi_1^{(1)}+\Phi_2^{(1)}),$$

which does not involve H, and is the same as the equation obtained when the magnetic field is absent. The rate of diffusion parallel to the magnetic field is therefore unaltered by the presence of the field.

Suppose next that d_{12} is perpendicular to H; then

$$d_{12}.[(C_1\wedge H)\wedge H] = -H^2 d_{12}.C_1,$$

and so $\Phi_1^{(1)}$ can be expressed in the form

$$\Phi_1^{(1)} = -d_{12}.(D_1^{\mathrm{I}}C_1+D_1^{\mathrm{II}}C_1\wedge H-D_1^{\mathrm{III}}H^2C_1),$$

or (as the term in D_1^{III} can now be absorbed into the term in D_1^{I}) in the form

$$\Phi_1^{(1)} = -d_{12}.(D_1^{\mathrm{I}}C_1+D_1^{\mathrm{II}}C_1\wedge H). \quad (19.41, 2)$$

Hence in this case the velocity of diffusion possesses two components, one parallel to d_{12}, the other parallel to $H\wedge d_{12}$, i.e. perpendicular to both d_{12} and H. These are the velocities of direct and transverse diffusion of 19.31.

19.42. The coefficients of diffusion

In the general case, when d_{12} is neither parallel nor perpendicular to H, the velocity of diffusion is the vector sum of the velocities of diffusion produced by the components of d_{12} parallel and perpendicular to H when acting separately. Thus only the cases when $\boldsymbol{d}_{12}$ is parallel or perpendicular to $\boldsymbol{H}$ need be considered; and as the first case is covered by the ordinary theory of diffusion, it is necessary to discuss only the case in which $\boldsymbol{d}_{12}$ is perpendicular to $\boldsymbol{H}$. Then $\Phi_1^{(1)}$ and $\Phi_2^{(1)}$ are given by (19.41, 2) and a similar equation, and $\boldsymbol{j}^{(1)}$ is expressible in the form

$$\boldsymbol{j}^{(1)} = -(L^{\mathrm{I}}\boldsymbol{d}_{12}+L^{\mathrm{II}}\boldsymbol{H}\wedge\boldsymbol{d}_{12}), \tag{19.42, 1}$$

where L^{I}, L^{II} are constants. Also (19.4, 10) takes the form

$$x_1^{-1}f_1^{(0)}\boldsymbol{d}_{12}.\boldsymbol{C}_1 = -f_1^{(0)}\left\{\frac{m_1}{\rho kT}\boldsymbol{C}_1.(\boldsymbol{j}^{(1)}\wedge\boldsymbol{H})+\frac{e_1}{m_1}(\boldsymbol{C}_1\wedge\boldsymbol{H}).\frac{\partial\Phi_1^{(1)}}{\partial\boldsymbol{C}_1}\right\}$$
$$-n_1^2 I_1(\Phi_1^{(1)})-n_1n_2I_{12}(\Phi_1^{(1)}+\Phi_2^{(1)}).$$

We substitute in this the known expressions for $\Phi_1^{(1)}$, $\Phi_2^{(1)}$ and $\boldsymbol{j}^{(1)}$; then, equating coefficients of $\boldsymbol{d}_{12}$ and of $\boldsymbol{H}\wedge\boldsymbol{d}_{12}$, we obtain the two equations

$$x_1^{-1}f_1^{(0)}\boldsymbol{C}_1 = f_1^{(0)}\left\{\frac{m_1}{\rho kT}H^2L^{\mathrm{II}}-\frac{e_1}{m_1}H^2D_1^{\mathrm{II}}\right\}\boldsymbol{C}_1+n_1^2I_1(D_1^{\mathrm{I}}\boldsymbol{C}_1)$$
$$+n_1n_2I_{12}(D_1^{\mathrm{I}}\boldsymbol{C}_1+D_2^{\mathrm{I}}\boldsymbol{C}_2),$$

$$0 = f_1^{(0)}\left\{-\frac{m_1}{\rho kT}L^{\mathrm{I}}+\frac{e_1}{m_1}D_1^{\mathrm{I}}\right\}\boldsymbol{C}_1+n_1^2I_1(D_1^{\mathrm{II}}\boldsymbol{C}_1)+n_1n_2I_{12}(D_1^{\mathrm{II}}\boldsymbol{C}_1+D_2^{\mathrm{II}}\boldsymbol{C}_2).$$

If $\quad D_1^{\mathrm{I}}+\iota HD_1^{\mathrm{II}} \equiv \zeta_1,\quad D_2^{\mathrm{I}}+\iota HD_2^{\mathrm{II}} \equiv \zeta_2,\quad L^{\mathrm{I}}+\iota HL^{\mathrm{II}} \equiv Z,\quad$ (19.42, 2)

these two equations can be combined into the single (complex) equation

$$x_1^{-1}f_1^{(0)}\boldsymbol{C}_1 = f_1^{(0)}\left\{-\frac{m_1}{\rho kT}Z+\frac{e_1}{m_1}\zeta_1\right\}\iota H\boldsymbol{C}_1+n_1^2I_1(\zeta_1\boldsymbol{C}_1)+n_1n_2I_{12}(\zeta_1\boldsymbol{C}_1+\zeta_2\boldsymbol{C}_2). \tag{19.42, 3}$$

Similarly from (19.4, 12) we derive the equation

$$-x_2^{-1}f_2^{(0)}\boldsymbol{C}_2 = f_2^{(0)}\left\{-\frac{m_2}{\rho kT}Z+\frac{e_2}{m_2}\zeta_2\right\}\iota H\boldsymbol{C}_2+n_2^2I_2(\zeta_2\boldsymbol{C}_2)$$
$$+n_1n_2I_{21}(\zeta_2\boldsymbol{C}_2+\zeta_1\boldsymbol{C}_1). \tag{19.42, 4}$$

Equations (19.42, 3 and 4) replace the equations (8.31, 5). It can be deduced from (19.42, 1), (19.4, 13) and (19.41, 2) that Z is connected with ζ_1, ζ_2 by the equation

$$Z = \tfrac{1}{3}e_1\int f_1^{(0)}\zeta_1 C_1^2\,d\boldsymbol{c}_1+\tfrac{1}{3}e_2\int f_2^{(0)}\zeta_2 C_2^2 d\boldsymbol{c}_2. \tag{19.42, 5}$$

To solve (19.42, 3, 4) we use a method similar to that of 8.51. We assume that ζ_1, ζ_2 are expressible in series of the functions $\boldsymbol{a}_1^{(p)}$, $\boldsymbol{a}_2^{(p)}$ of 8.51, the coefficients being now complex; the equations from which the coefficients are to be found are derived by multiplying (19.42, 3, 4) by $\boldsymbol{a}_1^{(q)}$ and $\boldsymbol{a}_2^{(q)}$, integrating over all values of $\boldsymbol{c}_1$ and $\boldsymbol{c}_2$ respectively, and adding. In particular, a first approximation to the solution is

$$\zeta_1\boldsymbol{C}_1 = d_0\boldsymbol{a}_1^{(0)},\quad \zeta_2\boldsymbol{C}_2 = d_0\boldsymbol{a}_2^{(0)}, \tag{19.42, 6}$$

where d_0 is determined from the equation

$$x_1^{-1}\int f_1^{(0)}\boldsymbol{C}_1.\boldsymbol{a}_1^{(0)}d\boldsymbol{c}_1 - x_2^{-1}\int f_2^{(0)}\boldsymbol{C}_2.\boldsymbol{a}_2^{(0)}d\boldsymbol{c}_2$$
$$= -\frac{\iota HZ}{\rho kT}\left[\int f_1^{(0)}m_1\boldsymbol{C}_1.\boldsymbol{a}_1^{(0)}d\boldsymbol{c}_1 + \int f_2^{(0)}m_2\boldsymbol{C}_2.\boldsymbol{a}_2^{(0)}d\boldsymbol{c}_2\right]$$
$$+d_0\,\iota H\left[\frac{e_1}{m_1}\int f_1^{(0)}\boldsymbol{a}_1^{(0)}.\boldsymbol{a}_1^{(0)}d\boldsymbol{c}_1 + \frac{e_2}{m_2}\int f_2^{(0)}\boldsymbol{a}_2^{(0)}.\boldsymbol{a}_2^{(0)}d\boldsymbol{c}_2\right] + n^2 d_0\{\boldsymbol{a}^{(0)}, \boldsymbol{a}^{(0)}\}.$$

It was shown in 8.51 that

$$\int f_1^{(0)}m_1\boldsymbol{C}_1.\boldsymbol{a}_1^{(0)}d\boldsymbol{c}_1 + \int f_2^{(0)}m_2\boldsymbol{C}_2.\boldsymbol{a}_2^{(0)}d\boldsymbol{c}_2 = 0;$$

also, by (8.51, 16),

$$\{\boldsymbol{a}^{(0)}, \boldsymbol{a}^{(0)}\} = \frac{3x_1x_2kT}{2nm_0[D_{12}]_1},$$

where $[D_{12}]_1$ denotes the first approximation to the coefficient of diffusion. Inserting these values and effecting the remaining integrations, we get

$$\tfrac{3}{2}n(2m_0kT)^{\frac{1}{2}} = \frac{3}{2}\frac{n_1n_2d_0\,\iota H}{\rho^2}(e_1m_2\rho_2 + e_2m_1\rho_1) + \tfrac{3}{2}kT\frac{n_1n_2d_0}{n[D_{12}]_1}$$
$$= \tfrac{3}{2}d_0\frac{\rho_1\rho_2}{\rho}\left(\iota\omega + \frac{1}{\tau}\right),$$

where
$$\omega \equiv \frac{H(e_1m_2\rho_2 + e_2m_1\rho_1)}{\rho m_1m_2}, \quad \tau \equiv \frac{m_1m_2n[D_{12}]_1}{\rho kT}. \tag{19.42, 7}$$

Thus
$$d_0 = \frac{(2m_0kT)^{\frac{1}{2}}\rho n}{\rho_1\rho_2}\frac{\tau(1-\iota\omega\tau)}{1+\omega^2\tau^2}.$$

Combining this with (19.42, 2, 6) and (19.41, 2), we see that

$$\Phi_1^{(1)} = -\frac{(2m_0kT)^{\frac{1}{2}}n\rho}{\rho_1\rho_2}\frac{\tau}{1+\omega^2\tau^2}\left\{\boldsymbol{d}_{12}.\boldsymbol{a}_1^{(0)} - \frac{\omega\tau}{H}\boldsymbol{d}_{12}.(\boldsymbol{a}_1^{(0)}\wedge\boldsymbol{H})\right\}.$$

The corresponding velocity of diffusion is given by

$$\overline{\boldsymbol{C}}_1 - \overline{\boldsymbol{C}}_2 = -\frac{(2m_0kT)^{\frac{1}{2}}n\rho\tau}{3\rho_1\rho_2}\left\{\frac{\boldsymbol{d}_{12} - (\omega\tau/H)\boldsymbol{H}\wedge\boldsymbol{d}_{12}}{1+\omega^2\tau^2}\right\}$$
$$\times\left\{\frac{1}{n_1}\int f_1^{(0)}\boldsymbol{C}_1.\boldsymbol{a}_1^{(0)}d\boldsymbol{c}_1 - \frac{1}{n_2}\int f_2^{(0)}\boldsymbol{C}_2.\boldsymbol{a}_2^{(0)}d\boldsymbol{c}_2\right\}$$
$$= -\frac{p\rho\tau}{\rho_1\rho_2}\frac{(\boldsymbol{d}_{12} - (\omega\tau/H)\boldsymbol{H}\wedge\boldsymbol{d}_{12})}{1+\omega^2\tau^2}. \tag{19.42, 8}$$

Thus the direct diffusion is reduced in the ratio $1/(1+\omega^2\tau^2)$, while the transverse diffusion is $\omega\tau$ times the direct. Similar results hold for the direct and transverse electric currents, and for the direct and transverse conductivities of the gas.

In the case when m_2/m_1 is negligibly small, equation (19.42, 8) is equivalent to (19.31, 7). This is a special case of a general result; for a mixture of any number of gases the formula (19.31, 1), with Dc_0/Dt replaced by D_0c_0/Dt, agrees with the diffusion formula derived from the first approximation to exact theory.*

The collision-interval τ in (19.42, 7) is the one effective in diffusion; by (14.2, 1), for rigid spherical molecules it is

$$\tau = \frac{3(m_1+m_2)}{8\rho\sigma_{12}^2}\left(\frac{m_1 m_2}{2\pi kT(m_1+m_2)}\right)^{\frac{1}{2}},$$

whereas in 5.21 the (differently defined) collision-intervals of molecules m_1, m_2 for collisions with molecules of opposite types were found to be

$$\tau_1 = \frac{1}{2n_2\sigma_{12}^2}\left(\frac{m_1 m_2}{2\pi kT(m_1+m_2)}\right)^{\frac{1}{2}}, \quad \tau_2 = \frac{1}{2n_1\sigma_{12}^2}\left(\frac{m_1 m_2}{2\pi kT(m_1+m_2)}\right)^{\frac{1}{2}}.$$

Thus $\tau = \frac{3}{4}(\rho_2\tau_1+\rho_1\tau_2)/\rho$, so that, apart from the factor $\frac{3}{4}$, τ is a weighted mean of τ_1 and τ_2. In the limiting case when m_2/m_1 is small, corresponding to an electron-ion mixture, $\tau = \frac{3}{4}\tau_2$. The electrical conductivity when there is no magnetic field (given by (19.11, 4)) is connected with τ by the equation

$$\vartheta = n_1 n_2(e_1 m_2 - e_2 m_1)^2 \tau/\rho m_1 m_2. \qquad (19.42, 9)$$

In 19.31 it was pointed out that, when a polarization electric field completely prevents transverse diffusion, direct diffusion in a binary gas is increased to its value in the absence of a magnetic field. This result is only approximate, since (19.42, 8) and (19.31, 7) are correct only to a first approximation. If further approximations to ζ_1 and ζ_2 are used it is found that the polarization field increases the direct diffusion, but not up to the value when there is no magnetic field. For the Lorentz approximation and for 'hard' molecules, the maximum reduction of the diffusion by the magnetic field when this electric force acts is in the ratio $9\pi/32 = 0{\cdot}88$, corresponding to rigid elastic spheres; however, for electrostatic interaction the reduction is in the ratio $3\pi/32 = 0{\cdot}29$.

19.43. Thermal conduction

The general theory of thermal conduction and thermal diffusion in an ionized gas subject to a magnetic field is similar to that of diffusion. Thus, for example, the rate of thermal conduction and thermal diffusion parallel to the magnetic field is not affected by the field; if ∇T is perpendicular to $\boldsymbol{H}$, on the other hand, the rates of thermal conduction and thermal diffusion in the direction of the temperature gradient are reduced, and additional conduction and diffusion occur in the direction of $\boldsymbol{H}\wedge\nabla T$. The general theory will not be given in detail, because of its complication; it involves

* T. G. Cowling, *Proc. R. Soc.* A, 183, 453 (1945).

combining the methods of 18.42, 18.43 and 19.42. We therefore merely quote the results for a binary mixture.

Suppose that the gas is at rest, and that the magnetic field is perpendicular to the temperature gradient. Then the contributions to the heat flux due to the two constituents in the mixture may be written

$$-a_1^{\mathrm{I}}\nabla T - a_1^{\mathrm{II}}\boldsymbol{H}\wedge\nabla T, \quad -a_2^{\mathrm{I}}\nabla T - a_2^{\mathrm{II}}\boldsymbol{H}\wedge\nabla T. \tag{19.43, 1}$$

Here a_1^{I}, Ha_1^{II} are the contributions from the first gas to the direct and transverse thermal conductivity, and similarly for a_2^{I}, Ha_2^{II}. If

$$\alpha_1 = a_1^{\mathrm{I}} + \iota H a_1^{\mathrm{II}}, \quad \alpha_2 = a_2^{\mathrm{I}} + \iota H a_2^{\mathrm{II}} \tag{19.43, 2}$$

the quantities α_1, α_2 satisfy, to the first approximation,

$$\alpha_1\left\{\frac{\mathrm{x}_1}{[\lambda_1]_1} + \iota\omega_1\frac{2m_1}{5kp} + \frac{T\mathrm{x}_2}{5p[D_{12}]_1}\{6M_1^2 + (5-4\mathrm{B})M_2^2 + 8M_1M_2\mathrm{A}\}\right\}$$
$$-\alpha_2\mathrm{x}_1\frac{M_1M_2T}{5p[D_{12}]_1}(11-4\mathrm{B}-8\mathrm{A}) = \mathrm{x}_1 \tag{19.43, 3}$$

and a similar equation (cf. (18.42, 6)).

These equations have not exactly the same form as those derived by relaxation theory, because of the appearance of α_2 in (19.43, 3), and α_1 in the corresponding equation for the second gas (compare 18.42). When m_2/m_1 is negligibly small, however, the results become similar to those given by the relaxation equations. Normally in this case α_2 is large compared with α_1, and is given by

$$\alpha_2\left[\left\{\frac{1}{[\lambda_2]_1} + \frac{T\mathrm{x}_1(5-4\mathrm{B})}{5p\mathrm{x}_2[D_{12}]_1}\right\} + \iota\omega_2\frac{2m_2}{5kp_2}\right] = 1. \tag{19.43, 4}$$

The expression in the brackets $\{\}$ in this case denotes $1/[\lambda_0]_1$, where $[\lambda_0]_1$ is the first approximation to the thermal conductivity in the absence of a magnetic field. Write

$$\tau_2 \equiv 2m_2[\lambda_0]_1/5kp_2. \tag{19.43, 5}$$

Then $\alpha_2 = [\lambda_0]_1/(1+\iota\omega_2\tau_2)$, and

$$a_2^{\mathrm{I}} = [\lambda_0]_1/(1+\omega_2^2\tau_2^2), \quad Ha_2^{\mathrm{II}} = -[\lambda_0]_1\,\omega_2\tau_2/(1+\omega_2^2\tau_2^2) \tag{19.43, 6}$$

in agreement with (19.32, 2, 3).

The thermal diffusion velocity is of the form

$$\overline{\boldsymbol{C}}_1 - \overline{\boldsymbol{C}}_2 = -(\mathrm{x}_1\mathrm{x}_2)^{-1}D_{12}\{\mathrm{k}_T^{\mathrm{I}}\nabla\ln T + \mathrm{k}_T^{\mathrm{II}}\boldsymbol{H}\wedge\nabla\ln T\}, \tag{19.43, 7}$$

where $\mathrm{k}_T^{\mathrm{I}}$, $H\mathrm{k}_T^{\mathrm{II}}$ are direct and transverse thermal diffusion ratios. Write

$$\mathrm{k}_T^{\mathrm{I}} + \iota H\mathrm{k}_T^{\mathrm{II}} \equiv \kappa_T. \tag{19.43, 8}$$

Then κ_T is given to a first approximation, in terms of the α_1, α_2 of (19.43, 3), by

$$[\kappa_T]_1 = T\mathrm{C}(\mathrm{x}_1\alpha_2M_1 - \mathrm{x}_2\alpha_1M_2)/p[D_{12}]_1 \tag{19.43, 9}$$

(cf. (18.43, 3)).

19.44. The stress tensor in a magnetic field

The effect of a magnetic field on the pressure distribution is more complicated, and less interesting to the physicist because the corresponding phenomena cannot be observed in metals. As with heat conduction, the results do not agree exactly with those of relaxation theory, because of the interaction of terms representing the viscous stresses of the different gases. However, they do possess a general resemblance to the relaxation results; and there is identity of form in certain special cases, notably that of a binary mixture in which m_2/m_1 is negligibly small. In this case the viscous stresses are due wholly to the first gas. Let the viscosity of this gas be μ_1, and write $\mu_1 = p_1\tau_1$. If $\boldsymbol{H}$ is in the x-direction, the stress-component p_{xx} has the same value as when the magnetic field is absent. The other stress-components are given to a first approximation by

$$p_{yy} = p - \frac{2\mu_1}{1+4\omega_1^2\tau_1^2}\{\mathring{e}_{yy} + 2\omega_1\tau_1\mathring{e}_{yz} + 2\omega_1^2\tau_1^2(\mathring{e}_{yy}+\mathring{e}_{zz})\},$$

$$p_{zz} = p - \frac{2\mu_1}{1+4\omega_1^2\tau_1^2}\{\mathring{e}_{zz} - 2\omega_1\tau_1\mathring{e}_{yz} + 2\omega_1^2\tau_1^2(\mathring{e}_{yy}+\mathring{e}_{zz})\},$$

$$p_{yz} = p_{zy} = -\frac{2\mu_1}{1+4\omega_1^2\tau_1^2}\{\mathring{e}_{yz} + \omega_1\tau_1(\mathring{e}_{zz}-\mathring{e}_{yy})\},$$

$$p_{xy} = p_{yx} = -\frac{2\mu_1}{1+\omega_1^2\tau_1^2}\{\mathring{e}_{xy} + \omega_1\tau_1\mathring{e}_{xz}\},$$

$$p_{xz} = p_{zx} = -\frac{2\mu_1}{1+\omega_1^2\tau_1^2}\{\mathring{e}_{xz} - \omega_1\tau_1\mathring{e}_{xy}\},$$

where $\mathsf{e} \equiv \overline{\overline{\nabla \boldsymbol{c}_0}}$. These results generalize those of (19.32, 4–6).

19.45. Transport phenomena in a Lorentzian gas in a magnetic field

The Lorentz approximation is of peculiar importance in considering the motion of electrons in a slightly ionized gas. The modifications in the theory when a magnetic field is present are readily made. Suppose that electrons or light molecules constitute the second gas of a binary mixture at rest; then, approximating as in 10.5, we can put (19.4, 12) in the form

$$\{(\mathscr{C}_2^2 - \tfrac{5}{2})\boldsymbol{C}_2.\nabla \ln T - x_2^{-1}\boldsymbol{d}_{12}.\boldsymbol{C}_2\}$$
$$= -\frac{e_2}{m_2}(\boldsymbol{C}_2 \wedge \boldsymbol{H}).\frac{\partial \Phi_2^{(1)}}{\partial \boldsymbol{C}_2} - n_1\iint(\Phi_2^{(1)} - \Phi_2^{(1)\prime})\,C_2 b\,db\,d\epsilon,$$

the first term in the bracket on the right of (19.4, 12) being omitted because of the smallness of m_2.

Substitute in this equation

$$\Phi_2^{(1)} = -(A_2^{\mathrm{I}}\boldsymbol{C}_2 + A_2^{\mathrm{II}}\boldsymbol{C}_2 \wedge \boldsymbol{H}).\nabla \ln T - (D_2^{\mathrm{I}}\boldsymbol{C}_2 + D_2^{\mathrm{II}}\boldsymbol{C}_2 \wedge \boldsymbol{H}).\boldsymbol{d}_{12}.$$

Then, integrating with respect to b and ϵ as in 10.5, and equating coefficients of ∇T, $\boldsymbol{H}\wedge\nabla T$, $\boldsymbol{d}_{12}$ and $\boldsymbol{H}\wedge\boldsymbol{d}_{12}$, we find

$$\mathscr{C}_2^2-\tfrac{5}{2} = -\omega_2 HA_2^{\mathrm{II}}+A_2^{\mathrm{I}}/\tau_2,$$
$$0 = \omega_2 A_2^{\mathrm{I}}+HA_2^{\mathrm{II}}/\tau_2,$$
$$-1/x_2 = -\omega_2 HD_2^{\mathrm{II}}+D_2^{\mathrm{I}}/\tau_2,$$
$$0 = \omega_2 D_2^{\mathrm{I}}+HD_2^{\mathrm{II}}/\tau_2.$$

Here $\omega_2 = -e_2H/m_2 = eH/m_2$, $-e$ being the electronic charge; also τ_2 is a function of C_2 given by

$$\tau_2 = \{n_1C_2\phi_{12}^{(1)}(C_2)\}^{-1} = l(C_2)/C_2 \tag{19.45, 1}$$

in terms of the 'free path' $l(C_2)$ of 10.5. From these equations it follows that

$$A_2^{\mathrm{I}} = -HA_2^{\mathrm{II}}/\omega_2\tau_2 = (\mathscr{C}_2^2-\tfrac{5}{2})\,\tau_2/(1+\omega_2^2\tau_2^2), \tag{19.45, 2}$$
$$D_2^{\mathrm{I}} = -HD_2^{\mathrm{II}}/\omega_2\tau_2 = -\tau_2/x_2(1+\omega_2^2\tau_2^2). \tag{19.45, 3}$$

Thus, for example, the rate of diffusion is given by

$$\begin{aligned} n_2\overline{\boldsymbol{C}}_2 &= \int f_2^{(0)}\Phi_1^{(1)}\boldsymbol{C}_2\,d\boldsymbol{c}_2 \\ &= \frac{1}{3x_2}\int f_2^{(0)}C_2^2\left\{\frac{\tau_2\boldsymbol{d}_{12}-\omega_2\tau_2^2(\boldsymbol{H}\wedge\boldsymbol{d}_{12})/H}{1+\omega_2^2\tau_2^2}\right\}d\boldsymbol{c}_2 \\ &\quad -\frac{1}{3T}\int f_2^{(0)}C_2^2(\mathscr{C}_2^2-\tfrac{5}{2})\left\{\frac{\tau_2\nabla T-\omega_2\tau_2^2(\boldsymbol{H}\wedge\nabla T)/H}{1+\omega_2^2\tau_2^2}\right\}d\boldsymbol{c}_2. \end{aligned} \tag{19.45, 4}$$

If τ_2 is independent of C_2, this shows that the direct diffusion is reduced in the ratio $1/(1+\omega_2^2\tau_2^2)$, and the transverse diffusion is $\omega_2\tau_2$ times as large as the direct. In general τ_2 is not independent of C_2, and it is impossible to evaluate the integrals in finite terms; for example, for rigid elastic spherical molecules it is $l(C_2)$ which is independent of C_2, and $\tau_2 \propto 1/C_2$. However, the formulae are still reminiscent of those of 19.31; they still indicate that for ordinary magnetic fields and normal densities the transverse diffusion is proportional to H and the reduction in conductivity to H^2, while for very large magnetic fields or very small densities the direct diffusion is proportional to $1/H^2$ and the transverse to $1/H$.*

Similar results hold for thermal conduction.

19.5. Alternating electric fields

In the theory given in earlier chapters it was assumed implicitly that the acceleration $\boldsymbol{F}$ of a molecule is a relatively slowly varying function of the time. When, as in the propagation of radio waves, an alternating electric field acts, whose period is comparable with the collision-interval for electrons, the theory requires considerable modification.

Consider a uniform gas under no forces other than the alternating electric

* For the evaluation of the integrals for rigid elastic spheres, see R. Gans, *Annln Phys.* **20**, 203 (1906), and L. Tonks and W. P. Allis, *Phys. Rev.* **52**, 710 (1937).

field and a constant magnetic field $\boldsymbol{H}$. To satisfy Maxwell's electromagnetic equations, the magnetic field should actually have an alternating part; but in practice this is small compared with the slowly varying terrestrial fields, and it will accordingly be ignored.

The electric field can be divided into components parallel and perpendicular to $\boldsymbol{H}$; the diffusion due to the first of these can be shown, by an argument similar to that of 19.41, to be the same as if the magnetic field were absent. Thus we have only to consider diffusion due to an alternating field in the two cases when $\boldsymbol{H}$ is perpendicular to the field, and when there is no magnetic field; and since the latter is a special case of the former, it is sufficient to work out the theory when $\boldsymbol{H}$ is perpendicular to the electric field. For reasons which will appear later, it is convenient to work with an electric field not always directed parallel to a given direction, but circularly polarized in a plane perpendicular to $\boldsymbol{H}$.

Consider the motion of a single particle of mass m_1 and charge e_1. For convenience let $\boldsymbol{H}$ be parallel to Ox, and let the electric field be given by

$$\boldsymbol{E} = (0, -E\sin st, E\cos st). \qquad (19.5, 1)$$

Then the part $\boldsymbol{F}_1$ of the acceleration of the molecule which is due to the electric field is

$$\boldsymbol{F}_1 = (0, -F_1 \sin st, F_1 \cos st)$$
$$= \left(0, -\frac{e_1 E}{m_1}\sin st, \frac{e_1 E}{m_1}\cos st\right). \qquad (19.5, 2)$$

The equations of motion of a particle are

$$\ddot{x} = 0, \quad \ddot{y} = \omega_1 \dot{z} - F_1 \sin st, \quad \ddot{z} = -\omega_1 \dot{y} + F_1 \cos st$$

(cf. (19.3, 2)), where $\omega_1 = e_1 H/m_1$. If $\dot{y} = v_1'$ and $\dot{z} = w_1'$ initially, when $t = 0$, it follows by integration that

$$\dot{y} = v_1' \cos \omega_1 t + w_1' \sin \omega_1 t + \frac{F_1}{(\omega_1 + s)}(\cos st - \cos \omega_1 t),$$

$$\dot{z} = -v_1' \sin \omega_1 t + w_1' \cos \omega_1 t + \frac{F_1}{(\omega_1 + s)}(\sin st + \sin \omega_1 t).$$

Now w_1' and v_1' denote the initial velocities of the particle in directions respectively parallel to $\boldsymbol{E}$, and perpendicular to both $\boldsymbol{E}$ and $\boldsymbol{H}$ (actually the direction is that of $\boldsymbol{E} \wedge \boldsymbol{H}$). If the velocities parallel to $\boldsymbol{E}$ and to $\boldsymbol{E} \wedge \boldsymbol{H}$ after time t are denoted by w_1 and v_1, then

$$w_1 = \dot{z}\cos st - \dot{y}\sin st$$
$$= w_1' \cos(\omega_1 + s)t - \left(v_1' - \frac{F_1}{\omega_1 + s}\right)\sin(\omega_1 + s)t,$$
$$v_1 = \dot{y}\cos st + \dot{z}\sin st$$
$$= \left(v_1' - \frac{F_1}{\omega_1 + s}\right)\cos(\omega_1 + s)t + w_1' \sin(\omega_1 + s)t + \frac{F_1}{\omega_1 + s}.$$

Comparing these equations with (19.3, 4), we see that the component velocities parallel to $\boldsymbol{E}$ and to $\boldsymbol{E}\wedge\boldsymbol{H}$ vary in the same way as if the acceleration $\boldsymbol{F}_1$ was constant in direction and of constant magnitude e_1E/m_1, and the angular velocity of spiralling in the magnetic field was increased from ω_1 to ω_1+s. Since the gas is uniform, diffusion in it depends on the velocities of the molecules, not on their positions; thus the velocity of diffusion at any instant is the same as if each molecule m_1 were subject to a force in a constant direction producing an acceleration $F_1 = e_1E/m_1$ and to a perpendicular magnetic field producing an angular velocity ω_1+s of spiralling.

All the results of 19.31, 19.41 and 19.42 can accordingly be modified to apply to the present problem. In addition to the direct diffusion current in the direction in which the electric field is at the moment acting, there is a second transverse current in a direction perpendicular to both the electric and the magnetic fields. By analogy with (19.31, 4) the relaxation method indicates that in a lightly ionized gas the velocity of diffusion of charged molecules m_s relative to the whole mass of gas has components $\overline{w}_s$, $\overline{v}_s$ in the direct and transverse directions, where

$$\overline{w}_s = \frac{e_sE}{m_s}\frac{\tau_s}{1+(\omega_s+s)^2\tau_s^2}, \quad \overline{v}_s = \frac{e_sE}{m_s}\frac{(\omega_s+s)\tau_s^2}{1+(\omega_s+s)^2\tau_s^2}. \qquad (19.5, 3)$$

Here τ_s is defined as in (19.31, 3); $\overline{v}_s$ is the velocity of diffusion in the direction which the electric force had one-quarter of a period earlier. Similarly in a binary mixture the velocities of relative diffusion in the direct and transverse directions are, to a first approximation,

$$\overline{w}_1-\overline{w}_2 = \frac{E(e_1m_2-e_2m_1)}{m_1m_2}\frac{\tau}{1+(s+\omega)^2\tau^2}, \qquad (19.5, 4)$$

$$\overline{v}_1-\overline{v}_2 = \frac{E(e_1m_2-e_2m_1)}{m_1m_2}\frac{(s+\omega)\tau^2}{1+(s+\omega)^2\tau^2}, \qquad (19.5, 5)$$

respectively, ω and τ having the same meaning as in (19.42, 7). No difficulties of interpretation similar to those met in 19.31 now arise, even when $(s+\omega)\tau$ is large.

If in place of the circularly polarized field (19.5, 1) we have to deal with a linearly oscillating field

$$\boldsymbol{E} = (0, 0, E_0\cos st), \qquad (19.5, 6)$$

we must represent this as the sum of the two circularly polarized oscillations

$$(0, -\tfrac{1}{2}E_0\sin st, \tfrac{1}{2}E_0\cos st) \quad\text{and}\quad (0, -\tfrac{1}{2}E_0\sin(-st), \tfrac{1}{2}E_0\cos(-st)).$$

Then, for example, to (19.5, 4, 5) correspond the equations

$$\overline{w}_1-\overline{w}_2 = E_0\frac{e_1m_2-e_2m_1}{2m_1m_2}\left\{\frac{\tau}{1+(s+\omega)^2\tau^2}+\frac{\tau}{1+(\omega-s)^2\tau^2}\right\} \qquad (19.5, 7)$$

and

$$\overline{v}_1-\overline{v}_2 = E_0\frac{e_1m_2-e_2m_1}{2m_1m_2}\left\{\frac{(s+\omega)\tau^2}{1+(s+\omega)^2\tau^2}+\frac{(\omega-s)\tau^2}{1+(\omega-s)^2\tau^2}\right\}, \qquad (19.5, 8)$$

the direction of the transverse velocity of diffusion being taken as that of the first circularly polarized oscillation one-quarter of a period earlier.

These results, based on relaxation theory, agree to a first approximation with those derived from the general theory, appropriately modified. The modification required consists of the retention of a first approximation to $\partial f_1^{(1)}/\partial t$ in the equation for $f_1^{(1)}$; if

$$f_1^{(1)} = \alpha(\mathbf{c}_1, t) \cos st + \beta(\mathbf{c}_1, t) \sin st,$$

the first approximation is taken to be

$$\frac{\partial f_1^{(1)}}{\partial t} = s[-\alpha(\mathbf{c}_1, t) \sin st + \beta(\mathbf{c}_1, t) \cos st].$$

PHENOMENA IN STRONG ELECTRIC FIELDS

19.6. Electrons with large energies

In certain circumstances the electrons conducting electricity in gases have energies far exceeding those appropriate to the temperature of the gas. It was one of the simplifying features of the Lorentz approximation that in the elastic encounter of two particles of widely different mass there is little interchange of energy; for example, if an electron of mass m_2 impinges on a molecule of mass m_1 at rest, and the relative velocity g is turned through an angle χ, the energy lost by the electron is

$$m_1 m_2^2 g^2(1 - \cos\chi)/(m_1 + m_2)^2,$$

which is a small fraction of its original energy, $\frac{1}{2}m_2 g^2$. Hence in the presence of a strong electric field a slowly moving electron gains, on an average, far more energy from the field during a free path than it can lose at an encounter. In consequence, the mean energy of electrons grows until the small fraction of it which is lost at collisions balances the energy gained during a free path. The mean energy in the steady state is thus much larger than the thermal energy $\frac{3}{2}kT$; the velocity-distribution function of the electrons differs widely from the Maxwellian form, and the theory of diffusion in the electric field differs considerably from that of 10.5.

If the mean energy of electrons is sufficiently large, inelastic collisions occur, in which part of the energy of an electron is used in exciting the quantum states of internal motion of a molecule. In a gas like helium, which has a high excitation potential, an inelastic collision can only occur if the electron has an energy some hundreds of times as large as the thermal energy, and the energy lost by the electron is large; in diatomic and polyatomic gases, on the other hand, a much smaller energy is able to stimulate the rotational and vibrational motions of the molecules, and inelastic collisions are important at much lower energies. These facts are of importance in considering the mobility of electrons in gases.

Another case in which the mean energy of electrons may considerably exceed the thermal energy occurs in the study of the upper atmosphere. The electrons are liberated from molecules by ultra-violet or corpuscular radiation; they remain free until they combine with positive ions to form neutral molecules, or with neutral molecules to form negative ions. The energy required to ionize a molecule is much larger than the thermal energy $\frac{3}{2}kT$, and the ionizing agent probably possesses energy exceeding this requisite energy by an appreciable fraction; the excess energy may be used, partly or wholly, in giving the liberated electron kinetic energy. The mean kinetic energy of electrons when liberated is therefore likely considerably to exceed the thermal value; and if electrons do not make too many collisions with molecules before recombination takes place, the mean kinetic energy of electrons present in the atmosphere is likely to be larger than the thermal.

19.61. The steady state in a strong electric field*

Suppose that electrons in a large mass of gas are subject to a strong electric field. After some time an approximately steady state is reached, in which the energy gained by an electron during a free path is, on an average, balanced by the energy lost at a collision with a gas-molecule. If collisions with the molecules are all elastic, these losses and gains of energy are small compared with the mean energy; and since the direction of motion of an electron undergoes a large deflection at encounter, the velocity-distribution function for electrons, though it may differ from the Maxwellian in form, must be nearly independent of the direction of the velocity.

Suppose first that the electrons are so few in number that their mutual encounters can be ignored. Consider the equilibrium of a mixture of neutral molecules, of mass m_1, and electrons, of mass m_2 and charge e_2, under the influence of an electric field $\boldsymbol{E}$. If the state is uniform as well as steady, the velocity-distribution function f_2 for electrons satisfies the equation

$$\boldsymbol{F}_2 \cdot \frac{\partial f_2}{\partial \boldsymbol{c}_2} = \iint (f_1' f_2' - f_1 f_2) g \alpha_{12} d\boldsymbol{e}' d\boldsymbol{c}_1, \qquad (19.61, 1)$$

where $\alpha_{12} d\boldsymbol{e}' = b\, db\, d\epsilon$ and α_{12} is a function of g and b; also

$$\boldsymbol{F}_2 = e_2 \boldsymbol{E}/m_2. \qquad (19.61, 2)$$

* Early investigations of the steady state of electrons in a strong electic field include those of F. B. Pidduck, *Proc. Lond. Math. Soc.* **15**, 89 (1916); M. J. Druyvesteyn, *Physica*, **10**, 61 (1930), and **1**, 1003 (1934); E. M. Morse, W. P. Allis and E. S. Lamar, *Phys. Rev.* **48**, 412 (1935); W. P. Allis and H. W. Allen, *Phys. Rev.* **52**, 703 (1937); and B. Davydov, *Phys. Z. SowjUn.* **8**, 59 (1935). Of these all save Pidduck and Davydov ignore the motion of the molecules; Druyvesteyn in addition takes an average of the energy-loss at collision of elastic spheres. More recent investigations have taken into account electron–electron encounters, and have considered the approach to the steady state, and 'runaway' effects (see 19.66).

The velocity-distribution function f_1 for the molecules can be supposed to have the Maxwellian form for a gas at rest, i.e.

$$f_1 = n_1\left(\frac{m_1}{2\pi kT}\right)^{\frac{3}{2}} \exp(-m_1 c_1^2/2kT). \tag{19.61, 3}$$

Since the distribution of electronic velocities is nearly isotropic, we put*

$$f_2 = f_2^{(0)} + \boldsymbol{F}_2 . \boldsymbol{c}_2 f_2^{(1)} + \boldsymbol{F}_2\boldsymbol{F}_2 : \overset{\circ}{\boldsymbol{c}_2\boldsymbol{c}_2} f_2^{(2)} + \ldots, \tag{19.61, 4}$$

where each of the functions $f_2^{(0)}, f_2^{(1)}, f_2^{(2)}, \ldots$ denotes a function of the scalars c_2, F_2 alone, and the quantities $\boldsymbol{c}_2$, $\overset{\circ}{\boldsymbol{c}_2\boldsymbol{c}_2}, \ldots$ are the vector and the tensors of second and higher orders whose components are solid harmonics in the components of $\boldsymbol{c}_2$, i.e. functions satisfying the equation

$$\frac{\partial^2\phi}{\partial u_2^2} + \frac{\partial^2\phi}{\partial v_2^2} + \frac{\partial^2\phi}{\partial w_2^2} = 0.$$

We substitute from (19.61, 4) into (19.61, 1), and equate terms involving scalars only, terms involving the vector $\boldsymbol{F}_2$, and terms involving the tensor $\overset{\circ}{\boldsymbol{F}_2\boldsymbol{F}_2}$, etc.: then we get

$$F_2^2\left(f_2^{(1)} + \tfrac{1}{3}c_2\frac{\partial f_2^{(1)}}{\partial c_2}\right) = \iint (f_1' f_2^{(0)\prime} - f_1 f_2^{(0)}) g\alpha_{12} de'\, dc_1, \tag{19.61, 5}$$

$$\frac{\boldsymbol{F}_2 . \boldsymbol{c}_2}{c_2}\frac{\partial f_2^{(0)}}{\partial c_2} + \tfrac{4}{15}F_2^2\frac{\boldsymbol{F}_2 . \boldsymbol{c}_2}{c_2^4}\frac{\partial (f_2^{(2)} c_2^5)}{\partial c_2}$$
$$= \iint \{f_1' f_2^{(1)\prime}(\boldsymbol{F}_2 . \boldsymbol{c}_2') - f_1 f_2^{(1)}(\boldsymbol{F}_2 . \boldsymbol{c}_2)\} g\alpha_{12} de'\, dc_1, \tag{19.61, 6}$$

and a series of similar equations. Since the second term of (19.61, 4) is small compared with the first, it is natural to expect that the third and subsequent terms are small compared with the second; they will accordingly be ignored in subsequent work. Thus we have to deal with (19.61, 5) and (19.61, 6) alone, and the second term on the left of (19.61, 6) can be omitted. Moreover, since m_2/m_1 is very small, only the least power of m_2/m_1 which appears on integration need be retained. In this connection it may be observed that $m_1\overline{c_1^2} < m_2\overline{c_2^2}$, and so c_1/c_2 can be regarded as at least as small as $(m_2/m_1)^{\frac{1}{2}}$.

In (19.61, 6) it is sufficient to put $c_2 = c_2' = g$, $c_1 = c_1'$, which is equivalent to ignoring m_2/m_1 completely. Then, transforming as in 10.5, we find

$$\frac{\boldsymbol{F}_2 . \boldsymbol{c}_2}{c_2}\frac{\partial f_2^{(0)}}{\partial c_2} = -n_1 f_2^{(1)}(\boldsymbol{F}_2 . \boldsymbol{c}_2)\int (1 - \cos\chi)\, c_2 \alpha_{12}(c_2)\, de'$$
$$= -n_1 f_2^{(1)}(\boldsymbol{F}_2 . \boldsymbol{c}_2)\, c_2 \phi_{12}^{(1)}(c_2).$$

* The notation here is rather different from that used earlier. A second approximation to f_2 was denoted earlier by $f_2^{(0)} + f_2^{(1)}$; it is now represented by $f_2^{(0)} + \boldsymbol{F}_2 . \boldsymbol{c}_2 f_2^{(1)}$: and so on.

As in (19.45, 1), an equivalent mean free path of electrons of speed c_2 is defined by

$$l(c_2) = 1/n_1\phi_{12}^{(1)}(c_2). \qquad (19.61, 7)$$

Then it follows that

$$\frac{\partial f_2^{(0)}}{\partial c_2} = -c_2^2 f_2^{(1)}/l(c_2). \qquad (19.61, 8)$$

If the same approximations are made in (19.61, 5), the integral on the right vanishes. It is therefore necessary to proceed to a further approximation; we retain all terms not small compared with m_2/m_1.

Multiply (19.61, 5) by $d\mathbf{c}_2$ and integrate over all values of $\mathbf{c}_2$ such that $c_2 < v$. Then we get

$$\iiint (f_1' f_2^{(0)\prime} - f_1 f_2^{(0)}) g\alpha_{12}\, d\mathbf{e}'\, d\mathbf{c}_1 d\mathbf{c}_2 = \tfrac{4}{3}\pi F_2^2 \int_0^v \frac{\partial}{\partial c_2}(c_2^3 f_2^{(1)})\, dc_2$$
$$= \tfrac{4}{3}\pi F_2^2 v^3 f_2^{(1)}(v). \qquad (19.61, 9)$$

Using a transformation similar to that of 3.53,* it can be proved that

$$\iiint f_1' f_2^{(0)\prime} g\alpha_{12}\, d\mathbf{e}'\, d\mathbf{c}_1 d\mathbf{c}_2,$$

integrated over all values of the variables such that $c_2 < v$, is identical with

$$\iiint f_1 f_2^{(0)} g\alpha_{12}\, d\mathbf{e}'\, d\mathbf{c}_1 d\mathbf{c}_2$$

integrated over all values such that $c_2' < v$. By (3.41, 6, 7), since $g' = g$,

$$c_2'^2 = c_2^2 - 2M_1\mathbf{G}.(\mathbf{g}_{21} - \mathbf{g}_{21}')$$
$$= c_2^2 - 2M_1(\mathbf{c}_1 + M_2\mathbf{g}_{21}).(\mathbf{g}_{21} - \mathbf{g}_{21}')$$

and so, correct to terms of order m_2/m_1,

$$c_2' = \sqrt{\{c_2^2 - 2(\mathbf{c}_1 + M_2\mathbf{g}_{21}).(\mathbf{g}_{21} - \mathbf{g}_{21}')\}}$$
$$= c_2 - (\mathbf{c}_1 + M_2\mathbf{g}_{21}).(\mathbf{g}_{21} - \mathbf{g}_{21}')/c_2 - \tfrac{1}{2}\{\mathbf{c}_1.(\mathbf{g}_{21} - \mathbf{g}_{21}')\}^2/c_2^3. \qquad (19.61, 10)$$

In this write $\mathbf{c}_2 = c_2\mathbf{a}$, where $\mathbf{a}$ is a unit vector (so that $d\mathbf{c}_2 = c_2^2 dc_2 d\mathbf{a}$). Then, by (19.61, 10), the condition $c_2' < v$ is equivalent to $c_2 < v + \Delta v$, where Δv is a function of v, $\mathbf{a}$, $\mathbf{c}_1$ and $\mathbf{e}'$; $\Delta v/v$ is a small quantity of order c_1/c_2 or $(m_2/m_1)^{\frac{1}{2}}$, and (since g/c_2 is nearly unity)

$$\Delta v = \mathbf{c}_1.(\mathbf{a} - \mathbf{e}') + \text{smaller terms}. \qquad (19.61, 11)$$

Thus (19.61, 9) is equivalent to

$$\tfrac{4}{3}\pi F_2^2 v^3 f_2^{(1)}(v) = \iiint \left(\int_v^{v+\Delta v} f_1 f_2^{(0)} g\alpha_{12} c_2^2 dc_2 \right) d\mathbf{e}'\, d\mathbf{a}\, d\mathbf{c}_1.$$

We substitute in this the Taylor expansion

$$f_2^{(0)}(c_2) = f_2^{(0)}(v) + (c_2 - v)\frac{\partial f_2^{(0)}(v)}{\partial v} + \tfrac{1}{2}(c_2 - v)^2 \frac{\partial^2 f_2^{(0)}(v)}{\partial v^2} + \ldots.$$

* The difficulties about convergence can be avoided if it is supposed as in 3.6 that encounters are neglected in which the deflection of the relative velocity is less than an assigned small quantity.

The third, fourth, ... terms of the expansion, after integrating with respect to c_2, give quantities of order $(\Delta v)^3$, $(\Delta v)^4$, ..., which can be neglected. Thus

$$\tfrac{4}{3}\pi F_2^2 v^3 f_2^{(1)}(v) = f_2^{(0)}(v)\iiint\left(\int_v^{v+\Delta v} f_1 g\alpha_{12} c_2^2 dc_2\right) d\mathbf{e}'\, d\mathbf{a}\, d\mathbf{c}_1$$

$$+\frac{\partial f_2^{(0)}(v)}{\partial v}\iiint\left(\int_v^{v+\Delta v}(c_2-v) f_1 g\alpha_{12} c_2^2 dc_2\right) d\mathbf{e}'\, d\mathbf{a}\, d\mathbf{c}_1. \quad (19.61, 12)$$

In evaluating the second term on the right we can approximate by putting

$$g\alpha_{12}(g)\, c_2^2 = \alpha_{12}(v)\, v^3, \quad \Delta v = \mathbf{c}_1.(\mathbf{a}-\mathbf{e}').$$

Then, since $\mathbf{a}.\mathbf{e}' = \cos\chi$ to the same approximation, the term becomes, using (19.61, 7),

$$v^3\frac{\partial f_2^{(0)}(v)}{\partial v}\iiint \tfrac{1}{2}\{\mathbf{c}_1.(\mathbf{a}-\mathbf{e}')\}^2 f_1 \alpha_{12}(v)\, d\mathbf{e}'\, d\mathbf{a}\, d\mathbf{c}_1$$

$$= n_1\frac{kT}{m_1} v^3 \frac{\partial f_2^{(0)}(v)}{\partial v}\iint(1-\cos\chi)\,\alpha_{12}\, d\mathbf{e}'\, d\mathbf{a}$$

$$= 4\pi n_1\frac{kT}{m_1} v^3\frac{\partial f_2^{(0)}(v)}{\partial v}\iint(1-\cos\chi)\, b\, db\, d\epsilon$$

$$= 4\pi\frac{kT}{m_1}\frac{v^3}{l(v)}\frac{\partial f_2^{(0)}(v)}{\partial v}.$$

In the calculation of the first term on the right of (19.61, 12) it is necessary to take into account the smaller terms on the right of (19.61, 11); the integration is very involved. It is, however, possible to determine the value of the integral indirectly.* Let it be $\psi(v)$; the value of ψ will not depend on F_2 or on $f_2^{(0)}$. Then (19.61, 12) becomes

$$\tfrac{4}{3}\pi F_2^2 v^3 f_2^{(1)}(v) = 4\pi\frac{kT}{m_1}\frac{v^3}{l(v)}\frac{\partial f_2^{(0)}(v)}{\partial v} + f_2^{(0)}(v)\,\psi(v).$$

If $F_2 = 0$, this equation is satisfied by

$$f_2^{(0)}(v) = n_2\left(\frac{m_2}{2\pi kT}\right)^{\frac{3}{2}} e^{-m_2 v^2/2kT}.$$

For this to be so, we must have

$$\psi(v) = 4\pi\frac{m_2}{m_1}\frac{v^4}{l(v)}.$$

Substituting this expression and replacing the variable v by c_2, we have finally

$$\tfrac{1}{3}F_2^2 f_2^{(1)}(c_2) = \frac{kT}{m_1 l(c_2)}\frac{\partial f_2^{(0)}(c_2)}{\partial c_2} + \frac{m_2 c_2}{m_1 l(c_2)} f_2^{(0)}(c_2). \quad (19.61, 13)$$

* A device similar to this was used by Davydov (*loc. cit.*).

This equation can be given a simple interpretation. The left-hand side, multiplied by $4\pi m_2 c_2^4 dc_2$, can be proved equal to the rate at which electrons of speeds between c_2 and c_2+dc_2 gain energy from the electric field per unit volume; the second term on the right, similarly multiplied, gives the loss of energy of these electrons in encounters if the molecules m_1 are at rest before encounter; and the first term represents the correction due to the relatively slow motions of the molecules.

From (19.61, 8, 13), it follows that

$$-\frac{F_2^2 l(c_2)}{3c_2^2}\frac{\partial f_2^{(0)}}{\partial c_2} = \frac{kT}{m_1 l(c_2)}\frac{\partial f_2^{(0)}}{\partial c_2} + \frac{m_2 c_2}{m_1 l(c_2)} f_2^{(0)},$$

whence

$$f_2^{(0)} = A\exp\left(-\int\frac{m_2 c_2 dc_2}{kT+m_1F_2^2 l^2/3c_2^2}\right), \quad f_2^{(1)} = \frac{m_2 c_2 l}{kTc_2^2+\frac{1}{3}m_1F_2^2l^2}f_2^{(0)},$$

(19.61, 14)

where A is a constant. For small values of F_2, these expressions approximate to those obtained by the Lorentz method; for large F_2, when the mean energy of an electron is large compared with $\frac{3}{2}kT$, they approximate to

$$f_2^{(0)} = A\exp\left(-\int\frac{3m_2 c_2^3 dc_2}{m_1 F_2^2 l^2}\right), \quad f_2^{(1)} = \frac{3m_2 c_2}{m_1 F_2^2 l}f_2^{(0)}. \qquad (19.61, 14')$$

The condition that (19.61, 14′) should apply is that kT should be small compared with $m_1 F_2^2 l^2/3c_2^2$ in the range of c_2 for which $f_2^{(0)}$ is appreciable; since $m_2\overline{c_2^2} > 3kT$ when this is so, the condition is that the mean value of $e_2 El$ shall be large compared with $3kT(m_2/m_1)^{\frac{1}{2}}$.

If the molecules are rigid elastic spheres, so that the free path is independent of c_2, (19.61, 14′) becomes

$$f_2^{(0)} = A\exp\left(-\frac{3m_2c_2^4}{4m_1F_2^2l^2}\right), \quad f_2^{(1)} = \frac{3m_2c_2}{m_1F_2^2 l}A\exp\left(-\frac{3m_2c_2^4}{4m_1F_2^2l^2}\right).$$

(19.61, 14″)

This result was originally derived by Druyvesteyn. It indicates that the number of electrons with energies large compared with the mean energy is much smaller than in a Maxwellian distribution with the same mean energy. Using this result, it follows that A is connected with the number-density n_2 by the relation

$$n_2 = 4\pi A\int_0^\infty \exp\left(-\frac{3m_2c_2^4}{4m_1F_2^2l^2}\right)c_2^2dc_2$$

$$= \pi A\left(\frac{4m_1F^2l^2}{3m_2}\right)^{\frac{3}{4}}\Gamma(\tfrac{3}{4}),$$

and that the mean energy of an electron is

$$\tfrac{1}{2}m_2\overline{c_2^2} = \frac{2\pi A m_2}{n_2}\int_0^\infty \exp\left(-\frac{3m_2 c_2^4}{4m_1 F_2^2 l^2}\right) c_2^4\, dc_2$$

$$= m_2\left(\frac{m_1 F_2^2 l^2}{3m_2}\right)^{\frac{1}{2}}\frac{\Gamma(\frac{5}{4})}{\Gamma(\frac{3}{4})} = 0{\cdot}427(m_1 m_2)^{\frac{1}{2}} F_2 l, \qquad (19.61,\ 15)$$

and the mean velocity of diffusion of the electrons is of magnitude

$$\frac{F_2}{3n_2}\int c_2^2 f_2^{(1)}\, dc_2 = \frac{4\pi m_2 A}{n_2 m_1 F_2 l}\int_0^\infty \exp\left(-\frac{3m_2 c_2^4}{4m_1 F_2^2 l^2}\right) c_2^5\, dc_2$$

$$= \left(\frac{4}{3}\right)^{\frac{3}{4}}\frac{\sqrt{\pi}}{2}\left(\frac{m_2}{m_1}\right)^{\frac{1}{4}}\frac{(F_2 l)^{\frac{1}{2}}}{\Gamma(\frac{3}{4})} = 0{\cdot}897\left(\frac{m_2}{m_1}\right)^{\frac{1}{4}}(F_2 l)^{\frac{1}{2}}. \qquad (19.61,\ 16)$$

This is small compared with the value

$$\tfrac{4}{3}F_2 l\left(\frac{m_2}{2\pi kT}\right)^{\frac{1}{2}}$$

given by the Lorentz approximation.

The formula (19.61, 14″), however, gives too few electrons of high energies for actual gases. It is clear that the actual distribution of velocities is very sensitive to the law of interaction between electrons and molecules; for example, if the force between them varies as the inverse fifth power of the distance, on the classical theory $l(c_2)$ is proportional to c_2, and the distribution function is Maxwellian, though with a 'temperature' considerably higher than that of the gas-molecules. It is possible to calculate $l(c_2)$ numerically for actual gases, from the experimental results on the angular distribution of electrons scattered by molecules, using (19.61, 7) and

$$\phi_{12}^{(1)}(c_2) = 2\pi\int(1-\cos\chi)\,\alpha_{12}(c_2,\chi)\sin\chi\, d\chi,$$

where $\alpha_{12}(c_2,\chi)$ is a scattering amplitude as in 17.2. From this the distribution function for any F_2 can be determined numerically, and the mean energy and rate of diffusion of the electrons can be calculated. Since both of these quantities are also measurable by experiment, a direct check on the theory is available.

Calculations of this type have been made by Allen* for the motion of electrons in helium, argon and neon. The calculated values of the rate of diffusion agree fairly well with experiment for values of $F_2 l$ which are not too large, though they are in all cases rather too low; the calculated mean energies are much larger than the experimental, but it is suggested that this may be due in part to an incorrect interpretation of experiment. When $F_2 l$ is very large, fast electrons are losing energy by inelastic impacts, and the above theory does not apply. In this case experiment shows that the mean

* H. W. Allen, *Phys. Rev.* 52, 707 (1937).

energy approaches a constant value as the electric field increases, and that the velocity of diffusion increases more rapidly than the above equations indicate.

The distribution of electronic speeds in helium is illustrated in Fig. 14. For this gas the quantity $l(c_2)$ has a minimum corresponding to electronic energies of about $2\frac{3}{4}$ electron-volts, and increases rapidly for energies in excess of this. For comparison, the distributions corresponding to Maxwell's and Druyvesteyn's formulae are also shown; the three curves (with the axis of abscissae) include equal areas, and correspond to the same mean energy (5·84 volts). It is clear from the figure that the calculated distribution is intermediate between those of Maxwell and Druyvesteyn; its maximum is sharper than that of Maxwell's curve (III), but it gives more high-energy electrons than curve II (Druyvesteyn).

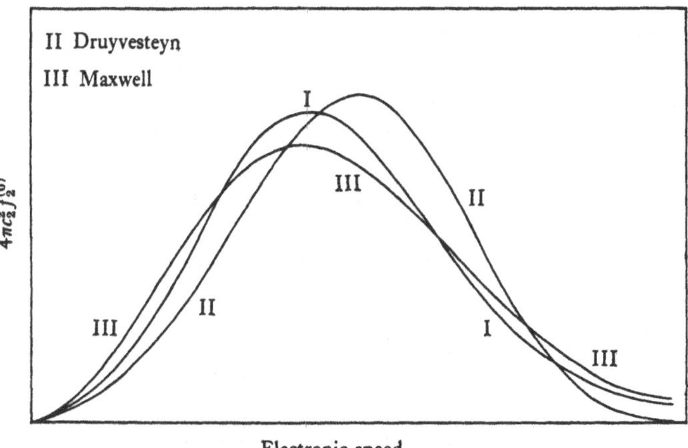

Fig. 14. The distribution function for electronic speeds ($4\pi c_2^2 f_2^{(0)}$): (I) calculated for electrons in helium, mean energy 5·84 V., (II) given by Druyvesteyn's formula, and (III) given by Maxwell's formula.

19.62. Inelastic collisions

The general theory of motion of electrons in a gas when inelastic collisions are possible is very complicated, and will not be given here. However, we can obtain some idea of the order of magnitude of the effects involved in the case of a gas like helium, which has a large excitation potential. An electron whose energy exceeds the excitation energy by an appreciable amount does not undergo many collisions before it loses energy inelastically, and so cannot obtain an energy much in excess of the excitation energy from the electric field before this happens; its energy after the collision is therefore small. Denote by $S(c_2)$ the number of electrons per unit volume and time which undergo inelastic collisions and whose speeds decrease from values above c_2 to values below c_2 because of the collisions. Then $S(c_2)$ is small if c_2 is either very large or very small; but there is an intermediate range of values of c_2 which is neither losing nor gaining electrons by inelastic collisions, and in which, therefore, $S(c_2)$ is constant. This range contains the majority of the electrons.

The left-hand side of (19.61, 9) represents the number of electrons per unit volume and time which enter the velocity-range for which $c_2 < v$ because of elastic collisions; the right-hand side gives the number leaving the same range because of the electric field. These quantities now differ by $S(v)$. Hence, integrating as before, we find that (19.61, 13) is replaced by

$$\tfrac{1}{3}F_2^2 f_2^{(1)} = \frac{S(c_2)}{4\pi c_2^3} + \frac{kT}{m_1 l}\frac{\partial f_2^{(0)}}{\partial c_2} + \frac{m_2 c_2}{m_1 l} f_2^{(0)}. \qquad (19.62, 1)$$

Here the term involving kT is unimportant for values of F_2 such that inelastic collisions are important, and so can be neglected.

It will be supposed that (19.61, 8) is unaffected by inelastic collisions; this amounts to assuming that the resultant momentum of electrons with speeds between c_2 and c_2+dc_2 is unaltered by inelastic collisions. This will be rigorously true in the range in which electrons are neither lost nor gained by inelastic collisions, and may be taken as a sufficient approximation elsewhere. Combining (19.61, 8) with (19.62, 1), we have

$$-\frac{F_2^2 l}{3c_2^2}\frac{\partial f_2^{(0)}}{\partial c_2} = \frac{S(c_2)}{4\pi c_2^3} + \frac{m_2 c_2}{m_1 l} f_2^{(0)}.$$

If $f^{(0)}$ denotes the value of $f_2^{(0)}$ in the absence of inelastic collisions, which is given by (19.61, 14′), the solution of this equation is

$$f_2^{(0)} = f^{(0)}\left\{B - \int_0^{c_2} \frac{3S(c_2)\,dc_2}{4\pi c_2 l F_2^2 f^{(0)}}\right\},$$

where B is a constant. Now when c_2 is large $f_2^{(0)}$ is small compared with $f^{(0)}$; also $S(c_2)$ denotes the number of electrons with speeds greater than c_2 which undergo inelastic collisions, and this, being roughly proportional to the total number of electrons with speeds greater than c_2, tends to zero more rapidly than $f^{(0)}$ as c_2 increases. For $f_2^{(0)}$ to be small compared with $f^{(0)}$ for large c_2, we must have

$$B = \int_0^{\infty} \frac{3S(c_2)\,dc_2}{4\pi c_2 l F_2^2 f^{(0)}},$$

and so

$$f_2^{(0)} = f^{(0)}\int_{c_2}^{\infty} \frac{3S(c_2)\,dc_2}{4\pi c_2 l F_2^2 f^{(0)}}. \qquad (19.62, 2)$$

If $f^{(0)}$ is small for energies approximating to the excitation energy, $f_2^{(0)}$ is not appreciably affected by inelastic collisions. For in this case the major part of the integral in (19.62, 2) comes from the range corresponding to energies close to the excitation energy; thus if the energy is much smaller than the excitation energy, (19.62, 2) approximates to

$$f_2^{(0)} = f^{(0)} \times \text{const.}$$

As F_2 increases, inelastic collisions begin to affect the values of $f_2^{(0)}$ corresponding to smaller energies.

A good approximation to $f_2^{(0)}$ is obtained if we take $S(c_2)$ as constant if $E_1 < \frac{1}{2}m_2c_2^2 < E_2$, where $E_2 - E_1$ is the energy of excitation of a molecule, and E_1 is small compared with this, and put $S(c_2) = 0$ elsewhere.

19.63. The steady state in a magnetic field

Consider now how the velocity-distribution of 19.61 must be modified when a magnetic field $\boldsymbol{H}$ acts perpendicular to the electric field. In this case (19.61, 1) is replaced by

$$\boldsymbol{F}_2 \cdot \frac{\partial f_2'}{\partial \boldsymbol{c}_2} + \frac{e_2}{m_2}(\boldsymbol{c}_2 \wedge \boldsymbol{H}) \cdot \frac{\partial f_2}{\partial \boldsymbol{c}_2} = \iint (f_1'f_2' - f_1f_2)g\alpha_{12}d\boldsymbol{e}'\,d\boldsymbol{c}_1. \tag{19.63, 1}$$

In this we insert

$$f_2 = f_2^{(0)} + (\boldsymbol{F}_2 \cdot \boldsymbol{c}_2)f_2^{(1)} + (\boldsymbol{H} \wedge \boldsymbol{F}_2) \cdot \boldsymbol{c}_2 \xi_2^{(1)}, \tag{19.63, 2}$$

where $f_2^{(0)}, f_2^{(1)}, \xi_2^{(1)}$ are functions of the scalars c_2^2, F_2^2, H^2 only, and we neglect terms involving $\overset{\circ}{\boldsymbol{c}_2\boldsymbol{c}_2}$ on the left. Then, equating terms involving only scalars, and terms involving the vectors $\boldsymbol{F}_2$ and $\boldsymbol{H} \wedge \boldsymbol{F}_2$, we get

$$\frac{F_2^2}{3c_2^2}\frac{\partial}{\partial c_2}(f_2^{(1)}c_2^3) = \iint (f_1'f_2^{(0)\prime} - f_1f_2^{(0)})g\alpha_{12}d\boldsymbol{e}'\,d\boldsymbol{c}_1, \tag{19.63, 3}$$

$$\frac{\boldsymbol{F}_2 \cdot \boldsymbol{c}_2}{c_2}\left(\frac{\partial f_2^{(0)}}{\partial c_2} - \frac{e_2H^2c_2}{m_2}\xi_2^{(1)}\right) = \iint \{f_1'f_2^{(1)\prime}(\boldsymbol{F}_2 \cdot \boldsymbol{c}_2') - f_1f_2^{(1)}(\boldsymbol{F}_2 \cdot \boldsymbol{c}_2)\}g\alpha_{12}d\boldsymbol{e}'\,d\boldsymbol{c}_1, \tag{19.63, 4}$$

$$\frac{e_2}{m_2}(\boldsymbol{H} \wedge \boldsymbol{F}_2) \cdot \boldsymbol{c}_2 f_2^{(1)} = \iint \{f_1'\xi_2^{(1)\prime}(\boldsymbol{H} \wedge \boldsymbol{F}_2) \cdot \boldsymbol{c}_2' - f_1\xi_2^{(1)}(\boldsymbol{H} \wedge \boldsymbol{F}_2) \cdot \boldsymbol{c}_2\}g\alpha_{12}d\boldsymbol{e}'\,d\boldsymbol{c}_1. \tag{19.63, 5}$$

The integrals are evaluated as in 19.61, giving

$$\tfrac{1}{3}F_2^2 f_2^{(1)} = \frac{kT}{m_1 l}\frac{\partial f_2^{(0)}}{\partial c_2} + \frac{m_2c_2}{m_1 l}f_2^{(0)}, \tag{19.63, 6}$$

$$\frac{\partial f_2^{(0)}}{\partial c_2} - \frac{e_2H^2c_2}{m_2}\xi_2^{(1)} = -\frac{f_2^{(1)}c_2^2}{l}, \tag{19.63, 7}$$

$$\frac{e_2}{m_2}f_2^{(1)} = -\frac{\xi_2^{(1)}c_2}{l}. \tag{19.63, 8}$$

We neglect the term involving kT in (19.63, 6); then elimination of $f_2^{(1)}$ and $\xi_2^{(1)}$ gives

$$-\frac{F_2^2 l}{3c_2^2}\frac{\partial f_2^{(0)}}{\partial c_2} \bigg/ \left(1 + \frac{e_2^2H^2l^2}{m_2^2c_2^2}\right) = \frac{m_2c_2}{m_1 l}f_2^{(0)},$$

the solution of which is

$$f_2^{(0)} = A \exp\left\{-\int \frac{3m_2 c_2^3}{m_1 F_2^2 l^2}\left(1+\frac{e_2^2 H^2 l^2}{m_2^2 c_2^2}\right) dc_2\right\}. \qquad (19.63, 9)$$

Comparing this with (19.61, 14′), we see that the effect of the magnetic field is to reduce the mean energy of electrons, the effect being the same as if F_2 were reduced by a factor equal to a mean value of $\sqrt{(1+e_2^2H^2l^2/m_2^2c_2^2)}$.

Again, from (19.63, 6, 8),

$$f_2^{(1)} = \frac{3m_2 c_2}{m_1 l F_2^2} f_2^{(0)}, \quad \xi_2^{(1)} = -\frac{3e_2}{m_1 F_2^2} f_2^{(0)}, \qquad (19.63, 10)$$

and so the velocity of diffusion of electrons is

$$\bar{c}_2 = \frac{1}{3n_2}\int (f_2^{(1)}\boldsymbol{F}_2 + \xi_2^{(1)}\boldsymbol{H}\wedge\boldsymbol{F}_2)\, c_2^2\, dc_2$$

$$= \frac{m_2}{m_1 n_2 F_2^2}\int f_2^{(0)} \frac{c_2}{l}\left(\boldsymbol{F}_2 - \frac{e_2 l}{m_2 c_2}\boldsymbol{H}\wedge\boldsymbol{F}_2\right) c_2^2\, dc_2.$$

Thus, if l/c_2 is constant and equal to τ_2 (so that τ_2 is an equivalent mean collision-interval), it is found that

$$\bar{c}_2 = \frac{\tau_2(\boldsymbol{F}_2 - \boldsymbol{H}\wedge\boldsymbol{F}_2 e_2\tau_2/m_2)}{1+e_2^2H^2\tau_2^2/m_2^2}, \qquad (19.63, 11)$$

which is identical with the result when the electric field is small. Equation (19.63, 11) can be taken as a first approximation when l/c_2 is not constant, τ_2 then denoting a mean value of l/c_2.

If l is constant and H is small the results can be expanded as series in H; if $\bar{c}_{20}$ denotes the mean value of c_2 when $H = 0$, it is found that the mean energy is reduced in the ratio

$$1 - 0{\cdot}618\, e_2^2 H^2 l^2/m_2^2 \bar{c}_{20}^2 + \ldots,$$

the direct diffusion is reduced in the ratio

$$1 - 0{\cdot}874\, e_2^2 H^2 l^2/m_2^2 \bar{c}_{20}^2 + \ldots,$$

and the ratio of transverse to direct diffusion is

$$1{\cdot}085\, e_2 H l/m_2 \bar{c}_{20} - \ldots.$$

It is to be observed that the direct diffusion-velocity is not much less than the ordinary velocity of diffusion in a gas subject to no magnetic force in which the electric force is adjusted to give the electrons the same energy.

19.64. Ionization and recombination

As indicated in 19.6, a steady state is possible in which electrons with high energies are produced by the ionization of molecules, the electrons losing energy elastically at collisions with molecules before recombination takes place. Suppose that the gas is uniform and at rest, and that no electric or magnetic fields act. Let $\alpha\, d\mathbf{c}_2$ denote the rate of production of electrons with velocities in the range $\mathbf{c}_2$, $d\mathbf{c}_2$ per unit time, by the ionization of molecules, and let $\beta f_2 d\mathbf{c}_2$ be the rate of loss of electrons to this velocity-range per unit volume by recombination; α, β are supposed to depend only on the magnitude of $\mathbf{c}_2$, not on its direction. Then the equation satisfied by the velocity-distribution function f_2 for electrons is

$$\frac{\partial f_2}{\partial t} = \alpha - \beta f_2 + \iint (f_1' f_2' - f_1 f_2) g\alpha_{12}\, d\mathbf{e}'\, d\mathbf{c}_1. \qquad (19.64, 1)$$

If the state is steady, $\partial f_2/\partial t = 0$. Again, if the integral on the right is multiplied by $d\mathbf{c}_2$, and integrated over all values of $\mathbf{c}_2$ such that $c_2 < v$, then, by the argument used in simplifying (19.61, 5), the resulting expression is equal to

$$4\pi\left\{\frac{kTv^3}{m_1 l(v)}\frac{\partial f_2(v)}{\partial v} + \frac{m_2 v^4}{m_1 l(v)} f_2(v)\right\}.$$

Thus (19.64, 1) is equivalent to

$$0 = \alpha - \beta f_2 + \frac{\partial}{c_2^2 \partial c_2}\left\{\frac{kTc_2^3}{m_1 l(c_2)}\frac{\partial f_2(c_2)}{\partial c_2} + \frac{m_2 c_2^4}{m_1 l(c_2)} f_2(c_2)\right\}. \qquad (19.64, 2)$$

It is, unfortunately, impossible to solve this equation completely. Approximate solutions can, however, be given in two limiting cases, and these serve to indicate the properties of the general solution.

Suppose first that an electron undergoes few collisions during its free life-time. Then the mean energy of an electron is large compared with $\frac{3}{2}kT$, and the term involving kT can be omitted from the right of (19.64, 2). The solution of the equation can then be found in terms of integrals; for example, if β is a constant, and $l(c_2)/c_2$ is a constant τ_2, the solution is

$$f_2 = c_2^{(m_1/m_2)\tau_2\beta - 3}\left\{\int_{c_2}^{\infty} \alpha \frac{m_1 \tau_2}{m_2} c_2^{2-(m_1/m_2)\tau_2\beta}\, dc_2 + A\right\},$$

where A is a constant. For the integral

$$\int f_2 d\mathbf{c}_2 \equiv 4\pi \int_0^{\infty} c_2^2 f_2\, dc_2$$

to converge for large values of c_2 it is necessary that $A = 0$, and so

$$f_2 = c_2^{(m_1/m_2)\tau_2\beta - 3}\int_{c_2}^{\infty} \alpha \frac{m_1 \tau_2}{m_2} c_2^{2-(m_1/m_2)\tau_2\beta}\, dc_2. \qquad (19.64, 3)$$

The number-density n_2 of electrons is given by

$$n_2 = 4\pi \int_0^{\infty} c_2^{(m_1/m_2)\tau_2\beta - 1}\left[\int_{c_2}^{\infty} \alpha \frac{m_1 \tau_2}{m_2} c_2^{2-(m_1/m_2)\tau_2\beta}\, dc_2\right] dc_2,$$

whence, by a partial integration,

$$n_2 = \frac{4\pi}{\beta}\left\{\left[c_2^{(m_1/m_2)\tau_2\beta}\int_{c_2}^{\infty}\alpha c_2^{2-(m_1/m_2)\tau_2\beta}dc_2\right]_0^\infty + \int_0^\infty \alpha c_2^2 dc_2\right\}$$

$$= \frac{4\pi}{\beta}\int_0^\infty \alpha c_2^2 dc_2, \qquad (19.64, 4)$$

as can also be proved by multiplying (19.64, 2) by dc_2 and integrating over all values of c_2. Again, the mean energy of an electron is

$$\frac{2\pi m_2}{n_2}\int_0^\infty f_2 c_2^4 dc_2 = \frac{2\pi m_2}{n_2\beta}\frac{1}{(1+2m_2/m_1\tau_2\beta)}\int_0^\infty \alpha c_2^4 dc_2, \quad (19.64, 5)$$

by a further partial integration. Comparing (19.64, 4 and 5), we see that the mean energy of an electron is $1/(1+2m_2/m_1\tau_2\beta)$ times the mean energy of a liberated electron; and so the approximation (19.64, 3) is valid if the mean energy of a liberated electron is large compared with

$$\tfrac{3}{2}kT(1+2m_2/m_1\tau_2\beta). \qquad (19.64, 6)$$

Next suppose that an electron undergoes so large a number of collisions during its free life-time that its mean energy is nearly equal to $\frac{3}{2}kT$. To a first approximation

$$f_2 = f_2^{(0)} \equiv n_2\left(\frac{m_2}{2\pi kT}\right)^{\frac{3}{2}} \exp(-m_2 c_2^2/2kT),$$

while a second approximation is given by the equation

$$-c_2^2(\alpha - \beta f_2^{(0)}) = \frac{\partial}{\partial c_2}\left(\frac{kTc_2^3}{m_1 l}\frac{\partial f_2}{\partial c_2} + \frac{m_2 c_2^4}{m_1 l}f_2\right).$$

On integration, this equation gives

$$-\int_0^{c_2}(\alpha - \beta f_2^{(0)})\,c_2^2 dc_2 = \frac{kTc_2^3}{m_1 l}\left(\frac{\partial f_2}{\partial c_2} + \frac{m_2 c_2}{kT}f_2\right), \qquad (19.64, 7)$$

the constant of integration being chosen so as to make f_2 finite when c_2 is small. If c_2 in (19.64, 7) is made to tend to infinity, it is found that

$$\int_0^\infty (\alpha - \beta f_2^{(0)})\,c_2^2 dc_2 = 0, \qquad (19.64, 8)$$

from which n_2 can be determined. This equation expresses the fact that ionizations and recombinations balance each other.

The general solution of (19.64, 7) is

$$f_2 = f_2^{(0)}\left\{B - \int_0^{c_2}\frac{m_1 l}{kTc_2^3 f_2^{(0)}}\left[\int_0^{c_2}(\alpha - \beta f_2^{(0)})\,c_2^2 dc_2\right]dc_2\right\}, \quad (19.64, 9)$$

where B is a constant, whose value can be determined from the condition

$$n_2 = \int f_2 d\boldsymbol{c}_2.$$

Equation (19.64, 9) can be shown to apply if $2m_2/m_1\tau_2\beta$ is large, and the mean energy of a liberated electron is small compared with (19.64, 6), τ_2 now denoting a mean value of $l(c_2)/c_2$.

The value of f_2 when electric or magnetic fields are present is very difficult to derive. It may be observed, however, that the relaxation results derived in 19.31 did not depend on any assumption that the distribution of velocities is nearly Maxwellian; and these results have been shown by the exact theory to be nearly accurate for a mixture of electrons and heavy molecules both when the electric force is small, and, with an altered collision-interval, when it is large. It is therefore not unreasonable to suppose that they apply also to a gas in which ionization and recombination are taking place, using the appropriate value of the mean collision-interval.

19.65. Strongly ionized gases

Encounters between pairs of charged particles can be neglected only in a very slightly ionized gas. The effective radius for electrostatic interactions at ordinary temperatures may be as much as a hundred times that for interactions of charged with uncharged particles. Thus encounters between pairs of charged particles are dominant when more than a small fraction of one per cent of the molecules is ionized.

To illustrate the case when electrostatic interactions are dominant, we consider the effect of a strong electric field in a fully ionized binary gas (plasma) composed of electrons and positive ions. The electric field predominantly accelerates the electrons, giving them energy of directed motion. Encounters with heavy ions quickly randomize the direction of this motion but (cf. 19.6) produce only a slow interchange of energy between ions and electrons; hence the mean energy of electrons may increase well above that of ions. Electron–electron encounters, though not altering the mean energy of electrons as a whole, make the electron velocity-distribution rapidly approach the Maxwellian form corresponding to this mean energy. The temperature slowly increases because of work done by the electric field.

The suffixes 1, 2 are taken to refer respectively to the ions and electrons. In a uniform gas at rest, the equation (19.61, 1) for electrons is replaced by

$$\frac{\partial f_2}{\partial t}+\boldsymbol{F}_2.\frac{\partial f_2}{\partial \boldsymbol{c}_2} = \sum_s \iint (f_s' f_2' - f_s f_2) g\alpha_{s2}\, d\boldsymbol{e}'\, d\boldsymbol{c}_s; \tag{19.65, 1}$$

the term $\partial f_2/\partial t$ is included to take account of the slow heating of the electron gas. As in 19.61, f_1 is assumed to have the Maxwellian form

$$f_1 = n_1\left(\frac{m_1}{2\pi kT_1}\right)^{\frac{3}{2}} e^{-m_1 c_1^2/2kT_1}, \tag{19.65, 2}$$

the mass-motion of the ions being ignored; also (cf. (19.61, 4)) f_2 is given by

$$f_2 = f_2^{(0)} + \boldsymbol{F}_2.\boldsymbol{c}_2 f_2^{(1)} + \boldsymbol{F}_2\boldsymbol{F}_2 : \overset{\circ}{\boldsymbol{c}_2\boldsymbol{c}_2} f_2^{(2)} + \dots, \tag{19.65, 3}$$

where $f_2^{(0)}, f_2^{(1)}, \ldots$ are functions of c_2. On substituting from (19.65, 3) into (19.65, 1), and approximating as in 19.61, we get the coupled equations

$$\frac{\partial f_2^{(0)}}{\partial t}+\frac{F_2^2}{3}\frac{\partial}{\partial \mathbf{c}_2}\cdot(\mathbf{c}_2 f_2^{(1)}) = \iint \{f_1' f_2^{(0)\prime} - f_1 f_2^{(0)}\} g\alpha_{12} de'\, dc_1 + \iint \{f^{(0)\prime} f_2^{(0)\prime} - f^{(0)} f_2^{(0)}\} g\alpha_2 de'\, dc, \quad (19.65, 4)$$

$$\frac{\mathbf{c}_2}{c_2}\frac{\partial f_2^{(0)}}{\partial c_2} = \iint \{f_1' f_2^{(1)\prime} \mathbf{c}_2' - f_1 f_2^{(1)} \mathbf{c}_2\} g\alpha_{12} de'\, dc_1 + \iint \{f^{(0)\prime} f_2^{(1)\prime} \mathbf{c}_2' + f_2^{(0)\prime} f^{(1)\prime} \mathbf{c}' - f^{(0)} f_2^{(1)} \mathbf{c}_2 - f_2^{(0)} f^{(1)} \mathbf{c}\} g\alpha_2 de'\, dc. \quad (19.65, 5)$$

In these $f^{(0)}$, $f^{(1)}$ stand for $f_2^{(0)}(c)$, $f_2^{(1)}(c)$; the term $\partial f_2^{(1)}/\partial t$ is omitted from (19.65, 5) as representing only the slow change of $f_2^{(1)}$, which is itself small.

The first term on the right of (19.65, 4) is small, since c_1', c_2' differ only slightly from c_1, c_2; the second term, which is dominant, ensures that $f_2^{(0)}$ has nearly the Maxwellian form

$$f_2^{(0)} = n_2\left(\frac{m_2}{2\pi k T_2}\right)^{\frac{3}{2}} \exp(-m_2 c_2^2/2kT_2), \quad (19.65, 6)$$

corresponding to an electron-temperature T_2 not necessarily equal to T_1. The temperatures T_1, T_2 are defined as such that

$$\tfrac{1}{2}m_1\overline{c_1^2} = \tfrac{3}{2}kT_1, \quad \tfrac{1}{2}m_2\overline{c_2^2} = \tfrac{3}{2}kT_2. \quad (19.65, 7)$$

In (19.65, 5) we may approximate by using (19.65, 6), and also by writing $c_1' = c_1$, $c_2' = c_2$ in the first integral on the right. Then this integral becomes

$$n_1 \int f_2^{(1)}(\mathbf{c}_2' - \mathbf{c}_2) g\alpha_{12} de',$$

independent of T_1, and the equation becomes identical with that giving $f_2^{(1)}$ (and hence the electric current and conductivity) in a gas in a *weak* field at temperature T_2. Thus $f_2^{(1)}$ can be regarded as known from our earlier work. In particular, by (14.1, 1) and 10.34, the velocity of diffusion $\overline{\mathbf{c}}_2$ is given to the first approximation by

$$\overline{\mathbf{c}}_2 = 2m_2 \mathbf{F}_2 \tau(T_2)/m_1, \quad (19.65, 8)$$

where
$$\tau(T_2) = 3(2kT_2)^{\frac{3}{2}} m_1/\{16\pi^{\frac{1}{2}} n_1 m_2^{\frac{1}{2}} e_1'^2 e_2'^2 \ln(1+v_{01}^2)\}. \quad (19.65, 9)$$

The charges on the particles are here denoted by e_1', e_2' to draw attention to their being measured in electrostatic units.

Equation (19.65, 4) may be used to determine not only the deviation of $f_2^{(0)}$ from the Maxwellian form (19.65, 6) but also the value of $\partial T_2/\partial t$. The latter* is found from the equation of energy of the electron gas, obtained by multiplying (19.65, 4) by $\frac{1}{2}m_2 c_2^2$ and integrating with respect to $\mathbf{c}_2$. The contribution to the energy equation from the second integral on the right

* See L. Landau, *Phys. Z. SowjUn.* **10**, 154 (1936).

of (19.65, 4) is zero, by (3.54, 5), corresponding to the fact that the total energy of the electrons is not altered by their mutual encounters. On using the transformation of 3.53, the first integral on the right contributes

$$\iiint f_1 f_2^{(0)} \tfrac{1}{2} m_2 (c_2'^2 - c_2^2) g\alpha_{12} \, de' \, dc_1 \, dc_2.$$

The first and second terms on the left yield respectively terms $\frac{3}{2} n_2 k \, dT_2/dt$ and (after transformations, including an integration by parts) $-m_2 n_2 \boldsymbol{F}_2 . \bar{\boldsymbol{c}}_2$. Hence finally

$$\tfrac{3}{2} n_2 k \frac{dT_2}{dt} = m_2 n_2 \boldsymbol{F}_2 . \bar{\boldsymbol{c}}_2 - \iiint f_1 f_2^{(0)} \tfrac{1}{2} m_2 (c_2^2 - c_2'^2) g\alpha_{12} \, de' \, dc_1 \, dc_2. \tag{19.65, 10}$$

The first term on the right of (19.65, 10) represents the rate of gain of electron energy from the electric field, the second the rate of loss of energy through encounters with ions. In the second term, because of the smallness of $c_2^2 - c_2'^2$, it is sufficient to use the approximation (19.65, 6) for $f_2^{(0)}$.

In order to effect the integration in this term, we express $\boldsymbol{c}_1$, $\boldsymbol{c}_2$ in terms of $\boldsymbol{\Gamma}$ and $\boldsymbol{g}_{21}$, where

$$\boldsymbol{\Gamma} = M_1' \boldsymbol{c}_1 + M_2' \boldsymbol{c}_2, \tag{19.65, 11}$$

and

$$M_1' = m_1/(m_1 + m_2'), \quad M_2' = m_2'/(m_1 + m_2'), \quad m_2' = T_1 m_2 / T_2. \tag{19.65, 12}$$

Then

$$\boldsymbol{c}_1 = \boldsymbol{\Gamma} - M_2' \boldsymbol{g}_{21}, \quad \boldsymbol{c}_2 = \boldsymbol{\Gamma} + M_1' \boldsymbol{g}_{21}, \tag{19.65, 13}$$

and (cf. (3.41, 8) and (3.52, 5))

$$\frac{m_1 c_1^2}{2kT_1} + \frac{m_2 c_2^2}{2kT_2} = \frac{(m_1 + m_2')}{2kT_1} (\Gamma^2 + M_1' M_2' g^2), \quad d\boldsymbol{c}_1 \, d\boldsymbol{c}_2 = d\boldsymbol{\Gamma} \, d\boldsymbol{g}_{21}.$$

Also, by (3.41, 6, 7),

$$\begin{aligned} c_2^2 - c_2'^2 &= 2M_1 \mathbf{G} . (\boldsymbol{g}_{21} - \boldsymbol{g}_{21}') \\ &= 2M_1 \{\boldsymbol{\Gamma} + (M_2 M_1' - M_1 M_2') \boldsymbol{g}_{21}\} . (\boldsymbol{g}_{21} - \boldsymbol{g}_{21}'). \end{aligned}$$

On substituting these values, the term in question becomes

$$n_1 n_2 \left(\frac{m_1 m_2}{4\pi^2 k^2 T_1 T_2}\right)^{\frac{3}{2}} \iiiint \exp\{-(m_1 + m_2')(\Gamma^2 + M_1' M_2' g^2)/2kT_1\}$$

$$\times \frac{m_1 m_2}{m_1 + m_2} \{\boldsymbol{\Gamma} + (M_2 M_1' - M_1 M_2') \boldsymbol{g}_{21}\} . (\boldsymbol{g}_{21} - \boldsymbol{g}_{21}') g b \, db \, d\epsilon \, d\boldsymbol{\Gamma} \, d\boldsymbol{g}_{21}$$

or, substituting for $M_2 M_1' - M_1 M_2'$ and integrating with respect to $\boldsymbol{\Gamma}$ and ϵ,

$$2\pi n_1 n_2 \left(\frac{m_1 m_2}{2\pi k T_2 (m_1 + m_2')}\right)^{\frac{3}{2}} \frac{m_1^2 m_2^2 (1 - T_1/T_2)}{(m_1 + m_2)^2 (m_1 + m_2')}$$

$$\times \iint \exp\{-(m_1 + m_2') M_1' M_2' g^2 / 2kT_1\} (1 - \cos\chi) g^3 b \, db \, d\boldsymbol{g}_{21}.$$

In this we now regard m_2, m_2' as negligible compared with m_1. Then the expression is equivalent to

$$\frac{8\pi^{\frac{1}{2}}n_1n_2(m_1m_2)^{\frac{3}{2}}(1-T_1/T_2)}{(2kT_2)^{\frac{3}{2}}(m_1+m_2)^{\frac{3}{2}}}\iint \exp\left\{-\frac{m_0M_1M_2g^2}{2kT_2}\right\}(1-\cos\chi)g^5b\,db\,dg$$
$$=16n_1n_2k(m_2/m_1)(T_2-T_1)[\Omega_{12}^{(1)}(1)]_{T_2}$$
$$=3n_2k(T_2-T_1)/2\tau(T_2),$$

using (19.65, 9), (9.33, 2) and the results of 10.34. Thus (19.65, 10) becomes

$$\frac{dT_2}{dt}=\frac{4m_2^2F_2^2\tau(T_2)}{3km_1}-\frac{(T_2-T_1)}{\tau(T_2)}. \qquad (19.65, 14)$$

Suppose that T_1 is kept constant. Then if $F_2=0$ equation (19.65, 14) indicates that any initial *small* difference between T_2 and T_1 relaxes to zero with relaxation time $\tau(T_1)$. If a small F_2 is suddenly switched on, T_2 rises in a time comparable with $\tau(T_1)$ to the steady value found by taking $dT_2/dt=0$ in (19.65, 14). Because $\tau(T_2)$ is (very nearly) proportional to $T_2^{\frac{3}{2}}$, no such steady value can be found if F_2 is large. As T_2 increases above T_1, the first term on the right of (19.65, 14) (representing Joule heating) also increases, whereas the second term (representing the loss of energy at encounters with ions) initially increases but ultimately decreases. Thus if m_2F_2 is sufficiently large, the loss of energy at encounters cannot balance the gain of energy from the electric field for any value of T_2, and T_2 steadily increases.

By considering the maximum ratio of the second term on the right of (19.65, 14) to the first, it can be shown that the maximum steady value of T_2 occurs when

$$T_2=\tfrac{3}{2}T_1,\quad m_2F_2=\left(\frac{\pi m_2}{2m_1}\right)^{\frac{1}{2}}\frac{8n_1e_1'^2e_2'^2\ln(1+v_{01}^2)}{9kT_1}. \qquad (19.65, 15)$$

The corresponding diffusion velocity $\bar{c}_2$ has magnitude $(3kT_1/2m_1)^{\frac{1}{2}}$, comparable with the mean thermal speed of ions. If m_2F_2 exceeds the value given by (19.65, 15), in the absence of any further cooling mechanism for electrons T_2 increases without limit, and with it $\bar{c}_2$; the electrons are then said to run away.

19.66. Runaway effects

The runaway of electrons in a strong electric field was first considered by Giovanelli.* The runaway process can be regarded as total when the velocity of diffusion $\bar{c}_2$ given by (19.65, 8) becomes comparable with the mean thermal velocity $(8kT_2/\pi m_2)^{\frac{1}{2}}$ of electrons. This gives

$$m_2F_2\sim\frac{8n_1e_1'^2e_2'^2\ln(1+v_{01}^2)}{3kT_2}, \qquad (19.66, 1)$$

* R. G. Giovanelli, *Phil. Mag.* (7), **40**, 206 (1949).

so that the product $F_2 T_2$ is then considerably greater than its limiting value for the runaway process to commence (cf. (19.65, 15)).

High-speed electrons are liable to run away before the electron-gas is able to accelerate as a whole. The mean rate of loss of momentum of an electron of velocity $\boldsymbol{c}_2$ due to encounters with ions (regarded as stationary) is

$$\begin{aligned} n_1 \iint m_2(\boldsymbol{c}_2 - \boldsymbol{c}_2') g b \, db \, d\epsilon &= 2\pi n_1 m_2 c_2 \boldsymbol{c}_2 \int (1 - \cos\chi) b \, db \\ &= n_1 m_2 c_2 \boldsymbol{c}_2 \phi_{12}^{(1)}(c_2) \\ &= \frac{2\pi n_1 e_1'^2 e_2'^2 \boldsymbol{c}_2}{m_2 c_2^3} \ln(1 + v_{01}^2) \end{aligned}$$

(cf. (10.34, 4)). The corresponding retardation of the electron has magnitude $2\pi n_1 e_1'^2 e_2'^2 \ln(1 + v_{01}^2)/m_2^2 c_2^2$. If c_2 is large compared with the mean thermal velocity of electrons, there is a further retardation $\pi n_2 e_2'^4 \ln(1 + v_{01}^2)/m_2^2 c_2^2$ resulting from encounters with electrons, since the 'field' electrons may be regarded as at rest in comparison with the fast particle. Thus the total retardation is

$$2\pi e_2'^2 (n_1 e_1'^2 + \tfrac{1}{2} n_2 e_2'^2) \ln(1 + v_{01}^2)/m_2^2 c_2^2.$$

If c_2 is so large that this is less than the acceleration $\boldsymbol{F}_2$ due to the electric field, encounters are unable to prevent that field from accelerating the electron to steadily increasing velocities. The critical c_2 for runaway is given by

$$\tfrac{1}{2} m_2 c_2^2 = \pi e_2'^2 (n_1 e_1'^2 + \tfrac{1}{2} n_2 e_2'^2) \ln(1 + v_{01}^2)/m_2 F_2, \qquad (19.66, 2)$$

so that (cf. (19.66, 1)) $\frac{1}{2} m_2 c_2^2$ is comparable with the value of kT_2 for total runaway of the electron gas.

The rate at which high-speed electrons attain runaway energies depends on the number of electrons in a high-velocity 'tail' of the distribution function f_2; this number is difficult to estimate, because of the failure of the expansion method used in 19.65.* Moreover, when electron runaway once starts, a discussion based solely on elastic encounters may be inadequate. A sufficiently energetic electron may also produce further ionization, so possibly increasing the number of runaways.

* See, for example, H. Dreicer, *Phys. Rev.* **115**, 238 (1959), and **117**, 329 (1960); A. V. Gurevich, *Soviet Phys. JETP*, **11**, 1150 (1960); M. D. Kruskal and I. B. Bernstein, *Phys. Fluids*, **7**, 407 (1964).

THE FOKKER–PLANCK APPROACH

19.7. The Landau and Fokker–Planck equations

As was noted in 10.34, the dominant encounters for electrostatic interaction are those producing small deflections. If it is assumed that in all encounters the deflections are small, alternative expressions for $\partial_e f/\partial t$ can be derived from the Boltzmann integral.

Assume that in the encounter of two molecules m_1, m_2 the velocities $\boldsymbol{c}_1$, $\boldsymbol{c}_2$, $\boldsymbol{g}_{21}$ alter by small quantities $\Delta\boldsymbol{c}_1$, $\Delta\boldsymbol{c}_2$, $\Delta\boldsymbol{g}_{21}$. Then by (3.41, 6, 7)

$$\Delta\boldsymbol{c}_1 = -M_2\Delta\boldsymbol{g}_{21}, \quad \Delta\boldsymbol{c}_2 = M_1\Delta\boldsymbol{g}_{21}. \qquad (19.7, 1)$$

By Taylor's theorem,

$$f_1' \equiv f_1(\boldsymbol{c}_1+\Delta\boldsymbol{c}_1) = f(\boldsymbol{c}_1)+\Delta\boldsymbol{c}_1 . \frac{\partial f_1}{\partial \boldsymbol{c}_1} + \tfrac{1}{2}\Delta\boldsymbol{c}_1\Delta\boldsymbol{c}_1 : \frac{\partial^2 f_1}{\partial \boldsymbol{c}_1 \partial \boldsymbol{c}_1} + \ldots$$

with a similar expression for f_2'. Thus if we retain only terms of first and second order in $\Delta\boldsymbol{c}_1$, $\Delta\boldsymbol{c}_2$, we find, using (19.7, 1)

$$\left(\frac{\partial_e f_1}{\partial t}\right)_2 \equiv \iiint (f_1'f_2' - f_1 f_2) g b\, db\, d\epsilon\, d\boldsymbol{c}_2$$

$$= \iiint \left\{\Delta\boldsymbol{g}_{21} . \left(M_1\frac{\partial}{\partial\boldsymbol{c}_2} - M_2\frac{\partial}{\partial\boldsymbol{c}_1}\right) f_1 f_2\right.$$

$$\left. + \tfrac{1}{2}\Delta\boldsymbol{g}_{21}\Delta\boldsymbol{g}_{21} : \left(M_1\frac{\partial}{\partial\boldsymbol{c}_2} - M_2\frac{\partial}{\partial\boldsymbol{c}_1}\right)^2 f_1 f_2\right\} g b\, db\, d\epsilon\, d\boldsymbol{c}_2. \qquad (19.7, 2)$$

Now $\boldsymbol{g}_{21}+\Delta\boldsymbol{g}_{21} (\equiv \boldsymbol{g}_{21}')$ can be expressed in the form

$$\boldsymbol{g}_{21}+\Delta\boldsymbol{g}_{21} = \boldsymbol{g}_{21}\cos\chi + g(\mathbf{h}\cos\epsilon + \mathbf{i}\sin\epsilon)\sin\chi,$$

where $\mathbf{h}$, $\mathbf{i}$ are unit vectors perpendicular to $\boldsymbol{g}_{21}$ and to each other; the neglect of terms of higher order than the second in $\Delta\boldsymbol{g}_{21}$ means that χ^3 is negligible, and that $\sin^2\chi$ can be replaced by $2(1-\cos\chi)$. Thus

$$\iint \Delta\boldsymbol{g}_{21}\, b\, db\, d\epsilon = -2\pi\boldsymbol{g}_{21}\int(1-\cos\chi)\, b\, db = -\boldsymbol{g}_{21}\phi_{12}^{(1)}(g), \qquad (19.7, 3)$$

$$\iint \Delta\boldsymbol{g}_{21}\Delta\boldsymbol{g}_{21}\, b\, db\, d\epsilon = \pi g^2(\mathbf{hh}+\mathbf{ii})\int \sin^2\chi\, b\, db = \phi_{12}^{(1)}(g)(g^2\mathsf{U} - \boldsymbol{g}_{21}\boldsymbol{g}_{21}), \qquad (19.7, 4)$$

using (1.3, 9). From these relations

$$2\iint \Delta\boldsymbol{g}_{21}\, g b\, db\, d\epsilon = \frac{\partial}{\partial \boldsymbol{g}_{21}} . \iint \Delta\boldsymbol{g}_{21}\Delta\boldsymbol{g}_{21}\, g b\, db\, d\epsilon$$

$$= \left(M_1\frac{\partial}{\partial\boldsymbol{c}_2} - M_2\frac{\partial}{\partial\boldsymbol{c}_1}\right) . \iint \Delta\boldsymbol{g}_{21}\Delta\boldsymbol{g}_{21}\, g b\, db\, d\epsilon, \qquad (19.7, 5)$$

since, applied to a function of $\mathbf{g}_{21}$, $-\partial/\partial\mathbf{c}_1 = \partial/\partial\mathbf{c}_2 = \partial/\partial\mathbf{g}_{21}$. Hence (19.7, 2) can be written

$$\left(\frac{\partial_e f_1}{\partial t}\right)_2 = \frac{1}{2}\int\left\{\left(M_1\frac{\partial}{\partial\mathbf{c}_2} - M_2\frac{\partial}{\partial\mathbf{c}_1}\right)\right.$$

$$\left.\cdot\iint\left(M_1\frac{\partial}{\partial\mathbf{c}_2} - M_2\frac{\partial}{\partial\mathbf{c}_1}\right)f_1 f_2 . \Delta\mathbf{g}_{21}\Delta\mathbf{g}_{21}\, gb\, db\, d\epsilon\right\} d\mathbf{c}_2.$$

The term involving $M_1\,\partial/\partial\mathbf{c}_2$ in the first of the brackets on the right vanishes on integration, by the divergence theorem. Thus

$$\left(\frac{\partial_e f_1}{\partial t}\right)_2 = -\frac{\partial}{\partial\mathbf{c}_1}\cdot\mathbf{Q}_{12}, \tag{19.7, 6}$$

where

$$\mathbf{Q}_{12} = \tfrac{1}{2}M_2\iiint\left\{\left(M_1\frac{\partial}{\partial\mathbf{c}_2} - M_2\frac{\partial}{\partial\mathbf{c}_1}\right)f_1 f_2\right\}.\Delta\mathbf{g}_{21}\Delta\mathbf{g}_{21}\, gb\, db\, d\epsilon\, d\mathbf{c}_2 \tag{19.7, 7}$$

$$= \frac{1}{2}\iiint\left\{\left(\frac{m_1}{m_2}\frac{\partial}{\partial\mathbf{c}_2} - \frac{\partial}{\partial\mathbf{c}_1}\right)f_1 f_2\right\}.\Delta\mathbf{c}_1\Delta\mathbf{c}_1\, gb\, db\, d\epsilon\, d\mathbf{c}_2. \tag{19.7, 7'}$$

Equations (19.7, 6 and 7′) (in a different notation) were first derived by Landau.* As Landau pointed out, they indicate that the changes in f_1 due to encounters with molecules m_2 are the same as if there were a flow $\mathbf{Q}_{12}$ of molecules m_1 in velocity-space.

Again using (19.7, 5) and the divergence theorem, we may further transform (19.7, 7) as follows:

$$\mathbf{Q}_{12} = \tfrac{1}{2}M_2\int\left(M_1\frac{\partial}{\partial\mathbf{c}_2} - M_2\frac{\partial}{\partial\mathbf{c}_1}\right)\cdot\left(\iint f_1 f_2\Delta\mathbf{g}_{21}\Delta\mathbf{g}_{21}\, gb\, db\, d\epsilon\right)d\mathbf{c}_2$$

$$-M_2\iiint f_1 f_2\Delta\mathbf{g}_{21}\, gb\, db\, d\epsilon\, d\mathbf{c}_2$$

$$= -\frac{1}{2}\frac{\partial}{\partial\mathbf{c}_1}\cdot\left\{f_1\iiint f_2\Delta\mathbf{c}_1\Delta\mathbf{c}_1\, gb\, db\, d\epsilon\, d\mathbf{c}_2\right\}$$

$$+\iiint f_1 f_2\Delta\mathbf{c}_1\, gb\, db\, d\epsilon\, d\mathbf{c}_2$$

$$= f_1\,\Sigma_2\Delta\mathbf{c}_1 - \frac{1}{2}\frac{\partial}{\partial\mathbf{c}_1}\cdot(f_1\,\Sigma_2\Delta\mathbf{c}_1\Delta\mathbf{c}_1), \tag{19.7, 8}$$

where

$$\Sigma_2\phi \equiv \iiint f_2\phi gb\, db\, d\epsilon\, d\mathbf{c}_2. \tag{19.7, 9}$$

Thus $\Sigma_2\phi$ denotes the sum of the values of ϕ for all encounters per unit time of a molecule m_1 of velocity $\mathbf{c}_1$ with molecules m_2. Combining (19.7, 6) and (19.7, 8), we get

$$\left(\frac{\partial_e f_1}{\partial t}\right)_2 = -\frac{\partial}{\partial\mathbf{c}_1}\cdot(f_1\,\Sigma_2\Delta\mathbf{c}_1) + \frac{1}{2}\frac{\partial^2}{\partial\mathbf{c}_1\partial\mathbf{c}_1}:(f_1\,\Sigma_2\Delta\mathbf{c}_1\Delta\mathbf{c}_1). \tag{19.7, 10}$$

* L. Landau, *Phys. Z. SowjUn.* 10, 154 (1936).

This is the Fokker–Planck expression for $(\partial_e f_1/\partial t)_2$, introduced into the theory of ionized gases by Cohen, Spitzer and Routly.* A direct derivation of this expression can readily be given; the present derivation displays the essential identity of the Fokker–Planck approach with that based on the usual Boltzmann integral, within the range in which both are valid.

19.71. The superpotentials

From (19.7, 1, 3, 4, 9)

$$\Sigma_2 \Delta \mathbf{c}_1 = M_2 \int f_2 \mathbf{g} g \phi_{12}^{(1)}(g)\, d\mathbf{c}_2, \qquad (19.71, 1)$$

$$\Sigma_2 \Delta \mathbf{c}_1 \Delta \mathbf{c}_1 = M_2^2 \int f_2 (g^2 \mathsf{U} - \mathbf{g}\mathbf{g}) g \phi_{12}^{(1)}(g)\, d\mathbf{c}_2. \qquad (19.71, 2)$$

Some care is required in applying these expressions to electrostatic interaction. Difficulties regarding convergence for large values of b have already been discussed in 10.34; in addition there are difficulties for small b, where the deflection is not small and the assumptions of the Fokker–Planck approach are inapplicable. Landau proposed taking a lower limit b' for b such that the electrostatic energy $|e_1' e_2'|/b'$ is equal to the mean kinetic energy of a particle; a quantum lower limit has also been proposed. In what follows we shall use the expression (10.34, 4) for $\phi_{12}^{(1)}$; this is roughly equivalent to using a lower limit b' such that $|e_1' e_2'|/b' = 4kT$.† Then very nearly

$$\phi_{12}^{(1)} = 4\pi (e_1' e_2' / m_0 M_1 M_2 g^2)^2 \ln v_{01} \qquad (19.71, 3)$$

and $\ln v_{01}$ is regarded as independent of g.

From (19.71, 1, 3), since $\partial g/\partial \mathbf{c}_1 = -\mathbf{g}_{21}/g$,

$$\Sigma_2 \Delta \mathbf{c}_1 = \frac{\partial h_{12}}{\partial \mathbf{c}_1}, \qquad (19.71, 4)$$

where $$h_{12} = 4\pi M_2 (e_1' e_2' / m_0 M_1 M_2)^2 \ln v_{01} \int f_2 g^{-1} d\mathbf{c}_2. \qquad (19.71, 5)$$

Similarly, since $\partial^2 g/\partial \mathbf{c}_1 \partial \mathbf{c}_1 = (g^2 \mathsf{U} - \mathbf{g}\mathbf{g})/g^3$,

$$\Sigma_2 \Delta \mathbf{c}_1 \Delta \mathbf{c}_1 = \frac{\partial^2 H_{12}}{\partial \mathbf{c}_1 \partial \mathbf{c}_1}, \qquad (19.71, 6)$$

where $$H_{12} = 4\pi M_2^2 (e_1' e_2' / m_0 M_1 M_2)^2 \ln v_{01} \int f_2 g\, d\mathbf{c}_2. \qquad (19.71, 7)$$

In terms of the 'superpotentials' h_{12}, H_{12},

$$\frac{\partial_e f_1}{\partial t} = -\frac{\partial}{\partial \mathbf{c}_1} \cdot \left(f_1 \frac{\partial h_{12}}{\partial \mathbf{c}_1} \right) + \frac{1}{2} \frac{\partial^2}{\partial \mathbf{c}_1 \partial \mathbf{c}_1} : \left(f_1 \frac{\partial^2 H_{12}}{\partial \mathbf{c}_1 \partial \mathbf{c}_1} \right). \qquad (19.71, 8)$$

The introduction of h_{12}, H_{12} is due to Rosenbluth, MacDonald and Judd.‡ The notation h_{12}, H_{12}, like the $\Sigma_2 \Delta \mathbf{c}_1$, $\Sigma_2 \Delta \mathbf{c}_1 \Delta \mathbf{c}_1$ in the Fokker–Planck

* R. S. Cohen, L. Spitzer and P. McR. Routly, *Phys. Rev.* **80**, 230 (1950).

† Note that in deriving (19.71, 2) the approximation $\phi_{12}^{(2)} = 2\phi_{12}^{(1)}$ has been used; cf. (10.34, 4, 5).

‡ M. N. Rosenbluth, W. M. MacDonald and D. L. Judd, *Phys. Rev.* **107**, 1 (1957).

expression, conceals integrations involving f_2, so that the Boltzmann equation remains an integro-differential equation. Rosenbluth and his co-workers gave equations for calculating h_{12}, H_{12} when f_2 is expanded in a series involving spherical harmonics in the components of $\boldsymbol{c}_2$; but often a relatively crude approximation is made by taking the velocity-distribution function f_2 of the 'field' particles in (19.71, 5 and 7) to be Maxwellian. However, as Spitzer and Härm* showed, the values of the transport coefficients calculated using this approximation are subject to fairly large errors.

In many problems the Fokker–Planck equation is at least as difficult to apply as that based on the usual Boltzmann integral. However, it sometimes has advantages when detailed information about the velocity-distribution function is required, e.g. in the region of high velocities.

COLLISIONLESS PLASMAS

19.8. The usual Enskog expansion of f_s,

$$f_s = f_s^{(0)} + f_s^{(1)} + f_s^{(2)} + \dots \qquad (19.8,\ 1)$$

is effectively an expansion in powers of the collision-interval τ_s. In 19.31 it was pointed out that in rarefied plasmas (strongly ionized gases) in the presence of a magnetic field the time ω_s^{-1} might be comparable with τ_s, and expressions attaching equal importance to ω_s^{-1} and τ_s were derived. For sufficiently rarefied plasmas in a magnetic field, an expansion of f_s in powers of ω_s^{-1} has been suggested, neglecting encounters completely.† This procedure regards the spiral-radius in the magnetic field as analogous to the free path in an ordinary gas, the magnetic deviations corresponding to collisions.

The analogy is not exact, since the magnetic 'collisions' affect only the motions perpendicular to $\boldsymbol{H}$. If the electric field $\boldsymbol{E}$ has components $\boldsymbol{E}_\parallel$, $\boldsymbol{E}_\perp$ respectively parallel and perpendicular to $\boldsymbol{H}$, the component $\boldsymbol{E}_\parallel$ accelerates charged particles freely; thus a solution for f_s resembling the earlier normal solutions can in general be obtained only if $\boldsymbol{E}_\parallel$ is small. Again, in the earlier theory it was possible to express a normal solution in terms of a finite number of parameters—partial densities and the mass-velocity and temperature. Because magnetic 'collisions' lead to no interchange of energy between particles, here f_s depends on an infinite number of parameters.

The equation satisfied by f_s is now the collisionless Boltzmann, or Vlasov, equation; assuming only electro-magnetic forces, this is

$$\frac{\partial f_s}{\partial t} + \boldsymbol{c}_s \cdot \frac{\partial f_s}{\partial \boldsymbol{r}} + \frac{e_s}{m_s}(\boldsymbol{E} + \boldsymbol{c}_s \wedge \boldsymbol{H}) \cdot \frac{\partial f_s}{\partial \boldsymbol{c}_s} = 0. \qquad (19.8,\ 2)$$

In (19.8, 2) we substitute an expansion of form (19.8, 1), in which the

* L. Spitzer and R. Härm, *Phys. Rev.* **89**, 977 (1953).

† G. F. Chew, M. L. Goldberger and F. E. Low, *Proc. R. Soc.* A, **236**, 112 (1956).

successive terms are supposed to be of order 1, $\omega_s^{-1}, \omega_s^{-2}, \ldots$; also $\partial_e f/\partial t$ is supposed divided into parts of similar orders, and $e_s \boldsymbol{E}_\perp/m_s$, $e_s \boldsymbol{E}_\parallel/m_s$ are arbitrarily regarded as of order ω_s and 1 respectively. Then on equating terms of successive orders, we get

$$\frac{e_s}{m_s}(\boldsymbol{E}_\perp + \boldsymbol{c}_s \wedge \boldsymbol{H}) . \frac{\partial f_s^{(0)}}{\partial \boldsymbol{c}_s} = 0, \quad (19.8, 3)$$

$$\frac{\partial_0 f_s^{(0)}}{\partial t} + \boldsymbol{c}_s . \frac{\partial f_s^{(0)}}{\partial \boldsymbol{r}} + \frac{e_s}{m_s}\left\{\boldsymbol{E}_\parallel . \frac{\partial f_s^{(0)}}{\partial \boldsymbol{c}_s} + (\boldsymbol{E}_\perp + \boldsymbol{c}_s \wedge \boldsymbol{H}) . \frac{\partial f_s^{(1)}}{\partial \boldsymbol{c}_s}\right\} = 0, \quad (19.8, 4)$$

$$\ldots\ldots\ldots\ldots\ldots\ldots\ldots\ldots\ldots,$$

where $\partial_0 f_s^{(0)}/\partial t$ and the other parts of $\partial f_s/\partial t$ are so far undetermined.

Let $\boldsymbol{c}_0$ here denote a velocity such that

$$\boldsymbol{E}_\perp + \boldsymbol{c}_0 \wedge \boldsymbol{H} = 0, \quad (19.8, 5)$$

and write $\boldsymbol{C}_s = \boldsymbol{c}_s - \boldsymbol{c}_0$. Then the general solution of (19.8, 3) is

$$f_s^{(0)} = F_s(C_s^2, \boldsymbol{C}_s . \boldsymbol{H}_s, \boldsymbol{r}, t), \quad (19.8, 6)$$

where F_s is arbitrary. This implies that in velocity-space $f_s^{(0)}$ is symmetric about an axis through $\boldsymbol{c}_s = \boldsymbol{c}_0$ parallel to $\boldsymbol{H}$. Thus to the first approximation the mean velocity $\bar{\boldsymbol{c}}_s$ has the same component perpendicular to $\boldsymbol{H}$ as $\boldsymbol{c}_0$; since the component of $\boldsymbol{c}_0$ parallel to $\boldsymbol{H}$ is not determined by (19.8, 5), $\boldsymbol{c}_0$ can be chosen to equal the mass velocity of the gas, to the first approximation. The function F_s may be chosen to equal, to a first approximation, the actual velocity-distribution function f_s, averaged over all directions of the component of $\boldsymbol{C}_s$ perpendicular to $\boldsymbol{H}$.

The first approximation to the pressure tensor is

$$\mathsf{p}^{(0)} = \sum_s \int m_s \boldsymbol{C}_s \boldsymbol{C}_s f_s^{(0)} \, d\boldsymbol{c}_s. \quad (19.8, 7)$$

It does not represent a hydrostatic pressure, but is symmetric about the direction of $\boldsymbol{H}$. If $p_\parallel$ and $p_\perp$ denote the pressures parallel and perpendicular to $\boldsymbol{H}$,

$$\mathsf{p}^{(0)} = p_\perp \mathsf{U} + (p_\parallel - p_\perp)\boldsymbol{H}\boldsymbol{H}/H^2. \quad (19.8, 8)$$

No simple general rule for assigning the value of $\partial_0 f_s^{(0)}/\partial t$ in (19.8, 4) can be given. If (19.8, 4) is multiplied by an arbitrary function $\chi_s(C_s^2, \boldsymbol{C}_s . \boldsymbol{H}, \boldsymbol{r}, t)$ and integrated over all $\boldsymbol{c}_s$, the 'collision term', involving $\boldsymbol{E}_\perp + \boldsymbol{c}_s \wedge \boldsymbol{H}$, can readily be seen to vanish; thus we find an infinite set of equations for the quantities $\partial_0 \bar{\chi}_s^{(0)}/\partial t$, where

$$n_s \bar{\chi}_s^{(0)} = \int f_s^{(0)} \chi_s \, d\boldsymbol{c}_s. \quad (19.8, 9)$$

This whole infinite set of equations has to be taken into account in defining $\partial_0 f^{(0)}/\partial t$, in place of the equations for $\partial_0 n/\partial t$, $\partial_0 \boldsymbol{c}_0/\partial t$ and $\partial_0 T/\partial t$ in the discussion of 7.14. Thus a general method of solving (19.8, 4) cannot be given; we must either assign $\partial_0 f_s^{(0)}/\partial t$ arbitrarily and test what are the consequences, or else content ourselves with such information as can be derived without explicitly solving (19.8, 4).

As an example of such information, if (19.8, 4) is multiplied by $m_s \boldsymbol{c}_s$ and integrated with respect to $\boldsymbol{c}_s$, we obtain on summation over all s the equation of motion

$$\rho \frac{D_0 \boldsymbol{c}_0}{Dt} = -\nabla . \mathsf{p}^{(0)} + \boldsymbol{j}^{(1)} \wedge \boldsymbol{H}, \tag{19.8, 10}$$

where $\boldsymbol{j}^{(r)}$ is the current density given by

$$\boldsymbol{j}^{(r)} = \sum n_s e_s \overline{\boldsymbol{C}}_s^{(r)} = \sum_s e_s \int f_s^{(r)} \boldsymbol{C}_s d\boldsymbol{c}_s, \tag{19.8, 11}$$

and the volume charge $\sum_s n_s e_s$ is regarded as negligible. The equation (19.8, 10) is the only one now available to determine the component of $\boldsymbol{j}^{(1)}$ perpendicular to $\boldsymbol{H}$;* for its effective use, we need to know the values of $p_{\parallel}$ and $p_{\perp}$.

Again, sum over all s the equation obtained by multiplying (19.8, 4) by $m_s \boldsymbol{C}_s \boldsymbol{C}_s$ and integrating with respect to $\boldsymbol{c}_s$. From the double products of the resulting equation into $\boldsymbol{HH}$ and into the unit tensor U, it is found that

$$\frac{D_0 p_{\parallel}}{Dt} = -p_{\parallel} \nabla . \boldsymbol{c}_0 - 2p_{\parallel} \frac{\boldsymbol{HH} : \nabla \boldsymbol{c}_0}{H^2} - (\nabla . \mathsf{Q}) : \frac{\boldsymbol{HH}}{H^2} + 2\boldsymbol{E}_{\parallel} . \boldsymbol{j}^{(0)}, \tag{19.8, 12}$$

$$\frac{D_0 p_{\perp}}{Dt} = -2p_{\perp} \nabla . \boldsymbol{c}_0 + p_{\perp} \frac{\boldsymbol{HH} : \nabla \boldsymbol{c}_0}{H^2} - \tfrac{1}{2}(\nabla . \mathsf{Q}) : \left(\mathsf{U} - \frac{\boldsymbol{HH}}{H^2}\right), \tag{19.8, 13}$$

where Q is a third-order tensor measuring the flux of $m\boldsymbol{CC}$, i.e. (in an obvious notation)

$$\mathsf{Q} = \sum_s \int f_s^{(0)} m_s \boldsymbol{C}_s \boldsymbol{C}_s \boldsymbol{C}_s d\boldsymbol{c}_s; \tag{19.8, 14}$$

$\boldsymbol{j}^{(0)}$ is the electric current corresponding to the first approximation $f_s^{(0)}$.

When $\boldsymbol{E}_{\parallel} = 0$, equations (19.8, 12, 13) can be simplified by using the equation of continuity $D_0 \rho / Dt = -\rho \nabla . \boldsymbol{c}_0$ and Maxwell's equation

$$\partial \boldsymbol{H} / \partial t = -\operatorname{curl} \boldsymbol{E};$$

since now $\boldsymbol{E} = -\boldsymbol{c}_0 \wedge \boldsymbol{H}$, this gives

$$\frac{D_0 \boldsymbol{H}}{Dt} = \boldsymbol{H} . (\nabla \boldsymbol{c}_0) - \boldsymbol{H}(\nabla . \boldsymbol{c}_0).$$

The simplified forms of (19.8, 12, 13) then are

$$\frac{D_0}{Dt}\left(\frac{p_{\parallel} H^2}{\rho^3}\right) = -(\nabla . \mathsf{Q}) : \frac{\boldsymbol{HH}}{\rho^3}, \tag{19.8, 15}$$

$$\frac{D_0}{Dt}\left(\frac{p_{\perp}}{\rho H}\right) = -\frac{1}{2\rho H^3}(\nabla . \mathsf{Q}) : (\mathsf{U} H^2 - \boldsymbol{HH}). \tag{19.8, 16}$$

* E. N. Parker, *Phys. Rev.* **107**, 924 (1957). Parker gave an explicit expression for this component of $\boldsymbol{j}^{(1)}$ in terms of $p_{\parallel}$ and $p_{\perp}$.

To use these equations we need to know **Q**. If it is assumed that **Q** is negligible, we obtain the 'double-adiabatic' equations

$$\frac{D_0}{Dt}\left(\frac{p_{\parallel} H^2}{\rho^3}\right) = 0, \quad \frac{D_0}{Dt}\left(\frac{p_{\perp}}{\rho H}\right) = 0. \tag{19.8, 17}$$

In general **Q** is unknown. Its rate of variation can be obtained by multiplying $\boldsymbol{H}$ by $m_s \boldsymbol{C}_s \boldsymbol{C}_s \boldsymbol{C}_s$, integrating and summing; the resulting equation involves the mean value of a function of fourth degree in the components of $\boldsymbol{C}_s$. Similarly the time-derivative of this mean value is given by an equation involving the mean value of a function of fifth degree; and so on. This type of difficulty is one often met; it can be overcome only by arbitrarily assuming a relation between moments of sufficiently high order.

Because of this difficulty a number of discussions of collisionless plasmas have been based on particle orbits instead of the Boltzmann equation.* The two approaches are fundamentally the same; a theorem due to Jeans† states that the most general solution of the collisionless Boltzmann equation is $f_s = \phi(\alpha_1, \alpha_2, \ldots, \alpha_6)$, where $\alpha_1, \alpha_2, \ldots, \alpha_6$ are functions of $\boldsymbol{c}_s$, $\boldsymbol{r}$, t such that the equations $\alpha_r =$ const. give six independent integrals of the equations of motion of a molecule m_s. For example, if $\partial_0 f_s^{(0)}/\partial t = 0$ and $\boldsymbol{c}_0$ is negligibly small, the equations for the $\partial_0 \overline{\chi}_s^{(0)}/\partial t$ of (19.8, 9) are satisfied by assuming that along any line of magnetic force $f_s^{(0)}$ is a unique function of α_s, β_s, where

$$\alpha_s = \tfrac{1}{2} m_s c_s^2 + e_s V_s, \quad \beta_s = |\boldsymbol{c}_s \wedge \boldsymbol{H}|^2/H^3,$$

and V_s is the electric potential. This agrees with the particle approach, since α_s is the total energy and β_s is proportional to the magnetic moment of a particle m_s, which is an adiabatic invariant of the motion. However, the development of the theory in terms of the subdivision indicated in (19.8, 1, 3, 4), is parallel to, rather than identical with, the theory based on particle orbits; and its connection with the theory given earlier in this book is not close.

* See, e.g., H. Alfvén, *Cosmical Electrodynamics*, chapter 2 (Oxford, 1947, 1963); M. N. Rosenbluth and C. L. Longmire, *Ann. Phys.* **1**, 120 (1957). The relation of the Boltzmann and particle approaches is discussed, e.g., by K. M. Watson, *Phys. Rev.* **102**, 12 (1956); S. Chandrasekhar, A. N. Kaufman and K. M. Watson, *Ann. Phys.* **2**, 435 (1957), and **5**, 1 (1958); E. N. Parker, *Phys. Rev.* **107**, 924 (1957); S. Chandrasekhar, *Plasma Physics* (Cambridge, 1960).

† J. H. Jeans, *Month. Not. R. astr. Soc.* **76**, 70 (1915).

HISTORICAL SUMMARY

Daniel Bernoulli (1700–82) in 1738, Herapath (1790–1868) in the years 1816–21, Waterston (1811–83) in 1845 (though his paper was not published until 1893) and Joule (1819–89) in 1848, made early significant contributions to the kinetic theory of gases. A new phase began in 1858, when Clausius (1822–88) introduced the concept of the mean free path; his work extended over nearly twenty years, from 1857. Maxwell (1831–79) in 1859 obtained formulae for the transport coefficients of a gas (viscosity, conduction and diffusion), using the mean free path and his own new concept of the velocity distribution function f. He found the equilibrium form $f^{(0)}$ of f, though his proof (both then and at a second attempt in 1866) was imperfect. In later years Meyer (1834–1909), Stefan (1835–93), Tait (1831–1901), Jeans (1877–1946) and others devoted much effort to improving the free-path theory; but already in 1866 Maxwell turned from this to base a new theory on equations of transfer that he formulated. From them, however, he could obtain formulae for the transport coefficients only for the special 'Maxwellian' gas, in which the molecules interact with an r^{-5} force; for such a gas he did not need to determine f.

Maxwell's new method inspired Boltzmann (1844–1906), who in 1872, by his H-theorem, much improved the proof for Maxwell's function $f^{(0)}$. In the same paper he gave his integro-differential equation for f, and inferred the specially simple form of f for a Maxwellian gas. In 1879 Maxwell used this form to show that in a very rare gas at non-uniform temperature, at rest, there are stresses not included in the Navier–Stokes equations. Again he discussed only a Maxwellian gas, 'for the sake of being able to effect the integrations'. In 1880/1 Boltzmann published three long memoirs in which he reviewed Maxwell's 1866 and 1879 papers. He said that the solution of his own equation for f is easy only for a Maxwellian gas; 'in all other cases, and especially for...elastic spheres, the solution...meets with great difficulties'. He added that one must almost despair of the general solution of his equation.

In 1887 Lorentz (1853–1928) further improved Boltzmann's discussion of $f^{(0)}$. He also, in 1905, achieved the first success, after Maxwell, in obtaining accurate formulae for the transport coefficients for another special type of gas. Hilbert (1862–1943) in 1912 discussed the theory of Boltzmann's equation, but without arriving at a satisfactory solution. Pidduck (1885–1952) in 1915, by numerical treatment based on Hilbert's discussion, obtained an accurate value for one special gas coefficient, the self-diffusivity, in a gas composed of elastic spheres. However, neither his work nor that of Lorentz gave any hint as to how to obtain general results.

Enskog (1887–1947) in 1911/12, and Chapman in 1912, wrote on the kinetic theory, but without success in solving the general problem. Not until 1916 could Chapman, using Maxwell's equations of transfer in an extended way, publish accurate general formulae for the transport coefficients (without needing to determine f). In 1917 Enskog's Uppsala inaugural dissertation gave the general solution for f, yielding transport coefficients identical with Chapman's. This identity of results independently obtained by different routes led to their acceptance by many physicists who might not wish to check the complicated mathematics.

The half-century between Maxwell's 1866 and Chapman's 1916 kinetic theory communications to the Royal Society, during which exact theory was limited almost entirely to a Maxwellian gas, left to Chapman and Enskog the discovery of the complete equations of diffusion and heat conduction in a mixed gas, revealing the phenomena of thermal diffusion and the diffusion thermoeffect. Thermal diffusion in liquids had been discovered experimentally by Ludwig in 1856 and studied by Soret in 1879–81. However, it vanishes for a Maxwellian gas and so, despite its appearance in Enskog's first 1911 paper, for a Lorentzian gas, its general occurrence in mixed gases remained unsuspected until in the summer of 1916 it was confirmed by Dootson in conjunction with Chapman. Even then, for another quarter-century it seemed to be little more than a scientific curiosity, until in the thermal diffusion column invented by Clusius it was applied on a large scale for the separation of uranium and other isotopes for a variety of medical and technological purposes.*

Since 1920 the ideas of Chapman and Enskog have been considerably developed and generalized. A variety of molecular models have been studied, including those with internal energy and those in which quantum effects are important. The theory has been extended to ionized gases and to very rare and very dense gases, each of which presents its own methodological difficulties. One further gas coefficient—the volume viscosity—has been isolated. These later developments are described in the body of this book; the present Historical Summary therefore deals mainly with the history of our subject prior to 1920.

Earlier editions of this book included a more extended Historical Summary, partly reproduced above.

HISTORICAL REFERENCES

Where page numbers are given in parenthesis, they refer to pages of this book where the reference is to be found.

L. Boltzmann, 1872 (p. 73); *Wien Ber.* **81**, 117 (1880), **84**, 40 and 1230 (1881); *Collected Works*, **2**, 388, 431, 523.

S. Chapman, *Phil. Trans. R. Soc.* A **213**, 433 (1912); **216**, 279 (1916); *Proc. R. Soc.* A **98**, 1 (Dec. 1916); S. Chapman and F. W. Dootson, 1917 (p. 268).

* Cf. L. B. Loeb, *The Kinetic Theory of Gases* (Dover edition, 1961), pp. xiii, xiv.

R. Clausius, *Annln Phys.* **105**, 239 (1858).
D. Enskog, *Phys. Zeit.* **12**, 56 and 533 (1911); *Annln Phys.* **38**, 731 (1912); *Inaugural Dissertation*, Uppsala, 1917.
J. H. Jeans, *Phil. Mag.* **8**, 692–9, 700–3 (1904).
H. A. Lorentz, 1887 (p. 73); 1905 (p. 188).
J. C. Maxwell, 1860 (p. 72); 1867 (p. 68); 1879 (p. 281).
O. E. Meyer, *Annln Phys.* **125**, 177, 401, 564 (1865).
F. B. Pidduck, 1915 (p. 195).
J. Stefan, *Wien Ber.* **65**, 323 (1872).
See also: S. G. Brush, *Kinetic Theory*, 2 vols. (Pergamon, 1965, 1966); (1957/8) *Ann. Sci.* **11**, 188 on Herapath, **13**, 273 on Waterston, **14**, 185 on Clausius, and **14**, 243 on Maxwell; also *Am. J. Phys.* **29**, 593 (1961), **30**, 269 (1962) on the later developments.
S. Chapman, in *Lectures in Theoretical Physics* vol. 9C (ed. by W. E. Brittin), pp. 1–13 (Gordon & Breach, 1967).

NAME INDEX

Note: where an author's name is used adjectivally (Boyle's law, Boltzmann's equation, etc.), only the first appearance is listed here. For later appearances, see the subject index

SUBJECT INDEX

REFERENCES TO EXPERIMENTAL DATA FOR PARTICULAR GASES

KEY. Simple gas: S = specific heat, V = viscosity, C = thermal conductivity, SD = self-diffusion, ITD = isotopic thermal diffusion.

Gas mixtures: MV = mixture viscosity, MC = mixture conductivity, D = diffusion, TD = thermal diffusion. In each case the table number of the second gas is given in parentheses following the page number.

Quantum results and those at high pressures are distinguished respectively by the addition of Q and P.

1 Acetylene, C_2H_2 *V*, 228, 232, 233; *C*, 249
2 Air *S*, 43; *V*, 228, 232, 233, 237; *C*, 249
3 Ammonia, NH_3 *S*, 43; *V*, 228, 232, 233, 234; *C*, 249; *D*, 263 (4, 11, 12, 15, 18, 24), 265 (4, 15); *ITD*, 274, 276; *TD*, 274 (18)
4 Argon, A *S*, 43; *V*, 228, 232, 233, 237, 238; *MV*, 241 (11); *C*, 249; *MC*, 254 (11); *D*, 261 (11), 263 (3, 6, 11, 12, 15, 18, 19, 22, 24), 265 (3, 11, 18); *SD*, 267; *ITD*, 276; *TD*, 275 (13), 278 (6, 11, 12, 15, 18, 19, 22, 24); *VP*, 310; *CP*, 310
5 Carbon monoxide, CO *S*, 43; *V*, 228, 232, 233, 237, 238; *C*, 249; *D*, 263 (6, 10, 12, 19, 22), 265 (22); *SD*, 266, 267, 268; *ITD*, 275; *TD*, 278 (12)
6 Carbon dioxide, CO_2 *S*, 43; *V*, 228, 232, 233, 237, 238; *C*, 249; *D*, 263 (4, 5, 8, 11, 12, 16, 18, 19, 21, 22), 265 (11, 12, 18, 19, 22); *SD*, 266, 267, 268; *ITD* 275; *TD*, 278 (4, 11, 12, 19, 22)
7 Chlorine, Cl_2 *S*, 43; *V*, 228, 232, 233, 237; *C*, 249
8 Deuterium, D_2 *S*, 43; *V*, 228, 232, 237; *C*, 249; *D*, 263 (6, 12); *TD*, 278 (19)
9 Ethane, C_2H_6 *S*, 43; *V*, 228, 233, 238; *C*, 249; *D*, 263 (12, 19)
10 Ethylene, C_2H_4 *S*, 43; *V*, 228, 233; *C*, 249; *D*, 263 (5, 12, 19); *TD*, 278 (12)
11 Helium, He S, 43; *V*, 228, 230, 231, 232, 233, 237, 238; *MV*, 241 (12, 4); *C*, 249; *MC*, 254 (4); *D*, 261 (4), 263 (3, 4, 6, 12, 15, 18, 19, 22, 24), 265 (4, 6, 18, 19); *SD*, 267; *ITD*, 276; *TD*, 278 (4, 6, 12, 15, 18, 19, 24); *VP*, 310; *VQ*, 331
12 Hydrogen, H_2 *S*, 43; *V*, 228, 232, 233, 237, 238; *MV*, 241 (11), 243 (13); *C*, 249; *D*, 261 (19), 263 (3, 4, 5, 6, 8, 9, 10, 11, 16, 19, 21, 22, 23), 265 (6, 22); *SD*, 266, 267; *ITD*, 275, 276; *TD*, 278 (4, 5, 6, 10, 11, 16, 18, 19, 22); *VP*, 310, *VQ*, 332
13 Hydrogen chloride, HCl *V*, 228, 232, 233, 237; *MV*, 243 (12); *C*, 249; *SD*, 267; *TD*, 275 (4)
14 Hydrogen sulphide, H_2S *S*, 43; *V*, 228, 233; *C*, 249
15 Krypton, Kr *V*, 228, 233, 237, 238; *C*, 249; *D*, 263 (3, 4, 11, 18, 24), 265 (3); *SD*, 267; *ITD*, 276; *TD*, 278 (4, 11, 18, 24)
16 Methane, CH_4 *S*, 43; *V*, 228, 232, 233, 237, 238; *C*, 249; *D*, 263 (6, 12, 22), 265 (22); *SD*, 267; *ITD*, 276; *TD*, 278 (12)
17 Methyl chloride, CH_3Cl *V*, 228, 233; *C*, 249
18 Neon, Ne *S*, 43; *V*, 228, 232, 233, 237, 238; *C*, 249; *D*, 263 (3, 4, 6, 11, 15, 24), 265 (4, 6, 11); *SD*, 267; *ITD*, 276; *TD*, 274 (3), 278 (4, 11, 12, 15, 24); *VP*, 310; *CP*, 310
19 Nitrogen, N_2 *S*, 43; *V*, 228, 232, 233, 234, 237, 238; *C*, 249; *D*, 261 (19), 263 (4, 5, 6, 9, 10, 11, 12, 22, 24), 265 (6, 11); *SD*, 267, 268; *ITD*, 276; *TD*, 278 (4, 6, 8, 11, 12, 21, 22); *VP*, 309; *CP*, 310
20 Nitric Oxide, NO *S*, 43; *V*, 228, 232, 233, 237; *C*, 249
21 Nitrous Oxide, N_2O *S*, 43; *V*, 228, 232, 233, 237; *C*, 249; *D*, 263 (6, 12); *SD*, 268; *TD*, 278 (19)
22 Oxygen, O_2 *S*, 43; *V*, 228, 232, 233, 237; *C*, 249; *D*, 263 (4, 5, 6, 11, 12, 16, 19, 24), 265 (5, 6, 12, 16); *SD*, 266, 267; *ITD*, 276; *TD*, 278 (4, 6, 12, 19)
23 Sulphur dioxide, SO_2 *S*, 43; *V*, 228, 232, 233, 237; *C*, 249; *D*, 263 (12)
24 Xenon, Xe *S*, 43; *V*, 228, 233, 237, 238; *C*, 249; *D*, 263 (3, 4, 12, 15, 18, 19, 22); *SD*, 267; *ITD*, 276; *TD*, 278 (4, 11, 15, 18); *VP*, 310